Agricultural Research for Sustainable Food Systems in Sri Lanka

Ranjith Premalal De Silva •
Gamini Pushpakumara • Pahan Prasada •
Jeevika Weerahewa
Editors

Agricultural Research for Sustainable Food Systems in Sri Lanka

Volume 2: A Pursuit for Advancements

Springer

Editors
Ranjith Premalal De Silva
Faculty of Agriculture
University of Peradeniya
Peradeniya, Sri Lanka

Gamini Pushpakumara
Faculty of Agriculture
University of Peradeniya
Peradeniya, Sri Lanka

Pahan Prasada
Faculty of Agriculture
University of Peradeniya
Peradeniya, Sri Lanka

Jeevika Weerahewa
Faculty of Agriculture
University of Peradeniya
Peradeniya, Sri Lanka

ISBN 978-981-15-3672-4 ISBN 978-981-15-3673-1 (eBook)
https://doi.org/10.1007/978-981-15-3673-1

This Springer imprint is published by the registered company Springer Nature Singapore Pte Ltd.
The registered company address is: 152 Beach Road, #21-01/04 Gateway East, Singapore 189721, Singapore

Contents

About the Editors

Ranjith Premalal De Silva is a Senior Professor in Agricultural Engineering in the Department of Agricultural Engineering, University of Peradeniya, Sri Lanka. He completed his B.Sc. degree in Agriculture from the University of Peradeniya; his M. Sc. degree from Asian Institute of Technology, Thailand; and his Ph.D. in Applied Remote Sensing and GIS from the Cranfield University, UK. He is the Founder President of the Geo-Informatics Society of Sri Lanka and is a Former Vice Chancellor of Uva Wellassa University, Sri Lanka, and Competent Authority of Trincomalee Campus. He has served in many national institutions and also contributed as a consultant for the FAO, WFP, and World Bank. He is presently serving as the Competent Authority of the University of Vocational Technology and Director/CEO of Hector Kobbekaduwa Agrarian Research and Training Institute.

Gamini Pushpakumara is a Senior Professor and Former Head of the Department of Crop Science and is the current Dean of the Faculty of Agriculture, University of Peradeniya. He obtained his B.Sc. degree in Agriculture from the University of Peradeniya and his M.Sc. and Ph.D. degrees in Forestry and its Relation to Land Use and Forest Genetics and Germplasm Conservation, respectively, from the University of Oxford, UK. He is an internationally recognized researcher with over 200 publications in peer-reviewed journals.

Pahan Prasada is a Senior Lecturer in the Department of Agricultural Economics and Business Management, Faculty of Agriculture, University of Peradeniya, Sri Lanka. He completed B.Sc. in Agriculture from the University of Peradeniya; M.Sc. in Agricultural Economics and International Development from the University of Guelph, Canada; and Ph.D. from the University of New South Wales, Australia. He is also a Fellow of the Adaptation Finance Fellowship Program of Frankfurt School-UNEP Centre and a certified expert in Climate Adaptation Finance (CECAF) at Frankfurt School of Management and Finance.

Jeevika Weerahewa is a Professor of Agricultural Economics in the Department of Agricultural Economics and Business Management, Faculty of Agriculture,

University of Peradeniya, Sri Lanka. She completed her B.Sc. and M.Phil. degrees in Agriculture from the University of Peradeniya and her Ph.D. in Agricultural Economics from the University of Guelph, Canada. She has served as President of the Sri Lankan Agricultural Economics Association.

Agriculture Scientist's Many Burdens: A Glimpse of Efforts in Land Use Planning, Waste Recycling, Food Storage Design, Managing Farmer Psychology and Other Eclectic Pursuits

Pahan Prasada, Gamini Pushpakumara, and Ranjith Premalal De Silva

Agriculture is too vast a field of study to generate a short systematic overview of current developments. The goals of agricultural sector are broad and generic, especially in low- and middle-income countries with an agricultural heritage. As a result, the scientific work in agriculture evolves in isolation and sometimes in distant parallels to the broader purpose of the sector. Such deviations of scientific pursuits from development goals are perhaps inevitable in rural socio-economic production systems (as in Sri Lanka), where thousands of smallholders have chosen agriculture as their livelihood, perpetuating routines that have been passed on through generations, while depending ever so little on the scientific frontier. This chapter attempts to preview quite a diverse collection of scientific work. Each of the contributions records the current status of a thematic area under agriculture while paying attention to how the findings relate to broader goals of sustainability. Within this diversity in both depth and breadth, four themes emerge. The first theme is best summarized as the science behind the environmental services relevant to agricultural ecosystems and technologies aimed at facilitating such services. The second is the wider theme of food production and its challenges, especially the contemporary ones. The science and practice behind evidence-based planning in agriculture is the third theme. The fourth is social, behavioural and management elements of agricultural sector.

P. Prasada (✉) · G. Pushpakumara · R. P. De Silva
Faculty of Agriculture, University of Peradeniya, Peradeniya, Sri Lanka

R. P. De Silva et al. (eds.), *Agricultural Research for Sustainable Food Systems in Sri Lanka*, https://doi.org/10.1007/978-981-15-3673-1_1

1 Interventions Towards Improving the Agricultural Ecosystems

Environmental goods and services have emerged as a co-output of agricultural production while being a crucial input to the same. That agriculture depends crucially on the ecosystem health for its sustenance is reason enough to pay attention to improving environmental services. While mono-cropping and intensive farming cannot be replaced due to ever-increasing food demand, the many corrective measures that will redress the damages caused by modern agriculture can be adopted. Spanning from microbial pesticides that could potentially replace industrial chemical equivalents, through shade trees in agricultural ecosystems, to recovering energy from waste, numerous are the scientific efforts towards 'greening' modern agricultural landscape.

Agroforestry is one of the more readily adoptable interventions towards enhancing the ecosystem health while offsetting the consequences of widespread mono-cropping in local agriculture. Agriculture scientists at the University of Peradeniya have contributed to this topic since as early as the 1970s. Current work focuses more on system and ecological benefits of agroforestry and trees outside the forest, including those in home gardens. With new commitments in greenhouse gas emission reductions in the Paris Climate Agreement, trees in agricultural systems have become very topical. In this context, Mohotti et al. (2019) consider the potential of shade trees in tea as a model of agroforestry capable of mitigation of climate change impacts among other functions. The merits and demerits of trees in plantations are discussed at length to highlight the need for planning appropriate species that would minimize the vulnerability of ecosystem while maximizing the carbon sequestration and other environmental benefits.

Soil quality (and fertility) is surfacing as a key concern after years of widespread mono-cropping. The modern agriculture has not really delivered any credible alternative to the model of mono-cropping under high chemical inputs. Dandeniya and Dharmakeerthi (2019) claim that more than 50% of the agricultural lands in Sri Lanka have been degraded based on the level of organic carbon reserves, nutrient content and soil acidity. Authors suggest green manure, animal manure, compost, biochar and biofertilizers as corrective measures to be adopted in an integrated nutrient management strategy to restore soil fertility. Application of agrochemicals is the common culprit in accounting for the deterioration of soils. Agrochemicals have many unintended consequences, not only to soil flora and fauna but also to the soil structure. The chapter on microbial pesticides (De Costa 2019a) appeals for incorporating microbiological solutions to pest control as a means of promoting eco-friendly agriculture. The authors highlight the relatively high use of microbial pesticides in advanced Asian economies such as China while pointing out relatively poor use of such applications in Sri Lanka. Over the last 27 years, only a handful of crops have seen microbial pesticide applications. In order of frequency, banana-, rice-, tomato-, pineapple- and chilli-related interventions top the list with at least two published findings for each over the said period. Nearly all the applications in Sri Lanka have been based on fungi, bacteria and combinations of each as the microbial antagonist used in the pesticides and almost all applications have targeted plant pathogens.

Agriculture appears greener, certainly compared to manufacturing. However, the environmental footprint of agriculture is significant. Globally, 32% of all greenhouse gas emissions are reported from agriculture, forestry and related land uses. While the Sri Lankan per capita emissions are substantively below the global average value, there is an urgency for improvement, especially since Sri Lanka is no less vulnerable to catastrophic consequences of climate change and natural disasters. Within the scope of corrective measures in improving agricultural ecosystems, the recycling, reuse and recovery concept is being recognized as an important paradigm. It is understood that ecosystems have within them important self-corrective and resilience-related functions. Optimization of such properties of natural systems with adjustments to our management practices appears the least intrusive of interventions. A good example is wetland restoration. Among the studies on environment in the present volume, Mowjood et al. (2019) consider environmental services rendered by wetland ecosystems in a comparative analysis of paddy fields and constructed wetlands as simulations of natural wetlands. Authors highlight the design details, biochemical processes, water management and habitat preservation as substantive challenges that confront establishment of constructed wetlands while pointing out the equivalent roles that paddy fields play in an agroecosystem.

On the resource recycling front, the fate of straw after paddy harvesting has long been a concern to agroecosystems. Burning is what farmers do for the most part in the absence of alternative reuse options. Burnt straw applied to soils have beneficial impacts, but the air pollution and the loss of material due to blowing away far outweigh the possible benefits of straw burning and application of ash to the soil. Biochar became topical in this context, first as a soil supplement and later as a method for trapping carbon in material form. Basnayake et al. (2019a) discuss the milestones of a research program since 1994 aimed at sustainable reuse of paddy straw, cow dung and green manure and culminating in work on biochar. Taking the theme of resource recovery, a step further into the larger operational scales, Basnayake et al. (2019b) narrate the process of designing and implementing municipal solid waste management using the example of Kandy landfill site at Gohagoda. The long journey towards energy recovery from waste to profits is laid out in detail to highlight the hurdles (both scientific and institutional). Lessons learnt by the authors clearly illustrate how difficult it is to find sustainable solutions to manage waste in an environmentally friendly and economically rewarding manner.

2 Food Sector and Its Contemporary Challenges

Food sector's perennial challenges have remained constant over half a century since the 'so-called' Green Revolution. Per capita food availability has risen due to increase in yields and partly due to imports. Postharvest losses, as a percentage, are no different to what they were few decades back. Regardless of numerous inefficiencies, the national preoccupation with breeding high-yielding varieties has served the goals of self-sufficiency well. While it was not utilized to the same degree as in rice, the diversity of crop germplasm in field crops remains one of the biggest potentials for future varietal development. Fonseka et al. (2019) report records of 410 species of food crop wild relatives (CWR), belonging to 47 families and 122 genera. In particular, crop wild relatives present desirable traits that are resilient to biotic stresses (pest and disease) and abiotic stresses (drought and salinity) likely in a warmer world that is inevitable. The authors argue for the need to include crop wild relatives in routine varietal development, citing the many benefits of the germplasm in various categories of domestically consumed food crops. It is also highlighted the need for in situ and field conservation of genetic resources of wild relatives.

Indigenous crops are a similar theme to CWR, but they are mostly underutilized ones or ones with limited spatial incidence. Ranil et al. (2019) consider the taxonomic diversity, distribution, medicinal properties, phytochemical aspects and conservation pressure of indigenous vegetable plants. The authors highlight the research needs in terms of germplasm collection, evaluation and conservation of agrobiodiversity, genetic improvement and use in plant breeding programs taking into account the fact that indigenous vegetables are well adapted to biotic and abiotic stresses.

The current challenges in safety and quality in food processing and consumption far outweigh the concerns regarding production methods. This is partly due to the mass consumption in urban settings where foods are sourced from both domestic and foreign sources. The expectations of quality and safety have risen in the widening middle-income class of consumers (Prasada 2016a). Functional and fortified foods have been a classic concern for health-concerned public (Prasada 2016b). In general, such concerns encompass a medley of food constituents varying from additives, antioxidants, antidiabetic compounds to hypocholesterolemic compounds. While there is substantive noise and uncertainty on the credibility of claims regarding nutritional and health implications of foods, there is a rising body of research aimed at identifying the naturally occurring edible material with functional properties. Madhujith and Wedamulla (2019) contribute to this literature highlighting the antioxidant potential of many local foods such as brinjal, guava, watermelon, dragon fruit, tamarind, himbutu (*Salacia chinensis*) and palu (*Manilkara hexandra*); the hypocholesterolemic activity of *Phyllanthus* species; and the antidiabetic properties of Thebu (*Costus speciosus*), Ranawara (*Cassia auriculata*) and Beli (*Agale marmelose*). Dietary implications of food constituents receive far more attention today than in any other period in the past. Fats and oils are the key concern. Madhujith and Sivakanthan (2019) argue the case for taking trans fats out of the diet. They caution on the use of partially hydrogenated vegetable oils for cooking traditional dishes in Sri Lanka, pointing out that partially hydrogenated vegetable fats contain *trans* fats up to 35–45%. Baked, confectionery, deep-fried products and many processed snack foods, all of which contain high levels of trans fats, are growing to be significant components of Sri Lankan diets. The alarming evidence of roadside fast food outlets reusing palm oil as many as 40 times is cited as another form of *trans* fat generation which leads to serious health implications in the consumers of these fast foods.

Biotechnology has remained an elusive frontier for Sri Lankan agricultural research for many years. It is not necessarily due to lack of expertise but due to a certain shortage of initiative at the institutional level. Except for work on rice, local research studies on other crops utilizing high-end biotechnology are comparatively sparse. The extent and the coverage of different types of local studies are summarized by Bandaranayake (2019). The author lists out 20 studies on marker-assisted selection (MAS) focused on rice and 15 studies using molecular characterization aiming identification of pests and diseases. Compared to the exclusive focus on rice observed in MAS, characterization studies have focused on pests and diseases of reasonably wide range of crops: banana, papaya, pineapple, cucurbits, potato and tomato. Approximately 27 studies have been conducted locally aiming at detecting genetic diversity of crops. While the majority of these have focused on deeper understanding of rice germplasm, field crops and fruit crops have also been considered in a few.

Perhaps the most pressing challenge of the day facing the crop scientist is climate vulnerability of existing cultivars. Drought tolerance, salinity tolerance and flood resistance are all priorities competing for scientists' attention. While the climate-related factors have hit the headlines quite recently, the efforts of agronomists towards enhancing the resilient properties of staple crops have seen many significant milestones in the post-Green Revolution era. De Costa (2019b) presents a comprehensive account of such work with examples and poses the remaining challenges in 'climate-proofing' the crop germplasm. The significant highlights in the above research program include germplasm screening for development of heat-tolerant rice varieties starting in 2008/2009 *Maha* season, varietal testing to screen heal-tolerant traits in chilli and tomato and investigation of flowering and fruit setting of coconut in the context of increased atmospheric temperatures and drought conditions. However, the research carried out is still deficient considering the vulnerability of Sri Lanka as a tropical island exposed to potential consequences of climate change scenarios such as a two-degree hotter world.

The burden of postharvest losses in the food value chain is a continuing challenge to the food scientist. Storage and transport stages of the supply chain in Sri Lanka remain quite rudimentary and backward in terms of possible technological improvements within our reach. Prasantha (2019) discusses the status of modified atmospheric (MA) technologies (or lack thereof) in food storage in local food value chains. The merits and demerits of high carbon dioxide, hermetic packaging, low pressure and high pressure modified atmospheric packaging are discussed for potential use of on-farm storage. While modified atmospheric storage involves expensive pressure chambers and control systems, the author argues that operational cost of MA is comparatively lower than that of chemical fumigation. Thus, high-value organic commodities like spices, nuts and herbs stand to benefit from MA techniques.

3 Evidence-Based Planning for Systematic Improvements

Sustainability of food supply systems is a strategic interest to a nation. However, the stability and continuity of Sri Lankan food supply is ensured by the involvement of a large smallholder community in staple crops, aided by public subsidization of farm inputs. As a result, there are no alarming signs of food deficits. Drought episodes generate the occasional shortage. In spite of temporary and infrequent setbacks of supply, the resilience of local food supply chain has been a vital part of the food security. This however is not evidence of an efficient food system, even in the case of staples. Several contributions to the present volume focus on several planning tools and support functions of agricultural system.

Input to output conversions are standard tools of planning in agricultural economics. The responsiveness of output to inputs is essentially an indicator of technological and behavioural context of input usage. These indicators are termed production elasticities in the jargon. A second category of measures of responsiveness are between market supply and prices (of inputs and output), known as supply elasticities. Weerasooriya and Hemachandra (2019) review evidence from local agricultural economic analyses in order to point out the relative efficiency of the local production technology and supply chains. In production relationships, with respect to paddy, fertilizer, land extent and labour appear inelastic and becoming more inelastic with time (later studies reporting more inelastic values). Further, fertilizer and labour are more inelastic compared to land. For plantation crops, fertilizer is the most inelastic input followed by labour and then land. In supply function relationships for paddy, elasticity estimates for fertilizer price ranged from relatively inelastic −0.009 to elastic −2.760. Elasticities for paddy price ranged from 0.100 to as large as 5.049. Authors claim that the direction of responsiveness in the above relationships is consistent with theory, but the wide variations of values leave many unanswered questions.

Modern land use planning is rarely performed without the help of geographic information systems (GIS) and remote sensing (RS). The output of spatial analysis generates vast collection of data requiring its own science of geo-informatics. The value of geo-informatics for agriculture and land use planning can hardly be overstated. Dayawansa and De Silva (2019) list out applications of GIS and RS in measurement and management of soil erosion and land degradation, as inputs in hydrological models, climate change mapping, land use change mapping and soil studies. GIS and RS have become indispensable tools in analytical toolkit of modern-day land use planner. Policy scientists also stand to gain from rapid assessments, and monitoring may be possible by GIS and RS on key resources in drafting relevant policies. Geo-informatics which is the data management arm of spatial sciences has become a specialized technique used in land suitability analysis, farm resource management, harvesting, supply-chain logistics and policy analyses such as evaluations of food security.

Reporting of research in agricultural sciences invariably involves statistical tools (except in the case of certain qualitative studies and reviews). There is a long-standing concern in academic literature on inappropriate use of statistics, as methods or as conclusions. Samita et al. (2019) look into 20 years of research published in two leading local journals to detect instance of inappropriate use/misuse of statistical tools and find that at the stages of experiment design and data analysis, there are clear deviations from correct use.

While statistics is indispensable for quantitative interventions, communication is integral to qualitative interventions. At multiple point of the food value chain, a key role exists for efficient, effective and timely communication interventions. From controlling epidemic pest attacks to cushion the shocks of price hikes, communication and information dissemination remain relevant in the local agricultural economy. In fact, efficient communication is a key element of reduced transaction costs in an agricultural system. Wijeratne (2019) discusses the evolution of communication strategies and their impact in various agricultural extension models that Sri Lanka has adopted over the years. The journey from the on-farm visits through field schools to remote communication is laid out in detail. On a related note, Dissanayeke et al. (2019) review the developments in ICT-based tools in agricultural extension focusing on the economic and efficiency implications that such developments entail. The emerging participation of large corporate telecommunication firms in providing mobile telephone-based extension is fast gaining ground marking a new paradigm in extension services. The relative impact of ICT interventions on product prices is discussed with respect to foreign experiences.

4 Organizational Structures, Individual and Group Psychology as Drivers of Modern Agriculture and Agribusiness

That smallholder sector dominates the world food supply is no secret. Sri Lanka has a similar experience with respect to the domestic food supply. While the role of imports in the local food basket is still significant in value, smallholder-generated agricultural produce dominate in terms of quantity. The smallholder is local agricultural sector's strength and its weakness too. Farmers acting as individual agents face a world of uncertainty unparalleled to those involved in any other routine economic activity. In the Sri Lankan setting, smallholder farming is an inherited livelihood. An average farmer is partially educated but well experienced, locked by limits on the land and capital. She cultivates two seasons if rains are good, but often fights with weather and outbreaks of pests and diseases to secure the odds of success. She often operates in credit-bound relationship with other players of agricultural value chain, with the hope of repayment based on a good harvest and favourable market prices. The latter is always an elusive prospect to the conventional operator with limited information and no market power.

Several contributions included in the volume investigate the limited, yet expanding, genre of literature on how farmers think, react and adapt to changes in the farming environment. In certain cases, the focus is on the psychological forces that govern individual behaviour. In others, the focus is on how institutions condition farmer responses in an ever-changing agribusiness landscape. The continuing struggle of the smallholder-dominated agricultural sector to flourish in a wider marketplace for cost-effective produce and an increasingly quality-conscious consumer characterizes many failures of development while offering glimpses of how farmer psychology, entrepreneurial tendencies and novel partnerships can shape the viability of agribusiness.

Entrepreneurship is often coined as the driving force of the market-driven economy. It unleashes the forces of value creation and finds expression in many economic enterprises and organizational settings. The lack of apparent success in the smallholder agriculture is attributed by many to lack of entrepreneurial traits in the average smallholder farmer. Kodithuwakku and Weerakoon (2019) discuss the level of integration of entrepreneurship in local agribusiness. An extensive survey of literature by authors highlights the relative neglect of agribusiness in the global and local work on entrepreneurship. It is argued that the production orientation eclipses the market sensing capabilities of the smallholder, thus creating an unfavourable setting for entrepreneurial success. Among potential solutions, social enterprises which have a social mission with a commercial mechanism, may prove a catalyst for entrepreneurship in agribusiness sector.

Two of the contributions (by Anuradha (2019) and by Jayawardena (2019), respectively) consider the psychology of farmer and the extension service provider. In the former, the context is the agricultural sector facing climate change. The article is motivated by the burgeoning literature on adaptation to climate change. In the latter, attention is drawn to emotional intelligence of the value chain participants in general and the extension service provider in particular. The two themes are closely related in that the challenges of the agricultural sector call for a psychological basis for adaptation[1] and efficient mediation. Several psychological constructs are highlighted in Anuradha (2019) as pivotal to understanding the crisis response and adaptation process. These are, namely, risk perception and appraisal, coping appraisal, subjective norms and social norms. In addition, affective responses, such as fear, worry, anxiety and hope, are also found to be significant determinants of adaptation behaviour. A corollary to coping, not necessarily beneficial, is the role of heuristics and habits. Nonetheless, these are also key bases of adaptive responses and will constitute the viability of adaptation response. The authors of the article on psychology of adaptation invoke the Protection Motivation Theory to argue that risk appraisals of threat and possible adaptation responses against the stressor can generate one of two opposing responses: adaptation and maladaptation. In particular, if the perceived risk and the adaptation capacity are high, it is likely that an individual will adapt to prevent damage from the threat, while low-risk perception and adaptation capacity may lead to maladaptation. The case for emotional intelligence is made in the other contribution which argues that the emotional style of extension or advisory service officers could impact the behaviour of farmers. The many uncertainties in agriculture call for high emotional maturity to manage these

[1] An operationalisation of adaptation as a latent construct in climate resilience can be found in Prasada (2020).

situations. The failed expectations and frustrations of the average farmer are cited as reasons for inculcating emotional intelligence to the advisory and extension services.

The individual-level drivers are only one part of the solution. The organizational innovations are common tools to circumnavigate the numerous inefficiencies of agricultural production and distribution processes. Public-Private (PP) partnerships came into vogue in the 1990s along with the economic structural adjustment processes, mainly at the behest of the World Bank and IMF. They were advertised as solutions to corruption, capital inadequacy and management lapses in large ventures. While the PP model worked for utilities and infrastructure, the barriers of agricultural sector posed complexities that were difficult to tackle. The Public-Private-Producer (PPP) partnership is the institutional model that emerged in response to agricultural sector's complexities. The PPP model aims to solve the same kind of issues that PP model addresses but through wider participation of the grassroots actors, mainly the farmers. The article on agricultural PPP in Sri Lanka by Prasada (2019) provides selective evidence from a recent program of PPP development in Sri Lanka. The wider context of agricultural PPPs is discussed, and the process of PPP development and its key indicators are discussed from the view of key stakeholders.

References (Cited from This Publication)

Anuradha JMPN (2019) A psychological lens to conceptualize Sri Lankan farmers' adaptation behavior in the face of significant environmental stressors

Bandaranayake PCG (2019) Use of biotechnology for crop improvement in Sri Lanka: current status and future prospects

Basnayake BFA, Ariyawansha RTK, Gamage DAS, Karunarathna AK (2019a) Experiences of biochar – applications for sustainable agriculture in Sri Lanka

Basnayake BFA, Ariyawansha RTK, Karunarathna AK (2019b) Compendium of principles applied and technologies developed for managing municipal solid wastes in Sri Lanka

Dandeniya WS, Dharmakeerthi RS (2019) Integrated plant nutrient management in major agricultural soils of Sri Lanka: a review of the current status and the way forward

Dayawansa NDK, De Silva RP (2019) Geo-informatics: contribution from spatial sciences for agricultural development and food security

De Costa DM (2019a) Microbial pesticides towards eco-friendly agriculture: present status and future prospects in Sri Lanka

De Costa WAJM (2019b) Increasing climate resilience of cropping systems in Sri Lanka

Dissanayeke U, Pahan Prasada DV, Wickramasuriya H (2019) ICT based information systems in agricultural extension and their economic implications: Sri Lankan perspectives

Fonseka RM, Fonseka HHD, Abhyapala KMRD (2019). Crop Wild Relatives (CWR): an underutilized genetic resource for improving agricultural productivity and food security

Jayawardena LNAC (2019) Emotional intelligence and interpersonal relations in managing the outcomes of agricultural extension and advisory services

Kodithuwakku SS, Weerakoon C (2019) Embracing entrepreneurship in Sri Lankan agribusiness research: a review and a research agenda

Madhujith T, Sivakanthan S (2019) Taking trans fats out of the food supply

Madhujith T, Wedamulla N (2019) Functional foods and health

Mohotti AJ, Pushpakumara DKNG, Singh VP (2019) Shade in tea plantations: a new dimension with an agroforestry approach for a climate smart agricultural landscape system

Mowjood MIM, Jinadasa KBSN, Basnayake BFA (2019) Paddy field and constructed wetland: the equivalencies
Pahan Prasada DV (2019) Public–private–producer (PPP) partnerships in Sri Lankan agriculture
Prasantha BDR (2019) Promising modified atmosphere storage methods to protect self-stable food commodities in Sri Lanka
Ranil RHG, Pushpakumara DKNG, Fonseka RM, Fonseka HHD, Bandaranayake PCG, Weerakkody WAP, Ariyaratne WMTP, De Silva AN, Gunawardhana NPT (2019) Utilizing neglected crop genetic resources for food and nutritional security: special reference to indigenous vegetables of Sri Lanka
Samita S, Sivanathawerl T, Suriyagoda LDB (2019) Statistics for furtherance of agricultural research in Sri Lanka
Weerasooriya S, Hemachandra D (2019) Empirical application of theory of firm in agriculture research in Sri Lanka: a review of the literature
Wijeratne M (2019) Communication in agricultural extension towards the achieving sustainable food Systems in Sri Lanka

References (Sources Outside this Volume)

Prasada DVP (2016a) Economic origins of dietary diseases: is obesity becoming a middle income problem?. Procedia Food Sci 6:113–116
Prasada DVP (2016b) Impact of legislature regarding ‘right to food’ and ‘wheat fortification’ on child malnutrition: cross-country estimates. Procedia Food Sci 6:108–112
Prasada DVP (2020) Climate resilience and varietal choice: a path analytic model for rice in Bangladesh. J Agribus Develop Emerg Econ. https://doi.org/10.1108/JADEE-09-2019-0135

Crop Wild Relatives: An Underutilized Genetic Resource for Improving Agricultural Productivity and Food Security

R. M. Fonseka, H. H. D. Fonseka, and K. M. R. D. Abhyapala

1 Introduction

Food production for human consumption is a challenge under the intense pressure of the increasing food demand, scarcity of natural resources, and the competition for arable land for the biofuel production. Increase in food production to meet the food demand is challenging mainly due to environmental degradation, climate change, and emerging threats of pests and diseases in animals and plants (Godfray et al. 2010; Garrett et al. 2011). During the last few decades, efforts have been made to increase the input (i.e., energy, water, agrochemicals) use efficiency in agriculture through various agronomic means. However, in the coming decades, higher yields are to be achieved with less inputs, as many of those agricultural inputs are becoming limited resources (Godfray et al. 2010; Ray et al. 2013). Further, already limited land resources are becoming more limited for cropping due to salinity, loss of fertility, and mismanagement (Fita et al. 2015). Hence, future food production should be planned in a way to produce "more from less" where low-input responsive crop varieties play a crucial role. In this context, boosting crop yields to meet rising demands is challenging, and achieving global food security would be one of the most pressing issues for mankind in the coming years (Godfray et al. 2010; Nayyar and Dreier 2012). Efficient crop, soil, and water management practices together with new crop varieties having increased resilience to stress factors will overcome the detrimental impacts of climate change and lead to improved food security, livelihoods, and environmental security (Dar and Gowda 2013; Fita et al. 2015). Moreover,

R. M. Fonseka (✉)
Department of Crop Science, Faculty of Agriculture, University of Peradeniya, Peradeniya, Sri Lanka

H. H. D. Fonseka
Onesh Agri Pvt Ltd., Colombo, Sri Lanka

K. M. R. D. Abhyapala
University of Sri Jayewardanepure, Colombo, Sri Lanka

R. P. De Silva et al. (eds.), *Agricultural Research for Sustainable Food Systems in Sri Lanka*, https://doi.org/10.1007/978-981-15-3673-1_2

expected climate change scenarios which may threaten the agricultural production recognized as more frequent extreme weather events such as heat waves, drought, storms, heavy precipitation, high temperature, and floods (Hansen et al. 2012; Parry et al. 2007) consequently lead to degradation of agricultural lands (Lindgren et al. 2011; Miraglia et al. 2009). By the year 2050, most parts of the globe will experience a hotter growing season than the hottest growing seasons of recent times (Battisti and Naylor 2009). Moreover, elevated temperatures and increased CO_2 may lead to decreased photosynthesis and biomass production in C_3 crops (Ruiz-Vera et al. 2013), while lower temperatures may prolong cropping seasons (Abhayapala et al. 2014, 2018). Such erraticism in weather patterns is expected to increase and negatively affect the production of major crops (Lobell et al. 2008); as a result the world could see significant production catastrophe in the future.

According to Yohannes (2016), “adapting” agriculture seems to be one of the possibilities to combat the problem, where shifting planting dates and switching to novel crop varieties with relevant traits are considered as some alternative strategies. Securing twenty-first-century global food demand and to cope up with the impact of climate change have created an increased need for effective crop breeding programs to deliver new crop varieties with higher productivity and resilience to the impacts of greater biotic and abiotic stresses in agriculture (Jarvis et al. 2008; Lobell et al. 2008; Henry and Nevo 2014; Fita et al. 2015). To address these complex and multifaceted problems, we need to transform current food systems into sustainable food systems, a paradigm shift, for which agricultural biodiversity are a key player (Bommarco et al. 2013). Crop wild relatives, an integral part of agricultural biodiversity found in forests, agro-forests, homegardens, etc., provide year-round healthy, nutrient-rich species and varieties, and some with the medicinal value which are often well adapted to local conditions. Moreover, they possess high nutritional value and the wide range of ecosystem plasticity; hence they can be cultivated with minimal external inputs (i.e., pesticides, fertilizers, etc.) and agronomic practices than the modern crop varieties (Castañeda-Álvarez et al. 2016; Singh et al. 2009).

In order to develop a new variety, plant breeder has to improve an already present trait or to add an entirely new trait. In this endeavor, the next critical step is to find sources of the appropriate gene(s) for making the desired change. From the breeders' point of view, the narrow genetic base of modern cultivars is the major bottleneck for crop improvement efforts, and, therefore, the use of crop wild relatives (CWR) is a promising approach to enhance genetic diversity of cultivated crops (Mammadov et al. 2018). In this context, crop wild relatives play a vital role in providing genes of interest outside the gene pool of cultivated species.

2 Crop Wild Relatives

A crop wild relative (usually abbreviated as CWR) is a plant species occurring in the “wild” that is the progenitor or closely related species in the same genus to a particular domesticated species. For instance, rice *Oryza rufipogon* and *Oryza nivara* are wild relatives of *Oryza sativa* L.; the domesticated species of rice, *Solanum incanum* and *Solanum insanum*, are wild relatives of *Solanum melongena* L.; the domesticated species of eggplant and *Abelmoschus angulosus*, *A. ficulneus*, and *A. moschatus* are the wild relatives of cultivated okra (*Abelmoschus esculentus* L.). A formal definition of a CWR, "A crop wild relative is a wild plant taxon that has an indirect use derived from its relatively close genetic relationship to a crop; this relationship is defined in terms of the CWR belonging to gene pools 1 or 2, or taxon groups 1 to 4 of the crop," has been proposed by Maxted et al. (2008).

Approximately a total of 50,000–60,000 crop and CWR species of which 10,739 considered to have of high potential to contribute to food security. About 1000 of these being very closely related to the important food crops such as cereals, vegetables, root and tubers, medicinal and fruits or fodder and forage crops (Maxted and Kell 2009). A survey carried out on the use of CWR for the crop improvement by Hodgkin and Hajjar (2008) revealed that, even though the use of CWR in plant breeding has certainly been done from early times, the commercial use of CWR is first recorded at the end of the nineteenth century and the potential significance of them in crop improvement programs was recognized by Vavilov and other pioneers of the genetic resources concept. Potential use of these CWR in plant breeding to produce new varieties with the ability to withstand adverse impacts of climate change, increasing scarcity of nutrients, water, and other inputs, and new pests and diseases and their use in sustainable agriculture are amply reported by many authors (Choudhary et al. 2017; Hodgkin and Hajjar 2008; Hopkins and Maxted 2011; Jarvis et al. 2008; Singh et al. 2009). Use of CWR in the past 20 years in some released cultivars and traits involved are shown in Table 1.

A comprehensive review published by Maxted and Kell (2009) citing 91 articles reported the identification and transfer of useful traits from 185 CWR taxa into 29 crop species. They also reported that the degree to which breeders had used CWR diversity varied markedly between crops.

Accordingly, the major actual and potential uses of crop wild relatives can be summarized as follows:

Table 1 Use of crop wild relatives in the past 20 years in released cultivars of some important crops (Hajjar and Hodgkin 2007)

Crop	Pest and disease resistance	Abiotic stress	Yield	Quality	Total number of contributed traits
Cassava	+	–	–	+	3
Wheat	+++++++++++	–	+	+	9
Rice	+++++++	+++	+	–	12
Banana	++	–	–	–	2
Tomato	++++++++++	++	–	++	55
Groundnut	+	–	–	–	1

+ indicates number of wild relatives that have contributed beneficial traits to crop varieties
– indicates wild relatives have not contributed beneficial traits in that category

1. As food crops, condiments, medicinal, forage and ornamental plants
2. Prevention of genetic uniformity and the consequences of genetic vulnerability (the susceptibility of most of the cultivated varieties of a crop species to biotic and abiotic stresses as a result of genetic uniformity) as CWR represent a large diverse gene pool
3. Crop yield increase (dramatic yield increase in major world food crops such as rice, wheat, and sorghum was accomplished through introgression of unadapt genes, e.g., dwarf genes)
4. Introduction of new quality traits (e.g., starch, protein and lycopene content, other bioactive compounds, etc.)
5. Introduction of resistance genes to biotic and abiotic stresses (e.g., resistance to major pests and diseases, salinity, drought, etc.)

As modern cultivars are highly genetically uniform, high genetic diversity found in CWR described above would be an attractive as well as an essential option for developing new crop varieties adapted to emerging challenges. The use of CWR in improvement programs for a wide range of crops did not, however, gain real prominence until the 1970s and 1980s (Hoyt and Brown 1988). Moreover, with a wealth of potentials, CWR remain a relatively low priority in collection and conservation efforts by the global gene banks and other regional gene banks due to financial and political impediments (Fielder et al. 2015; Maxted et al. 2013; Maxted and Kell 2009).

In addition to their value in plant breeding, CWR are of direct socioeconomic importance. They can be used as food, fodder and forage crops, medicinal plants, condiments, ornamental and forestry species, as well as plants used for industrial purposes, such as oils and fibers (Fielder et al. 2015). A number of wild cowpea species (*Vigna* spp.) in Africa contribute directly to food security through consumption of their tubers, fruits, and seeds, while wild yams (*Dioscorea* spp.) are an important source of carbohydrates and a source of income in Madagascar. In Sri Lanka also, tubers of number of *Dioscorea* spp. are consumed. Similarly, wild fruits such as apple, pistachio, etc. are harvested for food in Central Asia (http://www.cropwildrelatives.org/cwr/importance/).

Despite that CWR contain genes resistance to biotic and abiotic stresses, adaptability to seasonal differences, etc., they themselves are now under the threat of extinction. CWR, like other wild species, are subjected to a range of threats such as the change in land use, climate change, nitrogen deposition, and alien invasive species (Ford-Lloyd et al. 2011). As a result, over 70% of the total CWR species are facing the threat of extinction, and over 95% are insufficiently represented with respect to the full range of geographic and ecological variation in their native distributions (Castañeda-Álvarez et al. 2016) that demand immediate programs for collection and conservation. Therefore, in an era facing the challenges of genetic erosion of diversity and extinction of species, systematic collection, conservation, and characterization of the available genetic diversity of CWR are immensely important (Choudhary et al. 2017; Dempewolf et al. 2017; Maxted et al. 2012).

Table 2 Extent of genetic uniformity in selected crops

Crop	Country	Number of varieties
Rice	Sri Lanka	From 2000 in 1959 to fewer than 100 today; 75% descend from a common stock
Rice	Bangladesh	62% of varieties descend from a common stock
Rice	Indonesia	74% of varieties descend from a common stock
Wheat	USA	50% of crop in 9 varieties
Potatoes	USA	75% of crop in 4 varieties
Soybeans	USA	50% of crop in 6 varieties

Source: Thrupp (2000)

Table 3 Past global crop failures due to genetic uniformity

Year	Country	Crop	Effect
1800s	Sri Lanka	Coffee	Rust epidemic. Plantations destroyed
1846	Ireland	Potato	Famine
1940s	USA	Various crops	Loss to insects doubled
1943	India	Rice	Famine
1960s	Philippines, Indonesia	Rice	Tungo virus epidemic
1984	Florida, USA	Citrus	18 million trees destroyed

Source: Thrupp (2000)

2.1 Genetic Diversity and Value of Crop Wild Relatives

In the process of domestication, crops passed through a genetic bottleneck, at times losing some "undesirable traits," ending up with genetic uniformity in contrast to the wide genetic variability that exists in the wild species that maintained through generations (Tanksley and McCouch 1997; Dempewolf et al. 2012). Throughout the history of agriculture and mainly after the green revolution, selective breeding for commercial traits made the crop plants more genetically uniform (Thrupp 2000), and the depleted genetic diversity resulted more vulnerable to biotic and abiotic stresses of cultivated crops (Jarvis et al. 2008). Tables 2 and 3 show the extent of genetic uniformity and consequent global crop failures in the past due to the vulnerability of some selected crops in different countries including Sri Lanka. For example, more than half of the genetic variation has been lost in cultivated soybean (Hyten et al. 2006; Sedivy et al. 2017), 2–4% of maize genes experienced artificial selection (Wright et al. 2005), and genetic diversity has been significantly reduced in cultivated rice (Xu et al. 2011). In contrast, CWR have been exposed to natural selection in their native range that maintained a high degree of genetic diversity that can be effectively utilized in crop enhancement for novel traits. Coffee plantations in Sri Lanka during 1869 and potato cultivation in Ireland during 1846 were devastated due to epidemics of coffee leaf rust (*Hemileia vastatrix*) and potato blight (*Phytophthora infestans*), respectively. If scientists in the 1840s were aware that potato CWR (*Solanum demissum*) contain genes for resistance to the disease that triggered the Irish Potato Famine, they would have saved million Irish people who died from starvation and related causes.

Through the lessons learned over the decades, breeders were able to introduce traits from wild species into cultivated crops, mostly to overcome effects of genetic vulnerability due to biotic stress. Some of those revolutionary achievements include the introduction of late blight [caused by *Phytophthora infestans* (Mont.) de Bary] resistance from the wild potato *Solanum demissum* Lindl. and stem rust (caused by *Puccinia graminis* spp. *graminis* Pers.) resistances from the wild wheat, *Aegilops tauschii* Coss. (Prescott-Allen and Prescott-Allen 2009; Kilian et al. 2010). One of the remarkable achievements has been made in rice breeding using CWR. After screening more than 17,000 cultivated and wild rice samples, scientists of International Rice Research Institute found a wild relative of rice, *Oryza nivara*, growing in the wild in Sri Lanka and India with a single gene for resistance to the grassy stunt virus. This gene is now routinely incorporated in all new varieties of rice grown across Asian rice fields (Madurangi et al. 2011). IRRI reported that more than 30 million hectares in 30 countries were planted with varieties of virus-resistant rice bred to contain genes from *Oryza nivara*. Before the introduction of the resistant gene, at least 116,000 ha of rice fields were damaged, and the value estimated was more than US$ 2.5 billion (Kush and Virk 2005). Also, there are evidences for the use of several wild species of *Aegilops* to produce new varieties of wheat resistance to leaf rust, powdery mildew, and nematodes (Schneider et al. 2008), while commercial tomato varieties were produced using genetic variation that occurs in wild species (Rick and Chetelat 1995; Bai and Lindhout 2007; Robertson and Labate 2007).

Other than direct use in crop improvement programs for biotic stress tolerance or for higher yield potential, CWR have also been used to boost the nutritional value of foods. Chatzav et al. (2010) found wide genetic variation for grain nutrients, with the high concentrations of grain zinc, iron, and protein in wild species of wheat than domesticated genotypes. For example, a wild relative of wheat, *Triticum turgidum* var. *dicoccoides*, from the Eastern Mediterranean, was used to increase the protein content of bread and durum wheat (Kovacs et al. 1998), while a wild potato contributed to increased calcium content in potato (Spillman 2013). In addition, doubling of protein content in Brazilian cassava also attributed to crop wild relatives (Hoisington et al. 1999). By crossing cultivated broccoli with a wild Sicilian relative, scientists were able to develop a new variety of broccoli containing 100 times more sulforaphane, cancer-fighting chemical, an antioxidant that destroys compounds that can damage DNA (IPGRI and UNEP 2004). Some wild relatives of eggplant are reported to present high contents of phenolic acids, chlorogenic acid, and other bioactive compounds that are of interest for developing new eggplant cultivars with higher amounts of bioactive compounds (Kaushik et al. 2015; Plazas et al. 2016; Prohens et al. 2013). It was proven that eggplant is amenable to interspecific hybridization with a large number of wild species, from the primary and secondary and even with tertiary gene pool materials (Plazas et al. 2016). Hence, these wild species of eggplants could be utilized effectively in developing new eggplant cultivars with traits such as improved yield, fruit size, and quality as well as cultivars resilience to climate change. *S. insanum*, the closest wild relative of eggplant, which is naturally distributed in Southeast Asia, Madagascar, and Mauritius (Knapp et al. 2013; Vorontsova et al. 2013; Mutegi et al. 2015), can be utilized to improve functional quality (phenolic acids, chlorogenic acid, etc.) of *S. melongena* (Prohens et al. 2013). However, past evidences suggest that wild relatives of eggplant have not been utilized as compared to tomato in crop improvement programs (Diez and Nuez 2008). In Taiwan, two tomato varieties were bred to have six times more vitamin A than in average tomato with the characteristic orange color to promote its consumption as a functional food, especially to overcome widespread vitamin A deficiency of people in developing countries especially in Asia including Sri Lanka. The source of this gene called Beta was discovered in the wild tomato *Solanum habrochaites* (Zamir 2001). It was also reported that valuable genetic traits of wild rice had been used to increase domesticated rice yield, quality, and resistance to biotic and abiotic stresses (Madurangi et al. 2011; Brar and Khush 1997; Xiao et al. 1998).

Not only providing genes of interest for crop improvements, CWR possess many other direct uses in agriculture. There are evidences that CWR with characters such as high vigor, resistance to soilborne diseases, higher tolerance to environmental stresses, higher postharvest quality, etc. have been used as rootstocks in grafting fruit and vegetable crops (Fallik and Ilic 2014; Perera and Warshamana 1987; Rahman et al. 2002; Rouphael et al. 2010). For example, the wild eggplant, *Solanum torvum*, is commonly used for eggplant grafting due to its multiple disease resistance and high vigor (Rahman et al. 2002; Ranil et al. 2015, 2017). Further, it was reported that some CWR thrives well under marginal conditions (low rainfall, high temperatures, high pressure of pests and diseases, etc.) that would be highly stressful for elite varieties of crops (Dempewolf et al. 2014; Dwivedi et al. 2008; Warschefsky et al. 2014). Their high levels of genetic diversity can be utilized to expand a crop's range of cultivation into environments that are more extreme than those in which it was domesticated (Warschefsky et al. 2014). Besides, more importantly, crop wild relatives are of essential components of natural and agricultural ecosystems, where they are indispensable for ecosystem services as well as for maintaining ecosystem health (Wijesundara 2006).

The information presented above suggests that the crop wild relatives represent a "multi-trait" gene pool, hence an integrated approach involving multiple disciplines such as management and utilization of genetic resources, crop physiology, crop breeding, molecular biology and genomics, agronomy, stress tolerance, and reproductive/seed biology will help to address the future global challenges discussed earlier (Dempewolf et al. 2017; Dwivedi et al. 2017).

In terms of economic importance, it was reported that crop wild relatives have saved millions of dollars, both directly and indirectly, by improving crop resilience to biotic and abiotic stresses (Dwivedi et al. 2008). Estimates of the economic value of CWR vary widely. However, a business firm, PricewaterhouseCoopers, estimated the total potential value of wild gene pools of 32 major global crops up to US$ 196 billion (https://foodtank.com/news/2017/11/crop-trust-wild-relatives). An insight on the value and benefits of crop wild relatives gained through plant breeding was given in a joint press release by the International Plant Genetic Resources

Institute and the United Nations Environment Program (IPGRI and UNEP 2004). According to them, between 1976 and 1980, estimated contribution by wild relatives in yield and disease resistance to the farm economy of the USA alone was about 340 million US$ per year. Some other estimates of the economic impact of genetic transfers from CWR are the wild relative of tomato (*Lycopersicon peruvianum* (L.) Mill.) has made it possible to increase the total solids content of the tomato by 2.4%, worth 250 million US$ a year, and traits incorporated from wild relatives into sunflower are worth USD 267–384 million annually to the sunflower industry in the state of California, USA (Frison and Atta-Krah 2008). Similarly, wild relatives have contributed to increase the productivity of globally important food crops such as rice, wheat, barley, maize, oats, and potatoes. However, these treasurable genetic resources still remain grossly underutilized, and the above achievements represent no more than the proverbial tip of the iceberg (Bains et al. 2012).

3 Techniques for Using CWR in Crop Improvement

As described by Dempewolf et al. (2017) in their extensive literature review, a process of introducing genetic diversity from wild species into cultivars for food security requires a significant amount of time, resources, and human capacity. Developing a new crop variety with the desired new character using their wild relatives often involves crossing the crop with a wild relative that has the desired character (donor parent), obtaining the hybrid offspring, and then backcrossing this over several generations with the recurrent parent. This lengthy, time-consuming process, starting in the wild, locating the genetic resource, identification and collection by experts, botanists, and taxonomists and thereafter, passes through gene bank managers for the conservation and characterization process, and geneticists, agronomists, and pathologists take over for further characterization and evaluation of traits and finally breeders for variety development (Dempewolf et al. 2017). However, in order to achieve the desired product, plant breeders often have to overcome some major problems such as the issues associated with wide hybridization, i.e., infertility, negative linkage drag, and incompatibility. These issues should be dealt with appropriate techniques such as embryo rescue; for example, resistance to "little leaf" disease in eggplant is available in *Solanum gilo* and *S. integrifolium* which are not crossable with cultivated *S. melongena* due to interspecific barriers (Prohens et al. 2017). Similar incompatibility barriers were reported in crop improvement programs in *Abelmoschus* species. In many instances, the hybrids and subsequent generations of offspring continue to possess noncommercial, undesirable traits such as spines, saponins, etc., derived from the wild relatives. Efforts to remove these undesirable characteristics through further backcrossing can be a slow process and would delay the development of new varieties with the desired traits (Dempewolf et al. 2017; Hodgkin and Hajjar 2008). Inadequate phenotypic and genotypic data of CWR is one of the barriers to their increased use by the breeders to effectively select the accessions based on their genetic background (Dempewolf et al. 2017; Kadam et al. 2016). Similarly, lack of knowledge and data regarding how much diversity is in each accession, genetics of the traits of interest, how similar accessions are, how much duplication exists within and among collections, how to capture the most diversity across the gene pool, etc. limits their use in research and development programs. These deficiencies in collection, evaluation, and conservation of CWR are also reported by Maxted et al. (2012).

In addition to the issues mentioned above, perceptions on wild species' inferiority relative to elite material, lack of understanding of the genetic base of complex traits, poor management and maintenance of CWR in ex situ collections, lack of institutional support, and funding limitations are considered as major challenges that hinder the increased use of CWR in developing countries (Dempewolf et al. 2017).

Moreover, procedures which involved developing commercial cultivars with genes introduced from wild relatives can be expensive and time-consuming (e.g., some linkages with wild genes need to be broken, in tomato, it took 12 years to break the linkage between nematode resistance and undesirable fruit characters). Therefore, identifying, isolating, and transferring traits of interest from wild species into crop backgrounds, and then evaluating the resulting material, requires significant funding and long-term commitment (Dempewolf et al. 2017).

With the advances in novel biotechnological applications of omics-scale technologies and other advanced techniques such as high-throughput next-generation sequence technologies, transcriptomics, proteomics, genetic engineering, etc., the scope of efficient exploration of useful genetic variation of wild species for the genetic enhancement of cultivated crops seems more promising than before (Brar and Singh 2011, Brozynska et al. 2016). Recently, a new approach called "introgressiomics" has been proposed by Prohens et al. (2017) as a means to overcome "breeding barriers," which consists of mass-scale development of plant materials and populations introgressed with CWR into the genetic background of crops. The ultimate aim of introgressiomics is to provide a significantly enlarged genetic pool which enhances the efficiency of selection for complex traits. Subsequently, "introgressiomics" approach expected to deliver a new generation of cultivars with dramatically improved yield and resilience to climate change scenarios while contributing to sustainable agriculture.

However, the use of such techniques in plant breeding remains relatively expensive and technically challenging, and the tools and technical knowledge are not available to all plant breeders working on all crops especially in the developing countries like in Sri Lanka, where the use of such techniques are scarcely reported (Madurangi et al. 2011; Samarajeewa et al. 1998; Perera and Warshamana 1987). Therefore, the use of conventional breeding techniques for interspecies gene transfer between closely related species is likely to remain the global norm (Maxted and Kell 2009; Maxted et al. 2012).

3.1 CWR for Increase Yield

Efforts have been made to utilize the genetic diversity in crop wild relatives to improve yields of various crop species (Maxted et al. 2013; Nevo and Chen 2010).

Three independent yield-promoting genomic regions introduced from *Solanum pennellii*, a green-fruited wild relative of tomato, have led to developing hybrids with a 50% increased yield over a leading variety (Gur and Zamir 2004). Wild *Helianthus annuus* and *H. petiolaris* Nutt. have been used in the production of high-yielding commercial sunflower hybrids, and it significantly expands the sunflower industry (Hodgkin and Hajjar 2008).

These crop improvement efforts have been concentrated primarily on crop species including wheat, barley, rice, potato, and tomato (Foolad and Panthee 2012; Nevo and Chen 2010; Xiao et al. 1996). Zamir (2001) explained possible reasons for the greater use of CWR which has been restricted only to certain crops as (1) cross compatibilities, (2) the taxonomic relationship between crops and their close wild species, (3) low fertility in the F1 and subsequent progeny, (4) availability or conservation of CWR, (5) exploration and utilization of wild germplasm, and (6) regional financial support based on local need and geographical distribution of CWR.

3.2 Adaptation to Biotic and Abiotic Stresses

Crop yields are reduced due to various biotic stresses by pathogens such as fungi, viruses, bacteria, nematodes, and insect pests. Traditionally, in commercial agriculture insect control is done using insecticides. With the development of insecticide resistance in insect populations, due to the subsequent occurrence of pesticide residues in food and food products, and adverse effects of pesticide use on the environment, the alternative methods of pest control have received considerable attention. Host plant resistance is one of the most economical and environmentally friendly methods of keeping insect pest populations below economic threshold levels. However, continuous use of the limited number of resistant resources is not a long-term strategy for durable resistance since new pathogens and insects evolved very rapidly. To mitigate this problem, breeders have been exploiting genetic resources such as CWR, to develop varieties resistant to biotic stress (Hajjar and Hodgkin 2007; Lynch 1990; Xiao et al. 1996). Some important vegetable crop wild relatives as sources of resistance for biotic stress are shown in Table 4. The use of CWR may lead to potentially environmental-friendly methods of insect pest control and may be safe for beneficial soil living organisms (Sharma 2009). Currently, introgressing genes from the wild relatives into cultivated crops for incorporating resistance to different pests are common, for example, groundnut rust, *Puccinia arachidis* Speg., and late leaf spot, *Cercosporidium personatum*, and these derivatives are less susceptible than commercial cultivars. In rice, *Oryza nivara* genes provided resistance to grassy stunt virus on millions of hectares of rice fields in the south and Southeast Asia (Barclay 2004). In lettuce, downy mildew, *Bremia lactucae*, and lettuce aphid, *Nasonovia* spp., resistance was derived from the wild (Crute 1992; Eenink et al. 1982).

At present, salinity and drought are the most common abiotic factors limiting worldwide crop yields. These two stresses have been intensively studied in crops

Table 4 Important vegetable crop wild relatives as sources of resistance for biotic stress

Crop	Disease	Source
Tomato	Curly top, tomato mosaic, *Verticillium* wilt, powdery mildew, and leaf curl	*L. chilense*
	Leaf spot	*L. glandulosum*
	Leaf spot, bacterial canker	*L. hirsutum*
	Powdery mildew	*L. hirsutum* var. *glabratum*
	Nematode	*L. hirsutum* var. *typicum*
	Fusarium wilt, spotted wilt, bacterial canker, early blight	*L. peruvianum* and *L. pimpinellifolium*
Eggplant	Bacterial wilt	*Solanum insanum*, *S. nigrum*, *S. xanthocarpum*, *S. sisymbriifolium*, *S. toxicariu*, and *S. indicum*
	Root knot nematode	*S. sisymbriifolium* and *S. elaeagnifolium*
	Phomopsis blight	*S. xanthocarpum*, *S. nigrum*, *S. gilo*, *S. sisymbriifolium*, *S. torvum*
Okra	YVMV and powdery mildew	*Abelmoschus manihot*
Pumpkin	Powdery mildew	*C. lundellianai*
Cucumber	Cucumber green mottle mosaic virus	*C. anguria*

Source: Singh et al. (2009)

such as soybeans, tomato, and cereals as well as in the wild relatives (Munns et al. 2012; Placido et al. 2013; Qi et al. 2014). Six barley cultivars with drought tolerance derived from *Hordeum spontaneum* genes have been exploited for tolerance of soils with acidic sulfate content in Vietnam (Nguyen et al. 2003). In tomato, *Lycopersicon chilense* and *L. pennellii* genes have been used to increase drought and salinity tolerance (Rick and Chetelat 1995). Chickpea cultivar with introgressed wild genes for drought and temperature tolerance is derived from *Cicer reticulatum*, and it is already a leading cultivar in Northern India (Maqbool et al. 2017).

4 Options for Conservation of CWR Diversity

According to Maxted and Kell (2009), there are approximately 50,000–60,000 species of CWR, of which 10,000 are considered as high potential value to secure food security, among 1000 of these being closely related to the important food crops. Hence, the conservation of CWR is increasingly recognized as a high priority (Ford-Lloyd et al. 2011; Hunter et al. 2012), and the conservation of these species in their natural habitats is important for their continued evolution (Maxted and Kell 2009). Even though CWR are known to possess resilient genes to face the threats of climate change, climate change itself is likely to be a threat to crop wild relatives in addition to the severe impacts due to human activities such as deforestation and land clearance for agriculture and development purposes leading to habitat destruction

(Dempewolf et al. 2014; Ford-Lloyd et al. 2011, Jarvis et al. 2008; Norton et al. 2017). As they pointed out, these threats demand to have programs to collect, characterize, evaluate, and conserve traditional varieties (landraces) and wild relatives in order to have them available for use in future crop improvement programs. They also predicted that 16–22% of wild relatives of species with direct value to agriculture may be in danger of extinction. The International Union for Conservation of Nature announced that by the end of 2017, more than 20 wild relatives of agricultural crops are in danger. Among wild relatives of important food security crops such as banana and plantain, cassava, sorghum, and sweet potato are in urgent need of collection and conservation, along with those of pineapple, carrot, spinach, and many other fruits and vegetables. Even for the wild relatives of vital staples like rice, wheat, potato, and maize, which tend to be better represented in gene banks, there are still significant gaps in their collections according to the sources (https://www.sciencedaily.com/releases/2016/03/160321123704.htm). Jarvis et al. (2008), therefore, suggested that increased habitat conservation will be important to conserve most species, while gene banks should target as a priority for those predicted to extinct for collection and inclusion. According to the records of Commission on Genetic Resources for Food and Agriculture (2010), approximately seven million crop accessions are stored ex situ in gene banks worldwide; however, these gene collections are often incomplete, and critical information is lacking for most of these materials. Furthermore, a very little attention has been paid on the socioeconomic aspects on CWR collection and utilization which is of paramount importance to develop a benefit-sharing plan at the local level (Gunarathne 2010, unpublished data). In the light of the above critical issues, collection and conservation of this diversity are critical because the future of materials that is still in the wild is precarious.

There are two methods of conservation of crop wild relatives. One type of crop genetic conservation is ex situ maintenance of genetic resources in gene banks, botanical gardens, and agricultural research stations (Plucknett et al. 1987). Another type is in situ maintenance of genetic resources on-farm or in natural habitats (Brush 1991; Maxted et al. 2008). However, Maxted and Kell (2009) reported that CWR account for only 2–6% of global gene bank collections and of the total number of CWR species that exist, only about 6% have any accessions conserved ex situ. For the expansion of ex situ conservation activities, assistance is needed at the regional level for expansion of infrastructure facilities, technology improvements, and institutional and manpower development.

4.1 In Situ Conservation

There are two types of in situ conservation of crop wild relatives. First, in situ conservation refers to the persistence of genetic resources in their natural habitats, including areas where everyday practices of farmers maintain genetic diversity on their farms. This type is a historical phenomenon, but it is now especially visible in regions where farmers maintain local, diverse crop varieties (landraces), even though modern, broadly adapted, or higher-yielding varieties are available.

Second, in situ conservation refers to specific projects and programs to support and promote the maintenance of crop diversity, sponsored by national governments, international programs, and private organizations. In situ conservation practices and projects in agriculture theoretically can concern the wide spectrum of genetic resources relating to crops, from wild and weedy relatives of crop species to the intraspecific diversity within crop species (Maxted et al. 1997).

In situ conservation of cultivated plants requires novel approaches, while in situ conservation of wild crop relatives can draw on theories and methods developed for conserving many different species in their natural habitats. Focusing on variation within cultivated species is warranted by the fact that this type of diversity is arguably the most important one for the future viability of agricultural evolution, as it has been in the past (Maxted et al. 1997).

4.2 Ex Situ Conservation

Plant species and varieties can be preserved under artificial conditions away from the places where they naturally grow. Ex situ conservation is the conservation of components of biological diversity outside their natural habitats (CBD 1992). There are different methods of ex situ conservation. Among these methods, the storage of seeds in seed banks has some advantages for preserving species, but can only be used for species with seeds capable of remaining viable after long-term storage. In comparison to some other common methods of ex situ conservation, the advantages of seed storage can include low cost, less risk of disease, and more efficient use of space or land. Seed storage can be 50–500 times cheaper per collection than field gene banks or in vitro storage. However, seeds coming under the recalcitrant seed category cannot be stored in seed banks as they lose viability if their moisture contents are reduced to the required level (Maxted et al. 2008). Despite recent efforts to conserve important crops, CWR are underrepresented in gene banks, and a systematic effort is needed to conserve CWR for future food security. In addition, assessment of the diversity and relationships of the cultivated species will facilitate the establishment of conservation strategies and the use of genetic resources in breeding programs.

5 Crop Wild Relatives in Sri Lanka

Despite the small land size (6,524,540 ha) of the country, Sri Lanka is one of the most biologically diverse countries in Asia. Sri Lanka's biodiversity is considered to be the richest per unit area in the Asian region, overtaking several mega diversity countries such as Malaysia, Indonesia, and India (Sri Lanka's Fifth National Report to the Convention on Biological Diversity 2014). The global importance of the island's biodiversity has placed Sri Lanka together with the Western Ghats of

India among the 34 biodiversity hotspots in the world. Sri Lanka is characterized by a wide range of agroclimatic conditions, which account for the diversity of biological resources that exist in the country. Probably the most important of these resources is the immense genetic diversity of the various crop plants grown in the country. Some of the contributing factors for the rich biodiversity in the island are a wide range of topographic, climatic, edaphic, and hydrologic variation, biogeography and its influence on the evolutionary history of the biota of Sri Lanka, long history of agriculture in the country, and farmers' selection and cultivation of plant species over millennia and traditional farming practices to suit the diverse agricultural ecosystems. Prior to the year 2004, little attention was given to conserving and utilizing CWR, and a few had been comprehensively studied or researched. An inventory of food CWR in Sri Lanka was compiled using already published material on the Sri Lankan flora (Hasanuzzaman et al. 2003) and the records of the national herbarium (Table 5).

The list includes 410 species of food CWR, belonging to 47 families and 122 genera. Of these, 366 are native species and 77 are endemic relatives of food crops, while 44 species are naturalized exotics. Though ex situ conservation of plant genetic resources was started in Sri Lanka with the establishment of Royal Botanic Gardens, Peradeniya, in the year 1821, the nodal organization dealing with ex situ conservation of crop genetic resources, the Plant Genetic Resources Centre (PGRC), Gannoruwa, Peradeniya, was established in 1988. Records at PGRC show that the 308 accessions of CWR belong to 26 species of food crops that are stored at PGRC, of which about 2.7% of the total accessions are conserved. These figures are comparable with our neighboring countries such as India and Bangladesh those reported to have 320 and more than 300 species of CWR, respectively (FAO Country Report 2007). It is recorded that among the vegetables and fruits, there are about 10 species of wild *Vigna* out of 16 species present in the country (Liyanage et al. 2006; Samarasinghe et al. 2009; Liyanage 2010), 12 species of wild okra, 16 wild relatives of cucurbits, 7 wild relatives of orange and related crops, and 7 wild species of *Passiflora* and parent species of the cultivated banana (*M. acuminate* and *M. balbisiana*).

Hurtado et al. (2012) assessed the genetic diversity and relationships of 52 accessions of eggplant from three geographically distant secondary centers of diversity (China, Spain, and Sri Lanka) using 28 morphological descriptors and 12 highly polymorphic genomic simple sequence repeats (SSR). The total genetic diversity (H_T) within each origin was high, ranging between $H_T = 0.5400$ (Sri Lanka) and $H_T = 0.4943$ (China), and materials of Sri Lanka presented the highest number of alleles (63), followed by Chinese (52) and Spanish (50) accessions, which corresponds, respectively, to 57.3%, 47.3%, and 45.5% of the total number of alleles detected.

The number of species of CWR under different crop categories conserved at PGRC is shown in Table 6. For the effective utilization of these conserved materials, it is vital to ascertain the true genetic relationships of these species, valuable traits of interest, their genetics, nutraceutical properties, and barriers in using them in breeding programs must be studied in detail to facilitate the utilization process.

Table 5 List of available wild relatives of some selected food crops in Sri Lanka

Crop/plant	Crop wild relatives
Paddy	*Oryza eichingeri* A. Peter
Oryza sativa (cultivated species)	*Oryza nivara* Sharma et Shastry
	Oryza rhizomatis Vaughan
	Oryza rufipogon Griff.
	Oryza granulata
Eggplant	*Solanum torvum*
Solanum melongena L. (cultivated species)	*Solanum insanum*
	Solanum incanum
	Solanum violaceum
Pepper	*Piper zeylanicum* Miq.
Piper nigrum L. (cultivated species)	*Piper walkeri* Miq.
	Piper hymenophyllum Miq.
	Piper sylvestre Lam.,
	Piper trineuron Miq.,
	Piper siriboa L.,
	Piper chuvya (Miq.) C. DC.
	Piper betle L., (cultivated)
	Piper longum L.
Green gram	*Vigna aridicola* N. Tomooka and Maxted
Vigna radiata (cultivated species)	*Vigna dalzelliana*
	Vigna radiata var. *sublobata*
	Vigna stipulacea
	Vigna trilobata
	Vigna trinervia
Banana	*Musa acuminate* Colla.
Musa paradisiacal L. (cultivated species)	*Musa balbisiana* Colla.
Cinnamon	*Cinnamomum dubium* Nees
Cinnamomum verum J. Presl (cultivated species)	*Cinnamomum ovalifolium* Wight
	Cinnamomum litseaefolium Thw.
	Cinnamomum citriodorum Thw.
	Cinnamomum capparu-coronde Blume
	Cinnamomum sinharajaense Kostermans
	Cinnamomum rivulorum Kostermans
	Cinnamomum camphora (L.) Presl

Source: https://www.doa.gov.lk/index.php/en/ct-menu-item-9/cwr-project

Conserving wild relatives of crops – mainly *Oryza* spp. and *Vigna* spp. – was addressed through a project for the conservation of crop wild relatives, and 22 locations have been identified for in situ conservation of crop wild relatives in addition to the protected areas where they occur. The most important areas outside PA in this regard are the Manikdena Archaeological Reserve and Arboretum, Waulpane Forest, and Thumbathanna Forest (Liyanage 2009). Economic valuation

Table 6 Number of wild relative species conserved at PGRC, Sri Lanka, up to July 2018 (Unpublished data)

Crop group	No. of wild spp.	% of total species
Cereals	11	17
Legumes	17	26
Vegetables	26	40
Oil seed	03	5
Fruits	03	5
Root and tubers	00	–
Fiber crops	00	–
Spices and condiments	04	6
Others	01	2
Total	65	

for in situ conservation of *Oryza granulata* was undertaken in Wavulpane area indicating that the villagers of the adjacent area are willing to pay (WTP) Rs. 82.73 (~0.5US$) per annum for in situ conservation program as labor hours if the conservation program is executed only by the government-authorized body (Dissanayake et al. 2009).

In Sri Lanka most of its forests are rich in different CWR, and it is an important center for collection of crop wild relatives (Wijeratne and Piyasiri 2016). However, at present, CWR populations are under threat of extinction despite having 80 laws and regulations directly or indirectly relating to biodiversity conservation and signatory to seven international convections related to conservation of biodiversity. This may be attributed to the fact that most of their populations are not found in protected areas (PA) and are therefore vulnerable to deforestation, urbanization, and overextraction as food. Among the factors that posed serious threats to the preservation of natural floristic diversity in Sri Lanka are lack of public awareness as indicated by Dissanayake et al. (2009), heavy rate of deforestation due to various development projects and the selective felling of trees for timber and removal of plant species as wild harvested plants (WHP) particularly those with food and medicinal value and unplanned land use, pollution, and fragmentation. A decline in the natural forest cover over a century shows the gravity of the problem. At the beginning of the last century, natural forest cover was 70% of the total land area, while the latest figure, however, shows that the natural forest cover has decreased to about 29.7%.

A quarter of the total land area of Sri Lanka is reserved for forests and administered by the Department of Forests and the Department of Wildlife Conservation. Currently, the country's protected areas occupy around 26.5% of the total land area of the country (Table 7). The first environmental protection area (EPA) for wild rice relatives in Puttalam was declared under provincial law. Compared to Sri Lanka, India and Pakistan reported to have 659 and 233 EPAs, respectively (FAO Country Report 2007).

Table 7 Comparison of areas protected by countries participated in CWR project (2004)

Country	Land area (km^2)	Total protected area (km^2)	Total number of sites	% area protected
Armenia	29,800	2991	28	10
Bolivia	1,098,580	230,509	50	21
Madagascar	587,040	18,458	60	3
Sri Lanka	65,610	14,877	264	23
Uzbekistan	447,400	20,503	24	5

Source: https://www.bioversityinternational.org/2.crop_wild_relatives_project_countries.pdf

Table 8 National Red List status of endemic wild *Cinnamomum* species in Sri Lanka

Species name	Red list category
Cinnamomum dubium	NT
Cinnamomum ovalifolium	I
Cinnamomum litseaefolium	T
Cinnamomum rivulorum	T
Cinnamomum sinharajaense	T
Cinnamomum capparu-coronde	HT
Cinnamomum citriodorum	HT

NT not threatened, *I* intermediate, *T* threatened, *HT* highly threatened
Source: Kumarathilake et al. (2010)

Wild rice in Asia provides one of the classic examples for coupling wild species with breeding techniques for the genetic improvement of crops. Wild rice grown in the coastal areas of Sri Lanka may provide gene(s) that helps to develop a rice variety that can be cultivated in paddy fields affected by salinity due to possible sea level rise expected due to global warming. The cultivated species of cinnamon (*Cinnamomum verum*) is one of the important export agricultural crops in Sri Lanka and is well regarded as a high-value spice crop which attracts higher demand in the local as well as international markets and its uses for medicinal purposes (Wijesundara 2006).

It was reported that the endemic *Cinnamomum* species are encountered to a serious threat and therefore, both in situ and ex situ conservations were started in Kanneliya Forest Reserve. Moreover, community conservation groups were strengthened in the surrounding villages to mitigate illegal removal and conserve "Kapuru Kurundu" (*Cinnamomum camphora*) plant. Table 8 shows the National Red List status of endemic *Cinnamomum* species of Sri Lanka. The *Cinnamomum camphora* is a wild relative of cultivated cinnamon, an introduced plant to Sri Lanka, while other seven species are endemic to the country. Wijeratne and Piyasiri (2016) reported that the endemic *Cinnamomum* species are encountered to a serious threat of extinction as they provide a significant socioeconomic value to the communities and therefore, they have a tendency to rapid exploitation, especially by the people living in adjoining areas of forest reserves. According to Kumarathilake et al. (2010), lack of awareness, habitat destruction, urbanization, and poor agricultural practices were found to be the major threats to the wild cinnamon species, and due to the prevalence of the threats, those endemic *Cinnamomum* species have been recognized under the National Red List; for example, *C. capparu-coronde* is identified as highly threatened (HT) species in the National Red List.

To our knowledge, use of wild relatives of crop plants to incorporate valuable traits has not been systematically tried in many crop breeding programs though advance technologies and other facilities are available to undertake such research programs. Breeding methods used in vegetable improvement program in DOA show breeders prefer to use cultivated species as parents than CWR for the development of vegetable varieties. Of the total vegetable varieties released so far, 15% are local or foreign introductions; 71% are selections of landraces and farmer varieties including five varieties [*Thumbika*, *Golika*, *Visal*, *Kesara*, and *Parakum* (pollen parent)] of spine gourd (*Momordica dioica*), a wild relative of *M. charantia*, locally known as *Thumba Karawila*; and only 15% of the varieties developed using hybridization. Parents involved in most of the crossbred varieties were cultivated species (Unpublished data) with the exception of using *M. dioica* and *M. subangulata*, two wild relatives of *M. charantia*, as parents for the development of a F1 hybrid which were released in 2015 by the Department of Agriculture, Sri Lanka. Other than the use of *Momordica* spp., efforts to use CWR in wide hybridization are confined to a few instances, i.e., rice (Hemachandra 2008; Madurangi et al. 2011), okra (Samarajeewa et al. 1998), brinjal (Perera and Warshamana 1987), and spine gourd (Hitinayake et al. 2017). Except rice and spine gourd, there is no other single food crop variety developed using crop wild relatives.

6 Potential for Expanding the Use of CWR in Sri Lanka

Since the last few decades, Sri Lanka too are experiencing extreme weather events such as droughts, floods, cyclones, landslides, and issues such as saltwater intrusion owing to climate change. Sri Lanka was ranked fourth among ten most affected countries in the Global Climate Risk Index for 2016 (Eckstein et al. 2018), while the same ranking for 2015 was 98 and for the period of 1997–2016 was 45. As the report indicates, “Sri Lanka was hit by cyclone Roanu in May 2016, after already having experienced severe droughts during the beginning of the year. Floods and landslides took the lives of over 100 people and displaced half a million. The economic damages are estimated at US$ 2 billion.” In this context, there is a growing fear that these extreme climatic events can disturb agriculture in Sri Lanka. This necessitates being ready with mitigating measures, developing new crop varieties incorporating traits that are resilient to changing temperatures, increased pest and disease incidence, the higher level of salinity, and water shortages where crop wild relatives will be the key players. As indicated above, Sri Lanka has wide array of CWR for different types of crops such as rice, eggplant, banana, pepper, cinnamon, green gram, chilli, tomato, etc., which can provide adequate genetic resources for the development of new crop varieties with traits adaptable to drought, heat, salinity, water and nutrient limitations, and biotic stresses. Either through conventional breeding or using advanced technologies such as molecular breeding, genomics, and introgressiomics (Zhou et al. 2015; Brozynska et al. 2016; Prohens et al. 2017), plant breeders would be able to transfer desired traits from crop wild relatives to domesticated crops. However, it appears that this valuable genetic resource has not yet been properly utilized for crop variety improvement in Sri Lanka except for rice. Problems and the prospect of using CWR in national crop improvement programs were discussed by Fonseka and Fonseka (2010). In one initiative, Department of Agriculture (DOA) of Sri Lanka, in collaboration with the global project on crop wild relatives, implemented by the United Nations Environment Programme, making efforts through the Agriculture Information Park in Gannoruwa, Peradeniya, to increase public awareness on the potential role of wild relatives in crop improvement. So far, wild relatives of pepper, bean, okra, banana, and rice have been established along the banks of the Mahaweli River (Wijesekera 2006; Wijesekara et al. 2006). This initiative seems to be very effective since most of the national parks and other protected areas of Sri Lanka were not objectively established to conserve CWR but to conserve particular habitats or selected charismatic species. Meanwhile, the DOA has established the second Agriculture Information Park in Southern Sri Lanka, which will also feature a section devoted to crop wild relatives and expects to improve public awareness on CWR through these sites where people are able to observe and recognize the country's hidden genetic treasures and to protect them accordingly. Besides, these sites will also help to conserve the wild relatives outside their habitats.

Also, in the year 2013, a research project was initiated with a systematic approach to utilize the considerable genetic variation found in eggplant in Sri Lanka and some wild relatives from around the world to utilize eggplant wild relatives for pre-breeding. The project partners represent three geographical regions (Sri Lanka, Spain, and Ivory Coast) and are continuing to date, in which the University of Peradeniya is a partner. The scope of the project is to utilize wild relatives of eggplants belonging to primary, secondary, and tertiary gene pools through introgression and interspecific hybridization and backcrossing using cultivated local eggplant varieties as recurrent parents. Those pre-bred materials developed for desired traits, particularly drought and biotic and abiotic stresses, are then subjected to evaluation and incorporating them into breeding pipelines to improve cultivated eggplants targeting to develop new varieties which are of commercial interest and more resilient and adaptive especially to local conditions under climate change. So far a large number of backcross populations and introgressed materials involving parents of 9 cultivated *Melongena* species, 4 primary gene pool, 23 secondary gene pool, and 8 tertiary gene pool were produced, and further fixing of genetic material and characterization for traits of interest are going on with the participation of 4 international private seed companies (https://www.cwrdiversity.org/partnership/eggplant-pre-breeding-project/). The project is funded and managed by the Global Crop Diversity Trust with the Millennium Seed Bank of the Royal Botanic Gardens, Kew, England.

7 Conclusions

During the next few decades, food production needs to be increased substantially to meet the growing demand. Efforts have been made to increase the agricultural production by increasing the efficiency of inputs used in agriculture through various agronomic means. However, in the face of rapidly depleting resource, future food production endeavors should be planned to produce "more from less" where low-input responsive new crop varieties to be bred are not only resilient to biotic and abiotic stresses coupled with higher yields and enhanced quality but also to thrive under low-input farming environments. In this context, demand for effective crop breeding programs to deliver crops with higher productivity and resilience to the impact of greater biotic and abiotic stresses in agriculture are imperative. Breeders are now in the quest of probing for new genetic diversity to add entirely new traits required to face the challenges posed. CWR collectively constitute an enormous reservoir of novel genetic diversity which can be utilized to produce new varieties that withstand adverse impacts of climate change and new pests and diseases and to thrive well under input-limited growing conditions while ensuring enhanced production and food and nutrient security and sustaining productivity in an environment-friendly manner.

Though novel breeding techniques are available elsewhere, instances of application of them to develop new varieties using CWR are scarcely reported in Sri Lanka. This may be attributed to lack of human resource capacity and lack of understanding of the genetic base of complex traits, poor management and maintenance of CWR in ex situ collections, lack of institutional support, and funding limitations. Human resource capacity building, especially in the areas of plant breeding, cytogenetics, plant taxonomy, and curation of gene bank collections of wild material, would be supportive in effective utilization of these wild accessions. As suggested by many authors throughout, it is apparent that fixing of traits of interest from wild species into genetic backgrounds of cultivated crops and their improvement and evaluation process is time-consuming and requires state-of-the-art laboratory facilities and long-term commitment in addition to a committed human resource.

According to various reporters, Sri Lanka possesses a wide array of CWR for a large number of food crops, assuring adequate genetic resources and diversity for the production of new crop varieties. Even though the potential may be enormous, they remain as a greatly underutilized and neglected or yet to exploit resource, and at present, CWR populations are under threat due to habitat disturbances and human activities. These threats demand to have programs toward the systematic collection, characterization, and conservation of CWR in order to use them for future crop improvement programs where significant funding and strengthening of gene bank facilities play a crucial role. However, effective conservation and utilization of this "hidden treasure" would be able to deliver more diverse, resilient, and resource-efficient new crop varieties that can contribute to a more sustainable and productive agriculture despite the threats posed by the climate change.

8 Recommendations

Sri Lanka is blessed to have a large number of CWR of economically important crops. Majority of these CWR are victims of habitat destruction and are on the verge of extinction. In addition to the CWR listed in Tables 5 and 6 which were already conserved, concerted efforts should be extended for the collection, evaluation, conservation, and utilization of other important wild relatives available and to gather vital information necessary to identify and to protect the proprietary right of them. The collection and conservation efforts are progressing satisfactorily, but the progress of evaluation and characterization are not up to the expectations. Compared to other countries, very serious lagging can be observed in the utilization of crop wild relatives in local breeding programs. As remedial measures upgrading of the facilities at PGRC for maintenance, evaluation, characterization (both phenotypic and molecular), and documentation and to provide basic facilities at regional commodity research institutes to maintain regional collections, expanding the capacity of long-term storage, establishing a cryopreservation facility and developing a computer-based national database on plant genetic resources for use by the plant breeders and other researchers, strengthening the National Herbarium, and improving the quality of training programs related to plant genetic resource are some of the recommendations to upgrade the collection, maintenance, and evaluation process. Enhanced human capacity to employ better methods to maintain and use CWR is based on reestablishing the skill base and investing in strong breeding capacity. Though it is common to use wild species for developing commercial cultivars with genes introduced from wild relatives, it can be expensive and time-consuming. Hence, logical use of available molecular-assisted breeding techniques to enhance the efficiency of breeding methodologies and to develop new and improved varieties not only for pest and disease resistance but tolerance to extreme temperatures, tolerance to salinity, drought resistance, and nutritional quality should be encouraged. To facilitate this central facility should develop for molecular techniques. These targets can be achieved through the implementation of prioritized, nationally coordinated breeding programs to provide a national grid for breeding in collaboration with centers of excellence elsewhere. Perceptions of CWR's inferiority relative to the elite material, lack of understanding of the genetic base of complex traits, poor management and maintenance of CWR in ex situ collections, lack of institutional support, and funding limitations are some of the other challenges to be addressed under local conditions.

In order to protect these resources, the creation of public awareness through mass media and the inclusion of CWR in school, diploma and undergraduate curricula, farmer participatory programs, etc. should be considered. While such mechanisms to conserve CWR are vital, sustainable utilization of them is equally essential. However, exploiting this potential requires the capacity to improve varieties through plant breeding as well as partnerships and networks that encompass all relevant stakeholders, ranging from farmers to researchers to gene bank managers and policy makers to administrators. Economic valuation of these plant genetic resources, though not been calculated yet, shall shed light on "hidden" biodiversity values of both use (instrumental) and nonuse (intrinsic) for better decision-making by policy makers and administrators about their potential contribution to the economy and the benefits and costs to make conservation and utilization endeavors financially sustainable. Or else the "burden of proof" will be either on the conservationists or on the plant breeders.

References

Abhayapala KMRD, De Costa WAJM, Fonseka RM, Prasannath K, De Costa DM, Suriyagoda LDB, Abeythilakeratne PD, Nugaliyadde MM (2014) A response of potato (*Solanum tuberosum* L.) to increasing growing season temperature under different soil management and crop protection regimes in the Upcountry of Sri Lanka. J Trop Agric Res 25(4):555–569

Abhayapala KMRD, De Costa WAJM, Malaviarachchi MAPWK, Kumara JBDAP, Suriyagoda LDB, Fonseka RM (2018) Exploitation of differential temperature-sensitivities of crops for the improved resilience of tropical smallholder cropping systems to climate change: a case study with temperature responses of tomato and chilli. J Agric Ecosyst Environ 261:103–114

https://www.cwrdiversity.org/partnership/eggplant-pre-breeding-project/. Accessed on 1 Aug 2018

https://www.bioversityinternational.org/2.crop_wild_relatives_project_countries.pdf. Accessed on 25 July 2018

http://www.cropwildrelatives.org/cwr/importance/. Accessed on 25 July 2018

https://www.doa.gov.lk/index.php/en/ct-menu-item-9/cwr-project. Accessed on 25 July 2018

https://foodtank.com/news/2017/11/crop-trust-wild-relatives/. Accessed on 29 March 2020

Bai Y, Lindhout P (2007) Domestication and breeding of tomatoes: what have we gained and what can we gain in the future? Ann Bot 100(5):1085–1094

Bains NS, Singh S, Dhillon BS (2012) Enhanced utilization of plant genetic resources in crop improvement programmes. Indian J Plant Genet Resour 25(1):52–62

Barclay A (2004) Feral play: crop scientists use wide crosses to breed into cultivated rice varieties the hardiness of their wild kin. Rice Today:14–19

Battisti DS, Naylor RL (2009) Historical warnings of future food insecurity with unprecedented seasonal heat. Science 323(5911):240–244

Bommarco R, Kleijn D, Potts SG (2013) Ecological intensification: harnessing ecosystem services for food security. Trends Ecol Evol 28(4):230–238

Brar DS, Khush GS (1997) Alien introgression in rice. Plant Mol Biol 35:35–47

Brar DS, Singh K (2011) Oryza. In: Kole C (ed) Wild crop relatives: genomic and breeding resources. Springer, Berlin/Heidelberg. https://doi.org/10.1007/978-3-642-14228-4_7

Brozynska M, Furtado A, Henry RJ (2016) Genomics of crop wild relatives: expanding the gene pool for crop improvement. Plant Biotechnol J 14(4):1070–1085

Brush SB (1991) A farmer-based approach to conserving crop germplasm. Econ Bot 45:153–165

Castañeda-Álvarez NP, Khoury CK, Achicanoy HA, Bernau V, Dempewolf H, Eastwood RJ, Guarino L, Harker RH, Jarvis A, Maxted N, Müller JV, Ramirez-Villegas J, Sosa CC, Struik PC, Vincent H, Toll J (2016) Global conservation priorities for crop wild relatives. Nat Plants 2:16022. https://doi.org/10.1038/nplants.2016.22

CBD (1992) Convention on biological diversity: text and annexes. Secretariat of the Convention on Biological Diversity/Cornell University, Montreal/Urbana, pp 1–34

Chatzav M, Peleg Z, Ozturk L, Yazici A, Fahima T, Cakmak I, Saranga Y (2010) Genetic diversity for grain nutrients in wild emmer wheat: potential for wheat improvement. Ann Bot Preview

Choudhary M, Singh V, Muthusamy V, Wani S (2017) Harnessing crop wild relatives for crop improvement. Int J Life Sci 6:73. https://doi.org/10.5958/2319-1198.2017.00009.4

Commission on Genetic resources for Food (2010) The second report on the state of the world's plant genetic resources for food and agriculture. Food and Agriculture Org. https://scholar.google.com/scholar?cluter=1278742458347738512&hl=en&as_sdt=2005&sciodt=0,5

Crute IR (1992) From breeding to cloning (and back again?): a case study with lettuce downy mildew. Annu Rev Phytopathol 30:485–506

Dar WD, Gowda CLL (2013) Declining agricultural productivity and global food security. J Crop Improv 27(2):242–254

Dempewolf H, Hodgkins KA, Rummell SE, Ellstrand NC, Rieseberg LH (2012) Reproductive isolation during domestication. Plant Cell 24:2710–2717. https://doi.org/10.1105/tpc.112.100115

Dempewolf H, Eastwood RJ, Guarino L, Khoury CK, Müller JV, Toll J (2014) Adapting agriculture to climate change: A global initiative to collect, conserve, and use crop wild relatives. Agroecol Sustain Food Syst 38:369–377

Dempewolf H, Baute G, Anderson J, Kilian B, Smith C, Guarino L (2017) Past and future use of wild relatives in crop breeding. Crop Sci 57:1070. https://doi.org/10.2135/cropsci2016.10.0885

Diez MJ, Nuez F (2008) Tomato. In: Prohens J, Nuez F (eds) Vegetables II. Fabaceae, Liliaceae, Solanaceae, and Umbelliferae. Springer, New York, pp 249–323

Dissanayake DRRW, Jayasinghe-Mudalige UK, Gunawardena UADP, Randeni RPLC, Udugama JMM (2009) Economic Valuation of genetic resources of wild relatives: assessing the preferences of adjacent community for *in situ* conservation of *Oryza granulata* in Wavulpane area. In: Proceedings of 9th Agricultural Research Symposium, pp 76–81

Dwivedi SL, Upadhyaya HD, Stalker HT, Blair MW, Bertioli DJ, Nielen S, Ortiz R (2008) Enhancing crop gene pools with beneficial traits using wild relatives. Plant Breed Rev 30:179–280

Dwivedi SL, Scheben A, Edwards D, Spillane C, Ortiz R (2017) Assessing and exploiting functional diversity in germplasm pools to enhance abiotic stress adaptation and yield in cereals and food legumes. Front Plant Sci 8:1461. https://doi.org/10.3389/fpls.2017.01461

Eckstein D, Künzel V, Schäfer L (2018) Global climate risk index; Who suffers most from extreme weather events? Weather-related loss events in 2016 and 1997 to 2016, Chapman-Rose J, Baum D, Fuhrmann H, Kier G (eds). GERMANWATCH. http://germanwatch.org/en/cri. Accessed on 05/08/2018

Eenink AH, Groenwold R, Dieleman FL (1982) Resistance of lettuce (Lactuca) to the leaf aphid *Nasonovia ribisnigri* 1 transfer of resistance from *L. virosa* to *L. sativa* by inter-specific crosses and selection of resistant breeding lines. Euphytica 31:291–300

Fallik E, Ilic Z (2014) Grafted vegetables – The influence of rootstock and scion on postharvest quality. Folia Hort 26(2):79–90. https://doi.org/10.2478/fhort-2014-0008

FAO (2007) Status of plant genetic resources for food and agriculture. Country reports of Bangladesh, India, Nepal, Pakistan and Sri Lanka

Fielder H, Brotherton P, Hosking J, Hopkins JJ, Ford-Lloyd B, Maxted N (2015) Enhancing the conservation of crop wild relatives in England. PLoS ONE 10(6):e0130804. https://doi.org/10.1371/journal.pone.0130804

Fita A, Rodriguez-Burruezo A, Boscaiu M, Prohens J, Vicente O (2015) Breeding and domesticating crops adapted to drought and salinity: a new paradigm for increasing food production. Front Plant Sci 6:978

Fonseka HH, Fonseka RM (2010) Problems and prospects of using CWR in National Crop Improvement Programmes. In: Marambe B, Wijesekera A (eds) Conservation and utilization of crop wild relatives of Sri Lanka. Book of Abstracts. Department of Agriculture and Ministry of Environment and Natural Resources, p 34

Foolad MR, Panthee DR (2012) Marker-assisted selection in tomato breeding. Crit Rev Plant Sci 31:93–123

Ford-Lloyd B, Schmidt M, Armstrong SJ, Barazani O, Engels J, Hadas R, Hammer K, Kell SP, Kang D, Khoshbakht K, Li Y, Long C, Lu B-R, Ma K, Nguyen VT, Qiu L, Ge S, Wei W, Zhang Z, Maxted N (2011) Crop wild relatives – undervalued, underutilized and under threat? Bioscience 61:559–565

Frison E, Atta-Krah K (2008) Foreword. In: Maxted N, Ford-Lloyd BV, Kell SP, Iriondo JM, Dulloo E, Turok J (eds) Crop wild relatives, conservation and use. CAB International, Wallingford, pp xxiii–xxv

Garrett KA, Forbes GA, Savary S, Skelsey P, Sparks AH, Valdivia C (2011) Complexity in climate-change impacts: an analytical framework for effects mediated by plant disease. Plant Pathol 60 (1):15–30

Godfray HC, Beddington JR, Crute IR, Haddad L, Lawrence D, Muir JF (2010) Food security: the challenge of feeding 9 billion people. Science 327(5967):812–818

Gunarathne LHP (2010) Socio-economic aspects of conservation and utilization of crop wild relatives (Related to the CWR in the Central Region). Draft technical report. Unpublished

Gur A, Zamir RD (2004) Unused natural variation can lift yield barriers in plant breeding. PLOS Biol 2:1610–1615

Hajjar R, Hodgkin T (2007) The use of wild relatives in crop improvement: a survey of developments over the last 20 years. Euphytica 156:1–13

Hansen J, Sato M, Ruedy R (2012) Perception of climate change. Proc Natl Acad Sci U S A 109 (37):E2415–E2423

Hasanuzzaman SM, Dhillon BS, Saxena S, Upadhyaya MP, Joshi BK, Ahmad Z, Qayyum A, Ghafoor A, Jayasuriya AHM, Rajapakse RMT (2003) Plant genetic resources in SAARC countries: their conservation and management. SAARC Agricultural Information Centre, Dhaka

Hemachandra PV (2008) Collection of wild rice germplasm in Sri Lanka. In: Proceedings of the 64th Annual sessions, Sri Lankan Association of Advancement of Science

Henry RJ, Nevo E (2014) Exploring natural selection to guide breeding for agriculture. Plant Biotechnol 12:655–662

Hitinayake HMC, Sumanarathne JP, Abesekara WADS, Madushika KGN, Danushka WM, Sawarnalatha KG (2017) Yield improvement of spine gourd through gynomonoecious hybrid (*Momordica subangulata* sub spp. *renigera x Momordica dioica* Roxb. ex). Ann Dep Agric 19:71–78

Hodgkin T, Hajjar R (2008) Using crop wild relatives for crop improvement: trends and perspectives ves. In: Maxted N, Ford- Lloyd BV, Kell SP, Iriondo JM, Dulloo ME, Turok J (eds) Crop wild relative conservation and use. CAB International, Wallingford, pp 535–548

Hoisington D, Skovmand B, Taba S (1999) Plant genetic resources: what can they contribute towards increased crop productivity? PNAS 96:5937–5943

Hopkins J, Maxted N (2011) Crop wild relatives: plant conservation for food security. Natural England research report NERR037. ISSN 1754-1956

Hoyt E, Brown S (1988) Conserving the wild relatives of crops. IBPGR, IUCN, WWF, Rome/ Gland

Hunter D, Guarino L, Khoury C, Dempewolf H (2012) A community divided: lessons from the conservation of crop wild relatives around the world. In: Maxted N, Dulloo ME, Ford-Lloyd BV, Frese L, Iriondo JM (eds) Agrobiodiversity conservation: securing the diversity of crop wild relatives and landraces. CAB International, Wallingford, pp 298–304

Hurtado M, Vilanova S, Plazas M, Gramazio P, Fonseka HH, Fonseka RM, Prohens J (2012) Diversity and relationships of eggplants from three geographically distant secondary centers of diversity. PLoS One 7(7):e41748

Hyten DL, Song Q, Zhu Y, Choi IY, Nelson RL, Costa JM, Specht JE, Shoemaker RC, Cregan PB (2006) Impacts of genetic bottlenecks on soybean genome diversity. Proc Nat Acad Sci U S A 103:16666–16671

International Plant Genetic Resources Institute and the United Nations Environment Programme (IPGRI & UNEP): A Joint Press Release (2004). Every crop needs its wild relatives. http://www.umsl.edu/~naumannj/Geography%201001%20articles/ch%208%20agricultural%20geography/Every%20crop%20needs%20its%20wild%20relatives.doc

Jarvis A, Lane A, Hijmans R (2008) The effect of climate change on crop wild relatives. Agric Ecosyst Environ 126:13–23

Kadam S, Vuong TD, Qiu D, Meinhardt CG, Song L, Deshmukh R, Nguyen HT (2016) Genomic-assisted phylogenetic analysis and marker development for next generation soybean cyst nematode resistance breeding. Plant Sci 242:342–350

Kaushik P, Prohens J, Vilanova S, Gramazio P, Plazas M (2015) Phenotyping of eggplant wild relatives and interspecific hybrids with conventional and phenomics descriptors provides insight for their potential utilization inbreeding. Front Plant Sci 7:677

Kilian B, Martin W, Salamini F (2010) Genetic diversity, evolution and domestication of wheat and barley in the Fertile Crescent. In: Glaubrecht M (ed) Evolution in action. Springer, Berlin/Heidelberg, pp 137–166. https://doi.org/10.1007/978-3-642-12425-9_8

Knapp S, Vorontsova MS, Prohens J (2013) Wild relatives of the eggplant (*Solanum melongena* L.: Solanaceae): new understanding of species names in a complex group. PLoS One 8:e57039. https://doi.org/10.1371/journal.pone.0057039

Kovacs MIP, Howes NK, Clarke JM, Leisle D (1998) Quality characteristics of durum wheat lines deriving high protein from Triticum dicoccoides (6b) substitution. J Cereal Sci 27:47–51

Kumarathilake DMHC, Senanayake SGJN, Wijesekara GAW, Wijesundara DSA, Ranawaka RAAK (2010) Extinction risk assessment at the species level: national red list status of endemic wild cinnamon species in Sri Lanka. Trop Agric Res 21:247–257. https://doi.org/10.4038/tar.v21i3.3298

Kush GS, Virk PS (2005) IR varieties and their impact. IRRI Publication

Lindgren E, Albihn A, Andersson Y (2011) Climate change, water related health impacts, and adaptation: highlights from the Swedish Government's commission on climate and vulnerability. In: Ford JD, Berrang-Ford L (eds) Climate change adaptation in developed nations – from theory to practice, vol 42. Springer, Dordrecht, pp 177–188

Liyanage ASU (2009) Integrated in-situ system approaches for conservation of wild rice. The International Symposium on Wild Rice, Thailand

Liyanage ASU (2010) Eco-geographic survey of crop wild relatives. Plant Genetic Resources Centre, Gannoruwa

Liyanage ASU, Wasala WMD, Edirisinghe DK, Wijesekera A (2006) Eco-geographic survey of wild species of Vigna in Sri Lanka. Eleventh annual symposium proceedings Part 1. International Forestry and Environment Symposium, Department of forestry and Environmental Science, University of Sri Jayewardenepura, Sri Lanka

Lobell DB, Burke MB, Tebaldi C, Mastrandrea MD, Falcon WP, Naylor RL (2008) Prioritizing climate change adaptation needs for food security in 2030. Science 319(5863):607–610

Lynch R (1990) Resistance in peanut to major arthropod pests. Fla Entomol 73:360–363

Madurangi SAP, Samarasinghe WLG, Senanayake SGJN, Hemachandra PV, Ratnasekera D (2011) Resistance of *Oryza nivara* and *Oryza eichingeri* derived lines to brown plant hopper, *Nilaparvata lugens*. J Natl Sci Found Sri Lanka 39(2):175–181

Mammadov J, Buyyarapu R, Guttikonda SK, Parliament K, Abdurakhmonov IY, Kumpatla SP (2018) Wild relatives of maize, rice, cotton, and soybean: treasure troves for tolerance to biotic and abiotic stresses. Front Plant Sci 9:886. https://doi.org/10.3389/fpls.2018.00886

Maqbool MA, Aslam M, Ali H (2017) Breeding for improved drought tolerance in Chickpea (*Cicer arietinum* L.). Plant Breed 136:300–318

Maxted N, Kell SP (2009) Establishment of a global network for the in situ conservation of crop wild relatives: status and needs. FAO Commission on Genetic Resources for Food and Agriculture, Rome

Maxted N, Ford-Lloyd BV, Hawkes JG (1997) Complementary conservation strategies. In: Maxted N, Ford-Lloyd BV, Hawkes JG (eds) Plant genetic conservation: the in situ approach. Chapman & Hall, London

Maxted N, Kell SP, Ford-Lloyd BV (2008) Crop wild relatives conservation and use: establishing the context. In: Maxted N, Ford-Lloyd BV, Kell SP, Iriondo J, Dulloo E, Turok J (eds) Crop wild relative conservation and use. CABI Publishing, Wallingford, pp 3–30

Maxted N, Kell S, Ford-Lloyd B, Dulloo E, Toledo A (2012) Towards the systematic conservation of global crop wild relative diversity. Crop Sci 52:1–12

Maxted N, Kell S, Brehm JM, Jackson M, Ford-Lloyd B, Parry M (2013) Crop wild relatives and climate change. In: Jackson M, Ford-Lloyd BV, Parry M (eds) Plant genetic resources and climate change. CABI, Wallingford

Miraglia M, Marvin HJ, Kleter GA, Battilani P, Brera C, Coni E (2009) Climate change and food safety: an emerging issue with special focus on Europe. Food Chem Toxicol 47(5):1009–1021

Munns R, James RA, Xu B, Athman A, Conn SJ, Jordans C, Gilliham M (2012) Wheat grain yield on saline soils is improved by an ancestral Na+ transporter gene. Nat Biotechnol 30:360–364

Mutegi E, Snow AA, Mathu R, Pasquest R, Ponniah H, Daunay M, Davidar P (2015) Genetic diversity and population structure of wild/weedy eggplant (*Solanum insanum* L., Solanaceae) in Southern India: implications for conservation. Am J Bot 102(1):140–148

Nayyar D, Dreier L (2012) Putting the new vision for agriculture into action: a transformation is happening. Word Economic Forum. http://www.weforum.org/reports/putting-new-visionagriculture-action-transformation-happening. Accessed on 5 May 2018

Nevo E, Chen GX (2010) Drought and salt tolerances in wild relatives for wheat and barley improvement. Plant Cell Environ 33:670–685

Nguyen B, Brar D, Bui B, Nguyen T, Nguyen H (2003) Identification on and mapping of the QTL for aluminum tolerance introgressed from the new source, *Oryza rufipogon* Griff, into Indica rice (*Oryza sativa* L.). Theor Appl Genet 106:583–593

Norton SL, Khoury CK, Sosa Chrystian C, Castañeda-Álvarez Nora P, Achicanoy Harold A, Steven S (2017) Priorities for enhancing the *ex situ* conservation and use of Australian crop wild relatives. Aust J Bot 65:638–645. https://doi.org/10.1071/BT16236

Parry M, Canziani O, Palutikof J, van der Linden PJ, Hanson CE (2007) Climate change: impacts, adaptation and vulnerability. Cambridge University Press, Cambridge

Perera KDA, Warshamana IK (1987) Graft compatibility, wilt resistance and graft induced changes on *S. melongena* by two wild species, *S. torvum* and *S. indicum*. Trop Agric 143:27–35

Placido DF, Campbell MT, Folsom JJ, Cui XP, Kruger GR, Baenziger PS, Walia H (2013) Introgression of novel traits from a wild wheat relative improves drought adaptation in wheat. Plant Physiol 161:1806–1819

Plazas M, Vilanova S, Gramazio P, Rodrguez-Burruezo A, Fita A, Herraiz Instituto FJ, Rajakapasha R, Fonseka R, Niran N, Fonseka H, Kouassi B, Kouassi A, Kouassi A, Prohens J (2016) Interspecific hybridization between eggplant and wild relatives from different gene pools. J Am Soc Hortic Sci 141(1):34–44

Plucknett D, Smith N, Williams J, Murthi AYN (1987) Gene banks and the world's food. Princeton University Press, Princeton

Prescott-Allen R, Prescott-Allen C (2009) Genes from the wild: using wild genetic resources for food and raw materials. Earthscan Publications Limited, London

Prohens J, Whitaker BD, Plazas M, Vilanova S, Hurtado M, Blasco M, Gramazio P, Stommel JR (2013) Genetic diversity in morphological characters and phenolic acids content resulting from an interspecific cross between eggplant, *Solanum melongena* L, and its wild ancestor (*S. incanum*). Ann Appl Biol 162:242–257. https://doi.org/10.1111/aab.12017

Prohens J, Gramazio P, Plazas M, Dempewolf H, Kilian B, Díez MJ (2017) Introgressiomics: a new approach for using crop wild relatives in breeding for adaptation to climate change. Euphytica 213:158. https://doi.org/10.1007/s10681-017-1938-9

Qi XP, Li MW, Xie M, Liu X, Ni M, Shao GH, Lam HM (2014) Identification of a novel salt tolerance gene in wild soybean by whole-genome sequencing. Nat Commun 5:4340

Rahman MA, Rashid MA, Hossain MM, Salam MA, Masum ASMH (2002) Grafting Compatibility of Cultivated eggplant Varieties with Wild Solanum Species. Pak J Biol Sci 5:755–757. https://doi.org/10.3923/pjbs.2002.755.757

Ranil RHG, Niran HML, Plazas M, Fonseka RM, Fonseka HH, Vilanova S, Andújar I, Gramazio P, Fita A, Prohens J (2015) Improving seed germination of the eggplant rootstock Solanum torvum by testing multiple factors using an orthogonal array design. Sci Hortic 193:174–181. https://doi.org/10.1016/j.scienta.2015.07.030

Ranil RHG, Prohens J, Aubriot X, Niran HML, Plazas M, Fonseka RM, Vilanova S, Fonseka HH, Gramazio P, Knapp S (2017) *Solanum insanum* L. (subgenus Leptostemonum Bitter, Solanaceae), the neglected wild progenitor of eggplant (*S. melongena* L.): a review of taxonomy, characteristics and uses aimed at its enhancement for improved eggplant breeding. Genet. Resour. Crop Evol 64:1707–1722

Ray DK, Mueller ND, West PC, Foley JA (2013) Yield trends are insufficient to double global crop production by 2050. PLoS ONE 8(6):e66428

Rick CM, Chetelat RT (1995) Utilization of related wild species for tomato improvement. Acta Hortic 412:21–38

Robertson L, Labate J (2007) Genetic resources of tomato. In: Razdan MK, Mattoo AK (eds) Genetic Improvement of Solanaceous Crops, 2, Tomato. Science Publishers, Enfield

Rouphael Y, Schwarz D, Krumbein A, Colla G (2010) Impact of grafting on product quality of fruit vegetables. Sci Hort 127:172–179

Ruiz-Vera UM, Siebers M, Gray SB, Drag DW, Rosenthal DM, Kimball BA, Ort DR, Bernacchi CJ (2013) Global warming can negate the expected CO_2 stimulation in photosynthesis and productivity for soybean grown in the Midwestern United States. Plant Physiol 162(1):410–423. https://doi.org/10.1104/pp.112.211938

Samarajeewa PK, Attanayake P, Gamage NST (1998) Interspecific crosses between *A. esculentus* L. and *A. angulosus* L. Trop Agric 152:45–51

Samarasinghe WLG, Liyanage ASU, Jayaweera SLD (2009) Assessment of the threatened status of crop wild relatives of banana, rice and vigna in Sri Lanka. Ann Sri Lanka Dep Agric 11:121–130

Schneider A, Molnar I, Molnar-Lang M (2008) Utilization of *Aegilop*s (goatgrass) species to widen the genetic diversity of cultivated wheat. Euphytica 163:1–19

Sedivy EJ, Faqiang Wu F, Hanzawa Y (2017) Soybean domestication: the origin, genetic architecture and molecular bases. New Phytol 214(2):539–553. https://doi.org/10.1111/nph.14418

Sharma H (2009) Biotechnological approaches for pest management and ecological sustainability. CRC Press, New York

Singh B, Sanwal SK, Rai M, Rai AB (2009) Sources of biotic stress resistance in vegetable crops: a review. Veg Sci 36(2):133–146

Spillman A (2013) Calcium-rich potatoes: it's in their genes. Agric Res Washington 51(3):18–19

Tanksley SD, McCouch SR (1997) Seed banks and molecular maps: unlocking genetic potential from the wild. Science 277:1063–1066. https://doi.org/10.1126/science.277.5329.1063

Thrupp LA (2000) Linking agricultural biodiversity and food security: the valuable role of agro biodiversity for sustainable agriculture. Int Aff 76:283–297

Vorontsova MS, Stern S, Bohs L, Knapp S (2013) African spiny solanum (subgenus Leptostemonum, Solanaceae): a thorny phylogenetic tangle. Bot J Linn Soc 173:176–193

Warschefsky E, Penmetsa R, Cook DR, von Wettberg EJ (2014) Back to the wilds: tapping evolutionary adaptations for resilient crops through systematic hybridization with crop wild relatives. Am J Bot 101(10):1791–1800. https://doi.org/10.3732/ajb.1400116

Wijeratne M, Piyasiri KHKL (2016) Conservation of crop wild relatives: a Sri Lankan experience in community participation. Trop Agric Res Ext 18(2):87–93

Wijesekara A, Herath A, Illankoon J (2006) Base line study on public awareness on crop wild relatives. Ann Sri Lanka Dep Agric 8:251–262

Wijesekera GAW (2006) Bringing crop relatives to the public. In: Crop wild relatives. Bioversity International, p 7

Wijesundara S (2006) Spicy wild relatives get some respect. In: Raymond RD, Moore C (eds) Geneflow. Bioversity International, Rome, pp 26–28

Wright SI, Vroh Bi I, Schroeder SG, Yamasaki M, Doebley JF, McMullen MD, Gaut BS (2005) The effects of artificial selection on the maize genome. Science 308:1310–1314

Xiao JH, Grandillo S, Ahn SN, McCouch SR, Tanksley SD, Li JM, Yuan LP (1996) Genes from wild rice improve yield. Nature 384:223–224

Xiao J, Li J, Grandillo S, Ahn SN, Yuan L, Tanksley SD, McCouch SR (1998) Identification of trait-improving quantitative trait loci alleles from a wild rice relative, *Oryza rufipogon*. Genetics 150:899–909

Xu X, Liu X, Ge S, Jensen JD, Hu F, Li X, Dong Y, Gutenkunst RN, Fang L, Huang L, Li J, He W, Zhang G, Zheng X, Zhang F, Li Y, Yu C, Kristiansen K, Zhang X, Wang J, Wright M, McCouch S, Nielsen R, Wang J, Wang W (2011) Resequencing 50 accessions of cultivated and wild rice yields markers for identifying agronomically important genes. Nat Biotechnol 30:105–111. https://doi.org/10.1038/nbt.2050

Yohannes H (2016) A review on relationship between climate change and agriculture. J Earth Sci Clim Change 7:335. https://doi.org/10.4172/2157-7617.1000335

Zamir D (2001) Improving plant breeding with exotic genetic libraries. Nat Rev Genetics 2:983–989

Zhou Z, Jiang Y, Wang Z, Gou Z, Lyu J, Li W, Yu Y, Shu L, Zhao Y, Ma Y (2015) Resequencing 302 wild and cultivated accessions identifies genes related to domestication and improvement in soybean. Nat Biotechnol 33:408–414

Utilizing Neglected Crop Genetic Resources for Food and Nutritional Security: Special Reference to Indigenous Vegetables of Sri Lanka

R. H. G. Ranil, Gamini Pushpakumara, R. M. Fonseka, H. Fonseka, P. C. G. Bandaranayake, W. A. P. Weerakkody, W. M. T. P. Ariyaratne, A. N. De Silva, and N. P. T. Gunawardena

1 Introduction

Vegetables play an important role as a source of nutrients for human beings, and their consumption rate ensures the intake of various essential vitamins and minerals, mitigating possible malnutrition (Yamaguchi 1983). Indigenous vegetables (IVs) are an important source of food for the less privileged masses as well as the wealthy class in many parts of the developing world. They play a significant role in maintaining food and nutritional security for the underprivileged in both urban and rural areas (Weinberger and Msuya 2004). Most households use them as a primary food or a secondary condiment to vegetable dishes prepared from domestic varieties. They are a valuable source of energy and micronutrients in the diets of rural people in the past and to some extent at present. IVs are traded usually in the rural market

R. H. G. Ranil (✉) · G. Pushpakumara · R. M. Fonseka · W. A. P. Weerakkody · W. M. T. P. Ariyaratne
Department of Crop Science, Faculty of Agriculture, University of Peradeniya, Peradeniya, Sri Lanka

H. Fonseka
Onesh Agri Pvt Ltd., Colombo, Sri Lanka

P. C. G. Bandaranayake
Agricultural Biotechnology Centre, Faculty of Agriculture, University of Peradeniya, Peradeniya, Sri Lanka

A. N. De Silva
Department of Pharmaceutical Botany, Bandaranaike Ayurvedic Research Institute, Maharagama, Sri Lanka

N. P. T. Gunawardena
National Herbarium, Department of National Botanic Gardens, Peradeniya, Sri Lanka

R. P. De Silva et al. (eds.), *Agricultural Research for Sustainable Food Systems in Sri Lanka*, https://doi.org/10.1007/978-981-15-3673-1_3

stands and recently in urban and semi-urban supermarkets too. The indigenous vegetables are also popular among the ethnic groups, and some are cultivated for meeting the high-end markets as they fetch a high price than the local/rural markets. As they are either cultivated in rural areas or grown wild, many species of IVs are less popular and utilized only by the locals, without subjecting them to extensive agronomic improvements. In Sri Lanka, the value of IVs with regard to food security in local households has not been given sufficient attention, though their species diversity is very high within the country. The species diversity of vegetable crops in Sri Lanka comprises over 90 species from 34 families. Despite the high species diversity of vegetable crops, current vegetable production is heavily dependent on about 25–40 species. Many species are of exotic origin, and few species are endemic and indigenous to Sri Lanka (Pushpakumara et al. 2018). Due to commercial cultivation of vegetables in major growing areas, habitats of these species are being invaded by the improved vegetable varieties of local and exotic origin resulting in swift disappearance of IVs from the backyards and their natural habitats.

The term "indigenous vegetable" is highly subjective, and apparently there is no agreement on a widely accepted definition. IVs are often described with terms such as neglected and underutilized species (Jaenicke and Hoschle-Zeledon 2006), and at times they are termed as traditional or local vegetables as well. In Merriam Web master's dictionary, the term "Indigenous" is defined as having originated in and being produced, growing, living, occurring naturally in a particular region or environment. Therefore, it refers to the place of origin (Shanmugasundaram 2007). In this article, we define an indigenous vegetable as "a vegetable species or variety native to particular country or region or environment and its centre of diversity is known to be in the particular country or region of its origin." This is in agreement with the definitions proposed by Weinberger (2007), Engle and Faustino (2007), Chadha and Patel (2007), and Hossain and Razzaque (1999). Accordingly, the vegetable species that have been originated either in Sri Lanka or other countries in South and Southeast Asia are classified as indigenous vegetables in this article. This is because Sri Lankan flora keep strong phytogeographical affinities with South and Southeast Asian elements since origin of Gondwanaland (Gunatilleke et al. 2017). Moreover, this region has been identified as the Indian center [Indo-Burma (7) and Siam-Malaya-Java (7A)] in the Vavilo's center of origin for domesticated crops. Engle and Faustino (2007) have listed the naturalized species or varieties that had been evolved from materials introduced to the region from another geographical area over a long period of time. Even though there may be a possibility to develop new genotypes due to the adaptation to new habitats through the natural selection and selection made by famers, there are no strong scientific evidences for existence of such distinct species or varieties in Sri Lanka. Therefore, in this article, we exclude either the naturalized species or varieties by definition.

2 Taxonomic Diversity

Sri Lanka is particularly blessed with a wide range of ecosystem diversity, which is ideal for growing many species of tropical and subtropical vegetable crops. Due to long history of agriculture and wide range of eco-edaphic conditions present in the

Table 1 Taxonomic diversity of indigenous vegetable of Sri Lanka

Category	Number
Families	47
Genera	82
Species	105
Endemic species	07

country, considerable genetic and species diversity of vegetable crops and their farming systems exist in Sri Lanka. The species diversity of vegetables in Sri Lanka is represented by exotic (introduced), indigenous (native), and endemic species. Despite high species diversity of vegetable crops, current vegetable production is heavily dependent on a few species (about 25–40 species). Many of these species come under underutilized or neglected category, though they depict potent value of medicinal and nutritional properties. Here we are focusing and summarizing various aspects of indigenous vegetable crops of Sri Lanka.

According to the definition proposed earlier, all the available species lists (Alwis and Senanayake 2010; Herat et al. 1994; Pushpakumara et al. 2018; Rajapaksha 1998) were critically reviewed and analyzed. The proposed new list was further enriched and strengthened by information collected from Dassanayake et al. (1994), Dassanayake and Clayton (1995–2000), Dassanayake and Fosberg (1980–1991), Ilankoon and Wijesekara (2008), Jayaweera (1981a, b), MOE (2012), Senaratna (2001), Shaffer-Fehre (2006), Sugathadasa et al. (2008), and Yasapalitha and Rupasinghe (2017). Moreover, additional information was collected in consultation with indigenous medical practitioners, taxonomists, researchers, and selected rural communities. The newly proposed list of IVs of Sri Lanka is attached as Annex I (based on the APG III classification), while a summary of the list is given in Table 1. Moreover, we use the most updated botanical names for all treated species following recent taxonomic update of world flora (www.theplantlist.org). Though endemic species form a separate group within indigenous category, due to their unique biological value in a particular ecosystem, here we do not consider the endemic species as a separate group.

The correct species identification is a fundamental need of any plant-based research. For example, many local publications have erroneously named some of the vegetable species as indigenous or native. *Luffa cylindrica* (L.) M. Roemer [Niyan-wetakolu/sponge gourd], *Phaseolus lunatus* L. [Potu-dambala/lima bean], and *Canavalia ensiformis* (L.) DC. and *Canavalia gladiata* (Jacq.) DC. [sword bean/awara] are common examples of such misuse. Hence, the proposed list will be beneficial to the taxonomists, horticulturists, and researchers who are working on indigenous vegetables for correct identification of the species.

3 Geographical Distribution

Even though the IVs show an island-wide distributional pattern in its Distribution Map 1, the majority is concentrated in the wet zone of Sri Lanka. The map was developed based on the records available at the National Herbarium, Peradeniya,

Map 1 General distribution pattern of IVs species of Sri Lanka based on records at the National Herbarium, Peradeniya

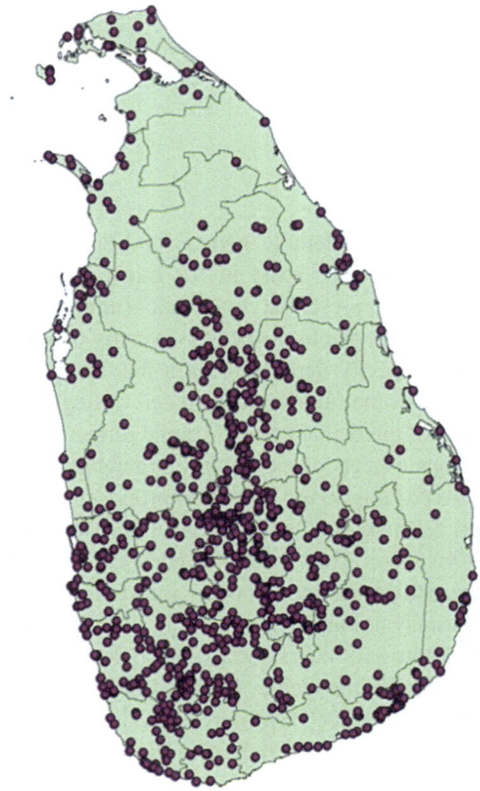

which is based on the specimens collected and conserved over the last century. It is a well-known fact that compared to the dry zone ecosystems, particularly North and East regions, the wet zone has a well-botanized and a large number of authenticated records for many species. Hence, it is necessary to focus on IVs available at other unexplored regions together with their potential domestication in the future. However, the general understanding is that major portion of IVs comes from cultivated small patches in domestic gardens or growing as weeds in marginal areas within farms or abandoned lands rather than in forests and related ecosystems.

4 Food and Medicinal Value

Vegetables are one of the most sustainable and affordable dietary sources of micronutrients and fiber (Palada et al. 2006). This will also empower the households to access a variety of fresh vegetable rich in nutrients and diversify the food basket. The main argument for the promotion of IVs is the supply of dietary fiber, essential vitamins, and minerals which are deficient in the commonly consumed commercial types of vegetables. There is an array of evidence in this regard that IVs contained minerals and vitamins required for a balance diet (Afolayan and Jimoh 2009; Glew et al. 2005; Maisuthisakul et al. 2007). The use of these vegetables could diversify the diets, improve nutrition, increase income, and improve the well-being of the people. Apart from that IVs were an integral component of indigenous medicine since the beginning of human civilization of many Asian and African countries. Their role in indigenous medicine system is significant due to their remarkable medicinal properties (Cheikhyoussef et al. 2011).

The eating habits and availability of IVs are changing from community to community and also according to their socioeconomic status. In general, urban populations have lesser access to IVs, and in the long run, it will help the Asian countries to increase farm income, improve nutritional status, and generate employment in the rural areas since most of these vegetables are high-value and labor-intensive crops. Apart from that their uses in indigenous medicine are well known. Therefore, it is mandatory to collect and document the scattered information on indigenous knowledge and medicinal uses of IVs. After referring the available literature and consultation with a wide range of audience, medicinal uses of few selected vegetable crop species are presented in Table 2.

The food and nutrient security of the rural population could be addressed by paying greater attention to indigenous vegetables which are familiar to them and commonly grown by the rural farmers. Considering the prevalence of over- and unbalanced nutrition-related noncommunicable diseases such as obesity, hypertension, diabetes, and cancer among the urban populations, these crops will play a major role in the future food supply chain. Therefore, the overall objective of promoting IVs is to increase the health, secure food and nutritional status among the rural communities, increase income, and improve the well-being of the people.

5 Phytochemical and Nutritional Properties of IVs

There is a large amount of literature available on nutritional and phytochemical properties of the indigenous vegetables, dating back to the 1940s. While some of them are available in detail, others are rather sketchy. For example, *Momordica charantia* is one of the most comprehensively studied species, even up to the types of sugar molecules present and their derivatives and other secondary metabolites (reviewed by Tan et al. 2016). On the other hand, very limited literature support is available on nutritional and phytochemical properties of *Blechnum orientale* (Lai et al. 2010). Some species, for example, *Hygrophila auriculata*, is phytochemically characterized by several groups, while limited work has been done on nutritional properties. In all the reports, the nutritional composition analyses were done following the standard methods of AOAC (2012). Either fresh or dried plant parts at different maturity stages have been used for analysis of moisture, ash, total available carbohydrate, lipids, fats and oils, amino acids, proteins, specific minerals, and vitamins. The nutritional and phytochemical properties of many IVs have been subjected to studies in other regions of the world of Sri Lanka (its unique geographical, climatic, and soil characteristics) indicating some unique chemical compound in some wild and cultivated species. Hence, phytochemical analysis should be a

Table 2 Medicinal uses of selected IVs in Sri Lanka

Species	Medicinal uses
Aerva lanata (Pol-pala)	Plant is used for treatment of coughs and as a vermifuge for children; roots used to treat headaches. It is also used for the treatment of other health problems such as anemia, Alzheimer, arthritis, cholesterol, lung problems, bone problems, and also blood circulation. Further, used as a remedy for kidney stones, sore throat, and indigestion. The plant is also used as diuretic and demulcent and for the treatment of sudden swellings (Jayaweera 1981a, b; Goyal et al. 2011; Rahman et al. 2013)
Alternanthera sessilis (Mukunuwenna)	It is a laxative and is useful in chronic congestion of the liver, biliousness, and dyspepsia associated with sluggish liver. Good for nursing mothers and used to treat snakebites. Also used to treat hepatitis, tight chest, bronchitis, asthma, and other lung problems, to stop bleeding, and as a hair tonic. Plant is used to reduce fever (Jayaweera 1981a, b; Rahman and Iffat Ara Gulshana 2014)
Amaranthus spinosus (Katu-thampala)	The plant is used as a sweetening and used to reduce fever and is recommended for eruptive fevers. Leaves have soothing effect and increase the milk secretion, good for colic. The root is considered as specific for gonorrhea. Proved to have antimalarial activity. Improves appetite; useful in burning sensation, leprosy, piles, bronchitis, leucorrhoea, constipation, and flatulence. Decoction of the herb is used as a mouthwash for toothache. The root is heating and expectorant; lessens the menstrual flow; useful in leucorrhoea, leprosy, and eczema (Jayaweera 1981a, b; Hilou et al. 2006; Rahman and Iffat AraGulshana 2014)
Amaranthus viridis (Kura-thampala)	Used for treatment of urinary problems and congestion of the liver and in gonorrhea. It is also used as an antidote for snakebite, stings of wasps, and bites of centipedes. The tender leaves are a popular vegetable. The plant is cooling, alexiteric, laxative, stomachic, appetizer, and antipyretic; used in burning sensation, hallucination, leprosy, bronchitis, piles, leucorrhoea, and constipation. The leaves are used as an emollient. The root is heating and expectorant; lessens the menstrual flow (Jayaweera 1981a, b; Rahman and Iffat Ara Gulshana 2014)
Amorphophallus paeoniifolius (Kidaran)	The corm is used externally to relieve pain in acute rheumatism. It is used for treatments of piles, acute dyspepsia, abdominal colic, and skin and blood diseases. The corm and the roots are useful for hemorrhoids. The crushed seed relieves toothache. The corm is eaten during periods of food scarcity (Jayaweera 1981a, b)
Aponogeton crispus (Kekatiya)	Used for feminine diseases, eye diseases, nausea, and diabetes. Nutritious food for weakness of the body (Anon 2002)
Basella alba (Niwithi)	The cultivated forms of this plant are used as a popular leafy vegetable. Medicinally, the roots are used as a poultice to reduce swellings. The plant is a demulcent, diuretic, and emollient. The leaves macerated into a pulp are used on boils, ulcers, and abscesses to hasten suppuration. With butter it is a soothing application on burns and scalds. A decoction of the leaves is a

(continued)

Table 2 (continued)

Species	Medicinal uses
	good laxative for pregnant women and for children (Jayaweera 1981a, b)
Centella asiatica (Gotukola)	Whole plant good for the brain and increases memory revitalizing the nerves and brain cells. Used to treat catarrh, skin rashes, worms, and wounds. Has an antitumor effect and neuroprotective effect. Different ailments like asthma, skin disorders, ulcers, and body aches for improving memory, as a nervine tonic and in treatment of dropsy, elephantiasis, gastric catarrh, kidney troubles, leprosy, leucorrhoea, and urethritis in maternal health care, in treatment of stomach disorders and also as a vegetable (Babu et al. 1995; Brinkhaus 1998; Anon 2002; Singh et al. 2010; Orhan 2012)
Cheilocostus speciosus (Thebu)	Leaves used for eye diseases, worm infections, and burning sensation. The rhizomes and roots are ascribed to be bitter, astringent, acrid, cooling, aphrodisiac, purgative, anthelmintic, depurative, febrifuge, expectorant, tonic, improves digestion and stimulant herb that clears toxins. Juice of the rhizome is applied to head for cooling and relief from headache (Anon 2004; Nadkarni and Nadkarni 2007)
Colocasia esculenta (Gahala)	Besides being a starchy food, the tubers of this plant are laxative, diuretic, lactagogue, and styptic. Juice of petioles is used to treat hemorrhage. It is also used for earache and otorrhoea and also as an external stimulant and rubefacient (Brown and Valiere 2004; Jayaweera 1981a, b)
Cordia dichotoma (Lolu)	The juice of the bark with coconut milk relieves severe colic. In Java, the bark is given for dysentery together with pomegranate rind. The mucilage in the fruit is used for treating coughs and diseases of the chest, uterus, urethra, etc. In larger doses, it is given for bilious ailments as a laxative. It is also used in gonorrhea (Jayaweera 1981a, b)
Cyanthillium cinereum (Monara-kudumbiya)	Especially good for improvement of immunity system. Used for skin diseases and stones in the bladder and gallbladder. Stem and leaf: Decoction for diuretic, kidney disorders, swellings, inflammation, lower abdominal pains, and menstrual pains; also to expel the placenta and as an abortifacient (Anon 2003; Gunjan et al. 2011)
Dregea volubilis (Anguna)	Good for feeding mothers, ulcers, and respiratory diseases. Also used for snakebites. Leaves have anti-inflammatory and anti-rheumatic effects and are used for treating pain, cough, fever, severe cold, and dyspepsia. Studies have proven its therapeutic effects such as antidiabetic, antileukemic, antiulcer, and antioxidant effects (Hossain et al. 2010; Nandi et al. 2012)
Eclipta prostrata (Kikirindiya)	Specially used for hair growth and color and treatment for hair fall. Good for eye diseases, skin diseases, headaches, urine troubles, and gastritis (Anon 2002)
Ehretia microphylla (Heen-thambala)	In South India the root of this plant is used for cachexia and syphilis and as an antidote for vegetable poisons. A decoction of the leaves is used to cure diarrhea accompanied with discharge

(continued)

Table 2 (continued)

Species	Medicinal uses
	of blood and also for cough. The leaves are used for the same ailments in the Philippine Islands (Jayaweera 1981a, b)
Ipomoea aquatica (Kan-kun)	The leaves are used as a vegetable. The plant is mildly laxative, and owing to the presence of insulin-like principle, it is used against diabetes mellitus (Jayaweera 1981a, b)
Lasia spinosa (Kohila)	The leaves, stems, and roots are used as a common remedy for piles. Possess a profound anticestodal effect (Jayaweera 1981a, b; Temjenmongla and Yadav 2005)
Senna auriculata (Ranawara)	The root is used as decoction for fevers, diabetes, diseases of the urinary system, and constipation. The leaves have laxative properties. The dried flowers and flower buds are used as a substitute for tea in the case of diabetic patients. It is also supposed to improve the complexion in women. The powdered seed is also used in diabetes and applied to the eye, in cases of chronic purulent conjunctivitis. Leaf juice is used to reduce body heat (Jayaweera 1981a, b; Sandhya et al. 2006)
Senna tora (Tora)	The leaves are laxative and are useful in habitual constipation and hemorrhoids. The seeds have antiparasitic properties. Seeds are used both externally and internally for all types of eye diseases, liver problems, and boils. The root of the plant is used as a bitter tonic and stomachic and the leaves as an antiperiodic, aperient, and anthelmintic (Jayaweera 1981a, b)
Trianthema portulacastrum (Heen-sarana)	Eaten as a pot herb. Root is abortion cautious and increases menstrual flow. Whole plant is applied as a dressing or poultice. Powdered root is given as a cathartic. Used to cure edema (Jayaweera 1981a, b; Kumar et al. 2004, 2005)
Zaleya decandra (Maha-sarana)	The root used for hepatitis and asthma. Juice of the leaf is dropped into the nostrils to relieve migraine. Whole plant extract is superior as a wound dressing, and leaf extract is used to reduce chronic pain in osteoarthritis. Antimicrobial activity has also been reported (Jayaweera 1981a, b; Balamurugan and Muthusamy 2008; Geethalakshmi et al. 2010)

mandatory component of future IVs studies in Sri Lanka. Summarized information on nutritional and phytochemical properties of few selected IVs species in Sri Lanka are presented in Tables 3 and 4.

6 Current Status of IVs Production in Sri Lanka

IVs provide a wide range of edible products that could be used in fresh form or as raw materials to prepare various products. Thus, cultivation of IVs can provide job opportunities and supports agribusiness and related service industries. Moreover, cultivation of IVs plays a fundamental role in biological, chemical, and hydrological cycles, protecting soils and providing non-timber forest products, and favorable ecological niches for other plants and animal species. Though it plays a vital role

Table 3 Nutritional properties of selected IVs in Sri Lanka

Species and plant part	Nutritional properties
Alternanthera sessilis (Mukunuwenna) (leaves, stems)	For fresh weight; energy (kcal) -39 mg/100 g, moisture 90 mg/100 g, ash 0.4 mg/100 g, protein 2.7 mg/100 g, fat 0.5 mg/100 g, total available carbohydrate 0.5 mg/100 g, calcium 7.03 mg/100 g, potassium 199 mg/100 g, sodium 0.7 mg/100 g, zinc 0.5 mg/100 g, copper 0.85 mg/100 g (Othman et al. 2016)
Centella asiatica (Gotukola) (leaves, stems)	Protein 2.4%, carbohydrate 6.7%, fat 0.2%, moisture 87.7%, insoluble dietary fiber 5.4%, 0.49% soluble dietary fiber, 17.0 mg/100 g phosphorus, 14.9 mg/100 g iron, sodium 107.8 mg/100 g, potassium 345 mg/100 g, calcium 171 mg/100 g. Total calories in 100 g 37.0 kcal. Vitamin C -48.5 mg/100 g, B1 0.09 mg/100 g, B2 0.19 mg/100 g, niacin 0.1 mg/100 g, carotene 2649 μg/100 g, and vitamin A 442 μg/100 g (Chandrika et al. 2011, 2015; Das 2011; Hashim 2011; Hashim et al. 2011; Joshi and Chaturvedi 2013)
Ipomoea aquatica (leaves, stems)	Vitamins A, B1, B2, B6, B12, C, E, K, "U" (S-methylmethionine), aliphatic pyrrolidine amides, carotenoids, hentriacontane, β-sitosterol and its glycosides, prostaglandin, leukotriene, N-trans- and N-cis-feruloyltyramines. Amino acids (aspartic acid, threonine, serine, glutamic acid, proline, glycine, alanine, leucine, tyrosine, lysine, histidine, arginine). Sugars (glucose, fructose, sucrose). Organic acids (malic acid, citric acid, oxalic acid). Minerals (sodium, potassium, calcium, iron, magnesium, zinc) (Bruemmer and Roe 1979; Candlish et al. 1987; Chen and Chen 1992; Igwenyi et al. 2011; Rao et al. 1990; Snyder et al. 1981; Tofern et al. 1999; Wills et al. 1984; Wills and Rangga 1996)
Lasia spinosa (Kohila) (rhizome)	17.6 kcal/100 g protein, 83 kcal/100 g moisture, 1.16 kcal/100 g fats, 34 kcal/100 g ash, 17 kcal/100 g total solids, 35.7 kcal/100 g carbohydrate dietary fiber with 40–75%. Micronutrients – zinc 7.44 ppm, magnesium 6.22 ppm, molybdnum1.18 ppm, copper 0.31 ppm, iron 17.06 ppm, and manganese 1.33 ppm (Brahma et al. 2014; Shefana and Ekanayake 2009)
Momordica charantia (Karavila) (fruit, flesh, pericarp seeds)	Vitamin A 471 mg/100 g, potassium 296 mg/100 g, vitamin C 84 mg/100 g, phosphorus 31 mg/100 g, iron 0.43 mg/100. Total protein content 1.2–2. 4 mg/100 g. Eight of the nine essential amino acids (histidine, isoleucine, leucine, lysine, methionine, phenylalanine, threonine, and valine) in the pericarp and seeds. Similar to other plant proteins, bitter melon appears to be lack of one essential amino acid, tryptophan. A particular polypeptide, named polypeptide-p or p-insulin, isolated from the whole bitter melon fruit, the flesh, and the seeds (Deng et al. 2014; Fonseka et al. 2007; Horax et al. (2010); Khanna et al. 1981; Liu et al. 2014; Orlovskaya and Chelombit'ko (2007); Panda et al. 2015; Paul and

(continued)

Table 3 (continued)

Species and plant part	Nutritional properties
	Raychaudhuri 2010; Ullah et al. 2011; Xu et al. 2015; Yawai et al. 1991; Zhang et al. 2016)
Senna auriculata (Ranawara) (whole plant)	Sodium 56.1 mg/100 gm, potassium 266 mg/100 gm, calcium 27.93 mg/100 gm, copper 4.87 mg/100 gm, zinc 0.44 mg/100 gm, iron 7.03 mg/100 gm 7. Vitamins (vitamin C50.08 mg/100 gm, riboflavin 102 (μg)/100 gm, folic acid 1 μg/100 gm, β-carotene 2.5 μg/100 gm) (Mukherjee and Datta 2017)

in crop diversification programs and agroforestry systems, it is not much popular among commercial farmers as a separate group of vegetables, as compared to exotic vegetables, low-country vegetables, or leafy vegetables. Even though indigenous food crop species play a vital role in agricultural biodiversity, extensive use of improved crop varieties has led to a situation where indigenous crop species are virtually disappearing from the ecosystem (genetic erosion) creating a vacuum in vegetable biodiversity. Although the country is virtually self-sufficient in vegetables, there is a big potential that exists to expand the commercial cultivation of IVs, targeting domestic consumption as well as export markets.

Besides that the major vegetable crops that are cultivated in large scale in a well-organized manner and their seeds and planting materials and cultivation technologies are available even at the commercial level, there is species diversity in domesticated IVs in various parts of the country. There are also many other species that grow in home gardens in small extents for household consumption, common lands, and fallow areas and in the natural forests, which could be a good source for number of benefits, particularly to the rural community.

Out of the listed species in the Appendix, only a few species are cultivated on a commercial scale and are connected to near-perfect, well-established marketing channels (e.g., *Centella asiatica*, *Alternanthera sessilis*, *Ipomoea aquatica*, *Momordica charantia*, *M. dioica*, etc.). The Department of Agriculture has identified them as high-priority plant species, and they have been subjected to development of technological packages for production and also extensive studies on improvement of the quality and production of those crops. Except for bitter gourd (Fig. 1), the cultivation extent and production details are not readily available on other commercialized IVs. Apart from that, though statistics are not readily available, *Lasia spinosa*, *Dioscorea* species, and *Amaranthus* species are also cultivated in farmlands as small-scale ventures. As reported by Ketipearachchi et al. (2002), *Solanum violaceum* is successfully cultivated in both dry and intermediate zones, and the total cultivated extent was over 50 ha during 2001.

Table 4 Phytochemical composition of selected IVs in Sri Lanka

Species	Phytochemical composition
Alternanthera sessilis (Mukunuwenna) (leaves and stems)	Flavonoids, terpenoids, phenols, phytosterols, and alkaloids in all of them. However, methanolic extract contained all the ten constituents tested including tannins, saponins, steroids, cardiac glycosides, and reducing sugars. Beta-carotene (Chandrika et al. 2005; Kota et al. 2017; Saravanan et al. 2013)
Centella asiatica (Gotukola) (leaves and stems)	Ursane-type pentacyclic triterpenoids known as centelloids, mainly, asiaticoside, madecassoside (brahminoside), asiatic acid, and madecassic acid (brahmic acid). Other triterpenoids: asiaticosides C, D, E, F; centellasaponins B, C; isothankunic acid and oleanane-type saponins, e.g., terminolic acid; centellasaponin D. 0.1% essential oils with α-humulene, germacrene B/D, β-caryophyllene, flavonoids-sesquiterpenes, steroids, β-carotene, lutein, neoxanthin, and violaxanthin (Brinkhaus et al. 2000; Chandrika et al. 2011; 2015; James and Dubery 2011; Guo et al. 2004; Hussin et al. 2009; Huang et al. 2011; James and Duebery 2009; Nhiem et al. 2011; Won et al. 2010)
Lasia spinosa (Kohila) (dried leaves and rhizome)	Alkaloids, saponins, tannins, phenolic compounds, antioxidants (Dubey et al. 2014; Goshwami et al. 2012; Kumar et al. 2013; Maisuthisakul et al. (2008); Shefana and Ekanayake 2009)
Ipomoea aquatica (Kan-kun) (leaves, stems)	Polyphenols (myricetin, quercetin, luteolin, apigenin, and kaempferol). Twelve pigments (various types of chlorophylls, carotenoids, viz., lutein, antheraxanthin, flavoxanthin, auroxanthin, luteoxanthin, neoxanthin, B-carotene, violaxanthin, cryptoxanthin, neoxanthin A, and neoxanthin B, and polyphenols, viz., quercetin 3′-methyl ether, quercetin 4′-methyl ether, and anthocyanins) (Chen and Chen 1992; Chu et al. 2000; Daniel 1989; Miean and Mohamed 2001; Wills and Rangga 1996)
Momordica charantia (Karavila) (flesh and seeds)	Gallic acid, gentisic acid, catechin and epicatechin, caffeic acid and chlorogenic acid, various saponins, both steroids and triterpenoids, steroidal glycosides from bitter melons are β-sitosterylglucoside and 5,25-stigmasteryl glucoside, and both together (1:1) are often referred to as charantin. Triterpenoids (momordicosides A and B; momordicosides C, D, and E; momordicosides F and I; momordicosides G and F_2; momordicosides K and L; momordicosides Q and R; momordicosides S and T; goyaglycoside-a, goyaglycoside-b, goyaglycoside-c, goyaglycoside-d, goyaglycoside-e, goyaglycoside-f, goyaglycoside-g, and goyaglycoside-h; and goyasaponins I, II, and III (oleanane types)). Alkaloids (vicine) (Budrat and Shotipruk 2008; Horax et al. 2005; Kubola and Siriamornpun 2008; Murakami et al. 2001;

(continued)

Table 4 (continued)

Species	Phytochemical composition
	Raman and Lau 1996; Tan et al. 2008; reviewed by Tan et al. 2016)
Senna auriculata (Ranawara) (whole plant)	Phytosterols, tannins, carbohydrates, flavonoids, terpenoids, and sterols. Oil from the seeds (uronic, palmitic, stearic, oleic) and linoleic acids. Flowers (apigenin-7-O-glucuronide and apigenin-7-oglucoside). Plant (lupeol, betulin, and stigmasterol). Roots (alkaloids, steroids, tannins, proteins, flavonoids, carbohydrates, fats, and oils). Leaves (alkaloids, carbohydrates, proteins, steroids, glycosides, flavonoids, tannins, phenolic compounds, fats, and oils, phytosterols, namely, β-sitosterol and lupeol) (Godbole et al. 1941; reviewed by Kshirsagar et al. 2010; Mazumder et al. 1999; Misra et al. 2001; Patra et al. 2009; Rastogi and Mehrotra 1993; Sunita and Abhishek 2008; Usha et al. 2007)

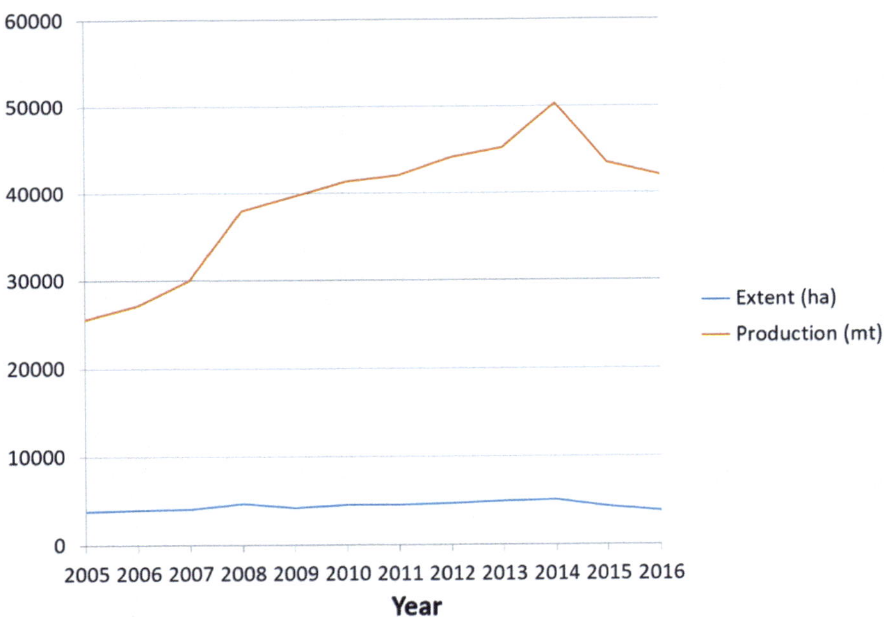

Fig. 1 Bitter gourd production in Sri Lanka. (Source: AgStat 2005–2016)

7 Conservation Perspectives

It is a well-known fact that not only IVs, but the entire biological diversity is subjected to various threats due to current unsustainable agricultural and development activities, climate change, and land/forest degradations over the last few

Table 5 Conservation status of indigenous vegetable based on the National Red List 2012

Categories	No. of taxa
Critically endangered (possibly extinct)	01
Critically endangered	01
Endangered	06
Vulnerable	07
Near threatened	14
Least concern	72
Data deficient	01

Source: MOE (2012)

decades. With the introduction of new improved varieties through selection and breeding, the indigenous varieties are threatened through extinction of their genetic resources. The erosion of genetic resources of indigenous vegetables is accelerated further due to the introduction of high-yielding hybrid varieties in the recent years. The hybrids are replacing not only the indigenous varieties but even the improved selections which were grown by the farmers in many countries. Unless action is taken, the genetic base of vegetables will be narrowed which may lead to serious catastrophe in the future. As Ketipearachchi et al. (2007) had highlighted, the existence of *Dioscorea* species inhabited to Sri Lanka is in danger due to urbanization, habitat destructions, unsustainable harvesting, damage by wild animals, and lack of protection, replanting, or cultivation programs. Out of the total IVs species listed here, 28% has been identified as threatened species (CR, EN, VU, NT), whereas the majority comes under the "least concern" category (Table 5, Map 2).

Currently, the indigenous vegetables have become important, and the attention of the scientists is being diverted to conserve and use them. A major reason for this is the danger of extinction, vis-a-vis the narrowing of the genetic base of the vegetables. It is a well-known fact that collection of germplasm is a fundamental tool in conservation of plant genetic resources and their utilization. Moreover, it provides good source of information for plant breeding and crop improvement programs. It increases the availability of germplasm for breeders, expands the genetic diversity, increases the variety yield potential and productivity, develops varieties with resistance to biotic and abiotic stresses, and develops varieties with desirable agronomic traits. Hence collection, conservation, and evaluation of IVs are urgently needed, and it should be a mandatory action of relevant institutes and authorities. Currently, the Plant Genetic Resources Center, Gannoruwa, is continuing exploration, evaluation, and conservation of IVs species. The National Herbarium is also maintaining herbarium specimens of IVs species. Table 6 shows number of collections of Sri Lankan IVs specimens maintained by various herbaria worldwide.

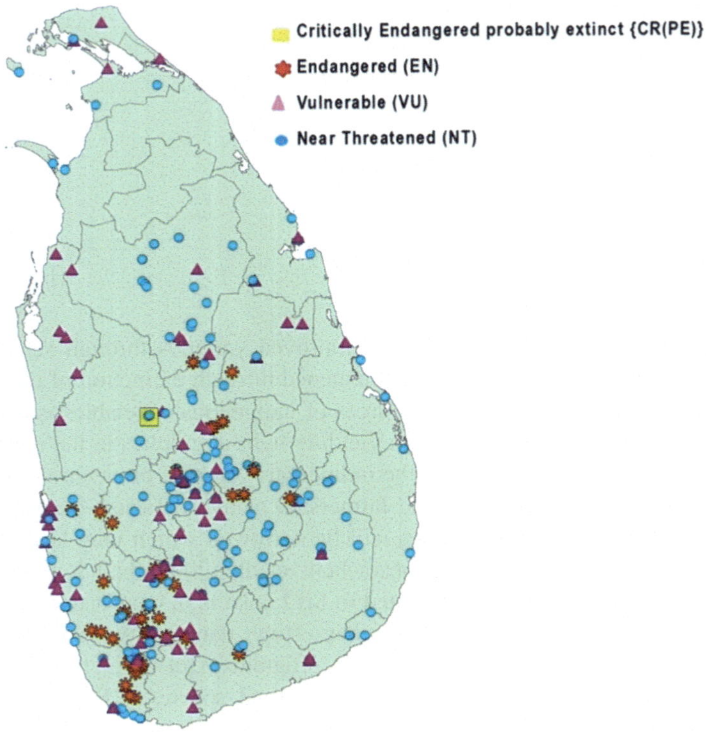

Map 2 Distribution of threatened IVs species in Sri Lanka based on records of the National Herbarium, Peradeniya

Table 6 Sri Lankan IVs specimens collection at selected herbaria

Institute/organization	No. of IVs specimens deposited
National Herbarium, Peradeniya, Sri Lanka (PDA)	2049
Royal Botanic Gardens, Kew, England (K)	512
Natural History Museum, London, England (BM)	125
Royal Botanic Gardens, Edinburgh, England (E)	28
Conservatoire et Jardin botaniques de la Ville de Genève, Switzerland (G)	8
Swedish Museum of Natural History (S), Sweden	32
Smithsonian Institution, USA (US)	624
Cambridge University, England (CGE)	3

Sources: Records at the National Herbarium, Peradeniya, and information from Flora of Ceylon volumes 1–14

8 Research Gaps and Development Needs

It is suggested that the following areas should be particularly explored and focused for the advancement of IV-based research in Sri Lanka.

1. Germplasm collection, evaluation and conservation

Conservation and utilization of plant genetic resources is a fundamental requirement in improvement program of any agricultural crop, and their importance has increased in the recent years with changing scenario of owners and legal regime with respect to biodiversity. With the introduction of new high-yielding varieties and hybrids, local crop genetic resources are being eroded at a rapid rate. Moreover, it is accelerated by unsustainable land use pattern related to agriculture and developmental activities. Hence collection, evaluation, and conservation is a primary need for management and utilization of IVs species in Sri Lanka.

2. Genetic improvement and use in plant breeding programs

IVs have immense potential for contribution to food production as they are well adapted to biotic and abiotic stress. Some of the IVs corps and their wild relatives can effectively be utilized for improvement of cultivated crops for future food and nutritional security. Thus IVs should be an integral component of varietal development programs.

3. Development of cultivation package

Except few IVs crops, there are no developed standards or recommendations for many of IVs crops in Sri Lanka. Therefore, agronomists need to contribute in formulations of new recommendations aiming for popularizing of IVs as commercial ventures among Sri Lankan farmers.

4. Collection of ethnobotanical information

The modern generation has little opportunity to use these indigenous food species since they are of limited supply as compared to attractive exotic vegetables with which even remote markets are always flooded with. This has created a situation where they have no chance to encounter the traditional knowledge gathered around these species. Though countries in Asia and African region are rich in IVs diversity, ethnobotanical information of such species has not been well studied and exploited or documented. With the unsustainable development activities, climate change, and forest/land degradation, world biodiversity is depleting at an alarming rate. Not only IVs, but all species are subjected to high risk of extinction from natural ecosystems. Moreover, modernization and changing of lifestyles have resulted in changing of food habits, reduction of dietary diversity, and loss of knowledge of edible plant species from the society. Therefore, extraction and documentation of ethnobotanical knowledge of IVs is a prime need.

9 Future Perspective

The IVs represent a large group of species which have the potential for ensuring food security while contributing to poverty alleviation both at national and household levels. The food base of the rural populations in Sri Lanka, especially those living in the marginal areas, has become narrower, causing shortage of food and nutrient deficiencies. The wild and weedy species which were used in the past have disappeared due to the changes to land use systems, use of improved hybrid seeds, and rapidly changing food habits. The lack of traditional knowledge on food values and cooking methods to minimize nutrient losses is also rapidly disappearing.

Currently, some of the IVs are becoming popular both among the rural and urban populations due to the therapeutic and food values. Mostly fresh green leaves of kangkung, *Gotukola*, *Amaranthus*, spinach, and Kathurumurunga are traded as commercial products in markets. There is a growing public awareness regarding the value of these green leaves as providers of dietary fiber, vitamins, and minerals that are essential for maintaining good health. Some specific examples in this regard are seen in Ayurvedic medicine, where it is claimed that the use of *Gotukola* improves memory and bitter gourd is a cure for diabetes (personal communication with Ayurvedic doctors). These are also available in value-added products form in Sri Lanka and in many other Western countries, indicating the economic and entrepreneurial potential of underutilized vegetables. The prices of some green vegetables are equal or sometimes higher than that of exotic leaves such as lettuce. These vegetables could also be grown in home gardens and urban agriculture for household consumption. Whereas the commercially produced green leaves may be contaminated with pesticide residues and heavy metals, those grown in home gardens are free of any contaminations. Therefore, the ideal strategy will be to promote these as home garden crops as well as large-scale commercial cultivations aiming at future food and nutritional security.

However, systematic research on indigenous vegetables and fruits are generally not been carried out by the commercial and research sector. Limited research has been carried out on these plant groups and information documented about their basic biology, management practices, and processing, while in some species, research is often carried out by isolated groups with their finding restricted to academic journals, making the dissemination of information to a wider audience being very low. It is essential to compile and use available information on such plant category for future studies. Effective development and utilization of indigenous vegetables require the collection and characterization of their germplasm and multiplication of identified superior germplasm, identification of effective means of propagation and agronomic practices, and harvest and postharvest and marketing studies.

Appendix: The List of Indigenous Vegetables of Sri Lanka

Family	Botanical name	Conservation status	Common name	Part used
Amaryllidaceae	*Allium hookeri* Thwaites	EN	Wal-lunu (S)	Leaves, bulb
Acanthaceae	*Hygrophila schulli* (Buch.-Ham.) M.R. & S.N. Almeida	LC	Neera-mulliya (S), Niramulli (T)	Leaves, tender shoots
Aizoaceae	*Sesuvium portulacastrum* (L.) L.	NT	Maha-sarana (S), Vankiruvilai (T)	Leaves, tender shoots
	Trianthema portulacastrum L.	LC	Heen-sarana (S)	Leaves, tender shoots
	Zaleya decandra (L.) Burm. f. (*Trianthema decandra* L.)	NT	Maha-sarana (S), Charania (T)	Leaves, tender shoots
Amaranthaceae	*Aerva lanata* (L.) Juss. ex Schult.	LC	Pol-pala (S)	Leaves, tender shoots
	Alternanthera sessilis (L.) R.Br. ex DC.	LC	Mukunuwenna (S), Ponankani (T)	Leaves, tender shoots
	Amaranthus spinosus L.	LC	Katu-kera, Katu-tampala (S), Mudkirai (T)	Leaves, tender shoots
	Amaranthus viridis L.	LC	Kura-tampala (S), Araikkirai (T)	Leaves, tender shoots
	Celosia argentea L.	LC	Kiri-henda (S), Cocks comb (E)	Leaves, flowers
Apiaceae	*Centella asiatica* (L.) Urb.	LC	Gotukola (S), Vallarai (T)	Leaves
Apocynaceae	*Dregea volubilis* (L.f.) Benth. ex Hook.f. (*Wattakaka volubilis* (L.f.) Stapf)	LC	Anguna, anguna-kola, Tiththa-anguna (S), Kodi-palai, Kurincha (T)	Leaves
	Gymnema lactiferum (L.) R.Br. ex Schult.	LC	Kurinnan (S), Kurinnan (T)	Leaves
	Tylophora sp.		Kiri-anguna (S)	Leaves
Aponogetonaceae	*Aponogeton crispus* Thunb.	VU	Kekatiya (S)	Inflorescence, petiole

(continued)

Family	Botanical name	Conservation status	Common name	Part used
	Aponogeton natans (L.) Engl. & K.Krause	VU	Wel-kekatita (S)	Inflorescence, petiole
	Aponogeton rigidifolius Bruggen	EN	Kekatiya (S)	Inflorescence, petiole
Araceae	*Amorphophallus paeoniifolius* (Dennst.) Nicolson	CR	Kidaran (S), Elephant foot yam (T)	Corm
	Colocasia esculenta (L.) Schott	LC	Taro (E), Gahala (S)	Stolen
	Lasia spinosa (L.) Thwaites	LC	Kohila (S)	Rhizome, immature leaf with petiole
	Pothos scandens L.	LC	Pota-wel (S)	Immature leaves
	Typhonium roxburghii Schott	NT	Panu-ala (S)	Whole plant
	Typhonium trilobatum (L.) Schott	LC	Panu-ala (S)	Whole plant
Asparagaceae	*Asparagus falcatus* L.	LC	Hatawariya (S)	Tuber
	Asparagus racemosus Willd.	LC	Hatawariya (S). Chattavari (T)	Tuber
Asteraceae	*Cyanthillium cinereum* (L.) H.Rob. (*Vernonia cinerea* (L.) Less.)	LC	Monara-kudumbiya (S), Neichatti-kirai (T)	Leaves
	Eclipta prostrata (L.) L.	LC	Kikirindi (S), Kaikechi (T)	Leaves
Basellaceae	*Basella alba* L.	EN	Nivithi (S), Pasalai (T)	Leaves, young shoots and stem
Begoniaceae	*Begonia cordifolia* (Wight) Thwaites	VU	Gal-ambala (S)	Leaves
	Begonia malabarica Lam.	NT	Hak-ambala (S)	Leaves
Boraginaceae	*Cordia dichotoma* G.Forst.	LC	Lolu (S), Sebesten plum (E), Naruvilli, Vidi (T)	Leaves, fruits
	Ehretia microphylla Lam. (*Carmona retusa* (Vahl) Masamune)	LC	Heen-tambala (S), Pakkuvetti (T)	Leaves, young shoots
Brassicaceae	*Cardamine africana* L.	?	Wel-aba-kola (S)	Leaves

	Cardamine hirsuta L.	?	Wel-aba-kola (S)	Leaves
Blechnaceae	*Blechnum orientale* L.	LC	Baru-koku (S)	Circinately coiled Leaves
Caryophyllaceae	*Drymaria cordata* subsp. *diandra* (Blume) J.A. Duke	LC	Kukulu-pala (S)	Leaves
Commelinaceae	*Commelina benghalensis* L.	LC	Diya-meneiya (S)	Leaves, young shoots
	Commelina diffusa Burm.f.	LC	Gira-pala (S)	Leaves, young shoots
	Murdannia esculenta (Wall. ex C.B.Clarke) R.S. Rao & Kammathy	NT	Kiri-bada-tel (S)	Leaves, young shoots
	Murdannia nudiflora (L.) Brenan	LC	Kiri-bada-tel (S)	Leaves, young shoots
Convolvulaceae	*Argyreia populifolia* Choisy	LC	Giri-tilla (S)	Leaves
	Ipomoea aquatic Forssk.	LC	Kankun (S)	Leaves, young shoots and stem
	Ipomoea littoralis Blume	NT	Tel-kola (S)	Leaves
Costaceae	*Cheilocostus speciosus* (J.König) C. Specht (*Costus speciosus* (J.König) Sm.)	LC	Tebu (S)	Leaves, young shoots
Cucurbitaceae	*Coccinia grandis* (L.) J.Voigt	LC	Kowakka (S), Ivy gourd (E), Kovvai (T)	Leaves
	Diplocyclos palmatus (L.) C.Jeffrey	LC	Pasagilla-gedi (S)	Fruits
	Momordica charantia L.	LC	Karavila (S), Bitter gourd (E), Pakal (T)	Fruits
	Momordica denudata (Thw.) C.B.Clarke	LC	Batu-karavila (S)	Fruits
	Momordica dioica Roxb. ex Willd	LC	Tumba-karavila (S), Spine gourd (E), Paluppakal (T)	Fruits
	Mukia maderaspatana (L.) M.Roem.	NT	Gon-kekiri (S), Mochumochukka (T)	Fruits
	Solena amplexicaulis (Lam.) Gandhi	LC	Kawdu-kekiri (S), Peyppudal (T)	Fruits
	Trichosanthes cucumerina L.	LC	Dum-mella, Kunu-mella (S), Pudal (T)	Fruits

(continued)

Family	Botanical name	Conservation status	Common name	Part used
Cycadaceae	*Cycas nathorstii* J.Schust.	VU	Madu (S)	Immature stem
Dilleniaceae	*Dillenia retusa* Thunb.	LC	Goda-para (S)	Fruits
Dioscoreaceae	*Dioscorea bulbifera* L.	LC	Udala, Panu-kondol (S), Aerial yam (E), Mothaka (T)	Stem tuber
	Dioscorea koyamae Jayasuriya	EN	Gonala, Kahata-gonala (S)	Stem tuber
	Dioscorea oppositifolia L.	NT	Hiritala, Kitala (S)	Stem tuber
	Dioscorea pentaphylla L.	LC	Katu-ala (S), Allai (T)	Stem tuber
	Dioscorea spicata Roth	VU	Gonala (S)	Stem tuber
	Dioscorea tomentosa Koenigh ex Spreng.	LC	Uyala (S)	Stem tuber
	Dioscorea trimenii Prain & Burkill	EN	Dahiya-ala (S)	Stem tuber
Euphorbiaceae	*Acalypha indica* L.	LC	Kuppameniya (S), Kuppameni (T)	Leaves
Fabaceae	*Canavalia rosea* (Sw.) DC.	LC	Mudu-awara (S)	Immature pod, seed
	Cyamopsis tetragonoloba (L.) Taub.	DD	Kotaranga (S), Koth-averay (T), Cluster bean (E)	Immature pod
	Lablab purpureus (L.) Sweet	LC	Lablab bean (E), Kos-eta-dambala (S)	Immature pods, seed
	Mucuna pruriens (L.) DC.	LC	Cowage (E), Wanduru-me (S), Chunao-varai (T)	Seeds
	Senna auriculata (L.) Roxb. (*Cassia auriculata* L.)	LC	Ranawara (S), Matara tea (E), Avarai (T)	Leaves
	Senna tora (L.) Roxb. (*Cassia tora* L.)	LC	Peti-tora (S), Vaddutakarai (T)	Leaves
	Vigna marina (Burm.) Merr.	EN	Field bean (E), Karal-li-me (S), Kodippayaru (T)	Leaves, Pods
Gleicheniaceae	*Dicranopteris linearis* (Burm.f.) Underw.	LC	Kekilla (S)	Circinately coiled Leaves
Lamiaceae	*Leucas zeylanica* (L.) W.T.Aiton	LC	Geta-tumba (S), Mudi-tumpai (T)	Leaves
	Pogostemon heyneanus Benth.	LC	Gas-kollan-kola (S)	Leaves

	Premna procumbens Moon	LC	Le-kola-pala (S), Mulla, Mullai (T)	Leaves
	Rotheca serrata (L.) Steane & Mabb. (*Clerodendrum serratum* (L.) Moon)	LC	Kan-henda (S), Chiru-dekku (T)	Leaves
Lecythidaceae	*Careya arborea* Roxb.	LC	Patana-oak (E), Kahata (E), Kachaddai (T)	Flowers, Fruits
Malvaceae	*Abelmoschus moschatus* Medik.	NT	Kapu-kinissa (S), Katukkasturi (T)	Immature fruits
	Hibiscus surattensis L. (*Hibiscus furcatus* Roxb.)	LC	Hin-napiritta (S)	Leaves
Marantaceae	*Stachyphrynium spicatum* (Roxb.) K.Schum. (*Stachyphrynium zeylanica* (Benth.) K.Schum.)	CR(PE)	Hulan-kiriya (S)	Tuber
Marattiaceae	*Angiopteris evecta* (Forst.) Hoffm.	NT	Wal-menda (S)	Circinately coiled Leaves
Melastomataceae	*Osbeckia octandra* (L.) DC.	LC	Heen-bovitiya (S)	Leaves
Molluginaceae	*Glinus oppositifolius* (L.) Aug.DC	LC	Heen-pala (S), Kachantarai (T)	Leaves
Moraceae	*Ficus racemosa* L.	LC	Attikka (S), Atti (T)	Leaves, fruits
Nelumbonaceae	*Nelumbo nucifera* Gaertn.	LC	Nelum (S), Tamarai (T), Lotus (E)	Rhizome
Nyctaginaceae	*Boerhavia diffusa* L.	LC	Pita-sudu-pala (S), Karichcharanai (T)	Leaves, young shoot
	Pisonia grandis R.Br.	LC	Watha-banga (S), Chandi (T), Lettuce tree (E),	Leaves
Nymphaeaceae	*Nymphaea nouchali* Burm.f.	VU	Maneal (S), Water lily (E)	Rhizome, flower stalk
	Nymphaea pubescens Willd.	LC	Et-olu (S), Water lily (E)	Rhizome
Olacaceae	*Olax imbricata* Roxb.	NT	Telatiya (S)	Young shoot, fruits
Phyllanthaceae	*Antidesma bunius* (L.) Spreng.	LC	Karawala-kebella	Leaves
	Aporosa cardiosperma (Gaertn.) Merr. (*Aporosa lindleyana* (Wight) Baill.)	LC	Kebella (S)	Leaves
	Sauropus androgynus (L.) Merr.	LC	Japan batu (S)	Leaves
Plantaginaceae	*Bacopa monnieri* (L.) Wettst.	LC	Lunu-wila (S)	Leaves

(continued)

Family	Botanical name	Conservation status	Common name	Part used
Pteridaceae	*Acrostichum aureum* L.	LC	Karen-koku (S)	Circinately coiled Leaves
Pontederiaceae	*Monochoria hastata* (L.) Solms	NT	Diya-habarala (S)	Leaves, young shoot
	Monochoria vaginalis (Burm.f.) C.Presl	LC	Diya-habarala (S)	Leaves, young shoot
Portulacaceae	*Portulaca oleracea* L.	LC	Genda-kola (S), Pulikkirai (T)	Leaves, young stem
Rubiaceae	*Canthium coromandelicum* (Burm.f.) Alston	LC	Kara (S), Pulikkirai (T)	Leaves
	Hedyotis fruticosa L.	LC	Weraniya (S)	Leaves
	Hedyotis neesiana Arn.	LC	Pita-sudu-pala (S)	Leaves
	Psychotria sarmentosa Blume	NT	Wal-gonika (S)	Leaves
	Tamilnadia uliginosa (Retz.) Tirveng. &Sastre	VU	Wadiga (S)	Fruits
Sapindaceae	*Cardiospermum halicacabum* L.	LC	Penela (S)	Leaves
Solanaceae	*Solanum violaceum* Ortega	LC	Thitta-tibbatu (S)	Fruits
	Solanum trilobatum L.	LC	Wel-tibbatu (S)	Fruits
Woodsiaceae	*Diplazium esculentum* (Retz.) Sw.	NT	Miyena-dalu (S)	Circinately coiled leaves

The botanical name with bold letters indicates the endemic species

Note: *CR* critically endangered, *CR(PE)* critically endangered (possibly extinct), *EN* endangered, *VU* vulnerable, *NT* near threatened, *LC* least concern, *DD* data deficient

References

Afolayan AJ, Jimoh FO (2009) Nutritional quality of some wild leafy vegetables in South Africa. Int J Food Sci Nutr 60(5):424–431

Alwis SD, Senanayake G (2010) Underutilized food crops of Sri Lanka (in Sinhala). Biodiversity Secretariat, Ministry of Environment, Sri Lanka

Anon (2002) Compendium of medicinal plants. A Sri Lankan study, vol 2. Department of Ayurveda, Sri Lanka, pp 16–91

Anon (2003) Compendium of medicinal plants, A Sri Lankan study, vol 3. Department of Ayurveda, Sri Lanka, pp 67–69

Anon (2004) Compendium of medicinal plants, A Sri Lankan study, vol 4. Department of Ayurveda, Sri Lanka, pp 81–83

Babu TD, Kuttan G, Padikkala J (1995) Cytotoxic and anti-tumor properties of certain taxa of Umbelliferae with special reference to *Centella asiatica* (L.) urban. J Ethnopharmacol 48 (1):53–57

Balamurugan B, Muthusamy P (2008) Observation of the hepatoprotective and antioxidant activities of *Trianthema decandra* Linn. (Vallai sharunnai) roots on carbon tetrachloride-treated rats. Bangladesh J Pharmacol 3(2):83–89

Brahma J, Chakravarty S, Rethy P (2014) Qualitative estimation of the presence of bioactive and nutritional compound in *Lasia spinosa*: an important vegetable plant used by the Bodos of Kokrajhar District. Int J ChemTech Res 6(2):1405–1412

Brinkhaus B (1998) *Centella asiatica* in traditional and modern phytomedicine-a pharmacological and clinical profile-part I: botany chemistry preparations. Perfusion 11:466–474

Brinkhaus B, Lindner M, Schuppan D, Hahn EG (2000) Chemical, pharmacological and clinical profile of the East Asian medical plant *Centella asiatica*. Phytomedicine 75:427–448

Brown AC, Valiere A (2004) The medicinal uses of poi. Nutr Clin Care 7(2):69–74

Bruemmer JH, Roe R (1979) Protein extraction from water spinach (*Ipomoea aquatica*). Proc Fla State Hort Soc 92:140–143

Budrat P, Shotipruk A (2008) Extraction of phenolic compounds from fruits of bitter melon (*Momordica charantia*) with subcritical water extraction and antioxidant activities of these extracts. Chiang Mai J Sci 35:123–130

Candlish JK, Gourley L, Lee HP (1987) Dietary fiber and starch contents of some South East Asian vegetables. J Agric Food Chem 35:319–321

Chadha KL, Patel VB (2007) Prospect of indigenous perennial plants as source of vegetable. In: Chandha ML, Kuo G, CLL G (eds) Proceeding of the 1st international conference of Indigenous vegetable and legumes prospectus for fighting poverty, hunger and malnutrition. Acta Horticulture 752: 49–54

Chandrika UG, Jansz ER, Warnasuriya ND (2005) Analysis of beta carotene and lutein in some leafy vegetables and effect of cooking on serum vitamin A formation from *Alternanthera sessilis* (Sinh. Mukunuwanna). J Natl Sci Found 33(2):141–145

Chandrika UG, Salim N, Wijepala GDDJ, Perera KSU, Goonetilleke AKE (2011) Carotenoid and mineral content of different morphotypes of *Centella asiatica* L. (Gotu kola). Int J Food Sci Nutr 62:552–557

Chandrika UG, Peramune AAS, Kumara P (2015) Gotu Kola (*Centella asiatica*): nutritional properties and plausible health benefits. Adv Food Nutr Res 76:1043–4526. https://doi.org/10.1016/bs.afnr.2015.08.001

Cheikhyoussef A, Mapaure I, Shapi M (2011) The use of some indigenous plants for medicinal and other purposes by local communities in Namibia with emphasis on Oshikoto region: a review. Res J Med Plants 5:406–419

Chen BH, Chen YY (1992) Determination of carotenoids and chlorophylls in water convolvulus (*Ipomoea aquatica*) by liquid chromatography. Food Chem 45:129–134

Chu YH, Chang CL, Hsu HF (2000) Flavonoid content of several vegetables and their antioxidant activity. J Sci Food Agric 80:561–566

Daniel M (1989) Polyphenols of some Indian vegetables. Curr Sci 58:1332–1333

Das AJ (2011) Review on nutritional, medicinal and pharmacological properties of *Centella asiatica* (Indian pennywort). J Biol Active Prod Nat 4:216–228

Dassanayake MD, Clayton WD (1995–2000) A revised handbook to the flora of Ceylon, vol X–XIV. Oxford & IBH Publishing Company Private Limited, New Delhi

Dassanayake MD, Fosberg FR (1980–1991) A revised handbook to the flora of Ceylon, vol I–VII. Amerind Publishing Company Private Limited, New Delhi

Dassanayake MD, Fosberg FR, Clayton WD (1994) A revised handbook to the flora of Ceylon, vol VIII–IX. Oxford & IBH Publishing Company Private Limited, New Delhi

Deng YY, Yi Y, Zhang LF, Zhang RF, Zhang Y, Wei ZC, Tang XJ, Zhang WM (2014) Immunomodulatory activity and partial characterisation of polysaccharides from *Momordica charantia*. Molecules 19(9):13432–13447

Dubey MK, Das S, Yadav S, Gupta PC, Jaiswal SK, Sharma (2014) Gastroprotective potential of bioactive fraction from *Lasia spinosa* rhizome on experimentally induced gastric ulceration. Int J Pharm Sci Res 5(12):5209–5215

Engle LM, Faustino FC (2007) Conserving the indigenous vegetable germplasm of South East Asia. In: Chandha ML, Kuo G, Gowda CLL (eds) Proceeding of the 1st international conference of indigenous vegetable and legumes prospectus for fighting poverty, hunger and malnutrition. Acta Horticulture 752:55–60

Fonseka H, Chandrasekara A, Fonseka M, Wickramasinghe IP, Kumara KPDRSP, Wickramarchchi WNC (2007) Determination of anti-amylase and anti-glucosidase activity of different bitter gourd (*Momordica charantia* L.) genotypes and *Thumba Karavila* (*M. dioica*). Proceedings of the 1st international conference on indigenous vegetables and legumes prospectus for fighting poverty, hunger and malnutrition. Acta Hortic 752:131–137

Geethalakshmi R, Sarada DVL, Marimuthu P (2010) Evaluation of antimicrobial and antioxidant potentials of *Trianthema decandra* L. Asian J Biotechnol 2(4):225–231

Glew R, Vanderjagt D, Chuang LT, Huang YS, Millson M, Glew R (2005) Nutrient content of four edible wild plants from West Africa. Plant Foods Hum Nutr (Formerly Qualitas Plantarum) 60 (4):187–193

Godbole NN, Gunde BG, Shrivastav PD (1941) An investigation of oil from seed of *Hygrophila spinosa*. J Am Oil Chem Soc 18:206–207

Goshwami D, Rahman Md M, Muhit Md A, Islam Md S, Ansari M (2012) Antioxidant property, cytotoxicity and antimicrobial activity of *Lasia spinosa* leaves. Nepal J Sci Technol 13 (2):215–218

Goyal M, Pareek A, Nagori BP, Sasmal D (2011) Aerva lanata: A review on phytochemistry and pharmacological aspects. Pharmacogn Rev 5(10):195–198

Gunatilleke N, Gunatilleke S, Ashton PS (2017) South-West Sri Lanka: a floristic refugium in South Asia. Ceylon J Sci 46(5):65–78. https://doi.org/10.4038/cjs.v46i5.7454

Gunjan G, Rajkumar V, Kumar RA (2011) Therapeutic potential of polar and non-polar extracts of *Cyanthillium cinereum* in vitro. In: Evidence-based complementary and alternative medicine, vol 2011. Article ID 784826, 10 pages, 2011. https://doi.org/10.1093/ecam/nep155

Guo JS, Cheng CL, Koo MW (2004) Inhibitory effects of *Centella asiatica* water extract and asiaticoside on inducible nitric oxide synthase during gastric ulcer healing in rats. Planta Med 70:1150–1154

Hashim P (2011) *Centella asiatica* in food and beverage applications and its potential antioxidant and neuroprotective effect. Int Food Res J 18:1215–1222

Hashim P, Sidek H, Helan MHM, Sabery A, Palanisamy UD, Ilham M (2011) Composition and bioactivities of *Centella asiatica*. Molecules 16:1310–1322

Herat TR, Somaratne L, Perera PPDC (1994) Common vegetables of Sri Lanka. The National Resources, Energy and Science Authority of Sri Lanka, Colombo 7, Sri Lanka

Hilou A, Nacoulma OG, Guiguemde T (2006) In vivo antimalarial activities of extracts from *Amaranthus spinosus* L. and *Boerhavia erecta* L. in mice. J Ethnopharmacol 103(2):236–240

Horax R, Hettiarachchy N, Islam S (2005) Total phenolic contents and phenolic acid constituents in 4 varieties of bitter melons (*Momordica charantia*) and antioxidant activities of their extracts. J Food Sci 70:275–280

Horax R, Hettiarachchy N, Kannan A, Chen P (2010) Proximate composition and amino acid and mineral contents of *Momordica charantia*. L. pericarp and seeds at different maturity stages. Food Chem 122:1111–1115

Hossain SM, Razzaque MA (1999) Collection, conservation and utilization of indigenous vegetable in Bangladesh. In: Engle LM, Altoveros NC (eds) Proceedings of a workshop. Asian Vegetable Research and Development Center, Shanhua, Taiwan

Hossain E, Sarkar D, Maiti A, Chatterjee M, Mandal SC, Gupta JK (2010) Anti-inflammatory effect of a methanolic extract of leaves of *Dregea volubilis*. J Ethnopharmacol 132(2):525–528

Huang SS, Chiu CS, Chen HJ, Hou WC, Sheu MJ, Lin YC, Shie PH, Huang GJ (2011) Antinociceptive activities and the mechanisms of anti- inflammation of asiatic acid in mice. Evid Based Complement Alternat Med 2011:895857

Hussin M, Hamid AA, Mohamad S, Saari N, Bakar F, Dek SP (2009) Modulation of lipid metabolism by *Centella asiatica* in oxidative stress rats. J Food Sci 74:H72–H78

Igwenyi IO, Offor CE, Ajah DA, Nwankwo OC, Ukaomah JI, Aja PM (2011) Chemical compositions of Ipomea aquatica (Green Kangkong). Int J Pharm Bio Sci 2(4):593–598

Ilankoon J, Wijesekara A (2008) Crop wild relatives of Sri Lanka and their conservation. Crop wild relative project, Ministry of Environment and Natural Resources, Battaramulla, Colombo

Jaenicke H, Hoschle-Zeledon I (2006) Strategic framework for underutilized plant species research and development, with special reference to Asia and Pacific, and to Sub-Sharan Africa. International Center for Underutilized Crops/Global Facilitation Unit for Underutilized Species, Colombo/Rome

James J, Dubery IA (2011) Identification and quantification of triterpenoid centelloids in *Centella asiatica* (L.) Urban by densitometric TLC. J Planar Chromatogr 24:82–87

James JT, Duebery IA (2009) Pentacyclic triterpenoids from medicinal herb, *Centella asiatica* (L) Urban. Molecules 14:3922–3941

Jayaweera DMA (1981a) Medicinal plants (indigenous & exotic) used in Ceylon, part 1. The National Science Council of Sri Lanka, Colombo, pp 33, 35, 41–42, 43, 49, 53, 127, 131, 135, 181, 199

Jayaweera DMA (1981b) Medicinal plants (indigenous & exotic) used in Ceylon, part 3. The National Science Council of Sri Lanka, Colombo, p 177, 187

Joshi K, Chaturvedi P (2013) Therapeutic efficiency of *Centella asiatica* (L.) Urb. An underutilized green leafy vegetable: an overview. Int J Pharm Bio Sci 4:135–149

Ketipearachchi Y, Ketipearachchi KW, Palipana WMU, Athauda AATR, Hettiarchchi HAPIP (2002) Characteristics and potential of Tibbatu, *Solanum violaceum*. Ann Sri Lanka Dep Agric 4:435

Ketipearachchi Y, Ketipearachchi KW, Hettiarchchi HAPIP (2007) Underutilized wild Dioscorea species found in the Dry zone forests of Sri Lanka. In: Proceeding of the international forestry and environment symposium, Department of Forestry and Environmental Science, University of Sri Jayewardenepura, Sri Lanka

Khanna P, Jain SC, Panagariya A, Dixit VP (1981) Hypoglycemic activity of polypeptide-p from a plant source. J Nat Prod 44:648–655

Kota S, Govada VR, Anantha RK, Verma MK (2017) An investigation into phytochemical constituents, antioxidant, antibacterial and anti-cataract activity of *Alternanthera sessilis*, a predominant wild leafy vegetable of South India. Biocatal Agric Biotechnol 10:197–203

Kshirsagar AD, Ingale KG, Vyawahare NS, Thorve VS (2010) *Hygrophila spinosa*: a comprehensive review. Pharmacogn Rev 4(8):167–171. https://doi.org/10.4103/0973-7847.70912

Kubola J, Siriamornpun S (2008) Phenolic contents and antioxidant activities of bitter gourd (*Momordica charantia* L.) leaf, stem and fruit fraction extracts in vitro. Food Chem 110:881–890

Kumar G, Sharmila Banu G, Vanitha Pappa P, Sundararajan M, Rajasekara Pandian M (2004) Hepatoprotective activity of *Trianthema portulacastrum* L. against paracetamol and thioacetamide intoxication in albino rats. J Ethnopharmacol 92(1):37–40

Kumar G, SharmilaBanu G, Pandian MR (2005) Evaluation of the antioxidant activity of *Trianthema portulacastrum* L. Indian J Pharm 37(5):331–333

Kumar M, Mondal P, Borah S, Mahato K (2013) Physico-chemical evaluation, preliminary phytochemical investigation, fluorescence and TLC analysis of the plants *Lasia spinosa* (Lour) Thwaites. Acad Sci 5(2):306–310

Lai HY, Lim YY, Kim KH (2010) *Blechnum orientale* Linn – a fern with potential as antioxidant, anticancer and antibacterial agent. BMC Complement Altern Med 10:15

Liu X, Chen T, Hu Y, Li KX, Yan LS (2014) Catalytic synthesis and antioxidant activity of sulfated polysaccharide from *Momordica charantia* L. Biopolymers 101(3):210–215

Maisuthisakul P, Suttajit M, Pongsawatmanit R (2007) Assessment of phenolic content and free radical scavenging capacity of some Thai indigenous plants. Food Chem 100(4):1409–1418

Maisuthisakul P, Pasukb S, Ritthiruangdejc P (2008) Relationship between antioxidant properties and chemical composition of some Thai plants. J Food Compos Anal 21:229–240

Mazumder UK, Gupta M, Maiti S (1999) Chemical and pharmacological evaluation of *Hygrophila spinosa* root. Indian J Exp Biol 61:181–183

Miean KH, Mohamed S (2001) Flavonoid (myricetin, quercetin, kaempferol, luteolin, and apigenin) content of edible tropical plants. J Agric Food Chem 49(6):3106–3112

Misra TN, Singh RS, Pandey HS, Singh BK, Pandey RP (2001) Constituents of *Asteracantha longifolia*. Fitoterapia 72:194–196

MOE (2012) The national red list 2012 of Sri Lanka: conservation status of the fauna and flora. Ministry of Environment, Colombo

Mukherjee C, Datta S (2017) Estimation of micronutrients in fresh Kulekhara leaves (*Hygrophila auriculata*). Int J Sci Res (IJSR) 6(2):838–840

Murakami T, Emoto A, Matsuda H, Yoshikawa M (2001) Medicinal foodstuffs. XXI. Structures of new cucurbitane-type triterpene glycosides, goyaglycosides-a, −b, −c, −d, −e, −f, −g, and -h, and new oleanane-type triterpene saponins, goyasaponins I, II, and III, from the fresh fruit of Japanese *Momordica charantia* L. Chem Pharm Bull (Tokyo) 49:54–63

Nadkarni KM, Nadkarni AK (2007) Indian materia medica, vol 1. Bombay Popular Prakashan Pvt Ltd, Mumbai, pp 385–286

Nandi D, Elizabeth Besra S, Vedasiromoni JR, Giri SV, Prince R, Parasuraman J (2012) Anti-leukemic activity of *Wattakaka volubilis* leaf extract against human myeloid leukemia cell lines. J Ethnopharmacol 144(2):466–473

Nhiem NX, Tai BH, Quang TH (2011) A new ursane-type triterpenoid glycoside from Centella asiatica leaves modulates the production of nitric oxide and secretion of TNF-α in activated RAW 264.7 cells. Bioorg Med Chem Lett 15:1777–1781

Orhan IE (2012) *Centella asiatica* (L.) Urban: from traditional medicine to modern medicine with neuroprotective potential. Evidence-based complementary and alternative medicine 2012. Article ID 946259, 8 pages. https://doi.org/10.1155/2012/946259

Orlovskaya T, Chelombit'ko V (2007) Amino acid composition of *Momordica charantia* seeds and pericarp. Chem Nat Compd 43:237–238

Othman A, Smail A, Hassan FA, Yusof BNM, Khatib A (2016) Comparative evaluation of nutritional compositions, antioxidant capacities, and phenolic compounds of red and green sessile joyweed (*Alternanthera sessilis*). J Funct Foods 21:263–271. https://doi.org/10.1016/j.jff.2015.12.014

Palada MC, Kalb TJ, Lumpkin TA (2006) The role of AVRDC–The world vegetable center in enhancing and promoting vegetable production in the tropics. Hort Sci 41(3):556–560. https://doi.org/10.21273/HORTSCI.41.3.556

Panda BC, Mondal S, Devi KSP, Maiti TK, Khatua S, Acharya K, Islam SS (2015) Pectic polysaccharide from the green fruits of Momordica charantia (Karela): structural characterization and study of immunoenhancing and antioxidant properties. Carbohydr Res 401:24–31

Patra A, Murthy NP, Jha S (2009) Pharmacognostical standardization of leaves of *Hygrophila spinosa* T. Anders Phcog J 1:82–87

Paul A, Raychaudhuri SS (2010) Medicinal uses and molecular identification of two *Momordica charantia* varieties-a review. Electron J Biol 6(2):43–51

Pushpakumara DKNG, Gunasena HPM, Wahundeniya WMKB (2018) Exploring the potential of underutilized vegetable of Sri Lanka. Unpublished report on ICRAF Sri Lanka Program, ICRAF South Asia, New Delhi, India

Rahman AHM, Iffat Ara Gulshana M (2014) Taxonomy and medicinal uses on Amaranthaceae Family of Rajshahi, Bangladesh. Appl Ecol Environ Sci 2(2):54–59

Rahman AHMM, Nitu SK, Ferdows Z, Islam AKMR (2013) Medico-botany on herbaceous plants of Rajshahi, Bangladesh. Am J Life Sci 1(3):136–144

Rajapaksha U (1998) Traditional food plants in Sri Lanka. Hector Kobbekaduwa Agrarian Research and Training Institute, Colombo

Raman A, Lau C (1996) Anti-diabetic properties and phytochemistry of *Momordica charanti*a L. (Cucurbitaceae). Phytomedicine 2:349–362

Rao SK, Dominic R, Singh K, Kaluwin C, Rivett DE, Jones GP (1990) Lipid, fatty acid, amino acid, and mineral composition of five edible plant leaves. J Agric Food Chem 38:2137–2139

Rastogi RP, Mehrotra BN (1993) Compendium of Indian medicinal plants. I. Publication and Information Directorate, CSIR, New Delhi, p 220

Sandhya B, Thomas S, Isabel W, Shenbagarathai R (2006) Ethnomedicinal plants used by the Valaiyan community of Piranmalai hills (reserved Forest), Tamil Nadu, India – a study. Afr J Tradit Complement Med 3:101–114

Saravanan S, Kumar SS, Vijay S (2013) Evaluation of antioxidant and thrombolytic potential of *Alternanthera sessilis*. IOSR J Environ Sci Toxicol Food Technol 2(5):01–04

Senaratna LK (2001) A checklist of the flowering plants of Sri Lanka. National Sciences Foundation, Maitland Place, Colombo 07, Sri Lanka

Shaffer-Fehre M (2006) A revised handbook of the flora of Ceylon. Pteridophyta (ferns and Fern allies), vol XV. CRC Press

Shanmugasundaram S (2007) Past, present and future of indigenous vegetables. In: Chandha ML, Kuo G, Gowda CLL (eds) Proceeding of the 1st international conference of Indigenous vegetable and legumes prospectus for fighting poverty, hunger and malnutrition. Acta Horticulture 752:377–380

Shefana G, Ekanayake S (2009) Nutritional aspects of *Lasia spinosa* (kohila). Vidyodaya J Sci 14:49–64

Singh S, Gautam A, Sharma A, Batra A (2010) *Centella asiatica* (L.): a plant with immense medicinal potential but threatened. Int J Pharm Sci Rev Res 4(2):9–17

Snyder GH, Morton JF, Genung WG (1981) Trials of Ipomoea aquatica, nutritious vegetables with high protein and nitrate extraction potential. Proceedings of the Florida State Horticultural Society 94:230–235

Sugathadasa KSS, Jeevandara PM, Devanarajana A, Pushpakumara DKNG (2008) A checklist of medicinal plants of Sri Lanka. Bandaranayake Ayurvedia Research Institute, Department of Ayurveda and ICRAF Sri Lanka Program, Sri Lanka

Sunita S, Abhishek S (2008) A comparative evaluation of phytochemical fingerprints of Asteracantha longifolia Nees. using HPTLC. Asian J Plant Sci 7:611–614

Tan M, Ye J, Turner N, Behrens C, Ke C, Tang C, Chen T, Weiss H, Gesing E, Rowland A, James D, Ye Y (2008) Antidiabetic activities of triterpenoids isolated from bitter melon associated with activation of the AMPK pathway. Chem Biol 15:263–273

Tan SP, Kha TC, Parks SE, Roach PD (2016) Bitter melon (*Momordica charantia* L.) bioactive composition and health benefits: A review. Food Rev Int 32(2):181–202. https://doi.org/10.1080/87559129.2015.1057843

Temjenmongla T, Yadav AK (2005) Anticestodal efficacy of folklore medicinal plants of Naga tribes in North –East India. Afr J Tradit Complement Altern Med 2(2):129–133

Tofern B, Mann P, Kaloga M, Jenett-Siems K, Wigge L, Eich E (1999) Aliphatic pyrrolidine amides from two tropical convolvulaceous species. Phytochemistry 52:1437–1441
Ullah M, Chy FK, Sarkar SK, Islam MK, Absar A (2011) Nutrient and phytochemical analysis of four varieties of bitter gourd (*Momordica charantia*) grown in Chittagong Hill tracts. Bangladesh Asian J Agric Res 5:186–193
Usha K, Kasturi GM, Hemalatha P (2007) Hepatoprotective effect of *Hygrophila spinosa* and *Cassia occidentalis* on carbon tetrachloride induced liver damage in experimental rats. Indian J Clin Biochem 22:132–135
Weinberger K (2007) Are indigenous vegetables underutilized crops? Some evidences from Eastern Africa and South East Asia. In: Chandha ML, Kuo G, Gowda CLL (eds) Proceeding of the 1st international conference of Indigenous vegetable and legumes prospectus for fighting poverty, hunger and malnutrition. Acta Horticulture 752: 29–33
Weinberger K, Msuya J (2004) Indigenous vegetables in Tanzania, significance and prospects. Technical Bulletin 1. The World Vegetable Center, Taiwan
Wills RB, Rangga A (1996) Determination of carotenoids in Chinese vegetables. Food Chem 56:451–455
Wills RB, Wong AW, Scriven FM, Greenfield H (1984) Nutrient composition of Chinese vegetables. J Agric Food Chem 32:413–416
Won JH, Shin JS, Park HJ, Jung HJ, Koh DJ, Jo BG, Lee JY, Yun K, Lee KT (2010) Anti-inflammatory effects of madecassic acid via the suppression of NF-kappaB pathway in LPS-induced RAW 264.7 macrophage cells. Planta Med 76:251–257
Xu X, Shan B, Liao CH, Xie JH, Wen PW, Shi JY (2015) Anti-diabetic properties of *Momordica charantia* L. polysaccharide in alloxan-induced diabetic mice. Int J Biol Macromol 81:538–543
Yamaguchi M (1983) World vegetables. Principle, production and nutritive value. Ellis Horwood Limited Publishers, Chichester
Yasapalitha TGGA, Rupasinghe RPS (2017) Ancient foods of Sri Lanka (in Sinhala). Wasana Publishers Pvt (Ltd.), Dankotuwa
Yawai KE, Rao KS, Kaluwin C, Jones GP, Rivett DE (1991) Chemical composition of *Momordica charantia* L. fruits. J Agric Food Chem 39:1762–1763
Zhang F, Lin L, Xie J (2016) A mini-review of chemical and biological properties of polysaccharides from *Momordica charantia*. Int J Biol Macromol 92:246–253

Shade in Tea Plantations: A New Dimension with an Agroforestry Approach for a Climate-Smart Agricultural Landscape System

A. J. Mohotti, Gamini Pushpakumara, and V. P. Singh

1 Introduction

Tea (*Camellia sinensis* L. (O) Kuntze) is a commercially important beverage crop, of which origin is reported to be Southeast Asia (Wilson and Clifford 2012). At present, ten countries produce over 90 % of world tea output, namely, China, India, Kenya, Sri Lanka, Vietnam, Turkey, Iran, Indonesia, Argentina, and Japan, of which China and Japan mainly produce green tea, and the other countries are mainly black tea producers. In the world tea exports, Kenya secured the first position in 2017, followed by China, Sri Lanka, and India (Sri Lanka Tea Board 2017). Over 50% of the world tea production is consumed in China and India. Sri Lankan tea, which is world famous as "Ceylon tea," fetches the highest average auction prices at present owing to its unique quality. More than 70% of tea in Sri Lanka is produced by smallholders. Turkey, Iraq, Russia, Iran, and UAE remained as the largest export destinations of Ceylon tea in 2017.

Tea is known as a crop of wide adaptability, grown in a range of soils and climates. However, for successful tea cultivation, many management practices are considered as essential. In its original habitat, tea plants grew under natural shade, and hence, the establishment of shade trees in tea plantations was adopted to mimic the natural conditions (Eden 1976). Thus, the establishment and management of shade are considered as vital and integral components in tea cultivation. Heavy shade is considered essential during the nursery period, which is gradually removed before field planting. However, requirement of shade in field-grown tea has been a controversial subject since inception, and at times there have been differences of opinion.

A. J. Mohotti (✉) · G. Pushpakumara
Department of Crop Science, Faculty of Agriculture, University of Peradeniya, Peradeniya, Sri Lanka

V. P. Singh
World Agroforestry Centre, South Asia Office, New Delhi, India

R. P. De Silva et al. (eds.), *Agricultural Research for Sustainable Food Systems in Sri Lanka*, https://doi.org/10.1007/978-981-15-3673-1_4

In tea plantations, shade is provided by planting different species of shade trees. To date, there are many scientific records with information supporting the shade requirement for tea (Owen 1886; Harler 1960; Baura and Gogoi 1979; Hudson and Muraleetharan 1999; Mohotti and Lawlor 2002; Karunaratne et al. 2003; Mohotti 2004; Panda 2016) and vice versa (Owen 1886; McCulloch et al. 1965; Aoki 1982; Grice 1989). Much of the information on shade has been recorded as its effects on three components, i.e., quantity and quality of yield, tea bush physiology, and pest incidences, but not as a system approach. Hence, this paper reviews the use of shade and shade trees in tea plantations and discusses the importance of tea plantations with shade tree as an agroforestry system for harnessing maximum economic, ecological, and environmental benefits from a tea-based land management system.

2 Types of Shade Trees: High vs Medium Shade

The type and level of shade depends largely on the practices and policies adopted by different countries and estates, the extent of the lands and ownership, etc.

In Sri Lanka, shade is managed at two canopy levels, i.e., high shade and medium shade in large plantations by planting specific tree species. This may be different in the case of majority of the smallholders, where tea is grown as an agroforestry/mixed cropping system with a large mix of different perennials. In the large plantations, shade is mainly provided by high shade trees, which are maintained at a height of about 20 m. The other canopy, which is lower and called the medium shade, is provided by shorter trees planted in between the high shade trees, the main function of which is the provision of green manure with periodical lopping (Ekanayake 2008).

This dual layer of shade provided by temporary and permanent shade tree species is also found at lower elevations with relatively higher temperatures in South India (Barua 1961). In North India however, shade trees are known to provide a more or less continuous cover for the tea bushes at a 6–15 m (Barua 1961; Arunachalam 1995). For higher elevations above 600 m on southern and western slopes and above 500 m elevations on northern and eastern slopes in places (i.e., Darjeeling in Northeast India), shade is not considered as necessary (Arunachalam 1995). However, shade is recommended for tea in Nilgiris and Anamallais, where the intensity of sunshine is very high. In Bangladesh, about 50% diffused sunlight is provided by a canopy of permanent and semipermanent shade trees. *Grevillea robusta* is a common shade tree in Kenyan tea plantations (McCulloch et al. 1965).

3 Desirable Characters of Shade Trees

It is clear from the literature that the characteristics of the shade stratum depend on the botanical composition, number of vertical layers, degree of canopy closure, tree crown characteristics, and shade tree management (lopping, pollarding and thinning out of plants). Such aspects vary among major ecological regions and landholding sizes in response to both biophysical and socioeconomic factors. It is clear from observations that highly diverse shade tree assemblies are present in smallholdings, whereas simple shade tree structures are often found in large-scale commercial holdings. Inclusion of shade trees is less controversial in the case of smallholders since they usually maintain multiple species in their lands. At early stages of introducing shade trees in India, size and shape of the canopy, leaflet size, economic life span, pest and disease resistance, resistance to wind damage, depth of the root system, and suitability to local soil and climatic conditions had been considered as important in selection criteria (Arunachalam 1995). Recently, many characters have been considered in selecting and recommending a particular species for tea plantations as a shade tree. Characters such as adaptability to soil and climatic conditions of a given region, compatibility with tea, growth rate and biomass production, plant height, strong and deep root system capable of trapping nutrients not reached by tea roots, tolerance to repeated pollarding/lopping, stem and branch characteristics, canopy architecture, leaf characteristics, competitiveness with tea for water and nutrients, ability to fix atmospheric nitrogen, not harboring tea pests and diseases and potential of becoming a weed due to self-seeding, etc. need to be carefully evaluated when selecting a shade tree for tea plantations in Sri Lanka (Beddage and Mohotti 2005). Further, supply of good-quality timber, fuelwood, and any other product are considered as added advantages in selecting a shade tree species.

Under Sri Lankan and South Indian conditions, *Grevillea robusta* is considered as an ideal shade tree for the tea plantations. Hudson and Muraleetharan (1999) described its beneficial attributes such as quick growing ability, finely dissected leaves which facilitate filtering of light, the deep root system with high root cation exchange capacity which does not compete with tea for nutrients and moisture, high amount of organic litter it produces, coppicing ability and fast regeneration, tolerance to wind, and generally less susceptibility to pests and diseases.

4 Species of Shade Trees

In most tea-growing countries, the species used for the purpose of shade are quite similar. In Sri Lanka, the tea-growing areas are classified into three elevation categories (above mean sea level), viz., low country (below 600 m), mid-country (600–1200 m), and upcountry (above 1200 m). For the low country, *Falcataria moluccana* (Miq.) Barneby & J.W.Grimes (syn. *Albizia moluccana* Miq.) is recommended as a high shade tree and *Gliricidia sepium* as a medium shade tree. For the mid-country wet zone, *Grevillea robusta* (Fig. 1a) and *F. moluccana* species are recommended as a high shade, while *Calliandra calothyrsus*, *Erythrina lithosperma*, and *G. sepium* (Fig. 1b) are recommended as medium shade trees. For the mid-country semidry zone, *G. robusta* and *F. moluccana* species are recommended as high shade trees, and *C. calothyrsus*, *E. lithosperma*, *Gliricidia sepium*, and *Acacia pruinosa* are recommended as medium shade trees. For the

Fig. 1 (**a**) Tea plantation with *Grevillea robusta* and (**b**) *Gliricidia* grown for shade in Sri Lanka

upcountry, *G. robusta* is recommended as a high shade tree, while *C. calothyrsus*, *E. lithosperma*, *Acacia pruinosa*, and *Acacia decurrens* are recommended as medium shade trees (Ekanayake 2008). Most of the large plantations usually follow these recommendations. With a preliminary screening, the Tea Research Institute has made an effort to identify new shade tree species for tea plantations. Consequently, 84 species were identified as potential species based on morphological characteristics in a preliminary screening (Beddage and Mohotti 2005). However, in smallholder tea lands, the shade tree species are not established systematically as recommended, and tea is usually mixed cropped with many perennial species such as spices, timber trees, fruit trees, etc.

In old tea plantations in India, planting of tea under the shade of *A. chinensis* was found to be beneficial (Owen 1886; Gogoi 1976). Subsequently, many other species were used, i.e., *Albizia odoratissima* Benth (Syn. *Mimosa odoratissima*, Roxb.), *Albizia procera*, and *Dalbergia assamica* (Gogoi 1976). *Albizia odoratissima* and *Albizia chinensis* were the best shade trees suited to conditions in Northeast India (Harler 1960).

In Bangladesh *Albizia sinensis* and *F. moluccana* are grown as semipermanent shade trees, while *Albizia odoratissima*, *Albizia procera*, *Albizia lebbeck*, and *G. robusta* are used as permanent shade trees. About 75% of the total tea-shade trees in Bangladesh are *A. odoratissima* (Sana 1989).

China has the largest extent of tea in the world. Tea is being planted in association with many tree species in China (Shuobo et al. 1997), i.e., forest species, strawberry tree (*Myrica*), *Citrus unshiu*, *Eriobotrya*, peach (*Prunus persica*), *Diospyros*, pear (*Pyrus* spp.), grape (*Vitis* spp.), Chinese sapium (*Sapium sebiferum*), tung oil tree (*Aleurites fordii*), *Pinus*, *Paulownia*, and rubber (*Hevea*). Tea is also grown in association with many food crops such as corn, potato, barley, etc. In China, the presence of trees in tea is as a part of an agroforestry system, rather than for the provision of shade for the tea per se.

Under the conditions of Malawi, the species *Grevillea robusta* and *Albizia adianthifolia* were tried as shade trees for tea (Grice 1989). The shade was found to be beneficial only during the 2 months of hot and dry weather in Malawi.

5 Propagation and Planting of Shade Trees

5.1 *Grevillea robusta* and *Albizia* spp.

Both *Grevillea* and *Albizia* are propagated through seeds. They are first raised in poly bags and then field-planted with the onset of monsoons. *Grevillea* seeds are membranous and of light weight, approximately containing 120,000 seeds per kg. As the viability of seeds is short, fresh seeds are used for propagation (Hudson and Muraleetharan 1999; Ekanayake 2008).

In India, seeds are sown on raised beds of 1 m width and of convenient length, made of soil and sand mixture in the ratio of 3:1. The seeds should be sown in rows spaced at 5 cm, and 2–3 cm apart in the row, covered with a thin layer of soil-sand mixture and then with a suitable grass/straw mulch. Overhead shade provided at a 1 m height should be removed when the seeds germinate in about 3 weeks' time. When the seedlings have the first pair of leaves, they may be transplanted into polythene bags filled with soil/soil-sand mixture (preferable pH 6.0). Until the seedlings establish well, temporary shade has to be provided with ferns, which should be removed in stages. The seedlings are ready to be field-planted in about 6–9 months.

5.2 *Calliandra calothyrsus*

Calliandra could be directly established in new clearings by sowing three to four seeds at a point, on a soil mound, and thinned out leaving one vigorous plant after germination. Seeds can also be sown in poly bags and 4–6-month-old seedlings planted in the field (Ekanayake 2008).

5.3 Dadap (*Erythrina lithosperma*) and Gliricidia (*Gliricidia sepium*)

Both these species are established by using 6–8-month-old poles, having a girth of 5–7.5 cm and 1.8–2 m length, obtained from trees at the point of anticipated cut, with the onset of rains (Ekanayake 2008). They should be ring-barked 6–8 weeks before poles are taken. Prior to planting, both ends of poles should be shaved off with a slanting cut using a sharp knife. Planting holes should be 30 cm wide and 45–60 cm deep. Dolomite 225 g should be mixed with the soil from the planting hole and should be put back to the hole. The poles should be planted after making a hole with an "alawangu." The basal end of the stump should be buried to a depth of 30–40 cm.

The top end of the pole should be covered with mud or polythene to prevent desiccation (Ekanayake 2008).

6 Management of Shade

In Sri Lanka, *Grevillea* and *Albizia* seedlings are planted at an initial spacing of 6 m × 6 m with medium shade trees interplanted midway between two high shade trees at the spacing of 3 m × 3.6 m. By the third year, high shade is finally thinned down to a spacing of 12 m × 12 m and medium shade to 6 m × 7.2 m (Tea Research Institute 2003). In tea fields on southern and western slopes, final stand of high shade should be at a spacing of 12 m × 6 m and medium shade at 6 m × 7.2 m. Under South Indian conditions, *Grevillea robusta* is recommended to be planted at the time of tea planting at a spacing of 6.1 m × 6.1 m, which is to be thinned out first at the time of the formative pruning, to a spacing of 12.2 m × 6.1 m. Trees will be further thinned out to a spacing of 12.2 m × 12.2 m at the time of second pruning and at high elevations to a final spacing of 12.2 m × 24.4 m at the third pruning (Hudson and Muraleetharan 1999).

As tea is considered to be a shade-loving plant, it needs to be grown in association with shade trees. However, too much shade by unpollarded trees may be detrimental to the growth, yield, and quality of tea (Hudson and Muraleetharan 1999). Therefore, it is extremely important that the shade is regulated in tea lands depending on the weather pattern. Normally, the monsoonal periods are accompanied by high rains and a thick cloud cover, which cuts off the incident solar radiation. During such times, the shade should be reduced by lopping or pollarding of shade trees. The shade trees should also be trained to provide an even and a good cover for tea especially during dry/hot months.

Pollarding is done by cutting off the upper branches of the crown of the tree at the desired height, which also induces lateral growth of the branches. In Sri Lanka, pollarding of *Grevillea robusta* is done at a height of 5 m when they are 10 years old, and in *Albizia* at a height of 5.8–6 m when they are 3 years old (Ekanayake 2008). Under south Indian conditions, at high elevations, pollarding is done at a height of 7.6 m, and development of three tiers of lateral branches between 6.1 and 7.6 m height is recommended (Hudson and Muraleetharan 1999). At mid-elevations, pollarding at a height of 9.1 m and promotion of lateral branches between 7.6 and 9.1 m is recommended. Annual lopping/de-suckering of the pollarded trees is also essential, to remove the clusters of branches arising at the pollarded level. While pollarding, it is essential to leave at least 20% of the foliage in the remaining laterally spreading branches.

7 Wind Belts

Wind belts should be established to protect tea in windswept areas. In the upcountry of Sri Lanka, a mixed stand of *Grevillea robusta* and *Hakea saligna* or *Acacia* spp. is recommended at right angles to the direction of the wind (Ekanayake 2008). In the low country, a mixed stand of *Fragrarea fragrans* and *Gliricidia sepium* is used. The tea itself also is used in certain areas as a wind break, by allowing two to three adjacent tea rows to grow freely. Wind barriers in the form of physical structures made from grass or other materials are recommended only for young tea in Malawi, for areas prone for winds (Grice 1989). Windbreaks established by trees are not recommended for young or mature tea in Malawi, as crop loss is anticipated due to the shade cast by such windbreaks.

8 Effects of Shade Trees on Tea

Presence of shade trees in tea plantations is known to impart multifold beneficial effects for tea, such as conservation of soil and soil moisture, reduction of incident solar radiation, protection from high temperature, reducing humidity injuries, maintaining soil fertility and recycling of nutrients, addition of organic matter to the soil, and addition of nitrogen by the root nodules of leguminous trees (Obaga 1984; Hudson and Muraleetharan 1999). Shade also helps in achieving better distribution of crop and minimizing incidence of pests.

As drawbacks, the shade trees also transpire a considerable amount of water and also use up the nutrients in the soil. Further, heavy shade creates undesirable environmental conditions for growth and yield of tea. Hence, it is important to consider these interactions in establishment and management of shade.

8.1 Modification of Microenvironment

Tea grown with shade trees forms a mini-forest system, with a few canopy strata, modifying the microclimate. Increment of the relative humidity and lowering the sunlight intensity, ambient temperature, and wind velocity can be considered as some of such modifications. In the case of smallholder lands, there can be even more strata, modifying the microclimate even further. These modifications usually favor the growth of the tea plant. However, there can also be adverse impacts, such as increased incidence of blister blight disease (caused by the fungus *Exobasidium vexans* Massee) with high relative humidity under shade.

Ediriweera and Anandacoomaraswamy (1999) reported that in a tea stand interplanted with 8-year-old *Grevillea robusta* L. as the shade tree, on a clear day the shaded area ranged from 12% to 15%, highest during the early morning and lowest during the noon. The canopy temperature under shade was 4–6 °C less than the unshaded area. The relative humidity of the shaded area was marginally higher than the unshaded area in the afternoon. The organic carbon content ranged from 3% to

4%, with a high value around near the base of the tree and a low value midway in between the shade trees. The mean leaf area index of the *Grevillea* canopy was 6.9 and the canopy extinction coefficient was 0.21. In a study carried out in China, in tea interplanted with tree species, the light intensity was 20–30% less, the temperature variations during the day and seasons were much less, the humidity was 2–5% higher, and the wind speed was 10–20% less than in unshaded tea lands (Shuobo et al. 1997).

8.1.1 Light Reduction Effects

Predominant effect of shade is to reduce the incident radiation flux at the tea canopy level, but there can be other effects. With low radiation under shade, reductions in transpiration have also been observed (Ripley 1967). Mohotti (1998) has reported that field-grown tea gets light-saturated around 700–900 μmol m^{-2} $s^{-1}$of photosynthetically active radiation (PAR). On bright, clear days, unshaded tea receive over 1700 μmol m^{-2} s^{-1}PAR around midday (Mohotti and Lawlor 2002; Wijeratne et al. 2008) under upcountry conditions of Sri Lanka. When tea is under about 35% shade, the canopy receives about 1100 μmol m^{-2} s^{-1}PAR, which is closer to light saturation range, and hence the tea leaves recorded higher rates of photosynthesis than that of unshaded plants (Wijeratne et al. 2008). Unshaded tea can receive high PAR, resulting in photoinhibition of photosynthesis (Mohotti 1998). The shade decreases the incident PAR flux, thus decreasing photoinhibition while maintaining a high rate of photosynthesis. Karunaratne et al. (2003) observed a significant photoinhibition on clear, sunny days, but not on cloudy days, which supports the relationship between shade and photoinhibition. Therefore, the extent of photoinhibition in a given agroecological region is determined by its proportion of clear, sunny days per year. In environments characterized with only a small proportion of clear, sunny days per year, giving too much shade could cause yield reductions due to a low radiation budget of the tea canopy. In a field study across a range of locations with different levels of natural shading, Gamage et al. (2007) have shown that the optimum shade level for tea at lower altitudes (<600 m amsl) in the humid zone of Sri Lanka is between 30% and 40%, which agrees with findings of Mohotti (1998) and Wijeratne et al. (2008). However, the best shade level for India has been reported to be about 50% of incident radiation (Harler 1960).

Shade trees minimized sun scorch in tea leaves and branches, particularly after pruning (Obaga 1984). In the absence of shade, the bush frame is suddenly exposed to direct solar radiation, causing sun scorch of branches that are often prone to wood rot, pathogen and pest attacks, etc.

8.1.2 Soil and Leaf Temperature and Relative Humidity

Shelterbelts reduced the wind speed and increased the diurnal range of temperature in East Africa (Ripley 1967). Artificial and natural shade decreased diurnal temperature variation, slightly increased humidity, and slightly reduced wind speed in a mature tea stand (Ripley 1967). He reported that the decrease in leaf transpiration was mainly due to the reduction in light intensity.

Hadfield (1968) showed that in Malawi, when the ambient temperature reached 30–32 °C, leaves in full sunshine may reach temperatures of 40–45 °C, but shaded leaves remained 1–2 °C below the ambient temperature. The temperatures of fully exposed horizontal leaves were generally 2–4 °C warmer than erect or semi-erect leaves. Mohotti and Lawlor (2002) reported that in high altitudes of Sri Lanka, the leaf temperature reached about 30–32 °C around midday. When the tea was shaded with nylon netting which cut down incident radiation by about 30%, the midday leaf temperature was about 1.5 °C below that of the unshaded plants. Wijeratne et al. (2008) showed that with increasing shade the leaf temperature significantly decreased. The leaf temperature also decreased with increasing depth within the canopy.

An experiment conducted for 2 years with different shade levels showed that shade had a protecting effect on soil temperature fluctuations at different times of the day and different seasons (Gogoi 1976). During summer, the soil temperature was much higher in the full sunlight than under shade.

Under the environmental conditions of Malawi, shade was not considered as essential, as the maximum temperatures reach 30 °C only during 2-month-long dry period (Grice 1989). Hence, shade was considered to be beneficial only during this time period. Furthermore, the tea yield under shade was found to be lower than those unshaded, and hence, having shade trees was discouraged.

8.1.3 Wind Speed/Hail Damage

As the canopy of shade trees is much above that of tea plants, the shade trees can minimize damage to tea by wind. In certain tea-growing areas of the world, i.e., Kericho and Nandi Hills areas of Kenya, East Pakistan, and Northeast India, hail damage and soil erosion especially to young tea fields can cause major crop losses (Obaga 1984). Therefore, establishment of shade trees in these areas is a means of reducing crop losses. In tea-growing areas of the eastern slopes of Sri Lanka, too, tea is protected from winds by shade trees and wind belts.

8.1.4 Interactions with Moisture and Nutrients

8.1.5 Moisture Relations Under Shade

The information available seems to be diverse in terms of the effects of shade on moisture relations of soil. In a high altitude in Kenya, soil drying during the dry season was more in soils with no shade trees than in tea soils interplanted with *Grevillea* (McCulloch et al. 1966). However, Visser (1961) observed that at the beginning of a drought, *Grevillea* and *Erythrina* conserved moisture but adversely affected the tea during prolonged drought. Soil moisture available under *Grevillea* and *Gliricidia* was slightly less than under *Albizia gummifera* and markedly less than when no shade was present. Young shade trees compete with tea for soil moisture (Gogoi 1976), possibly due to similar root depths. Properly managed mature stands of shade trees conserve soil moisture, owing to reduction of solar radiation, lowering soil temperature, and soil organic matter addition. The degree of competition with tea also varies with the species of shade trees. Mature stands of *Albizia chinensis*,

Albizia odoratissima, and *Erythrina lithosperma* are known to conserve soil moisture in tea lands when compared with those unshaded.

8.1.6 Interactions with Nutrients

Direct effects of nutrient recycling may occur due to leaf fall, lopping and pollarding, uptake by different depths of soil, and biological nitrogen fixation by roots of some species (i.e., *Albizia* spp.). There are also indirect effects of shade on nutrient recycling.

In experiments carried out during the 1940s, it was found out that the yield increase due to inorganic fertilizers was less under shade (Gogoi 1976; Obaga 1984). In a trial carried out in northeast India on 12 tea jats, the interactions of shade with nitrogen were highly significant (Gogoi 1976). In Assam, growing tea under shade without N fertilizer gave the same yield as tea grown with adequate N supply under full sunlight (Harler 1960). However, dense shade is also reported to impair photosynthesis and dry matter production and affect the uptake, utilization, and response to nutrients, especially nitrogen (Harler 1960; Grice 1989; Hudson and Muraleetharan 1999). Beer (1987) reported that for many perennial crops grown in inferior soils, shade is beneficial, especially as shade slows down the metabolic activities of the plants and therefore decreases the demand for soil nutrients.

Smith et al. (1993) observed an interaction between photoinhibition and N nutrition only in treatments that were receiving 225 kg N ha^{-1} $year^{-1}$ or less. Although treatments receiving higher N rates (e.g., 375 kg N ha^{-1} $year^{-1}$) showed light saturation around 1400 μmol m^{-2} s^{-1} of PAR, photoinhibition was not observed (i.e., up to 2000 μmol m^{-2} s^{-1}). This is in agreement with the results of Mohotti and Lawlor (2002) who showed that photoinhibition of tea is minimized, but not completely eliminated, by abundant nitrogen supply. With abundant N inputs, the rates of photosynthesis increased in comparison with low N and unshaded conditions (Mohotti et al. 2000). The unshaded plant photoinhibition, during midday of bright, clear days, was minimized with abundant N supply. In a study conducted in China, the nitrogen metabolism was shown to be more efficient under 50% and 90% shade than under unshaded conditions (Shuobo et al. 1997).

8.2 Shade Effects on Physiology of Tea

Investigations on photosynthetic partial processes (i.e., light capture, electron transport, photochemical and non-photochemical energy quenching, and carboxylation) by Mohotti et al. (2000) and Mohotti and Lawlor (2002) have shown that the entire photosynthetic apparatus of tea is shade-adapted.

Leaves of glass-house and field-grown tea plants under shaded conditions showed a 40% higher maximum rate of photosynthesis compared with unshaded leaves in Sri Lanka (Mohotti 2004), which was attributed to the higher efficiency of some partial processes of photosynthesis. Under shaded conditions, along with a decrease in leaf temperature, an increase in rate of total linear electron transport, direction of a larger proportion of the electron transport products to CO_2 fixation, and a smaller proportion to photorespiration occurred. Studies with young tea plants under controlled environmental conditions have shown that shade increased photosynthesis by increasing the efficiency of photosystem II and the rate of linear electron transport (Mohotti et al. 2000). Higher N inputs improve the performance of unshaded plants, as shown by increased rates of photosynthesis, compared with low N in unshaded conditions. Several studies have shown that photosynthesis of tea decreases due to photoinhibition when the light intensity increases beyond 1400–1500 μmol m^{-1} s^{-1} (Smith et al. 1993; Mohotti and Lawlor 2002). Karunaratne et al. (2003) indicated that the midday depression of photosynthesis observed around 09:00 to 11:00 h local time on clear days was less noticeable under shade.

Photoinhibition of the unshaded tea plants is a result of increasing irradiance, leaf temperature, and vapor pressure deficit of the air, which decreased the stomatal conductance. Mohotti and Lawlor (2002) observed that in field-grown mature tea, the irradiance, leaf temperature, and vapor pressure deficits of the air increase with the progress of the day toward midday with a correspondingly decrease of the stomatal conductance and sub-stomatal CO_2 concentration, which lead to reduced RuBP carboxylation and photosynthesis. With low carboxylation, a greater percentage of excitation energy was directed towards non-photochemical quenching, and as a result photoinhibition increased (Mohotti et al. 2000). Shade also increased the stomatal conductance, thereby allowing greater CO_2 influx, which resulted in directing a greater proportion of excitation energy towards carboxylation (i.e., increased photochemical quenching), which in turn decreased photoinhibition (Mohotti et al. 2000; Mohotti and Lawlor 2002; Karunaratne et al. 2003).

Several studies have shown that photosynthesis of tea decreases due to photoinhibition when the light intensity increases beyond 1400–1500 μmol m^{-2} s^{-1} (Smith et al. 1993; Mohotti and Lawlor 2002; Wijeratne et al. 2008). In a study carried out in China, the photosynthetic efficiency in tea increased up to 50% shade during the summer and autumn and up to 30% in the spring (Shuobo et al. 1997). The photosynthetic efficiency gradually decreased thereafter with further increase in shade level. Owing to the lowered light intensity and temperature, shading decreases respiration and transpiration. Wijeratne et al. (2008) showed that the photosynthesis of tea was highest with 35% shade (approximately 12.2% higher than in unshaded leaves), followed by 65% shade and unshaded plants. The stomatal conductance was highest in the medium shade, followed by high shade (65% shade) and unshaded plants. Accordingly, the photosynthetic apparatus of the tea plant is shown to adapt to different shade conditions. The radiation use efficiency of tea was lowest when the leaves were unshaded and it significantly increased with increasing shade as well as with the depth of the canopy, showing its flexibility in adaptation to different light environments. As tea is thought to have originated as an understory plant in the tropical rainforests (Eden 1976), it is likely that its photosynthetic apparatus is adapted to function with maximum capacity under shade.

The shade also decreases the air and leaf temperatures, which can have direct effects on photosynthesis. Hadfield (1968) reported that the net photosynthesis sharply declined at leaf temperatures above 35 °C. Between 39 and 42 °C, there was no net photosynthesis. Respiration continued up to 48 °C above which temperature the leaf tissue was irrevocably damaged. Wijeratne et al. (2008) showed that the leaf temperature significantly decreases with increasing shade treatment as well as with increasing depth of the canopy.

8.3 Effects of Shade Trees on Anatomy of Tea Plants

In a study carried out in China, the thickness of the whole leaf, upper cuticle layer, palisade, and spongy tissues decreased under shade but increased the ratio of palisade to spongy tissue (Shuobo et al. 1997). In a study carried out in Sri Lanka with mature, field-grown tea, unshaded plants exhibited typical "sun" characteristics, i.e., significantly higher canopy depths and specific leaf weight (Wijeratne et al. 2008). Typical "shade" characteristics were observed in leaves of the shaded plants as well as in the leaves in lower layers of the canopy. Leaf thickness and palisade layer thickness decreased significantly with increased shade level and canopy depth. However, the stomatal density and none of the photosynthetic pigments showed a significant difference with the increased shade level and depth of the canopy in this study.

8.4 Effects of Shade Trees on Growth and Yield

Besides the beneficial physiological effects of shade on tea, many workers have reported different effects of shade on shoot growth, yield, and yield components.

In India, a rapid yield decline was observed from the second year of shade removal although it has shown an increase in yield in the first year (Barua and Gogoi 1979). In a series of experiments conducted under Northeast Indian conditions with about 50% shade provided by bamboo screens, Barua (1961) showed that the growth and yield and the weight of prunings of tea significantly increased under shade. Hudson and Muraleetharan (1999) indicated that the optimal light intensity that promotes growth of tea shoots is about 60% sunlight. In an experiment carried out in Kenya, shoot extension, leaf water potential, and shoot dry mass were highest under artificial shade compared to unshaded and under a stand of *Grevillea robusta* planted at 12.1 m × 12.1 m spacing (Othieno and Ng'etich 1992). Aoki (1982) reported that the photosynthetic rates increased with leaf development in unshaded leaves, but shading retarded this increase in growth. However, the maximum leaf area was found to be under 35–50% light intensity and minimum under full sunlight. In a study carried out in China, bud density (no. of buds/ 1000 cm^2 plucking table) decreased by 15.2% and 36.4% under 50% and 90% shade, respectively, while the average bud weight was much higher with 50% shade than with unshaded and 90% shaded conditions (Shuobo et al. 1997). The buds under shaded conditions were much more vigorous than those under sun-lit conditions. On the contrary, under the conditions of Malawi, the annual yield without shade was found to be higher than that with shade (Grice 1989).

In a tea plantation with *Grevillea robusta* at a high elevation (2130 m) in Kenya, varying degrees of shade, yields have shown highly significant negative correlations with shade (McCulloch et al. 1966). Increasing shade level decreased terminal bud activity, the number and duration of active phases, the number and dry weight of leaves, the number and length of side shoots, and girth of stem (Kulasegaram and Kathiravetpillai 1980).

8.5 Effects of Shade Trees on Pests and Diseases of Tea

Among the multifold effects of shade trees, Hudson and Muraleetharan (1999) emphasized that it also helps minimizing the incidence of pests, especially mites and thrips. Shade decreased the red spider mite attack in India (Harler 1960). Having shade could reduce the building up of mite populations in the dry season, although some shade trees such as *Grevillea* are known to harbor some mite species attacking tea, such as scarlet mite (*Brevipalpus californicus*) (Senaratne 2008). Banerjee (1969) reported that removal of shade increased the numbers of red spider mite (*Oligonychus coffeae* [Neitner]), scarlet mite (*Brevipalpus phoenicis* [Geijskes]), pink mite (*Acaphylla theae* [Watt] Keifer), and purple mite (*Calacarus carinatus* [Green]).

The branches of tea are highly vulnerable for sun scorch soon after pruning, which provide good entry points for fungi causing wood rot (Arulpragasam 2008). Such stems with wood rot can be invaded by pests such as live-wood termites (i.e., *Glyptotermes dilatatus*) or scavenging termites (Vitarana 2008). Therefore, shade effects on preventing sun scorch of pruned branches help reducing pest infestations such as live-wood termite. At lower elevations of Sri Lanka, the shade tree species *Gliricidia sepium* can also act as a diversionary host for the low-country live-wood termite (Ekanayake 2008) reducing the chances for termites to attack tea bushes.

There is a direct relationship between the shade and the incidence of blister blight disease caused by the fungus *Exobasidium vexans* (Arulpragasam 2008). During heavy rains and monsoon periods, presence of heavy shade can cause severe economic losses due to the environmental conditions favoring the disease.

Despite its suitability as an ideal shade tree in tea plantations in higher elevations of Sri Lanka, *Grevillea robusta* has recently been identified as quite vulnerable to infestation by parasitic plants, particularly the mistletoe *Dendrophthoe falcata* (Yapa et al. 2018). Further, decline and death of *Grevillea* trees have been recently reported from many tea estates in Sri Lanka (Niranjan and Sinniah 2016). It is not yet clear at this stage whether the primary cause of the decline is due to parasitic plants, but declines of woody perennials have been reported due to the occurrence of parasitic plants elsewhere (Dolezal et al. 2016).

8.6 Effects of Shade Trees on Tea Quality

Flavonoids, which are secondary metabolites that possess protective roles in the tea plant, play a major role in quality determination. Flavonoid biosynthesis often increases in response to external stress factors such as excessive UV light, drought, cold temperature, and pathogen infection (Wang et al. 2012). Light is an important environmental parameter that drives photosynthesis and regulates plant growth and development. It also can have profound effects on flavonoid biosynthesis and subsequently on tea quality. However, the opinions on the effects of shade on tea quality too seem to be divided.

Shade had notable effects on both flavonoid and lignin biosynthesis but had no significant effect on anthocyanin accumulation (Wang et al. 2012). Color and strength of the liquor was found to be better under shade in Assam, India (Rahman et al. 1978). With orthodox tea manufacturing, artificial shade improved the tip content, while infused leaf, quality, and briskness were better under unshaded condition. With CTC manufacturing, none of the attributes were changed with shade. Tea manufactured from tea bushes highly shaded by *Grevillea robusta* at a high elevation (2130 m) in Kenya fetched much lower prices than that from unshaded tea bushes (McCulloch et al. 1966). In China, shading increased the contents of total amino acid, caffeine, chlorophyll, and nitrite compounds and decreased the catechin and polyphenol contents. The shaded tea also produced more succulent leaf with higher moisture contents in all the seasons. The overall result was a decrease in quality (Shuobo et al. 1997). Owuor et al. (1988) reported that chemical parameters such as theaflavin, thearubigins, and flavor index of CTC tea in Kenya have been changed favorably with the presence of shade in tea.

9 Tea-Shade Tree Intercrop as an Agroforestry System

Although at the inception shade trees were considered important merely for provision of shade per se, it is clear from the above discussion that shade trees in tea plantations can provide multitudinal benefits for physiological processes of tea, yield, yield quality, etc. as summarized in Fig. 2. Further, with increasing contribution to the national tea production by smallholders who maintain their tea with many tree species in combination, it is more appropriate to consider tea systems as agroforestry systems. The tree species planted in smallholdings may serve multipurposes, such as providing food/fruits [e.g., mango (*Mangifera indica*), avocado (*Persea americana*), jak (*Artocarpus heterophyllus*), citrus, etc.], timber [e.g., mango, jak, toona (*Toona ciliata*), mahogany (*Swietenia macrophylla*), etc.], fuelwood (e.g., *Gliricidia*), medicinal [e.g., neem (*Azadirachta indica*), etc.], cash crops [e.g., cloves (*Syzygium aromaticum*), nutmeg (*Myristica fragrans*), etc.], and wood for other purposes such as fencing, etc. in addition to providing shade and green manure. In a world with increasing population and decreasing land availability for agriculture, increasing productivity of a unit land area (land productivity) is more sustainable and the way forward. In that light, growing trees with multiple uses as

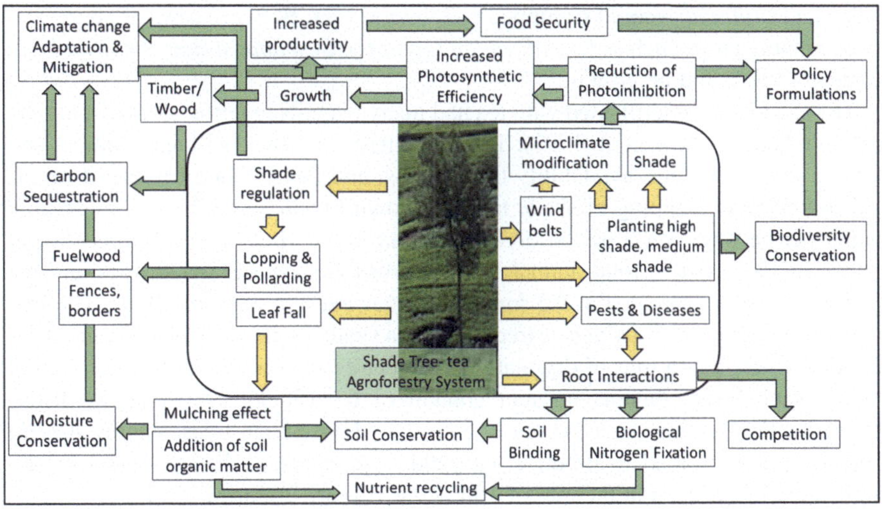

Fig. 2 Components/interactions/benefits in a tea-shade tree system. The inner box shows direct interactions, and the outer box, indirect interactions

shade trees in tea plantations will be an inevitable option in the future for ensuring much needed ecosystem services.

However, other advantages of maintaining shade trees in tea plantations such as the income generation from their products, for example, timber, fuelwood, and bioenergy, level of carbon sequestration in the field, carbon trading opportunities, and ecosystem services are little studied and estimated, yet extremely important. This has been further stressed by increasing awareness of the environmental costs associated with high-input monocultures and environmental hazards such as soil erosion on sloping lands. Thus, in addition to direct benefits to the tea industry, attention should be given to the contribution of shade trees for ecosystem services such as maintaining biodiversity, carbon sequestration, carbon trading opportunities, soil fertility management and rainwater harvesting, etc., as comprehensive assessment of all such is currently lacking. This is mainly due to the lack of integrated and holistic approach on the tea and shade tree system as one unit in many parts of the world. Thus, it is timely to reevaluate the system using the agroforestry principles to capture all such benefits.

9.1 Shade Trees as a Source of Nutrients/as a Source of Green Manure/Organic Carbon

The lopping of shade trees provides large quantities of organic matter, which improves the physical, chemical, and biological properties of soil (Ekanayake 2008). The medium shade trees serve a dual purpose by providing both shade and green manure to the tea fields. *Gliricidia* could be lopped 2–3 times a year, and dadap and *Calliandra* could be lopped 1–2 times a year. Between rows of young tea, green manure crops such as *Crotalaria anagyroides*, *Sesbania sesban*, or *Crotalaria usaromoensis* could be grown.

The shade trees interplanted with tea had many benefits in terms of improving the soil, in China (Shuobo et al. 1997). The shade trees improved soil aggregation, soil organic matter content, availability of nutrients, and fixed N and decreased runoff and soil erosion. Lopping of shade trees is known to add about 30 tons of organic matter per hectare over a period of 4–5 years in India (Hudson and Muraleetharan 1999). In East Africa, total annual leaf fall from *G. robusta* shade trees and tea bushes has been estimated at 6.5 tons dry matter per acre (Gogoi 1976). The leaf litter from a stand of *A. chinensis* trees casting a shade of 50–60% of full sunlight is reported to add 2.5–5 tons of dry matter per hectare per year (Hadfield 1974; Gogoi 1976). Such litter contains nutrient content of 63–126 kg N, 18–36 kg P_2O_5, 22–44 kg K_2O, 32–64 kg CaO, and 16–32 kg MgO. Shade trees, mainly *Albizia chinensis*, are assumed to produce tea yields equivalent to that produced by the application of nitrogen at the rate of 100 kg ha^{-1} $year^{-1}$ (Obaga 1984).

Planting of shade trees and application of sulfate of ammonia to tea lands in Assam, Northeast India, has shown markedly high outturns (Harler 1960). It has been reported that the planting of shade trees was preferable and safer than continued application of large doses of sulfate of ammonia in the absence of shade.

9.2 Shade Trees in Climate Change Mitigation and Carbon Sequestration

Climate change effects are experienced by all crops across the globe, and all major tea-growing countries, i.e., Sri Lanka (Wijeratne et al. 2007), Kenya (Elbehri 2015), India (Duncan et al. 2016), and China (Nemec-Boehm et al. 2014), have claimed vulnerability of their tea lands to climate change (Marx et al. 2017). The tea-growing agroecological regions of Sri Lanka with average temperatures above 22 °C, annual rainfall below 2000 mm $year^{-1}$ and high variability, and poor soil conditions are considered as particularly vulnerable (Wijeratne et al. 2007). Further, tea yields especially at low elevations are likely to reduce due to climate change.

Among many adaptation measures to minimize climate change impacts, diversification of marginal lands as energy/timber plantations, soil and moisture conservation, improvement of soil organic matter content, and planting/management of shade trees have been identified as of prime importance. Of these, planting and management of shade trees can be considered as extremely important, since it has many other benefits such as soil and moisture conservation, organic matter addition, carbon sequestration, etc. Shade trees help in mitigating climate change effects mainly by reducing temperature and addition of soil organic matter. Further, shade trees have a great carbon sequestration potential. When they are added to the tea ecosystem, carbon sequestration potential of the overall tea plantation increases tremendously (Wijeratne et al. 2014). The tea plantations in low country, mid-country, upcountry, and Uva regions of Sri Lanka had the potential of sequestering 6659, 3497, 2344, and 5085 kg of C ha^{-1} $year^{-1}$, respectively.

9.3 Classification and Estimation of Carbon Stocks of Shade Tree-Based Agroforestry Systems

In order to visualize the importance of shade tree-based tea agroforestry system, it is appropriate to classify the systems (i.e., high-density vs low-density shade, two storied vs single storied, mixed vs line planting) using GIS-based techniques (Premakantha et al. 2008) and to estimate its timber and carbon stocks using large area inventory with GIS techniques (Premakantha et al. 2012). This helps estimating the distribution of different tea-shade-tree-based agroforestry systems and their potential benefits for necessary policy implications for the future. This will in turn enhance the value of tea-based agroforestry systems.

For example, Premakantha et al. (2012) estimated a total wood volume of 2,935,791 m^{-3} in tea-based systems in the Nuwara Eliya district of Sri Lanka, which comprised of 52.9% of the total wood volume of the district. The shade trees provided a considerable amount of wood volume, of which *G. robusta* provided 20.39 m^3 ha^{-1}, *G. sepium* 0.11 m^3 ha^{-1}, and *C. calothyrsus* 0.02 m^3 ha^{-1}. Estimated total carbon content in tea-based systems was 1524 Gg in 79,182 ha (19.2 t ha^{-1}).

9.4 Shade and Ecosystem Services

Shade trees also can be considered as important agents in providing many ecosystem services. For example, in a study done in coffee plantations across six agroforestry and tree cover transition studies spanning in tropical/subtropical forest zones in three continents, the absence of shade trees resulted in loss of native earthworm populations, which in turn caused a reduction of 76% of soil macroporosity. Increased tree cover contributed to 53% increase in tea crop yield, maintained 93% of crop pollinators found in the natural forest and nearby forest fragments, and contributed to as much as 86% lower incidence of coffee berry borer (Barrios et al. 2017). In a study carried out in Yunnan, China, where tea is traditionally grown under forest, biodiversity, soil and water conservation, natural pest regulation, climate regulation, contribution to carbon storage in trees, etc. were proven to be significantly high in comparison with modern tea terraces (Liang et al. 2013). However, detailed studies in tea-shade tree systems are still lacking.

10 Conclusion and Way Forward: Shade Tree-Tea System as a Climate-Smart Agricultural Landscape System

Present-day agriculture holds many challenges ahead in view of the ever-changing needs of humans, rising populations, land fragmentation, limitation and competition for resources, environmental degradation, climate change and its consequences, etc. Although tea was traditionally grown in large monoculture plantations, present trend across the world is to increase smallholder contribution. With these challenges ahead, the tea landscape needs revisiting by following agroforestry principles and adding value to the tea-based ecosystem with shade trees in climate-smart agriculture (CSA) landscape models as the way forward. CSA has been considered for many sectors in Sri Lanka (World Bank and CIAT 2015), and such models have also been discussed and implemented in Kenya (Milder et al. 2015; Minang et al. 2015). In CSA, integrated landscape management principles are applied to incorporate climate change adaptation and mitigation goals into multifunctionality in rural landscapes. It includes many field- and farm-scale widely used agricultural practices, such as agroforestry, crop residue management, water harvesting and irrigation, indigenous crop and agrobiodiversity conservation and use, etc. This is done by carrying out landscape-scale planning, policy, land management, support activities, improving coordination, and alignment of activities, policies, and investments among sectors and scales.

Taking the ideal smallholder as the model, with all components in place, country- and region-specific CSA models may have to be developed in the future. However, as mentioned above, this will require proper planning, government policy-level interventions, land management, and support activities such as development of knowledge management systems. Evaluation of such existing models also will provide wealth of information for development of perfect tea-shade tree systems for future.

Acknowledgments Authors wish to thank Dr. Keerthi Mohotti of the Tea Research Institute of Sri Lanka for the comments on the manuscript and photographs.

References

Aoki S (1982) Changes in the photosynthetic capacities of tea leaves by shading. Study of Tea 62:8–13

Arulpragasam PV (2008) Common diseases of tea and their management. In: Zoysa AKN (ed) Tea hand book. Tea Research institute, Talawakele, pp 173–209

Arunachalam K (1995) Shade for tea and its management. In: A hand book on Indian tea. All India Press, Pondicherry, pp 90–102

Banerjee B (1969) Shade and manurial experiments at Murmuria: a report on the incidence of mites. Two Bud 16(3):93–96

Barrios E, Valenciab V, Jonssonc M, Braumand A, Hairiahe K, Mortimerf PE, Okubo S (2017) Contribution of trees to the conservation of biodiversity and ecosystem services in agricultural landscapes. Int J Biodivers Sci Ecosyst Serv Manage 14(1):1–16

Barua DN (1961) Effect of shade and shade trees. Two Bud 8(4):4–12

Barua DN, Gogoi BN (1979) Effects of shad removal. Two Bud 26(1):40–42

Beddage HP, Mohotti AJ (2005) A compilation of potential alternate shade tree species for tea plantations. In: Proceedings of the tenth annual forestry and environmental symposium, Department of Forestry and Environmental Science, University of Sri Jayawardenapura, Sri Lanka, pp 23–24. https://doi.org/10.31357/fesympo.v0i0.1244

Beer J (1987) Advantages, disadvantages and desirable characteristic of shade trees for coffee, cacao and tea. Agrofor Syst 5:3–13

Dolezal J, Leheckova E, Sohar K, Altman J (2016) Oak decline induced by mistletoe, competition and climate change: a case study from Central Europe. Presila 88:323–346

Duncan JMA, Saikia SD, Gupta N, Biggs EM (2016) Observing climate impacts on tea yield in Assam, India. Appl Geogr 77:64–71

Eden T (1976) Tea. Longmans, London

Ediriweera EPSK, Anandacoomaraswamy A (1999) Study of high shade trees (*Grevillea robusta*) on micro climate of a tea field. In: Proceedings of the fifth annual forestry and environment symposium of the Department of Forestry and Environmental Science, University of Sri Jayewardenepura, Sri Lanka, p 32

Ekanayake PB (2008) Planting and management of shade trees, green manure crops and wind belts. In: Zoysa AKN (ed) Tea hand book. Tea Research institute, Talawakele, pp 86–93

Elbehri A (ed) (2015) Kenya's tea sector under climate change: an impact assessment and formulation of a climate-smart strategy. Food and Agriculture Organization of The United Nations Rome, 184pp

Gamage AJ, Wijeratne MA, De Costa WAJM (2007) Role of shade trees in tea: a low-country perspective. In: Proceedings of the 215th meeting of the experiments and extension forum. Tea Research Institute of Sri Lanka, Talawakelle, Sri Lanka

Gogoi BN (1976) A review of research in shade tree in tea. Two Bud 23:67–73

Grice WJ (1989) Shade trees and windbreaks. In: Tea planter's handbook. Tea Research Foundation of Central Africa, pp 1–4

Hadfield W (1968) Leaf temperature, leaf pose and productivity of the tea bush. Nature 219:282–284

Hadfield W (1974) Shade in North East India tea plantations: I. The shade pattern. J Appl Ecol 11:151–178

Harler CR (1960) Shade for tea. World Crops 12:464–466

Hudson JB, Muraleetharan N (1999) Shade for tea. Planter's Chronicle:551–555

Karunaratne PMAS, Mohotti J, Nissanka SP, Wadasinghe WADSK (2003) Effect of shade in minimizing photoinhibition of photosynthesis of high grown tea (*Camellia sinensis* L. (O.) Kuntze) in Sri Lanka. Trop Agric Res 15:133–143

Kulasegaram S, Kathiravetpillai A (1980) Effect of level of shade and of nitrogen on growth of young tea (*Camellia sinensis* L.). Tea Q 49(3/4):112–124

Liang L, Xiang Y, Takeuchi K (2013) Harnessing ecosystem services for local livelihoods: the case of tea forests in Yunnan, China. TEEBcase, Retrieved from TEEBweb.org

Marx W, Haunschild R, Bornmann L (2017) Global warming and tea production-the bibliometric view on a newly emerging research topic. Climate 5(46):1–14. https://doi.org/10.3390/cli5030046

McCulloch JSG, Pereira HC, Kerfoot O, Goodchild NA (1965) Effect of shade trees on tea yields. Agric Meteorol 2(6):385–399

McCulloch JSG, Pereira HC, Kerfoot O, Goodchild NA (1966) Shade tree effects in tea gardens. World Crops:26–27

Milder JC, Moroge M, Shames S (2015) Operationalizing climate-smart agricultural landscapes: the case of a tea producing landscape in Kericho, Kenya. In: Minang PA, van Noordwijk M, Freeman OE, Mbow C, de Leeuw J, Catacutan D (eds) Climate-smart landscapes: multifunctionality in practice. World Agroforestry Centre (ICRAF), Nairobi, pp 319–333

Minang PA, van Noordwijk M, Freeman OE, Mbow C, de Leeuw J, Catacutan D (eds) (2015) Climate-smart landscapes: multifunctionality in practice. World Agroforestry Centre (ICRAF), Nairobi

Mohotti AJ (1998) Effect of irradiance and N nutrition on photosynthesis of tea (*Camellia sinensis* L. (O.) Kuntze) in comparison with sunflower (*Helianthus annuus* L.). PhD thesis, Department of Agricultural Botany, University of Reading, UK, 195pp

Mohotti AJ (2004) Shade in tea: is it beneficial? Sri Lankan J Tea Sci 69(1&2):27–39

Mohotti AJ, Lawlor DW (2002) Diurnal variation of photosynthesis and photoinhibition in tea: effects of irradiance and nitrogen supply during growth in the field. J Exp Bot 53(367):313–322

Mohotti AJ, Dennett MD, Lawlor DW (2000) Electron transport as a limitation to photosynthesis of tea (*Camellia sinensis* (L.) O. Kuntze); a comparison with sunflower (*Helianthus annuus* L.) with special reference to irradiance. Trop Agric Res 12:1–10

Nemec-Boehm RL Cash SB, Anderson BT, Ahmed S, Griffin TS, Orians CM, Robbat A Jr, Stepp RA, Han W (2014) Climate change, the monsoon, and tea yields in China. In: Proceedings, Agricultural & Applied Economics Association's 2014 AAEA annual meeting, Minneapolis, MN, July 27–29, 2014

Niranjan M, Sinniah GD (2016) Investigation of die back of *Grevillea robusta* in tea plantations of the Uva region in Sri Lanka. In: Proceedings of sixth symposium on plantation crop research, 02nd–04th November 2016, BMICH, 2, Colombo, Sri Lanka, pp 133–142

Obaga SO (1984) Shade trees in tea – a review. Tea 5(1):39–47

Othieno CO, Ng'etich WK (1992) Studies on the use of shade in tea plantations in Kenya: II effects on yields and components. Tea 13(2):282–292

Owen TC (1886) Tea planter's manual. A.M.J. Ferguson, Colombo

Owuor PO, Othieno CO, Howard GE, Robinson JM, Cooke RD (1988) Studies on the use of shade in tea plantations in Kenya: effects on chemical composition and quality of made tea. J Sci Food Agric 46:63–70

Panda H (2016) Historical perspective of shade trees. In: The complete book on cultivation and manufacture of tea (2nd rev edn). Asia Pacific Business Press Inc, New Delhi, pp 226–244

Premakantha KT, Pushpakumara DKNG, Dayawansa NKD (2008) Identification of tree resources outside forests in the upcountry of Sri Lanka using medium resolution satellite imagery. Trop Agric Res 20:354–365

Premakantha KT, Pushpakumara DKNG, Dayawansa NKD, Sivananthawerl T (2012) Estimation of wood volume of trees outside forests (TROF) in Nuwara Eliya district of Sri Lanka as a measure of carbon stocks. Ceylon Forester 34:21–36

Rahman F, Bhagavathy HN, Basu RP, Choudhury R, Das AK, Dutta SK, Gilchrist RCJH, Sharma KN, Trinick JM (1978) Effect of different field management practices on cup characters and valuation of teas. Two Leaves Bud 25(2):86–89

Ripley EA (1967) Effects of shade and shelter on the micro-climate of tea. East Afr For J 33:67–80

Sana DL (1989) Tea science. BTRI, Moulvibazar

Senaratne KADW (2008) Management of insect and mite pests. In: InZoysa AKN (ed) Tea hand book. Tea Research Institute, Talawakele, pp 210–230

Shuobo H, Genshen P, Renjum G (1997) Physiological and biochemical characteristics of tea plants interplanted with trees. IDRC: library: documents: agroforestry systems in China. Retrieved from http://www.idrc.ca/library/document/090916/chap26_e.html, on 29/07/2002

Smith BG, Stephens W, Burgess PJ, Carr MKV (1993) Effects of light, temperature, irrigation and fertilizer on photosynthetic rate in tea (*Camellia sinensis*). Exp Agric 29(3):291–306

Sri Lanka Tea Board (2017) Tea market update, October – December 2017, 13(4), 8 pp. Retrieved from file:///C:/Users/HP-PC/Downloads/Fourth%20Quarter%202017.pdf

Tea Research Institute (2003) Shade in tea. Advisory Circular No SI-2. Retrieved from http://www.tri.lk/userfiles/file/Advisory_Circulars/SI/TRI_SI02(e).pdf

Visser T (1961) Interplanting in tea. 1. Effects of shade trees, weeds, and bush crops. Tea Q 32:69–82

Vitarana SI (2008) Management of tea termites. In: Zoysa AKN (ed) Tea hand book. Tea Research institute, Talawakele, pp 231–240

Wang YS, Gao LP, Shan Y, Liu YJ, Tian YW, Xia T (2012) Influence of shade on flavonoid biosynthesis in tea (*Camellia sinensis* (L.) O. Kuntze). Sci Hortic 141:7–16

Wijeratne MA, Anandacoomaraswamy A, Amarathunga MKSLD, Ratnasiri J, Basnayake BRSB, Kalra N (2007) Assessment of impact of climate change on productivity of tea (*Camellia sinensis* L.) plantations in Sri Lanka. J Natl Sci Found Sri Lanka 35(2):119–126

Wijeratne TL, Mohotti AJ, Nissanka SP (2008) Impact of long term shade on physiological, anatomical and biochemical changes in tea (*Camellia sinensis* (L.) O. Kuntze). Trop Agric Res 20:376–387

Wijeratne TL, De Costa WAJM, Wijeratne MA (2014) Carbon sequestration potential of tea plantations in Sri Lanka as an option for mitigating climate change; a step towards a greener economy. In: Proceedings of fifth symposium on plantation crops research, held in Colombo, Sri Lanka, pp 205–212

Willson KC, Clifford MN (2012) Tea: cultivation to consumption. Springer Science & Business Media, Dordrecht

World Bank, CIAT (2015) Climate-smart agriculture in Sri Lanka. CSA country profiles for Africa, Asia and Latin America and the Caribbean Series. The World Bank Group, Washington, DC

Yapa SS, Mohotti AJ, Seneviratne MAPK, Peiris BL, Tennakoon KU (2018) Prevalence of mistletoes in fruit and timber trees in the wet and intermediate zones of Sri Lanka. Trop Agric Res 29(3):330–340

Use of Biotechnology for Crop Improvement in Sri Lanka: Current Status and Future Prospects

P. C. G. Bandaranayake

1 Introduction

Recent crop breeding efforts and advances in agricultural practices have contributed significantly to the annual gain of 0.8–1.2% in crop productivity in the world. Nevertheless, further improvement is essential to keep up with demand for feeding 9.7 billion people in 2050. A linear increase of productivity is needed to achieve the projected 100% increase in global demand for food, feed, and fiber by 2050 (Tilman et al. 2011). For example, global production of major cereals, maize, wheat, and rice, needs to be increased at the rate of between 1.16% and 1.31% annually (Hall and Richards 2013; Fischer et al. 2014) to satisfy the 2% yearly increase in the world population (FAO 2011). Nevertheless, such achievements are very challenging since there is limited room to improve the harvest index further (Fischer and Edmeades 2010). The anticipated yield increments of crop plants are also challenged by the escalating occurrence of adverse environmental consequences due to climate change and the manifestation of more aggressive strains of pests and pathogens (Nelson et al. 2014; Garnett et al. 2013).

Sri Lankan population is projected to increase by about 18%, from 20.3 to 23.9 million before plateauing it by 2050 (DCS 2012; UN 2012). Rice is the staple food of Sri Lanka, and it comprises of approximately 40% of the total crop production (FAO 2018a). Sri Lanka has been self-sufficient in rice, producing 100% domestic consumption since 2005 (DCS 2018; FAO 2018a). Several initiatives including high-yielding varieties, paddy expansion, and increased use of irrigation and fertilizer contributed immensely to increase rice production to meet the target (DCS 2018; FAO 2018a). Nevertheless, a recent analysis has predicted that Sri Lanka may need to increase water use and nitrogen fertilizer application by as much as 69% and 23%,

P. C. G. Bandaranayake (✉)
Agricultural Biotechnology Centre, Faculty of Agriculture, University of Peradeniya, Peradeniya, Sri Lanka
e-mail: pradeepag@pdn.ac.lk; pgunathilake@ucdavis.edu

R. P. De Silva et al. (eds.), *Agricultural Research for Sustainable Food Systems in Sri Lanka*, https://doi.org/10.1007/978-981-15-3673-1_5

respectively, to achieve self-sufficiency in rice in mid-century (Davis et al. 2016). Currently, 90% of the total water withdrawals in Sri Lanka are used for agriculture (FAO 2018b), and much of it are utilized for rice production (FAO 2018a). Therefore, the challenge would be to maintain self-sufficiency in rice while utilizing lesser amount of resources. Situation is more or less similar for other food crops. Therefore, new crop varieties are needed to achieve higher productivity with less amount of resources.

Genetic improvement or success of crop breeding has been described by the concept of genetic gain. A linear progress of 2% genetic gain has to be achieved to sustain the productivity (Ray et al. 2012). It is a huge challenge when crop yields have reached a plateau in some regions in the world, due to the narrow genetic base and lack of potential to increase harvest index within elite breeding stocks (Ray et al. 2012). Basically, genetic gain in a crop breeding program is determined by the equation $\Delta G = ih\sigma A/L$, where i is the selection intensity, h is the square root of the heritability in the narrow sense, σA is the square root of the additive genetic variance, and L is the length of breeding cycle interval or generation (Heffner et al. 2010; Ray et al. 2012). Accordingly, introduction of new promising alleles through rapid breeding cycles seems an efficient mechanism to improve the rate of gain significantly, and it could help to achieve the goal of feeding the projected population in 2050 (Reviewed by Li et al. 2018). There are many biotechnological tools available and applied around the world for introduction of favorable alleles and to speed up the breeding efforts. These can be broadly categorized into two major approaches, viz., marker-assisted selection (MAS) and genetic modifications (GMs) (Moose and Mumm 2008). This review discusses application of such key biotechnological tools in current food crop improvement programs in Sri Lanka. Further, the discussion will be extended to the future prospects and major challenges faced in the crop biotechnological research and development.

The research references included here are on major food crops, conducted in Sri Lanka and published as full papers in peer-reviewed international or local journals and proceedings. International journal articles and proceedings were accessed through Web of Science, Google Scholar, and Scopus, while the local publications were retrieved through the National Science Library and Resource Centre maintained by the National Science Foundation of Sri Lanka. Further, publications were collected from all science and agriculture libraries in major universities in Sri Lanka and the Department of Agriculture, Sri Lanka.

2 Molecular Markers

Farmers first and the plant breeders later have been artificially selecting traits for thousands of years aiming to shape plants for better growth and yield performance in cultivated environments. These bulk breeding processes relied on transgressive segregation, in which two genotypes would be crossed and breeders would select offspring with better performance than their parents. This classical breeding process has been repeated multiple times to produce the current elite varieties for most crops with many favorable alleles. Particularly, this process relied on the selection of plants with desired phenotypes that have high heritability. Selection is done manually under field conditions across a number of generations. Generally, large breeding populations are needed to achieve homozygosity of loci that encode desired traits, and often more than six generations are required. It can therefore take more than 10 years to bring a new variety of crop to the market, and there is considerable uncertainty about predicting the effects of combining phenotypes, which limits the application of classical breeding to address global food security and meet the challenges of food demands (Tester and Langridge 2010).

Molecular plant breeding on the other hand could improve the efficiency of the classical breeding by using the marker-assisted selection (MAS) (Moose and Mumm 2008). For example, MAS is extensively used in crop breeding for indirect selection of traits in progenies by monitoring genetic regions as markers associated with trait of interest (Barabaschi et al. 2016). As a critical and effective method, MAS has widely been applied to enhance crop yield, quality, and tolerance to biotic or abiotic stresses (reviewed by Collard and Mackill 2008; Kaur et al. 2015; Kumar et al. 2011). Molecular markers also have other multiple roles to play in plant genetics and breeding. For example, markers are vital for gene mapping and tagging, segregation analysis, genetic diagnosis including pest and pathogen detection, genetic diversity analysis, forensic examination, phylogenetic analysis, and many other biological applications (Semagn et al. 2006; Lam et al. 2010; Singh et al. 2010; Sonah et al. 2011).

Nevertheless, in Sri Lanka, over 95% of the crop breeding programs still rely on classical breeding approaches. Applications of markers are limited and used predominantly on genetic diversity assessments (Table 1) and pathogen detections and identifications (Table 2). Identification of the genetic diversity in the existing germplasm is the essential first step in plant breeding programs. Past breeding efforts were mainly based on morphological characterization and screening of the existing germplasm for particular trait. Molecular characterization can be done without growing the plants in the field saving time and cost substantially. So far, majority of genetic diversity studies were on rice (Table 1). This may be both due to existence of a larger germplasm and the dynamic rice breeding program in the country. Detection and identification of pathogens are important for crop improvement programs. It is essential to know the pathogen or pest before designing efficient resistance breeding program. Further, their population structures and evolutionary mechanisms are important for developing resistant or tolerant crops. Molecular detection and identification methods bypass tedious pathogen culturing steps, and long procedures follow traditionally. This allows use of technology for routinely and release of quality planting materials for breeding programs, commercial cultivations, and exportation.

Several types of informative markers for DNA sequence polymorphisms have been developed and used for decades (reviewed by He et al. 2014; Agarwal et al. 2008; Sonah et al. 2011). Among them, the most widely used techniques in Sri Lanka include random amplification of polymorphic DNA (RAPD) introduced by Williams et al. (1990), inter-simple sequence repeats (ISSRs) described by Bornet

Table 1 Genetic diversity and characterization of germplasm

Crop	Purpose	Technique	References
Capsicum (*Capsicum* spp.)	Diversity analysis	SSR	Chamikara et al. (2014)
Carica papaya	Identification of diseases	Universal primers for detection of phytoplasma	Nadeeshani et al. (2011)
Cashew (*Anacardium occidentale*)	Genetic diversity	RAPD – five primers	Ranathunge et al. (2010)
Commercial cultivars of capsicum	Diversity analysis	SSR	Chamikara et al. (2015)
Finger millet (*Eleusine coracana*)	Diversity analysis	SSR	Wakista et al. (2015)
Mountain papaya	Disease resistance to papaya mosaic virus	Specific primers	Rajapakshe et al. (2017)
Mountain papaya and *Carica papaya*	Characterization	ISSR (2 primers) and SSR (11 primers)	Warnakula et al. (2014)
Mountain papaya and *Carica papaya*	Genetic diversity assessment	SSR and ISSR	Warnakula et al. (2017)
Mung bean (*Vigna radiata*)	Genetic diversity in relation to flowering behavior	SSR (3 primer sets)	Thilakarathna et al. (2016)
Onion (*Allium cepa*)	Characterization	RAPD – five primers	Sumanasinghe et al. (2005)
Papaya ring spot virus	Detection, germplasm evaluation for resistance	PRSV-specific primers	Amalka et al. (2015)
Pineapple (*Ananas comosus*)	Genetic purity	RAPD – four primers	Kondasinghe et al. (2010)
Pomegranate (*Punicia granatum*)	Genetic diversity assessment	ISSR – twenty primers	Attanayake et al. (2017)
Rice (*Oryza sativa*)	Novel allele and protein structure of mutated fragrant gene	Gene-specific primers	Dissanayaka et al. (2013)
Rice (*Oryza sativa*)	Identification and characterization of disease-resistant genes	R gene-specific primers	Wijayalath et al. (2005)
Rice (*Oryza sativa*)	Screening for trehalose gene	Trehalose gene-specific primers	Perera et al. (2013)
Rice (*Oryza sativa*)	Brown plant hopper resistance and aroma	SSR	Jayathilaka et al. (2014)
Rice (*Oryza sativa*)	Fragrance gene	Allele-specific primers	Anushka et al. (2008)

(continued)

Table 1 (continued)

Crop	Purpose	Technique	References
Rice (*Oryza sativa*)	Seed storage	SDS-PAGE	Vithyashini and Wickramasinghe (2015)
Rice (*Oryza sativa*)	Fragrant gene	Allele-specific primers	Wettewa et al. (2011)
Rice (*Oryza sativa*)	Fragrant gene – novel allele	Allele-specific primers	Dissanayaka et al. (2013)
Rice (*Oryza sativa*)	Characterization and selection of phosphorus deficiency tolerance	SSR	Aluwihare et al. (2016)
Rice (*Oryza sativa*)	Fingerprinting and characterization	SSR	Bandara et al. (2009)
Rice (*Oryza sativa*)	Diversity of cis-acting elements of abscisic acid-responsive element-binding protein	Bioinformatics	Senavirathne et al. (2017)
Rice (*Oryza sativa*)	Salt tolerance	RAPD	Safeena et al. (2003)
Yam (*Dioscorea* sp.)	Identification of different species	RAPD – five primers	Jayamaha et al. (2006)

and Branchard (2001) and Reddy et al. (2002) and simple sequence repeats (SSRs) introduced by Litt and Luty in 1989 (Tables 1 and 2). This could mainly be due to simplicity of those techniques and cost-effectiveness.

In the past few decades, thousands of genes and tightly linked markers for various agronomic traits have been isolated and characterized in many crops (Takeda and Matsuoka 2008; Varshney et al. 2006; Collard and Mackill 2008; Kaur et al. 2015; Kumar et al. 2011). However, the MAS breeding efforts in Sri Lanka are mostly limited to rice (Table 3) and still at the research level: selection of markers from published work elsewhere, optimization of PCR and screening process, and development of breeding populations. Only few efforts are reported on identification of new quantitative trait loci (QTLs) and marker development work (Dissanayaka et al. 2014; Gimhani et al. 2016). Single nucleotide polymorphism (SNP), introduced in 2002, is considered as the most abundant in a genome and suitable for analysis on a wide range of genomic scales (Rafalski 2002; Zhu et al. 2003) and routinely being used in plant breeding in other country lines (He et al. 2014). Surprisingly, the SNP-based markers are not widely used so far in Sri Lankan breeding programs.

The genomic selection has been introduced as one of the most promising crop breeding technologies for expansion of genetic gain (Crossa et al. 2017). Among them, genotyping-by-sequencing (GBS) offers a definitive MAS tool to fast-track plant breeding and crop improvement (reviewed by He et al. 2014). Latest development of next-generation sequencing (NGS) technologies has led to whole genome sequencing remarkably cheaper and efficient, enabling ultra-throughput sequences to revolutionize plant genotyping and breeding including GBS (Crossa et al. 2017). The GBS procedure starts with restriction digestion of genomic DNA, followed by

Table 2 Detection and identification of pests and pathogens

Crop	Purpose	Technique	References
16SiXI and 16SrXIV phytoplasmas	Refinement of taxonomy	Multilocus sequence typing	Abeysinghe et al. (2016)
Banana bunchy top virus	Identification of strains	Specific primers for virus	Basnayake et al. (2005)
Banana streak virus	Identification of strains	Specific primers for virus	Gimhani et al. (2005)
Begomoviruses associated with cucurbits	Molecular detection and characterization	Specific primers	Bandaranayake et al. (2014)
Bipartite begomoviruses associated with cucurbits	Molecular detection and characterization	Specific primers	Bandaranayake et al. (2016)
Magnaporthe grisea	Identification	Repetitive element-based PCR (rep-PCR)	Wimalasena et al. (2009)
Papaya phytoplasma	Identification – dieback disease of papaya	16S rDNA universal primers for phytoplasma	Abeysinghe et al. (2014)
Papaya ring spot virus	Detection, germplasm evaluation for resistance	PRSV-specific primers	Amalka et al. (2015)
Papaya ring spot virus	Detection	PRSV-specific primers	Rajapakse et al. (2016)
Phytoplasma in crops	Detection	Universal primers for detection of phytoplasma	Sarathchndra et al. (2009)
Pineapple bacilliform virus	Investigation of host range	PCR with PBV-specific primers	Kularathne et al. (2005)
Seed potatoes	Detection of quarantine plant pathogens	16S rRNA primers	Fernando et al. (2015)
Tomato spotted wilt	Detection	TSWV-specific primers	Widana Gamage et al. (2014)
Virus in pineapple	Virus indexing	Mealybug wilt-associated virus-specific primers	Punchihewa et al. (2009)

the ligation of barcode adapter, PCR amplification, and sequencing of the amplified DNA pool using Illumina sequencing technology and bioinformatics analysis. As an efficient MAS tool and an economical technique, GBS has widely been used in many counties for genome-wide association studies (GWAS), genomic diversity studies, genetic linkage analysis, molecular marker discovery, and genomic selections. Introduction of such efficient and cost-effective technologies (He et al. 2014) would greatly benefit Sri Lankan crop breeding programs.

It is clear that Sri Lanka has more to explore in MAS, both in applications and marker development research. Most of the applications so far are in rice breeding

Table 3 Marker-assisted selection

Crop	Trait	Type of marker(s)	References
Rice (*Oryza sativa*)	Fragrant rice and decreased salt tolerance	SSR	Wijerathna et al. (2011)
Rice (*Oryza sativa*)	Dehydration tolerance	SSR	Ranawaka and Nakumara (2012)
Rice (*Oryza sativa*)	Root traits	SSR and SNP	Saliha et al. (2016)
Rice (*Oryza sativa*)	Root traits	SSR	Aberathna et al. (2014)
Rice (*Oryza sativa*)	PUP 1	SSR	Wijesekara et al. (2013)
Rice (*Oryza sativa*)	Fragrant rice	Fgr gene-specific primers	Gunarathna et al. (2009)
Rice (*Oryza sativa*)	Bacterial leaf blight	pTA248 marker for Xa21	Gokarella et al. (2009)
Rice (*Oryza sativa*)	Fragrant rice	Fgr gene-specific primers	Anushka et al. (2008)
Rice (*Oryza sativa*)	Development of marker for the badh2.7 allele – fragment rice	Cleaved Amplified Polymorphic Sequences (CAPS)	Dissanayaka et al. (2014)
Rice (*Oryza sativa*)	QTL mapping of salinity tolerance	SSR	Dahanayaka et al. (2017)
Rice (*Oryza sativa*)	Salinity tolerance – hybridity assessment	SSR	Gimhani et al. (2014)
Rice (*Oryza sativa*)	Salinity tolerance Saltol QTL	SSR	Dahanayaka et al. (2015)
Rice (*Oryza sativa*)	Submergence tolerance – BC1F1	SSR	Nawarathna et al. (2014)
Rice (*Oryza sativa*)	Blast-resistant genes, Pikh, Pit (p), and Pita	SSR	Jayawardana et al. (2015)
Rice (*Oryza sativa*)	Submerge tolerance – Sub-1	SSR	Rathnayake et al. (2012)

(continued)

Table 3 (continued)

Crop	Trait	Type of marker(s)	References
Rice (*Oryza sativa*)	Phosphorus deficiency tolerance Pup1	SSR	Aluwihare et al. (2016)
Rice (*Oryza sativa*)	Validation amylase content and gel consistency	SSR	Fernando et al. (2015)
Rice (*Oryza sativa*)	Anaerobic germination-tolerant QTL AG1	SSR	Sartaj et al. (2016)
Rice (*Oryza sativa*)	Discovery of salinity-tolerant QTLs in a bi-parental population	SNP	Gimhani et al. (2016)
Rice (*Oryza sativa*)	Fragrant gene and growth performances of fragrant rice	SSR	Wijerathna et al. (2014)

efforts. Currently, Sri Lanka is marginally self-sufficient in rice production and needs to use 69% more water and 23% more nitrogen fertilizer to achieve self-sufficiency of rice production in 2050 (Davis et al. 2016). However, there are limited applications of MAS breeding in rice for water use efficiency and fertilizer use efficiency (Table 3). Therefore, the rice breeders need to consider such traits and accelerate breeding efforts to cater the demand. While other food crop breeders also need to pay attention on application of MAS and improving the precision and the efficiency of the breeding efforts, numerous challenges have been encountered in adopting technology. For example, only several plant breeding stations at the Department of Agriculture have at least basic biotechnology facilities such as a PCR machine, equipment to run agarose gels, and a gel documentation system. No lab in the country has robotic systems and high-throughput genotyping platforms. Manual DNA extraction procedures (cetyltrimethylammonium bromide (CTAB) or sodium dodecyl sulfate (SDS) based) with a large number of individuals, PCR with several SSR markers, and loading and running polyacrylamide gels with those samples are less attractive to the plant breeders. Therefore, such infrastructure development would be the first and foremost effort toward application of MAS. Further, most of the plant breeders in the country have not gained enough exposure to new technologies.

3 Genetic Engineering

The advances in plant biotechnology have made it possible to identify and modify genes responsible for specific characteristics. Crop genetic engineering, also called GM, is a direct manipulation of crop genome using biotechnological tools (reviewed by Schaart et al. 2016; Prado et al. 2014). The first GM crops were developed via transgenesis – the process by which plants are engineered by inserting foreign DNA sequences from different plant or other species that are sexually incompatible. In this process, a gene or genes are inserted into a selected plant genome using precise technologies. Transgenic crop plants are such genetically engineered plants with one or more genes from another species (Francis et al. 2017). As this technology allows the movement of genes from any source into any target plant, tremendous opportunities exist to expand the genetic base used for improvement of quantity and quality of yield. While crop genetic engineering started in 1983 by introducing a bacterial gene to tobacco, the first transgenic food crop (i.e., Flavr Savr tomato) was commercialized in 1994. Over the past 25 years, many transgenic crops and traits have been commercialized. Most of the commercially successful GMs are transgenic field crops, which account for greater than 90% of the soybean, 85% of the corn, and 90% of cotton, grown in the USA (Fernandez-Cornejo et al. 2014).

However, some critics and groups of general public were concerned about integration of genetic materials into crops that cannot happen in natural ways. To meet such concerns, two alternative genetic engineering methods, cisgenesis and intragenesis, were developed (Schouten and Jacobsen 2008). In both methods, plants are engineered only with genetic material derived from the same species or from sexually compatible close relatives. However, unlike in cisgenesis, intragenesis applications use hybrid genes, with genetic elements from different genes and loci combined in vitro. Crops derived from both methods should not consist of foreign sequences such as selectable makers and vector backbone. Currently, intragenic and cisgenic crops are regulated as transgenic crops worldwide (European Food Safety Authority (Committee 2012)). However, consumer surveys done with several countries show higher public acceptance of intragenic and cisgenic crops compared to transgenic crops (Delwaide et al. 2015).

RNA interference (RNAi) is a technique used to shut off or silence the transcription of any native gene or genes. While the gene to be downregulated always belongs to the same crop, the regulatory sequences may come from the same species, any cross-breedable species, or any other living organism. Therefore, depending on the sources of the regulatory sequences, the RNAi-based crops could be transgenic, cisgenic, or intragenic. Non-browning apple and non-browning potato are two gene downregulated crops recently entered to the US market (reviewed by Schaart et al. 2016; Parmar et al. 2017).

While our neighboring country, India, has been trying to resale Bt (*Bacillus thuringiensis*) brinjal since 2009, Bangladesh deregulated Bt brinjal in 2014 (Choudhary et al. 2014), whereas Bt cotton has been released for commercial cultivation in India. Sri Lanka is still developing the Biosafety Act and the necessary legal framework for safe use of living modified organisms (LMO). Nevertheless, few laboratory studies are being done in Sri Lanka on optimization of methodologies and proof of concepts (Ratnayake and Hettiarachchi 2010; Ratnayaka et al. 2016; Peiris et al. 2008). Genetic engineering research and development (R&D) is costly and time-consuming. Intellectual property-related issues are also involved with already patented technologies. Since the country has not materialized the legal framework for R&D and future deregulation procedures, currently, the industry and government entities are less likely to invest on GM crop development in Sri Lanka. Further, Sri

Lankan consumers perceive GM food is having negative effects on human health and prefer labeling (Senarath and Karunagoda 2012; Wijeratna and Bandaranayake 2014).

4 Gene Editing

The most recent tool added to the toolbox, gene editing technologies, allows modification of a plant genome in a very precise manner. Often a single or few bases in a genome is replaced or deleted and it is not different any means from the frequently happening natural random mutations casing changes in the genome. Similar to natural random mutations, consequences of genome editing are indistinguishable in the genome (Woo et al. 2015; Arora and Narula 2017). Genome editing also allows for both intragenesis and cisgenesis, allowing scientists to replace genes of one crop with another variety or close relatives, an essential goal in conventional crop breeding as well. Therefore, genome editing allows to achieve the same alterations as conventional breeding but much faster (Arora and Narula 2017; Hussain et al. 2018; Schaart et al. 2016).

In the USA and many other countries, GM crops must be approved by regulatory agencies before reaching the market. Since there are no physical, chemical, or biological differences of random genetic mutations and gene-edited crops using techniques such as CRISPR-Cas, scientists are of the view that there is no reason for considering genome-edited crops as GMs (Huang et al. 2016). However, the plants resulting from genome editing are analyzed carefully to document the precise changes occurred and to confirm that no leftovers of foreign DNA remain, if such an approach was used for genome editing process. Therefore, under the current US regulation, when few nucleotides were edited in a crop with CRISPR-Cas or a similar technique, to change one or more endogenous genes or pathways, the resulting plant is not considered as a genetically modified organism (GMO). However, if the same technique was used to introduce an entire exogenous gene sequence into a plant, the resulting plant would be considered a GMO. One such regulatory body, viz., US Department of Agriculture (USDA), does not make the regulations mandatory if crops do not contain "foreign" DNA. One such crop, a CRISPR/Cas-edited white button mushroom developed by the Pennsylvania State University, USA, became the headline of several leading publications (Waltz 2016a, b). White button mushroom, which usually turns brown quickly, was engineered to resist browning by making a small deletion in the specific polyphenol oxidase gene. Many other crops are in the pipeline. Nevertheless, no published literature is found in Sri Lanka, except a few at laboratory level.

Further, over time, introduction of genetically uniform crop varieties and selection for better traits would lead to deterioration of genetic diversity of crop plants. With the available limited gene pool, conventional breeding alone would not satisfy the growing demand for quality food and to face with the global challenges such as climate change, as well as biotic and abiotic stresses. All crop plants have been domesticated from wild relatives, and those close relatives of domesticated species, the crop wild relatives (CWRs), are a rich source of readily accessible genetic material for crop improvement (reviewed by Brozynska et al. 2016; Kilian et al. 2011; Warschefsky et al. 2014). Sri Lanka being a country with rich biological diversity including endemic wild relatives of crops such as rice has unique opportunity to explore and use primary source of diversity for plant improvement. Recent advances in NGS technologies, with a three million-fold increase in sequencing throughput since 1975 and a million-fold decrease in cost over a decade (Bevan et al. 2017), provide us a venture for such initiatives.

Further, if the CRISPR-engineered crops are classified, as non-GM, the public perception and distrust around GMOs would change. However, education of the lay populace and communication with the public in an open and engaging way would be the trail for success of integrating advance technologies for crop improvements.

5 Way Forward

To the best of my knowledge, only few research publications listed and discussed above are gone beyond the research labs so far. For example, some pathogen detection and identification systems developed and published by the Plant Virus Indexing Centre are used for their regular testing purposes (Punchihewa et al. 2009). While a considerable number of markers are validated for several traits in rice (Table 3), only submerge tolerance (Nawarathna et al. 2014; Rathnayake et al. 2012) breeding program currently uses MAS. Further, no GM crop has gone beyond proof of concepts. It is clear that Sri Lanka has barely exploited the potential of biotechnology for crop improvement while the world is moving faster with new technologies. There may be multiple reasons and limitations for less adaptability of biotechnology in crop improvement efforts in Sri Lanka.

One of the major limitations is lack of trained personnel in the field. Even the trained scientists with PhD or similar qualifications do not get further exposure to technologies through continuous training to keep up with the fast growth in the field. Therefore, the Sri Lankan government and the relevant authorities, such as the University Grants Commission, the Department of Agriculture, the Department of Export Agriculture, and other relevant line ministries should take this issue seriously and invest more on human capacity building in agricultural biotechnology and related areas. At the institutional level, the relevant leaders can implement activities such as training of trainers, journal clubs, lab meetings, seminar presentations, and online discussion groups to have continuous dialog with trained personnel and to provide opportunities for others.

Physical capacity building is another important step toward application of new technologies in crop breeding. For example, some regional-level research institutions do not even have basic molecular biology lab facilities to implement MAS. General molecular biology equipment such as thermal cyclers, gel documentation systems, qPCR machines, freezers, etc. are needed for every molecular biology lab. However, some high-end equipment such as high-throughput genotyping systems, next-generation sequencing systems, metabolomics systems, etc. may be established in identified central facilities. It will also allow sustainable use and good maintenance of high-end equipment.

Bureaucratic procurement system may also hinder the application of biotechnology in crop improvement. In general, molecular biological chemicals and consumables are expensive compared to regular analytical grade chemicals. Currently, all the biotechnological chemicals and consumables are imported to Sri Lanka. There are several local businesses involved with importation. The government research institutions and the universities are supposed to follow the established procurement procedures. The local companies will only import after placing a purchase order, which is the last step of the said procurement procedure. This process can take several weeks to months. Options for addressing the above issue would be to cut down lengthy procedures and have local storage at least for the essential and most commonly used items.

Another noticeable factor is less interest and low investment of the private sector in application of biotechnological tools for their breeding programs. The main reason could be not having a proper legal framework for application and usage of the technology. Establishing the Biosafety Act for safe use of recombinant DNA technology would encourage private sector involvement. Further, the National Science Foundation, the Coordination Secretariat for Science and Technology, and the Ministry of Science and Technology have come up with the idea of establishing a "biotechnology park," with all the relevant industries and a government lab to facilitate private-public partnerships. One more step forward, molecular biology and recombinant DNA technology has recently been added to the high school curriculum, and the applications such as crop improvement are highlighted (NIE 2017).

Taken together, biotechnology sector in Sri Lanka in general and application of the technology in crop improvement specifically are growing slower than expected. Much efforts and investments are needed to cope up with the demand for crop improvement as well as regional and global technological advancements.

References

Aberathna M, Dahanayaka B, Kottearachchi NS (2014) Genetic analysis of root traits using recombinant inbred line populations of rice. Proceedings of 13th agricultural research symposium, pp 165–169

Abeysinghe S, Kumari W, Arachchi I, Dickinson M (2014) First report of the molecular identification of a phytoplasma associated with dieback disease of papaya in Sri Lanka. New Dis Rep 29:13–13

Abeysinghe S, Abeysinghe PD, Kanatiwela-De Silva C, Udagama P, Warawichanee K, Aljafar N, Kawicha P, Dickinson M (2016) Refinement of the taxonomic structure of 16srxi and 16srxiv phytoplasmas of gramineous plants using multilocus sequence typing. Plant Dis 100:2001–2010

Agarwal M, Shrivastava N, Padh H (2008) Advances in molecular marker techniques and their applications in plant sciences. Plant Cell Rep 27:617–631

Aluwihare Y, Ishan M, Chamikara M, Weebadde C, Sirisena D, Samarasinghe W, Sooriyapathirana S (2016) Characterization and selection of phosphorus deficiency tolerant rice genotypes in Sri Lanka. Rice Sci 23:184–195

Amalka B, Wickramaarachchi W, Kottearachchi NS (2015) Molecular detection of papaya ring spot virus and evaluation of potential resistant papaya germplasm. Proceedings of 14th agricultural research symposium, pp 145–149

Anushka K, Kottearachchi NS, Attanayaka D (2008) Identification of fragrance gene (Fgr) in Sri Lankan rice varieties using polymerase chain reaction based molecular markers, pp 182–186

Arora L, Narula A (2017) Gene editing and crop improvement using CRISPR-Cas9 system. Front Plant Sci 8:1932

Attanayake SRMR, Kumari SASM, Weerakkody WAP, Ranil RHG, Damania AB, Bandaranayake PCG (2017) Molecular diversity and genetic relationships among Sri Lankan pomegranate landraces assessed with inter simple sequence repeat (ISSR) regions. Nord J Bot 35(4):385–394

Bandara Y, Weerasooriya W, Wickramasinghe H (2009) DNA fingerprinting and phylogenetic study of some indica rice varieties using SSR markers. Int J Biol Chem Sci 3:1–6

Bandaranayake W, Wickramarachchi W, Wickramasinghe H, Rajapakshe R, Dissanayake D (2014) Molecular detection and characterization of begomoviruses associated with cucurbitaceae vegetables in Sri Lanka. J Natl Sci Found 42:265–271

Bandaranayake W, Wickramaarachchi W, Wickramasinghe H, Rajapakshe R, Dissanayake D (2016) Molecular detection and characterisation of bipartite begomoviruses associated with cucurbitaceous vegetables in Sri Lanka. J Natl Sci Found 44:379–384

Barabaschi D, Tondelli A, Desiderio F, Volante A, Vaccino P, Valè G, Cattivelli L (2016) Next generation breeding. Plant Sci 242:3–13

Basnayake B, Dassanayake E, Jayaweera S (2005) Identification of different genotypes of Banana Bunchy Top Virus (BBTV) in Sri Lanka. Ann Sri Lanka Dep Agric 7:57–66

Bevan MW, Uauy C, Wulff BB, Zhou J, Krasileva K, Clark MD (2017) Genomic innovation for crop improvement. Nature 543:346

Bornet B, Branchard M (2001) Nonanchored Inter Simple Sequence Repeat (ISSR) markers: reproducible and specific tools for genome fingerprinting. Plant Mol Biol Report 19:209–215

Brozynska M, Furtado A, Henry RJ (2016) Genomics of crop wild relatives: expanding the gene pool for crop improvement. Plant Biotechnol J 14:1070–1085

Chamikara M, Ishan M, Karunadasa S, Perera M, Rajapaksha P, Lelwala R, Kasthuriarachchi V, Jeyakumar D, Weebadde C, Sooriyapathirana S (2014) Morphological and DNA marker analysis of fruit size and shape in selected accessions and commercial cultivars of capsicum in Sri Lanka. Veg Sci 41:101–115

Chamikara M, Ishan M, Karunadasa S, Perera M, Rajapaksha P, Lelwala R, Kasthuriarachchi V, Jeyakumar D, Weebadde C, Sooriyapathirana S (2015) Morphological and microsatellite marker analysis of fruit size and shape in selected accessions and commercial cultivars of capsicum species in Sri Lanka. Int J Multidiscip Stud 2(1):27

Choudhary B, Nasiruddin KM, Gaur K and Mymensingh B (2014) The status of commercialized Bt brinjal in Bangladesh. ISAAA Brief 47, Ithaca

Collard BC, Mackill DJ (2008) Marker-assisted selection: an approach for precision plant breeding in the twenty-first century. Philos Trans R Soc Lond B Biol Sci 363(1491):557–572

Committee ES (2012) Guidance on selected default values to be used by the EFSA scientific committee, scientific panels and units in the absence of actual measured data. EFSA J 10:2579

Crossa J, Pérez-Rodríguez P, Cuevas J, Montesinos-López O, Jarquín D, De Los CG, Burgueño J, Camacho-González JM, Pérez-Elizalde S, Beyene Y (2017) Genomic selection in plant breeding: methods, models, and perspectives. Trends Plant Sci 22(11):961–975

Dahanayaka B, Gimhani D, Kottearachchi N, Samarasighe W (2015) Assessment of salinity tolerance and analysis of SSR markers linked with Saltol QTL in Sri Lankan rice (Oryza sativa) genotypes. Am J Exp Agric 9(5):1–10

Dahanayaka B, Gimhani D, Kottearchchi N, Samarasjighe W (2017) QTL mapping for salinity tolerance using an elite rice (Oryza sativa) breeding population. SABRAO J Breed & Genet 49 (2):123–134

Davis KF, Gephart JA, Gunda T (2016) Sustaining food self-sufficiency of a nation: the case of Sri Lankan rice production and related water and fertilizer demands. Ambio 45(3):302–312

DCS (2018) Paddy statistics. Agriculture and environment statistics division. Retrieved 11 Nov 2018, from http://www.statistics.gov.lk/agriculture/Paddy%20Statistics/PaddyStats.htm

DCS (Department of Census and Statistics) (2012) Census of population and housing—2012. Retrieved November 11, 2018, from www.statistics.gov.lk

Delwaide AC, Nalley LL, Dixon BL, Danforth DM, Nayga JRRM, Van Loo EJ, Verbeke W (2015) Revisiting GMOs: are there differences in European consumers' acceptance and valuation for cisgenically vs transgenically bred rice? PLoS One 10(5):e0126060

Dissanayaka D, Kottearachchi NS, Weerasinghe O, Peiris W (2013) Detection of novel allele and protein structure of mutated fragrant gene in Sri Lankan aromatic rice. Proceedings of 12th agricultural research symposium, pp 11–15

Dissanayaka S, Kottearachchi NS, Weerasena J, Peiris M (2014) Development of a CAPS marker for the Badh2. 7 allele in Sri Lankan fragrant rice (Oryza sativa). Plant Breed 133:560–565

FAO (2011) The state of the world's land and water resources for food and agriculture. Food and Agriculture Organization of the United Nations, Rome

FAO (2018a) FAOSTAT database. Retrieved November 11, 2018, from http://faostat.fao.org/

FAO (2018b) AQUASTAT database. Retrieved November 11, 2018, from http://www.fao.org/nr/water/aquastat/main/index.stm

Fernandez-Cornejo J, Wechsler S, Livingston M, Mitchell L (2014) Genetically engineered crops in the United States. United States Department of Agriculture, Economic Research Service, Washington, DC

Fernando H, Kajenthini T, Rebeira S, Bamunuarachchige T, Wickramasinghe H (2015) Validation of molecular markers for the analysis of genetic diversity of amylase content and gel consistency among representative rice varieties in Sri Lanka. Trop Agric Res 26(2):317

Fischer RA, Edmeades GO (2010) Breeding and cereal yield progress. Crop Sei 50:S85–S98. https://doi.org/10.2135/cropsci2009.10.0564

Fischer RA, Byerlee D, Edmeades G (2014) Crop yields and global food security. ACIAR, Canberra, pp 8–11

Francis D, Finer JJ, Grotewold E (2017) Challenges and opportunities for improving food quality and nutrition through plant biotechnology. Curr Opin Biotechnol 44:124–129

Garnett T, Appleby MC, Balmford A, Bateman IJ, Benton TG, Bloomer P, Burlingame B, Dawkins M, Dolan L, Fraser D (2013) Sustainable intensification in agriculture: premises and policies. Science 341:33–34

Gimhani D, Basnayake B, Dassanayake E (2005) Identification of different strains of banana streak virus in Sri Lanka, pp 73–77

Gimhani D, Kottearachchi N, Samarasinghe W (2014) Microsatellite marker based hybridity assessment for salinity tolerance in rice. J Agric Sci 9(2):96

Gimhani D, Gregorio GB, Kottearachchi N, Samarasinghe W (2016) SNP-based discovery of salinity-tolerant QTLs in a bi-parental population of rice (Oryza sativa). Mol Genet Genomics 291:2081–2099

Gokarella R, Samarasighe W, Kottearachchi NS, Abayawickrama A, Weerasinghe W (2009) Screening of rice breeding lines for bacterial leaf blight by Pta248 marker for Xa21 gene and field evaluation. Proceedings of 9th agricultural research symposium, pp 244–248

Gunarathna S, Abayawickrama A, Kottearachchi NS, Samarasighe W (2009) Screening of a rice breeding population and diverse germplasms using Fgr specific markers and KOH sensory test. Proceedings of 09th agricultural research symposium, pp 285–298

Hall AJ, Richards RA (2013) Prognosis for genetic improvement of yield potential and water-limited yield of major grain crops. Field Crop Res 143:18–33

He J, Zhao X, Laroche A, Lu ZX, Liu H, Li Z (2014) Genotyping-by-sequencing (GBS), an ultimate marker-assisted selection (MAS) tool to accelerate plant breeding. Front Plant Sci 5:484

Heffner EL, Lorenz AJ, Jannink JL, Sorrells ME (2010) Plant breeding with genomic selection: gain per unit time and cost. Crop Sci 50(5):1681–1690
Huang S, Weigel D, Beachy RN, Li J (2016) A proposed regulatory framework for genome-edited crops. Nat Genet 48:109
Hussain B, Lucas SJ, Budak H (2018) Crispr/Cas9 in plants: at play in the genome and at work for crop improvement. Brief Funct Genomics 17(5):319–328
Jayamaha D, Samarajeewa P, Attanayaka D (2006) Identification of different yam (Dioscorea) species by random amplified polymorphic DNA 141–145
Jayathilaka K, Kottearachchi NS, Kekulandara D, Mandanayaka M (2014) Marker assisted selection for the traits of aroma and brown plant hopper resistance in rice derived from cross between Bg380x At306. Proceedings of 13th agricultural research symposium, pp 190–194
Jayawardana W, Jayasekera G, Wijesundera R, Dissanayake DM, Sooriyapathirana S, Weebadde C, Perera K, Gunapala K, Hettige P (2015) Evaluation of DNA markers linked to blast resistant genes, Pikh, Pit (P), and Pita, for parental selection in Sri Lankan rice breeding. Trop Agric Res 26(1):82
Kaur S, Panesar PS, Bera MB, Kaur V (2015) Simple sequence repeat markers in genetic divergence and marker-assisted selection of rice cultivars: a review. Crit Rev Food Sci Nutr 55(1):41–49
Kilian B, Mammen K, Millet E, Sharma R, Graner A, Salamini F, Hammer K, Özkan H (2011) Wild crop relatives: genomic and breeding resources. Springer, Berlin
Kondasinghe K, Fahim M, Attanayaka D, Samarajeewa D (2010) Confirmation of genetic purity of four pineapple (Ananas comosus L.) lines by morphological and molecular trait analysis. Proceedings of 10th agricultural research symposium, pp 15–19
Kularathne R, Dassanayake E, Attanayaka D (2005) Investigation of host range of pineapple mealy bug wilt associated Pineapple Bacilliform Virus (PBV) by Polymerase Chan Reaction (PCR) within the Bromeliaceae family, pp 22–26
Kumar J, Choudhary AK, Solanki RK, Pratap A (2011) Towards marker-assisted selection in pulses: a review. Plant Breed 130(3):297–313
Lam HM, Xu X, Liu X, Chen W, Yang G, Wong FL, Li MW, He W, Qin N, Wang B (2010) Resequencing of 31 wild and cultivated soybean genomes identifies patterns of genetic diversity and selection. Nat Genet 42:1053
Li H, Rasheed A, Hickey LT, He Z (2018) Fast-forwarding genetic gain. Trends Plant Sci 23:184–186
Litt M, Luty JA (1989) A hypervariable microsatellite revealed by in vitro amplification of a dinucleotide repeat within the cardiac muscle actin gene. Am J Hum Genet 44:397
Moose SP, Mumm RH (2008) Molecular plant breeding as the foundation for 21st century crop improvement. Plant Physiol 147:969–977
Nadeeshani V, Dassanayake E, Basnayake B, Vivehandanthan K (2011) Identification of disease affecting Carica papaya L. cultivar Ms100 and varietal reaction. Proceedings of 11th agricultural research symposium, pp 46–50
Nawarathna R, Perera A, Samarasinghe W (2014) Screening of Bc1f1 population (Bg 379-2/Ir 07f102//Bg 379-2) of rice (Oryza sativa L.) for submergence tolerance using molecular markers. J Agric Sci 9(3):147
Nelson GC, Valin H, Sands RD, Havlík P, Ahammad H, Deryng D, Elliott J, Fujimori S, Hasegawa T, Heyhoe E (2014) Climate change effects on agriculture: economic responses to biophysical shocks. Proc Natl Acad Sci 111:3274–3279
NIE (Department of Science, National Institute of Education Sri Lanka), General Certificate of Education (Advanced Level) Grades 12–13 BIOLOGY SYLLABUS (Implemented from 2017), Retrieved November 11, 2018, from www.nie.lk
Parmar N, Singh KH, Sharma D, Singh L, Kumar P, Nanjundan J, Khan YJ, Chauhan DK, Thaku AK (2017) Genetic engineering strategies for biotic and abiotic stress tolerance and quality enhancement in horticultural crops: a comprehensive review. 3 Biotech 7:239

Peiris R, Wickramasinghe T, Indrasena SM (2008) 127-a promising tomato variety developed through induced mutation technique. Induced plant mutations in the genomics era. Proceedings of an international joint FAO/IAEA symposium, pp 379–80

Perera K, De Silva D, Rathnayake N, De Silva S, Senaratne S (2013) Optimizing a screening protocol for Trehalose gene (OsTPS1) in different traditional and improved rice varieties (Oryza sativa L. SSP Indica) in Sri Lanka. Proceedings of 12th agricultural research symposium, pp 35–38

Prado JR, Segers G, Voelker T, Carson D, Dobert R, Phillips J, Cook K, Cornejo C, Monken J, Grapes L (2014) Genetically engineered crops: from idea to product. Annu Rev Plant Biol 65:769–789

Punchihewa D, Dassanayake E, Peiris H, Ubesekara N, Smanmalee L, Attanayaka D (2009) Virus indexing and evaluation of pineapple cultivar Marian gold 3 for its agronomic suitability and in-vitro multiplication. Proceedings of 9th agricultural research symposium, pp 254–259

Rafalski A (2002) Applications of single nucleotide polymorphisms in crop genetics. Curr Opin Plant Biol 5:94–100

Rajapakse B, Wickramaarachchi W, Kottearachchi NS, Dissanayaka D (2016) Detection of papaya ring spot virus infection by Rt_PCR together with an amplification of reference gene. Proceedings of 15th agricultural research symposium, pp 130–134

Rajapakshe BN, Kottearachchi NS, Wickramarachchi RT, Amalka T (2017) Molecular evaluation of potential resistance to papaya ring spot virus in mountain papaya (Vasconcellea cundinamarcensis). Pak J Phytopathol 29:89–95

Ranathunge R, Attanayaka D, Samarajeewa D, Jayasekera S (2010) Genetic diversity of cashew (Anacardium occidentale L.) germplasm collected from five districts of Sri Lanka as revealed by Random Amplified Polymorphic DNA (RAPD). Proceedings of 10th agricultural research symposium, pp 35–39

Ranawaka A, Nakumara C (2012) QTL analysis of dehydration tolerance at seedling stage in rice (Oryza sativa). Trop Agric Res Ext 15:113–117

Rathnayake N, Bentota A, Dissanayake D, Perera K, Sooriyapathirana S, Jayasekera G (2012) DNA markers RM 464A and RM 219 haplotypes are effective in selecting Sub-1 locus for the introgression of submergence tolerance into new rice varieties. Ceylon J Sci (Bio Sci) 41:125–136

Ratnayaka RMLK, Jayasekera GAU, Hettiarachchi GHCM (2016) Overexpression of arabidopsis BBX 21 gene in Bg 250 rice enhances its architecture and productivity. Int J Appl Sci & Biotechnol 4(2):240–246

Ratnayake R, Hettiarachchi G (2010) Development of an efficient agrobacterium mediated transformation protocol for Sri Lankan rice variety-Bg 250. Trop Agric Res 22:45–53

Ray DK, Ramankutty N, Mueller ND, West PC, Foley JA (2012) Recent patterns of crop yield growth and stagnation. Nat Commun 3:1293

Reddy MP, Sarla N, Siddiq E (2002) Inter Simple Sequence Repeat (ISSR) polymorphism and its application in plant breeding. Euphytica 128:9–17

Safeena M, Sumanasinghe V, Bandara D (2003) Identification of RAPD markers for salt tolerance in rice. Trop Agric Res 15:39–50

Saliha M, Gimhani D, Kottearachchi NS (2016) Progress towards QTL mapping for root traits in chromosome 2 and 10 of rice (Oryza sativa). Proceedings of 15th agricultural research symposium, pp 174–178

Sarathchndra K, Dassanayake E, Ubesekara N, Attanayaka D (2009) Detection of phytoplasma in different crops by polymerase chain reaction using specific primers. Proceedings of 9th agricultural research symposium, pp 260–264

Sartaj A, Udawela K, Herath H (2016) Marker assisted backcross breeding of Bg 358 (Oryza sativa L.) for the anaerobic germination tolerant QTL1 AG1. Trop Agric Res 27:287–294

Schaart JG, Van De Wiel CC, Lotz LA, Smulders MJ (2016) Opportunities for products of new plant breeding techniques. Trends Plant Sci 21:438–449

Schouten HJ, Jacobsen E (2008) Cisgenesis and intragenesis, sisters in innovative plant breeding. Trends Plant Sci 13(6):260–261

Semagn K, Bjørnstad Å, Ndjiondjop M (2006) An overview of molecular marker methods for plants. Afr J Biotechnol 5:2540–2568

Senarath SN, Karunagoda RP (2012) Consumer attitude towards labeling of genetically modified foods in Sri Lanka. Trop Agric Res 23(3):283–288

Senavirathne W, Jayatilake D, Herath V, Wickramasinghe H (2017) Evaluation of genetic diversity of cis-acting elements of Abscisic acid responsive element binding protein (ABRE-BP) in selected Sri Lankan rice varieties. Trop Agric Res 28:120

Singh H, Deshmukh RK, Singh A, Singh AK, Gaikwad K, Sharma TR, Mohapatra T, Singh NK (2010) Highly variable SSR markers suitable for rice genotyping using agarose gels. Mol Breed 25:359–364

Sonah H, Deshmukh RK, Sharma A, Singh VP, Gupta DK, Gacche RN, Rana JC, Singh NK, Sharma TR (2011) Genome-wide distribution and organization of microsatellites in plants: an insight into marker development in Brachypodium. PLoS One 6:E21298

Sumanasinghe J, Samarasighe W, Wanigadeva S, Gunasekara I, Ranjanee S, Sujeewa Kumari L (2005) Characterization of onions (Allium cepa L.) by morphological traits, isozymes and randomly amplified polymorphic DNA markers. Ann Sri Lanka Depart Agric 7:253–270

Takeda S, Matsuoka M (2008) Genetic approaches to crop improvement: responding to environmental and population changes. Nat Rev Genet 9:444

Tester M, Langridge P (2010) Breeding technologies to increase crop production in a changing world. Science 327:818–822

Thilakarathna D, Kottearachchi N, Kumararathna M, Gimhani D (2016) Genetic diversity analysis in relation to flowering behavior in mungbean (Vigna radiata L.). Proceedings of 15th agricultural research symposium, pp 184–188

Tilman D, Balzer C, Hill J, Befort BL (2011) Global food demand and the sustainable intensification of agriculture. Proc Natl Acad Sci 108(50):20260–20264

UN (2012). World population prospects: the 2012 revision. Population division of the department of economic and social affairs of the United Nations Secretariat. Retrieved November 11, 2018, from http://esa.un.org/unpd/wpp/index.htm

Varshney RK, Hoisington DA, Tyagi AK (2006) Advances in cereal genomics and applications in crop breeding. Trends Biotechnol 24:490–499

Vithyashini L, Wickramasinghe H (2015) Genetic diversity of seed storage proteins of rice (Oryza sativa L.) varieties in Sri Lanka. Trop Agric Res 27:49–58

Wakista P, Dasanayaka P, Illeperuma R, Perera S (2015) SSR marker based molecular characterisation of finger millet accessions of India and Anuradhapura district of Sri Lanka. University of Sri Jayewardenepura, Nugegoda

Waltz E (2016a) CRISPR-edited crops free to enter market, skip regulation. Nat Biotechnol 34 (6):582–582

Waltz E (2016b) Gene-edited CRISPR mushroom escapes US regulation. Nat New 532(7599):293

Warnakula W, Kottearachchi NS, Yakandawala K (2014) Characterisation of mountain papaya in comparison with Carica papaya cultivars, using morphological, Inter Simple Sequence Repeat (ISSR) and Simple Sequence Repeat (SSR) markers. Proceedings of 13th agricultural research symposium, pp 225–229

Warnakula W, Kottearachchi N, Yakandawala K (2017) Morphological, SSR and ISSR marker based genetic diversity assessment of mountain papaya germplasm in comparison with Carica papaya. J Natl Sci Found 45:255–264

Warschefsky E, Penmetsa RV, Cook DR, Wettberg EJ (2014) Back to the wilds: tapping evolutionary adaptations for resilient crops through systematic hybridization with crop wild relatives. Am J Bot 101:1791–1800

Wettewa W, Kottearachchi NS, Gimhani D (2011) Allelic variation in fragrant gene in Sri Lankan rice varieties. Proceedings of 11th agricultural research symposium, pp 6–10

Widana Gamage S, Hassani-Mehraban A, Peters D (2014) Tomato spotted wilt virus in Sri Lanka: Emerging problems of tospoviruses. Trop Agric Res Ext 16:66–72

Wijayalath W, Weerasena O, Attanayaka D (2005) Identification and characterization of potential disease resistant genes from Sri Lankan traditional, wild and weedy rice varieties (Oryza sativa L.), pp 1–5

Wijeratna RMS, Bandaranayake PCG (2014) GM foods: knowledge, awareness and acceptance among Sri Lankans, Proceedings of the One Health International Conference, 5–6 September, University of Peradenoya, Sri Lanka (57)

Wijerathna Y, Kottearachchi NS, Gimhani D, Sirisena D (2011) Sri Lankan fragrant rice (Oryza sativa L.) varieties are associated with decreased salt tolerance. Proceedings of 11th agricultural research symposium, pp 51–55

Wijerathna Y, Kottearachchi N, Gimhani D, Sirisena D (2014) Exploration of relationship between fragrant gene and growth performances of fragrant rice (Oryza sativa L.) seedlings under salinity stress. J Exp Biol Agric Sci 2:7–12

Wijesekara U, Kottearachchi NS, Dahanayaka B (2013) Germplasm survey of Sri Lankan rice for Pup1 gene based markers: an approach towards development of rice varieties for phosphorous deficient soil. Proceedings of 12th agricultural research symposium, pp 1–5

Williams JG, Kubelik AR, Livak KJ, Rafalski JA, Tingey SV (1990) DNA polymorphisms amplified by arbitrary primers are useful as genetic markers. Nucleic Acids Res 18:6531–6535

Wimalasena G, Wijesundara W, Vivehandanthan K (2009) Optimization of repetitive element based PCR to identify polymorphism in Magnaporthe grisea. Proceedings of 09th agricultural research symposium, pp 280–284

Woo JW, Kim J, Kwon SI, Corvalán C, Cho SW, Kim H, Kim S-G, Kim S-T, Choe S, Kim J-S (2015) DNA-free genome editing in plants with preassembled CRISPR-Cas9 ribonucleoproteins. Nat Biotechnol 33:1162

Zhu Y, Song Q, Hyten D, Van Tassell C, Matukumalli L, Grimm D, Hyatt S, Fickus E, Young N, Cregan P (2003) Single-nucleotide polymorphisms in soybean. Genetics 163:1123–1134

Increasing Climate Resilience of Cropping Systems in Sri Lanka

W. A. J. M. De Costa

Abbreviations

ABA	Abscisic acid
AMSL	Above mean sea level
ARC	Aerobic rice culture
AWD	Alternative wetting and drying
CFLR	Continuously flooded lowland rice
DI	Deficit irrigation
DoA	Department of Agriculture, Sri Lanka
ETT	Extension and technology transfer
FIM	First Inter-monsoon
GHGs	Greenhouse gases
GWP	Global warming potential
HSPs	Heat shock proteins
NAP	National Adaptation Plan for Climate Change Impacts in Sri Lanka
NCCAS	National Climate Change Adaptation Strategy for Sri Lanka 2011–2016
NCCP	National Climate Change Policy of Sri Lanka
NEM	North-East monsoon
OTCs	Open-top chambers
PRD	Partial root-zone drying
PDSI	Palmer Drought Severity Index
QTL	Quantitative trait locus
RDI	Regulated deficit irrigation
RH	Relative humidity
RWH	Rainwater harvesting
SDI	Sustained deficit irrigation

W. A. J. M. De Costa (✉)
Department of Crop Science, Faculty of Agriculture, University of Peradeniya, Peradeniya, Sri Lanka

R. P. De Silva et al. (eds.), *Agricultural Research for Sustainable Food Systems in Sri Lanka*, https://doi.org/10.1007/978-981-15-3673-1_6

SEM	Second Inter-monsoon
SOM	Soil organic matter
SSC	Saturated soil culture
SWM	South-West monsoon
T_b	Base temperature
T_{ce}	Ceiling temperature
TGCs	Thermal gradient chambers
Tmax	Daily maximum temperature
T_{opt}	Optimum temperature
WP	Water productivity

1 Introduction

Climate change is, generally, defined as a statistically significant change in the overall state of the climate or a given climatic variable that has occurred over a longer (usually >30 years) time scale (Stocker et al. 2013b). The state of the climate over a given period can be defined in terms of means of different climatic variables (e.g. temperature, precipitation, etc.) and their variability (e.g. coefficient of variation of temperature, etc.). Therefore, climate change can involve either a long-term shift in the mean or variability or both of a given climate variable. Measures initiated in response to climate change can take one of two forms, namely, mitigation and adaptation. Mitigation of climate change includes measures taken to reduce the probability of future climate change by decreasing the emission rates of greenhouse gases (GHGs) or by increasing the capacity of their storage sinks (e.g. forests). In contrast, adaptation to climate change includes measures taken to reduce its impacts (e.g. heat-tolerant crop varieties, modifications in crop management). Adaptation also involves preparing for and adjusting to climate change.

Resilience to climate change can be defined as the ability to *withstand* the impacts, particularly the adverse impacts, of climate change. In a stricter definition, resilience refers to the 'capacity of a system to absorb disturbances and still retain its structure and function, while also keeping open the options to develop and change' (Carpenter et al. 2011; Nelson 2011). With regard to cropping systems, a key requirement for withstanding the adverse impacts of climate change is to ensure long-term stability in crop yields while protecting their support systems. This includes maintaining soil fertility and ensuring an adequate supply of external inputs such as water and essential plant nutrients. As the availability of these inputs is predicted to decrease with long-term climate change (Beddington et al. 2012; Wheeler and Von Braun 2013), increasing the efficiency with which the crops use external inputs to produce their yield constitutes another important aspect of climate resilience in crops and cropping systems.

The two principal pathways of responding to climate change, adaptation and mitigation, contribute to building climate resilience. As adaptation involves minimizing the adverse impacts of climate change, effective implementation of

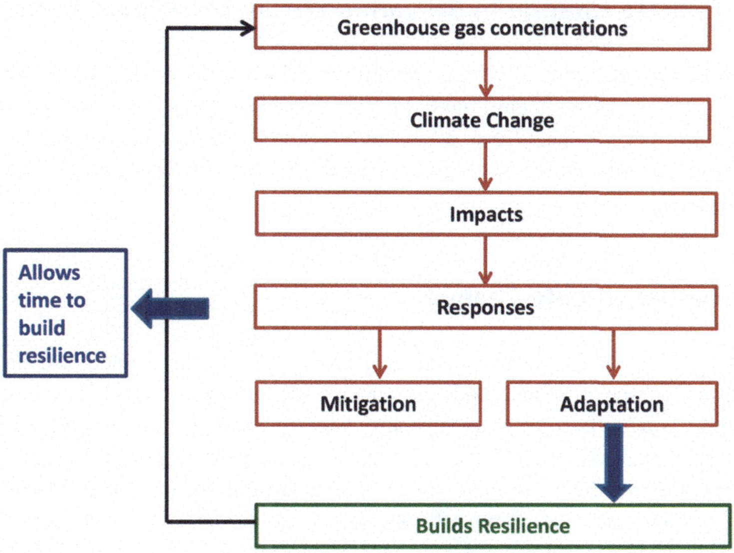

Fig. 1 Conceptual interrelationships between climate change mitigation, adaptation and increasing climate resilience. (Adapted from ACT Information Leaflet, December 2015. www.actiononclimate.today)

adaptation measures leads to increased resilience of crops and cropping systems to climate change. On the other hand, by reducing the drivers and causes of climate change, mitigation measures slow down the rate of climate change and thereby allow more time to develop effective adaptation measures and build resilience. The interrelationships between resilience, mitigation and adaptation are illustrated in Fig. 1.

2 Climate of Sri Lanka

The climate of Sri Lanka is essentially tropical, with three major climatic zones, the 'wet', 'intermediate' and 'dry' zones. In the Köppen-Geiger Climate Classification, these correspond approximately to 'equatorial rainforest, fully-humid (Af)', 'equatorial monsoonal (Am)' and 'equatorial savannah with dry summer (As)' climates (Kottek et al. 2006). Punyawardena (2009) and the website of the Department of Meteorology in Sri Lanka (Anonymous 2018) provide concise descriptions of essential features of the climate of Sri Lanka. Spatial and temporal variation of precipitation, altitude and temperature are the key determinants in classifying the agro-climate of Sri Lanka to 46 agro-ecological regions.

3 Climate Change in Sri Lanka Within the Global Context

Observed historical trends of climate change in Sri Lanka since the 1870s when the meteorological records began have largely followed the global trends for the corresponding period. Therefore, a brief description of the main features of global trends of past climate change is given to place the Sri Lankan trends within the global context.

3.1 Global Climate Change

The clearest signal of climate change has been the sustained increases in temperatures of the air, land and sea surface (Hartmann et al. 2013). Increased atmospheric concentrations of greenhouse gases caused by human activities have been the principal driver of increasing temperature in the earth's climate system (Rosenzweig et al. 2008; Shindell et al. 2009; Bindoff et al. 2013; Myhre et al. 2013). This increase has been shown not only in seasonal and daily mean temperatures but also in maximum and minimum temperatures, at different spatial scales such as global, regional and national (Karl et al. 2009; Sacks et al. 2010; Stott et al. 2011; Donat and Alexander 2012; Hansen et al. 2012; Lobell and Gourdji 2012). Moreover, there has been a shift in the frequencies of climate extremes with warm extremes becoming more frequent while cold extremes becoming less frequent (Alexander et al. 2006; Lobell et al. 2007, 2011; Meehl et al. 2009).

In comparison to the clear trends towards a warmer climate, the historical precipitation trends are less clear and show substantial spatial variation (IPCC 2013; Hegerl et al. 2015). While mid-latitudes in the Northern Hemisphere have shown increasing precipitation trends, other latitudes exhibit both increasing and decreasing trends, but with substantial inter-annual variability, thus decreasing the confidence level in the observed trends (Stocker et al. 2013a). Trends in ocean surface salinity have shown clear evidence that it has increased in areas where it is already higher and decreased in areas where it is already lower, thus pointing to a precipitation trend of wet areas getting wetter while dry areas getting drier (Trenberth 2011; Collins et al. 2013; Hegerl et al. 2015). Analyses of long-term precipitation data and simulation of coupled ocean-atmosphere climate models have confirmed this trend, both globally and regionally (Liu and Allan 2013; Chou et al. 2013). Furthermore, the range between wet and dry season precipitation has increased (Chou et al. 2013), owing to a greater increase in wet season precipitation than a decrease in dry season precipitation. Similarly, the annual range between maximum and minimum precipitation has increased in a majority of areas on a global scale during 1979–2008, and simulations using a range of coupled ocean-atmosphere models have predicted this trend to continue up to 2100 (Chou and Lan 2012).

The above asymmetry in the precipitation changes induced by global warming means that even if the annual total of precipitation does not change, the frequencies of excessively wet periods and droughts can increase. Furthermore, the frequency of excessive rainfall events has shown an increasing trend in many parts of the world, probably as a result of a warmer atmosphere (Easterling et al. 2000; Groisman et al. 2005; Alexander et al. 2006; Klein Tank et al. 2006; Allan et al. 2010). The above changes in the character of precipitation as induced by global warming have increased the frequency and severity of drought. Based on the Palmer Drought Severity Index (PDSI) (Palmer 1965), Dai et al. (2004) showed that, on a global scale, areas categorized as 'very dry (PDSI<-3)' increased from 12% to 30% during the last three decades of the twentieth century. Model simulations have shown clear indications of increased frequency and intensity of droughts with future climate change (Meehl et al. 2007; Sheffield and Wood 2008; Dai 2011).

3.2 Climate Change in Sri Lanka

There has been only limited work on the magnitudes of historical climate change in Sri Lanka and its predictions for the future. All analyses of historical temperature records, dating back to the second half of the nineteenth century, point to an increasing trend in air temperature, with the magnitudes and rates of increase varying in different climatic zones. As expected, precipitation shows much greater inter-annual variability than temperature. Analysis of long-term precipitation records shows declining trends in a majority of locations with a few showing increasing trends. However, because of the greater variability, many of these long-term trends are not statistically significant.

3.2.1 Temperature

De Costa (2008), in an analysis of the annual mean temperature (T_a) data from 1869 to 2007 from seven stations representing different climatic zones of Sri Lanka, showed that decadal mean air temperatures in six out of the seven locations had highly significant ($p < 0.001$) increasing trends for the nearly 140-year period (Table 1). Nuwara Eliya in the high-altitude (>900 m above mean sea level (AMSL)) wet zone (>1200 mm year^{-1} of rainfall) showed a continuous decadal warming at 0.106 °C per decade throughout the entire period. All locations except Nuwara Eliya showed significant warming during the last three decades of the nineteenth century, which was followed by a two- to three-decade period of cooling at the beginning of the twentieth century. Thereafter, all locations except Kandy, which is located in the mid-altitude (300–600 m AMSL) wet zone, showed a continuous decadal warming trend up to the first decade of the twenty-first century. Kandy showed a period of cooling during the two decades between 1950s and 1970s, which was followed by continuous warming, thus showing an overall cyclic variation pattern over the 140-year period. The rates of warming during the nearly 80-year period (60-year period for Kandy) of continuous decadal warming ranged from 0.065 (Ratnapura) to 0.195 °C (Anuradhapura) per decade. Notably, with the exception of Ratnapura, the above rate of warming exceeded the global average of 0.74 °C per decade for the period from 1906 to 2005 (Solomon et al. 2007). By analysis of frequency distributions, De Costa (2008) further showed that increases of

Table 1 Rates of increase of decadal mean air temperatures during the total period considered and during the period of almost continuous warming for selected locations in Sri Lanka

Location	Total period (1869–2007)		Period of almost continuous warming		
	Rate of increase of decadal mean T_a (°C decade^{-1}) ± std. error	R^2 value of linear regression	Rate of increase of decadal mean T_a (°C decade^{-1}) ± std. error	R^2 value of linear regression	Period
Anuradhapura	0.078 ± 0.019	0.58	0.195 ± 0.021	0.94	1930–2007
Kurunegala	0.053 ± 0.012	0.63	0.104 ± 0.015	0.89	1930–2007
Kandy	n.s.	–	0.096 ± 0.019	0.86	1950–2007
Ratnapura	0.049 ± 0.015	0.47	0.065 ± 0.013	0.75	1910–2007
Badulla	0.086 ± 0.018	0.66	0.191 ± 0.021	0.94	1930–2007
Nuwara Eliya	0.106 ± 0.008	0.93	0.141 ± 0.014	0.94	1930–2007
Colombo	0.040 ± 0.018	0.30	0.154 ± 0.018	0.93	1930–2007

Source: De Costa (2008)
Note: *n.s.* no significant linear relationship

T_a across its whole range rather than increases in the upper percentiles were responsible for the above rates of warming.

3.2.2 Rainfall

A parallel analysis of decadal mean annual total precipitation (P_a) showed statistically significant declining trends in six out of the seven locations (Table 2) with Nuwara Eliya showing the steepest decline at 51.6 mm decade^{-1}. Periods of continuous decadal precipitation could be identified in all locations, and the respective rates of decline ranged from 15.3 (Badulla, 1910–2007) to 120.6 (Kurunegala, 1970–2007) mm decade^{-1}. Furthermore, in all locations, mean annual total precipitation during 1990–2007 was lower than the corresponding means during 1950–1989, providing further evidence of precipitation decline. Analysis of frequency distributions of P_a revealed that precipitation reductions in the highest and lowest extremes contributed more to the overall reductions. Notably, a negative correlation was observed between the amount of precipitation and its variability (as measured by the coefficient of variation) across the whole range of locations, indicating that rainfall variability is greater in the lower rainfall environments, thus increasing the probability of drought. All seven locations showed significant negative correlations between annual mean temperature and annual total precipitation for the whole 140-year period, with Nuwara Eliya showing the strongest ($r = 0.483$, $p < 0.0001$) with a rate of precipitation decline of 371 mm per 1 °C warming.

There is evidence that the frequency and intensity of drought in Sri Lanka will increase with climate change (Eriyagama et al. 2010). De Costa (2009) showed significant declining trends in the decadal mean annual total water balance, calculated as the difference between decadal mean annual total precipitation and potential evapotranspiration, of six out of seven locations representing different climatic zones over a nearly 140-year period from 1869 to 2008 (Table 3). The rates of decline ranged from 22 (Kurunegala) to 55 mm (Nuwara Eliya) per decade. When the water balances were calculated for different rainfall seasons, it was shown that the declines in annual total water balance were primarily caused by significant declining trends in the water balance during the South-West monsoon (SWM) and particularly during the first 2 months of the SWM (i.e. May–June). Notably, water balance during the North-East monsoon (NEM) did not show a significant trend at any of the locations, while those during the First Inter-monsoon (FIM) and Second Inter-monsoon (SIM) showed significant declining trends in only two (Anuradhapura and Badulla) and one locations (Nuwara Eliya), respectively. In order to detect evidence of long-term changes in the water balance, De Costa (2009) compared the average water balance over three successive periods from 1869–1949 through 1950–1989 to 1990–2008. Highly significant reductions in the average annual water balance (Table 4) and the seasonal water balance during the SWM were shown with successive periods in all locations except Colombo (Table 5). In contrast, significant reductions in the seasonal water balance during the NEM over the three successive periods were shown in only two (Anuradhapura and Kurunegala) out of the seven locations. Notably, there was clear evidence that inter-annual variability in the annual and seasonal water balances had increased across the three successive periods.

Table 2 Rates of change of decadal mean rainfall during the total period considered and during periods of significant decline for selected locations in Sri Lanka

Location	Total period (1869–2007)		Periods of significant rainfall decline		
	Rate of change of decadal mean RF_a (mm decade^{-1}) ± std. error	R^2 value of linear regression	Rate of change of decadal mean RF_a (mm decade^{-1}) ± std. error	R^2 value of linear regression	Period
Anuradhapura	−12.92 ± 5.93	0.28	−45.52 ± 13.04	0.71	1940–2007
			−67.05 ± 10.47	0.91	1940–1990
Kurunegala	−9.56 ± 7.67	0.12	−120.57 ± 29.13	0.90	1970–2007
Kandy	−30.50 ± 6.81	0.63	−64.36 ± 11.04	0.87	1940–2007
			−85.76 ± 7.06	0.97	1940–1990
Ratnapura	−13.26 ± 8.82	0.16	−41.02 ± 18.39	0.42	1920–2007
Badulla	−19.16 ± 8.89	0.28	−15.29 ± 6.28	0.43	1910–2007
Nuwara Eliya	−51.60 ± 6.47	0.84	−67.18 ± 13.12	0.79	1920–2007
Colombo	+23.96 ± 16.99	0.14	−32.69 ± 33.18	0.14	1930–2007

Source: De Costa (2008)

Table 3 Long-term variation of decadal means of annual total water balance and the seasonal water balance in the South-West monsoon during the period from 1869 to 2008

Location	Annual total water balance (mm)		Seasonal water balance in South-West monsoon (mm)	
	Rate of change[a] ± SE (mm dec^{-1})	R^2	Rate of change ± SE (mm dec^{-1})	R^2
Anuradhapura	−34** ± 10	0.49	−15*** ± 3 (*44%*)[b]	0.64
Kurunegala	−22* ± 9	0.34	−12* ± 6 (*55%*)	0.27
Kandy	−31*** ± 6	0.67	−25*** ± 5 (*81%*)	0.67
Ratnapura	−24* ± 10	0.33	−19* ± 8 (*79%*)	0.33
Badulla	−29** ± 9	0.46	−9* ± 3 (*31%*)	0.36
Nuwara Eliya	−55*** ± 7	0.86	−41*** ± 6 (*75%*)	0.77
Colombo	+14[ns] ± 19	0.04	+13[ns] ± 10	0.13

Source: De Costa (2009)
ns, non-significant at $p = 0.05$; *significant at $p < 0.05$; **significant at $p < 0.01$; ***significant at $p < 0.001$
[a]Slope of the linear regression of water balance with time
[b]Reduction of seasonal water balance during the South-West monsoon as a percentage of the reduction of the annual total water balance

Table 4 Variation of annual total water balance of selected locations of Sri Lanka during three defined time periods

Location	Annual total water balance ± standard error (mm)			Probability for significance	CV (%)
	1869–1949 (Period 1)	1950–1989 (Period 2)	1990–2008 (Period 3)		
Anuradhapura	−448 ± 33 a	−647 ± 64 b	−890 ± 50 c (*99%*)	<0.0001	−52
Kurunegala	330 ± 53 a	305 ± 61 a	9 ± 74 b (*97%*)	0.0046	127
Kandy	780 ± 41 a	585 ± 52 b	420 ± 48 c (*46%*)	<0.0001	40
Ratnapura	2113 ± 57 a	1882 ± 76 b	1845 ± 104 b (*13%*)	0.0083	22
Badulla	701 ± 48 a	566 ± 51 ab	407 ± 70 b (*42%*)	0.0082	60
Nuwara Eliya	1593 ± 44 a	1268 ± 56 b	1119 ± 74 b (*30%*)	<0.0001	25
Colombo	378 ± 54 a	592 ± 84 a	382 ± 79 a	n.s.	115

Source: De Costa (2009)
Note: For any given location, means with the same letter are not significantly different at $p = 0.05$. Mean separation was done with Duncan's Multiple Range Test. *CV* coefficient of variation. Percentage reduction of annual total water balance from Period 1 to Period 3 is given in parenthesis

Table 5 Variation of seasonal water balance in the South-West monsoon in selected locations of Sri Lanka during three defined time periods

Location	Seasonal water balance ± standard error (mm) 1869–1949 (Period 1)	1950–1989 (Period 2)	1990–2008 (Period 3)	Probability for significance	CV (%)
Anuradhapura	−656 ± 21 a	−716 ± 22 a	−844 ± 27 b (*29%*)	0.0002	−24
Kurunegala	−85 ± 31 a	−89 ± 32 a	−254 ± 51 b (*199%*)	0.0143	−193
Kandy	287 ± 29 a	124 ± 30 b	−26 ± 24 c (*109%*)	<0.0001	87
Ratnapura	1239 ± 48 a	1097 ± 61 ab	1018 ± 91 b (*18%*)	0.0249	31
Badulla	−137 ± 17 a	−178 ± 18 ab	−238 ± 35 b (*74%*)	0.0197	−86
Nuwara Eliya	858 ± 35 a	607 ± 42 b	517 ± 48 b (*40%*)	<0.0001	39
Colombo	64 ± 41 a	178 ± 46 a	129 ± 61 a	n.s.	341

Source: De Costa (2009)

Note: For any given location, means with the same letter are not significantly different at $p = 0.05$. Mean separation was done with Duncan's Multiple Range Test. *CV* coefficient of variation. Percentage reduction of annual total water balance from Period 1 to Period 3 is given in parenthesis

4 Crops, Cropping Seasons and Cropping Systems in Sri Lanka

Organized agriculture in Sri Lanka dates back to more than 2000 years. Despite its relatively limited land area, a wide variety of crops and cropping systems can be observed in Sri Lanka. The composition and distribution of these crops and cropping systems are intimately linked to the highly heterogeneous climates that are found across Sri Lanka. Two of the principal perennial crops, tea and rubber, are grown mainly in the humid tropical climates of the wet zone. Coconut, the third major perennial crop, is distributed throughout all three climatic zones, with the highest concentration being in the intermediate zone. Because of its multiple products, coconut is widely grown in home gardens so that it is distributed inland from the coastal plains up to an altitude of ca. 300 m AMSL. Rice, the principal annual crop and the staple food of Sri Lanka, is distributed across all three of its climatic zones. Even though the well-distributed rainfall pattern of the wet zone can support rice cultivation throughout the year, a major portion of Sri Lanka's national rice production comes from the dry and intermediate zones, where an elaborate network of tanks and reservoirs makes rice cultivation possible during the drier period of the year. In addition, a wide variety of annual crops are grown in the different climatic zones of Sri Lanka. Grain legumes such as mung bean, cowpea, soybean and black gram are confined largely to the dry zone with limited extents in the low elevations of the intermediate zone. In contrast, the vegetable legumes such as common bean and yard-long bean are grown in the mid- and upper elevations of the wet zone. A wide range of vegetables belonging to families Cucurbitaceae and Solanaceae are grown in all climatic zones. Some such as tomato are grown across the whole range of climates, while others are more climate-specific. A majority of tropical vegetables are largely grown in the dry and intermediate zones, while a limited number of subtropical and temperate vegetables are grown in the mid- and upper elevations of the wet and intermediate zones. Limited extents of potato are grown in the upper elevations of the wet zone and in the extreme north during a 3-month window of the year when the day-night temperature differential is large enough for adequate filling of tubers. Sweet potato, cassava and a variety of indigenous yams are grown in lower and mid-elevations of the wet and intermediate zones. A limited extent of sugarcane is grown in the south-eastern and north-eastern intermediate and dry zones. The tropical climate of Sri Lanka supports a wide range of tropical fruits across the whole island. Except for a few widely consumed fruits such as banana and papaya, most fruit species are confined to specific climatic zones which suit their phenology and growth.

Cropping seasons in Sri Lanka are determined by its precipitation pattern, which is primarily based on the two monsoons. The North-East monsoon (NEM) starts in November and brings rain to almost the entire country and signals the beginning of the major cultivation season, which is locally known as the *maha* season. The South-West monsoon (SWM), beginning in May, brings substantial precipitation to the wet zone, but only limited rain to the dry and intermediate zones. Consequently, the *yala* season that begins with the SWM is a minor cropping season in the dry and intermediate zones.

The above-described variation in the climate of Sri Lanka and the range of crops grown therein gives rise to several specific cropping systems. The dry and intermediate zones, where a substantial portion of the production of rice and annual crops takes place, have a rice-based cropping system. Where supplementary irrigation is available during the *yala* season, it is a rice-rice cropping system. However, this is only possible in the areas under major irrigation schemes. In the large majority of the dry and intermediate zones, rainfed rice in the *maha* season alternates with a wide range of annual crops of varying durations in the *yala* season depending on the availability of rain and supplementary irrigation. The specific cropping system present in a given location is determined by the undulating land form which is a characteristic feature of the arable lands in Sri Lanka. The valley, which contains poorly drained, clayey soils, belonging to the Great Soil Group Tropaqualfs known as 'Low Humic Gley' in the local classification (Panabokke 1996), where water accumulates during the rainy season, has a rice-based cropping system, i.e. rice-rice or rice-annual crop. The upper catena containing the moderately well-drained soils of varying textural composition often has a mixed cropping system, varying from rice-rice to rice-annual crop, characterized by sequential cropping or crop rotations. At mid- and upper altitudes of the wet and intermediate zones, rice is confined to valleys, while most of the arable lands contain either perennial mono-cropping systems of tea and rubber or multiple cropping systems of short-duration annuals, particularly vegetables. The exact crop composition in a given season is often determined by the market demand and price for specific annual crops. In contrast to the largely annual crop-based systems in the dry zone, the wet and intermediate zones are characterized by perennial crop-based cropping systems. In the lower elevations, coconut is intercropped with fruit crops (e.g. pineapple), while in the mid-elevations, a wide variety of perennial spice crops (e.g. clove, cardamom, nutmeg, pepper etc.) are mixed with tropical forest tree species in complex multi-layered home gardens.

5 Approaches to Increase Climate Resilience of Crops and Cropping Systems

Attempts to increase the adaptive capacity of crops and cropping systems in Sri Lanka to anticipated climate change have taken two broad pathways. One pathway has focused on formulating policies aimed at promoting a broad range of adaptation options for different sectors with agriculture being one of the sectors (De Costa 2010a). The other pathway has focused on promoting research to generate new climate adaptive technologies or to modify the existing agro-technologies to increase the adaptive capacity of crops and cropping systems. The remainder of this chapter gives a comprehensive description of the specific measures initiated within these two principal approaches and a critical analysis of their strengths and weaknesses along with a prognosis and recommendations for the near- and medium-term future.

6 Policy Initiatives to Increase Climate Adaptive Capacity and Resilience

During the last decade, there have been three major policy initiatives in Sri Lanka to address issues on climate change, its mitigation and adaptation. These are the 'National Climate Change Adaptation Strategy (NCCAS) for Sri Lanka 2011 to 2016' formulated in 2010 (Anonymous 2010), the 'National Climate Change Policy (NCCP) of Sri Lanka' adopted in 2012 (Anonymous 2012) and the 'National Adaptation Plan (NAP) for Climate Change Impacts in Sri Lanka' published in 2016 (Anonymous 2016). In addition, there is a miscellaneous collection of sector-specific policy papers and action plans by various governmental, non-governmental and international donor agencies.

6.1 National Climate Change Adaptation Strategy (NCCAS) for Sri Lanka 2011–2016

The NCCAS contained an apparently detailed and prioritized framework for action and investment during the 2011–2016 period to systematically increase the adaptive capacity of different sectors of the economy and the people of Sri Lanka to both gradual long-term climate change (e.g. increasing temperatures, sea level rise) and increased frequency of extreme climatic events (e.g. droughts, floods, cyclones, landslides). Based on a sector-wise analysis, the overall adaptation strategy focused on an integrated framework consisting of five cross-cutting strategic thrusts: (1) mainstreaming climate change adaptation in to national planning and development; (2) enabling climate-resilient and healthy human settlements; (3) minimizing climate change impacts on food security; (4) improving climate resilience of key economic drivers; and (5) safeguarding natural resources and biodiversity from climate change impacts. In this overall strategy, increasing the climate resilience of cropping systems has been identified under the third strategic thrust, which included, in addition to agriculture, fisheries, irrigation and nutrition as well. Under this strategic thrust, four thematic areas for action were identified with each thematic area containing 3–5 priority adaptation measures. The four thematic areas were (A) ensure ability to meet food production and nutrition demand; (B) ensure adequate water availability for agriculture; (C) mitigate food security-related socio-economic impacts; and (D) increase awareness and mobilize communities for climate change adaptation. With the exception of the second thematic area, the other three are very generic and diffuse. Even the second, while focusing on water availability, is a naïve and ambitious goal given that the probability of increasing droughts is identified in the NCCAS itself as part of climate change. Despite the thematic areas being generic, the priority adaptation measures include some focused activities. Especially, those proposed for ensuring adequate water availability for agriculture include (1) promoting water-efficient farming methods and crops to improve water productivity; (2) improving maintenance of existing tanks and reservoirs including their watersheds and catchments; (3) adopting principles of Integrated Water Resources Management (IWRM); and (4) constructing new reservoirs and trans-basin diversions to meet demand.

6.2 National Climate Change Policy of Sri Lanka

The National Climate Change Policy (NCCP) of Sri Lanka was published in 2014 by the Climate Change Secretariat in the Ministry of Environment in Sri Lanka. This is a short document containing a total of 25 broad policy statements on action to be taken in 6 areas related to climate change given as (A) Vulnerability, (B) Adaptation, (C) Mitigation, (D) Sustainable consumption and production, (E) Knowledge management and (F) General statements. Policy Statement 5 under Section B aims to promote adaptation of cropping systems to climate change by taking timely action to address the adverse impacts of climate change on crops and minimize the impacts on food production to ensure food security. Policy Statement 5 also encourages climate-resilient, environmentally-friendly food production technologies. As part of 'Conservation of water resources and biodiversity', Policy Statement 6 promotes integrated watershed management and efficient water use, which has the potential to increase the adaptive capacity of cropping systems to climate change.

6.3 National Adaptation Plan (NAP) for Climate Change Impacts in Sri Lanka

Out of the three policy documents that deal with adaptation to climate change, the NAP is the most comprehensive and detailed. It has identified adaptation needs in nine key sectors vulnerable to climate change. In addition, it proposes a set of cross-cutting national adaptation needs. Food security is the first among the nine vulnerable sectors identified with the others being water, coastal sector, health, human settlements, biodiversity, tourism and recreation, export development and industry-energy-transportation. Because of the significant contribution of some of the agricultural crops to export earnings (e.g. tea, rubber, coconut, spices, etc.), the sector on estate development contains some adaptation measures which overlap with those proposed under food security. Introduction of new stress-tolerant cultivars and developing the capacity of research institutes are key common areas. In addition, adaptation measures specific to export agricultural crops such as improved nursery and plant management practices and sustainable cropping systems to increase climate resilience of plantation crops have been proposed.

It is notable that while being comprehensive and detailed, the NAP also pays limited attention to promotion of research that is essential for generation of most of the technologies that it proposes.

6.4 Strengths and Weaknesses of Policy Initiatives in Sri Lanka to Increase Climate Adaptive Capacity and Resilience

A major strength of two (i.e. NCCAS and NAP) of the three climate adaptation-related policy documents described above is that they bring together a range of relevant issues and information on climate change, climate vulnerability and possible adaptation pathways within the context of Sri Lanka and its different sectors. As such these documents serve the purpose of creating awareness among a broad range of stakeholders including law makers and governmental officials, professionals, civil society and an interested few in the general public.

However, all three policy documents have the common weakness of being too generic and thus not being able to provide sufficient guidelines for bringing about effective climate resilience 'on the ground' and to the masses that represent all sectors of the economy and layers of the society. Even though NCCAS has proposed a prioritized framework for action and NAP has proposed priority actions, for a subject specialist or a researcher, those proposals are too generic to be of much specific use except perhaps for justification statements in specific research proposals for funding. On the other hand, the NCCP does not even go as far as NCCAS and NAP in providing policy guidelines for initiating climate change adaptation measures. The 25 statements in the NCCP are even broader and more generic than those of NCCAS and NAP.

Apart from being too generic and lacking in specific guidance, a fundamental weakness in all three above-described climate change-related policies in Sri Lanka is that they include all possible options rather than providing guidance towards the most effective and viable options. Formulation of policies that recommend such viable and effective options among the many available requires an adequately comprehensive knowledge base and in-depth analysis of the available options (De Costa 2010a). These prerequisites for developing effective, high-quality policy documents do not exist in Sri Lanka. Generation of such policies of high quality needs a substantial investment in research to generate scientifically valid information, both in the 'hard sciences' (i.e. subject-specific technological input) and 'soft sciences' (i.e. a thorough analysis of the pros and cons of different policy options) (De Costa 2012).

The foregoing discussion highlights the fact that the limited strengths of the existing climate change-related policies are outweighed by fundamental weaknesses. Proof of this is evident from the fact that implementation of policies and action recommended therein has been very weak in Sri Lanka. Key decisions on climate change-related issues such as decreasing vulnerability and increasing adaptive capacity and resilience are often taken without referring to the relevant policy documents. Therefore, it is argued that it will be the technological advances through research, rather than elaborate policy discussions, that will bring about increased climate resilience to Sri Lankan agriculture. Accordingly, the next sections of this chapter will discuss the technological options that are available to increase climate resilience of crops and cropping systems.

7 Technological Advances and Research to Improve Adaptive Capacity of Cropping Systems in Sri Lanka and Increase Their Climate Resilience

Similar to almost all Third World developing countries, Sri Lanka has lagged behind the developed world in initiating research to increase climate resilience in its major crops and cropping systems (De Costa 2010a, b, 2012). As a result, it was only towards the end of the 1990s that research on assessing the impacts of climate change on crops grown in Sri Lanka started getting in to the agendas of crop research institutes and universities. Since then, research on the impacts of climate change on all important crops in Sri Lanka has gradually taken an important and often visible role during the last two decades. As development of new varieties through plant breeding is a major activity in the research programmes of all crop research institutes, screening to identify heat- and drought-tolerant germplasm has become a prominent feature.

7.1 Development of Heat-Tolerant Crop Varieties

Introduction of heat-tolerant crop varieties constitutes the adaptation option that has the greatest potential to increase climate resilience of crops and cropping systems in Sri Lanka. As heat tolerance is incorporated at the genetic level, this option is not overly dependent on crop management technology or the economic status of the farmers. At present, research programmes that aim to develop heat-tolerant varieties run along two parallel lines: (a) identifying the impacts of heat stress on the yield and yield-determining processes and (b) germplasm screening to identify tolerant genotypes. A brief description of crop-specific research on development of heat-tolerant varieties is given below.

7.1.1 Rice

Rice, the staple food of Sri Lanka and the cornerstone of its national food security, is highly sensitive to heat stress at its panicle development stage (Wassmann et al. 2009). Air temperatures above 34 °C during the spikelet development stage, even for a short period, have been shown to induce spikelet sterility in rice causing substantial yield reductions (Prasad et al. 2006; Jagadish et al. 2007, 2012). It is notable that this temperature threshold could be exceeded in many rice-growing areas in Sri Lanka, especially in the dry zone during the *yala* season (Punyawardena 2009). Furthermore, Weerakoon et al. (2008) have shown that the adverse impacts of high temperatures are exacerbated at the higher relative humidity (RH) levels by reducing transpirational cooling and increasing spikelet temperatures (Abeysiriwardena et al. 2002). Even though RH levels in Sri Lanka are generally in the higher range (i.e. >70%), the warm dry winds that occur during the period from June to August decrease the RH levels down to around 55%, especially during the day in the dry zone (Punyawardena 2009). As the June–August period coincides with the *yala* season, the above reduction in RH could have a mitigating effect on the adverse impacts of higher air temperatures on the spikelet fertility of rice.

Germplasm screening for development of heat-tolerant rice varieties in Sri Lanka started at the Rice Research and Development Institute (RRDI) at Batalagoda in the 2008/2009 *maha* season and was continued over the next three seasons. The screening was carried out in open-top chambers (OTCs) under field conditions. Air temperatures inside the OTCs were elevated by 2 °C above the prevailing ambient temperature by installing heating coils above the canopy level. In these experiments, which included cultivars of all age groups, significant inter-varietal variation was observed in the yield response to elevated temperature, with both positive and negative responses in different varieties. Furthermore, the response to elevated temperature by a given variety varied between different seasons, thus showing significant genotype × environment (G × E) interaction in the yield response to elevated temperature. This meant that currently cultivated rice varieties may not provide clear indications of heat tolerance, thus making them ineffective in being a genetic source for development of heat-tolerant varieties.

Subsequently, breeding lines known to contain genetic sources and traits of heat tolerance were obtained from the International Rice Research Institute (IRRI) and were screened for heat tolerance in thermal gradient chambers (TGCs). Breeders at the RRDI have been able to identify genotypes having relatively greater heat tolerance, which are being used as parents in crossing programmes.

7.1.2 Other Field Crops (OFCs)

Screening of OFCs for heat tolerance has started recently at the Field Crops Research and Development Institute (FCRDI), Mahailluppallama (MI). Initial screening has been started with chilli and soybean. However, this work has been hampered by the technical issue of controlling the temperature gradient in the TGC. This has been an issue encountered with the TGC at Batalagoda as well. Germplasm screening for heat tolerance in tomato and chilli has been carried out at the University of Peradeniya using OTCs where the temperature elevation above the ambient was 4–5 °C. In chilli, the varieties MI-2, *Galkiriyagama Selection* and *MI Varaniaya* were identified as relatively heat tolerant among the seven genotypes used in the screening. On the other hand, variety KA-2, which was found to be the most tolerant in the screening at MI, was classified as only 'moderately tolerant' in the screening at Peradeniya. These findings again highlight the significant genotype x phenotype interaction in the traits contributing to heat tolerance in crops. In tomato, varieties *Bathiya*, *Ravi* and *Tharindu* were identified as relatively heat tolerant from among 17 genotypes.

7.1.3 Coconut

The Coconut Research Institute (CRI) of Sri Lanka has made appreciable progress in investigating the impacts of heat and drought stress on the reproductive development of coconut. This work has identified the 3-month period prior to flower opening as the most critical period where heat and drought stress could significantly decrease fruit setting and the subsequent nut yield of coconut. As such, the influence of heat and drought stress on the quality of pollen and female flowers and its implications on fruit setting have been studied in a comprehensive experiment involving two hybrids, i.e. Sri Lanka Green Dwarf (SLGD) × Sri Lanka Tall (SLT) and SLGD × San Ramon (SR). Using controlled hand pollination, fruit setting of all possible combinations of stressed and unstressed pollen and female flowers was quantified. As expected, fruit setting was highest in the unstressed pollen × unstressed female flower combination and lowest in the stressed pollen × stressed female flower combination. Interestingly, fruit setting, while being lower than in the unstressed × unstressed combination, was greater in unstressed pollen × stressed female flower combination in comparison to the stressed × stressed and stressed pollen × unstressed female flower combination. This highlights the fact that maintaining the quality of pollen is more critically important than maintaining the quality of female flowers to minimize the adverse impacts of heat and drought stress on fruit setting and subsequent nut yield of coconut. Notably, in rice also, pollen has been shown to be more sensitive to heat stress than the female reproductive organ (Wassmann et al. 2009). Based on these findings, a pilot project has been initiated to increase the production of hybrid seed nuts in seed gardens during periods of heat and drought stress. Biochemical analysis of pollen and female flowers has shown that their carbohydrate contents varied with heat and drought stress intensity and are likely to play an important role in determining the quality of these reproductive organs.

Research conducted in the field and under semi-controlled environments has identified a threshold temperature of 33 °C above which the pollination and fertilization processes and consequently the nut setting and yield are significantly reduced. Because of the critical importance of pollen germination and pollen tube growth to ensure successful fertilization and nut setting, screening of coconut varieties for heat tolerance has been conducted under in vitro conditions in controlled environmental growth cabinets (Ranasinghe et al. 2018). Here, in vitro pollen germination and pollen tube growth in several coconut hybrids have been observed under a series of temperatures ranging from 16 to 38 °C varying at 2 °C intervals. This has enabled identification of cardinal temperatures, the base (T_b), optimum (T_{opt}) and ceiling (T_{ce}) temperatures, for pollen germination. The respective means and ranges of T_b, T_{opt} and T_{ce} were 14.8 °C (13.8–16.4 °C), 27.9 °C (27.2–28.7 °C) and 40.1 °C (38.4–43.0 °C). A principal component analysis showed T_{max} for pollen germination and T_{opt} for pollen tube growth contributing most to heat tolerance in these processes. Based on these two parameters, coconut hybrids with relatively greater and lower heat tolerance have been identified.

The CRI has also carried out detailed, long-term investigations on the flowering and fruit setting behaviour of mature coconut in the field. These studies have shown that the fruit set of an inflorescence opened in a given month was negatively correlated with the number of days on which the daily maximum temperature (T_{max}) exceeded 33 °C and positively correlated with the number of female flowers produced (Ranasinghe et al. 2015). It was also found that female flower production and fruit setting varied with month of the year and between different years and locations and that part of this variation was probably caused by the temporal and spatial variation of temperature and water availability.

7.1.4 Sugarcane

Research to quantify the impacts of climate change on sugarcane in Sri Lanka started only recently at the Sugarcane Research Institute, Uda Walawe. In 2017, an experiment started in open-top chambers (OTCs) to screen the response of eight cultivated varieties to elevated temperature and elevated atmospheric carbon dioxide, with the two factors varied both individually and in combination in different OTCs. Crops in this experiment are currently in their ripening phase.

7.1.5 Overall Assessment of the Current Status of Research on the Development of Heat-Tolerant Crop Varieties

The above description shows that work on the development of heat-tolerant crop varieties in Sri Lanka is still at an elementary level. Except in rice and coconut, work carried out so far has not progressed beyond initial screening of the available varieties. Notably, no research on heat tolerance has been initiated as yet in some of the agricultural crops of major economic importance such as tea, rubber and fruit crops. Fruit crops are especially vulnerable to climate change as flowering, fruit formation and their survival are highly sensitive to both higher temperatures and greater variability in precipitation, which could cause both droughts and extreme rainfall events.

The current screening methodology in Sri Lanka often lacks precision and needs a substantial technological input for improvement. The thermal gradient chambers (TGCs) and open-top chambers (OTCs) have the advantage of being able to grow the plants on natural soil under actual field conditions with adequate solar radiation. However, lack of precise temperature control is their biggest current weakness and requires improvement in temperature regulation technology. In particular, during hot, dry periods (e.g. *yala* season in the dry zone), air temperatures inside both TGCs and OTCs can exceed 38–40 °C during the hours around mid-day. While the crops growing under open field conditions would also experience higher temperatures during mid-day, it is rarely that the mid-day temperatures would reach 38–40 °C. Such unusually high screening temperatures could induce abnormal effects on the germplasm that is being screened, thus reducing the validity of results obtained from screening experiments. At present, Sri Lanka does not have dedicated screening facilities for screening for heat tolerance. In view of the urgent requirement for identification of heat-tolerant germplasm in a range of important crops in Sri Lanka, establishment of a field screening facility at a location in a high-temperature environment is a priority to accelerate development of heat-tolerant varieties. It would offer a relatively low-cost alternative to using expensive controlled environmental chambers and imprecise semi-controlled environmental facilities such as TGCs and OTCs. Hambantota in the south-east corner of Sri Lanka and Mannar in the north-west corner, both of which have warm, dry climates, are ideal locations for establishing a heat screening facility. Because plants at such a facility can be grown in the field under actual farming conditions, the studies on heat stress impacts and heat tolerance would be free from artefacts introduced by the un-natural experimental conditions in fully or semi-controlled environments. Furthermore, it can be noted that current research into development of heat-tolerant crop varieties in Sri Lanka involves only a limited number of available varieties or germplasm. In contrast, a dedicated heat screening facility at a high-temperature location would allow screening of a much greater number of current germplasm and/or breeding lines to be screened simultaneously thus enabling the deployment of modern high-throughput methods of phenotyping (Wahid et al. 2007).

There is an urgent need for research work on heat tolerance of the major crops in Sri Lanka to progress from elementary varietal screening to in-depth studies on mechanisms of heat tolerance and their physiological and genetic basis. In most screening studies, yield is the primary criterion that has been used to classify the degree of heat tolerance/susceptibility of the genotypes that are being screened. The only exception has been in rice and coconut, where the effects of heat stress on fertility of reproductive organs have been studied (Weerakoon et al. 2008; Ranasinghe et al. 2015). However, heat stress induces a wide range of physiological, biochemical and metabolic effects at different levels of plant organization ranging from cellular and tissue levels (Fig. 2) to whole plant and plant population levels (Hasanuzzaman et al. 2013). Therefore, multiple processes and functions have to be targeted when attempting to incorporate heat tolerance into existing varieties and developing new varieties.

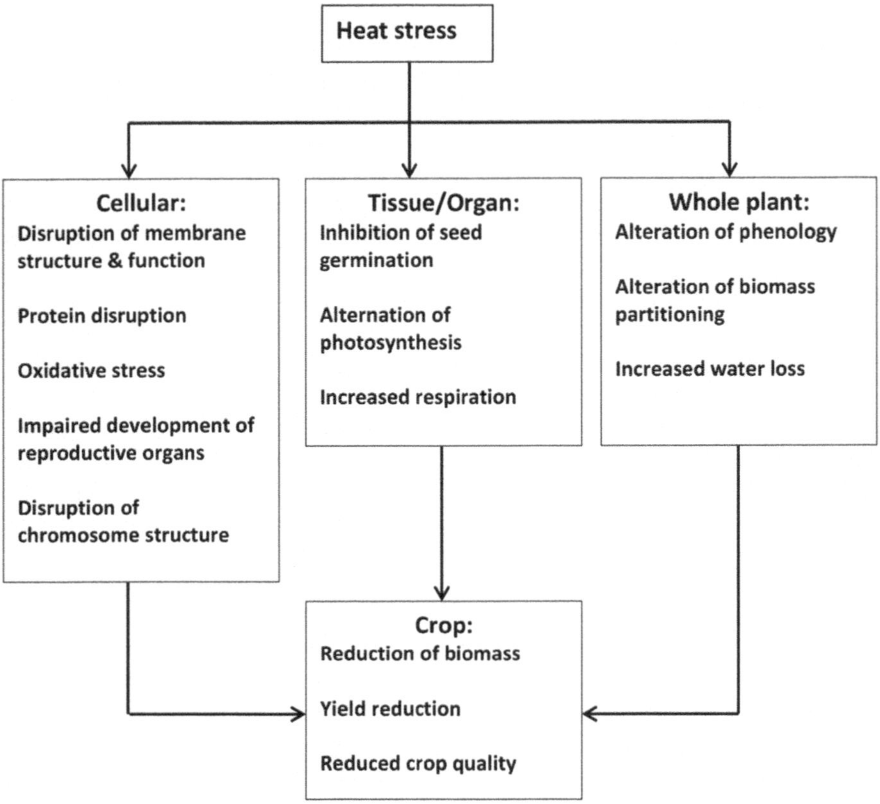

Fig. 2 Multiple effects of heat stress at different levels of plant organization. (Adapted from: Bita and Gerats (2013) and Hasanuzzaman et al. (2013))

7.1.6 Multiple Approaches to Develop Heat-Tolerant Crop Varieties

Development of heat-tolerant varieties should be done through a combination of classical/conventional and molecular approaches of plant breeding using all available mechanisms of heat tolerance (Wahid et al. 2007). Classical plant breeding primarily involves crossing of pre-identified relatively heat-tolerant genotypes among themselves or into existing varieties having other desirable traits (i.e. higher yield, acceptable grain characters and quality, adequate resistance to major pests and diseases, consumer acceptance, etc.), evaluation of their progeny and subsequent purification of identified heat-tolerant progeny over several generations until a 'pure' line is obtained. In classical/conventional plant breeding, yield is the major criterion that is used in evaluation of both existing germplasm and progeny arising from crossing. In breeding for heat tolerance, fertility of reproductive organs (e.g. pollen) has also been used as a simple screening criterion when it can be assessed relatively easily and rapidly. While being simple and straightforward, progress via conventional plant breeding could be slow, especially when trying to develop stress-tolerant varieties. Furthermore, when yield is used as the

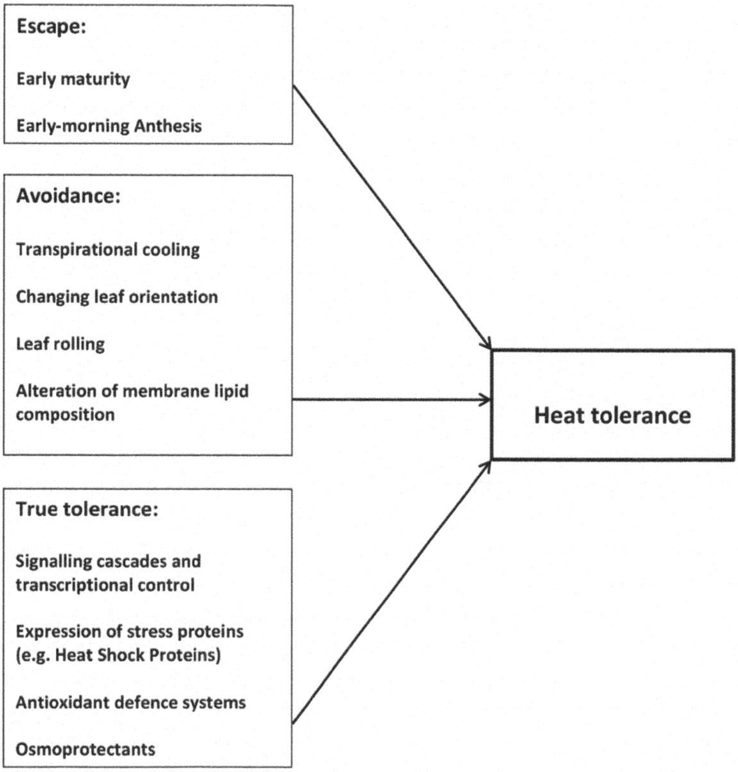

Fig. 3 Different mechanisms of heat tolerance. (Adapted from: Hasanuzzaman et al. 2013)

main criterion for germplasm screening, lower-yielding genotypes and progeny, which may be having higher levels of individual traits contributing to stress tolerance, could be discarded because of their lower yield performance. Therefore, adoption of ideotype breeding (Donald 1968), an approach based on identification of possible mechanisms and traits conferring stress tolerance and systematically combining ('pyramiding') them into a single tolerant genotype, could be an approach which could bring greater success in the development of heat-tolerant germplasm. In trying to bring together multiple traits of heat tolerance into a single genotype, both conventional and molecular approaches of plant breeding could be used. It can be noted that at present in Sri Lanka, the limited amount of work that is being carried out to develop heat-tolerant crop varieties employs the conventional plant breeding approach with yield as the primary screening criterion.

Plant breeding based on the mechanistic approach requires identification of processes and mechanisms of stress tolerance (Fig. 3) (Cossani and Reynolds 2012; Bita and Gerats 2013; Driedonks et al. 2016). Three different mechanisms of heat tolerance can be identified: (a) Heat escape includes mechanisms which enable heat-sensitive processes to occur during relatively cooler periods (e.g. flower opening and pollination during the cooler, early morning periods of the day); (b) heat avoidance includes mechanisms to maintain cooler tissue temperatures even when the external temperatures are higher (e.g. transpirational cooling, increased leaf reflectance); and (c) true heat tolerance includes mechanisms to maintain cellular functions even when the tissue temperatures increase beyond their optimum range (e.g. synthesis of heat shock proteins, antioxidant systems, osmoprotectants). While mechanistic plant breeding can proceed using the conventional tools of crossing and progeny evaluation based on individual traits or mechanisms of stress tolerance, research programmes in the developed world to produce stress-tolerant crop varieties employ a variety of modern approaches, which are based on the molecular-level processes and functions.

The term 'omics' approaches has been coined to describe the range of approaches currently used in investigating mechanisms of stress tolerance with the final aim of developing stress-tolerant genotypes. The 'genomics' approach determines the genes and the specific alleles present in the genome of a given genotype (e.g. a variety) via high-throughput DNA sequencing. This enables identification of polymorphism at specific locations of the genome in the germplasm of a crop. Bioinformatics enable information about genomic polymorphism to be correlated with variation of putative traits of stress tolerance in a varietal/germplasm screening under a given stress, thus enabling identification of genomic sequences (e.g. quantitative traits loci (QTLs)) containing genes conferring stress tolerance. The 'transcriptomics' approach investigates the changes in gene expression (as indicated by the specific mRNA levels present) in a known stress-tolerant genotype when it is subjected to a specific level of stress or at different levels of stress during gradual stress development. Accordingly, up- and downregulated genes in response to heat stress in a known heat-tolerant genotype (based on its yield performance in high-temperature environments) and their possible functions (found via homology search) provide indicators of pathways of stress tolerance. Expression of a specific DNA sequence of a genome (i.e. a gene or a QTL) via transcription into specific mRNA leads to synthesis of specific proteins via translation. They can be functional proteins such as enzymes, which regulate and control specific steps in the pathways of stress response and stress tolerance, or structural proteins, which may contribute to stress tolerance. The 'proteomics' approach identifies and quantifies the specific proteins that are synthesized under stress. Specifically, heat tolerance has been shown to be induced via heat shock proteins (HSPs) which are synthesized in response to heat stress in tolerant genotypes (Ahmad et al. 2016). Modern proteomic analytical techniques enable not only identification of specific proteins that are induced but also quantification of their amounts. When combined with knowledge on the functions of specific proteins in different physiological and metabolic processes and pathways, proteomics could lead to identification and elucidation of mechanisms of stress tolerance in a known ('phenotyped') stress-tolerant genotype. Accordingly, proteomic analysis of HSPs could identify key metabolic proteins involved in conferring heat tolerance. The large and varied amount of biochemical products (both primary and secondary) that are synthesized via the catalysis of biosynthesis pathways by proteins constitutes the

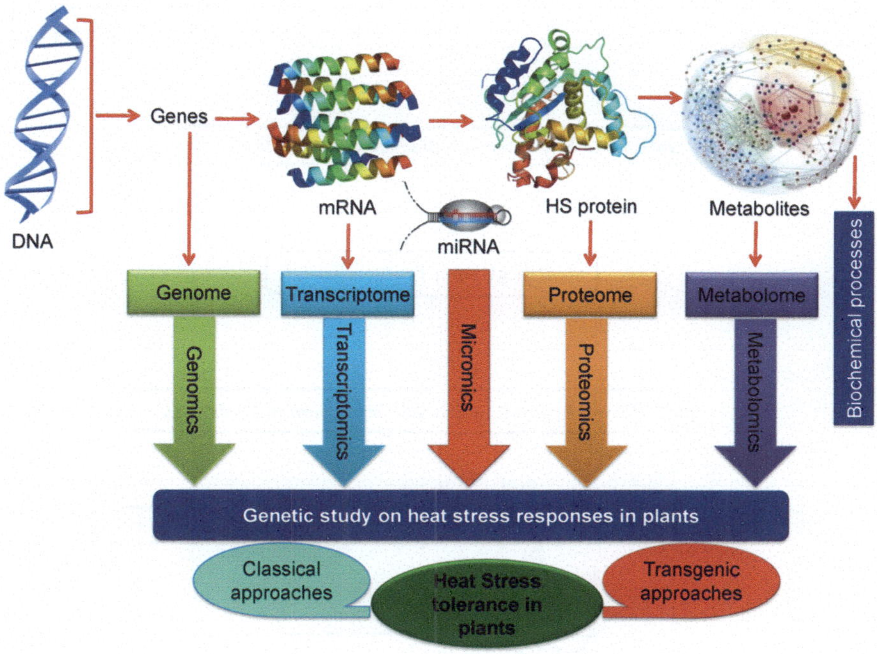

Fig. 4 Pathway for elucidation of mechanisms of heat tolerance using 'omics' approaches. (Source: Hasanuzzaman et al. (2013) Reproduced with permission)

'metabolome' and is analysed in 'Metabolomics'. It includes a range of methods to separate different primary (e.g. sugars, amino acids, etc.) and secondary (e.g. carotenoids, phenolic compounds, etc.) metabolites via techniques such as mass spectrometry and high-performance liquid chromatography (HPLC). Therefore, the 'omics' approaches offer exciting possibilities for elucidation of mechanisms of heat tolerance and determination of their genetic, physiological and biochemical basis (Fig. 4) (Hasanuzzaman et al. 2013), which lays a firm foundation for development of heat-tolerant crop varieties via biotechnology (Fig. 5) (Ahmad et al. 2016).

7.1.7 Institutional Capacity in Sri Lanka to Embark on the 'Omics' Approaches

Exploitation of the vast potential of 'omics' approaches to develop climate-resilient, heat-tolerant varieties of major crops in Sri Lanka requires a substantial improvement of its institutional infrastructure and trained manpower (De Costa 2010a). This is because the methodologies involved in 'omics' approaches require facilities for high-throughput DNA sequencing and RNA analysis and sequencing. They also require state-of-the-art facilities in electrophoresis, chromatography and mass spectrometry for detailed proteomic and metabolomics analysis. All 'omics' approaches require a significant input from Bioinformatics to process the large amount of

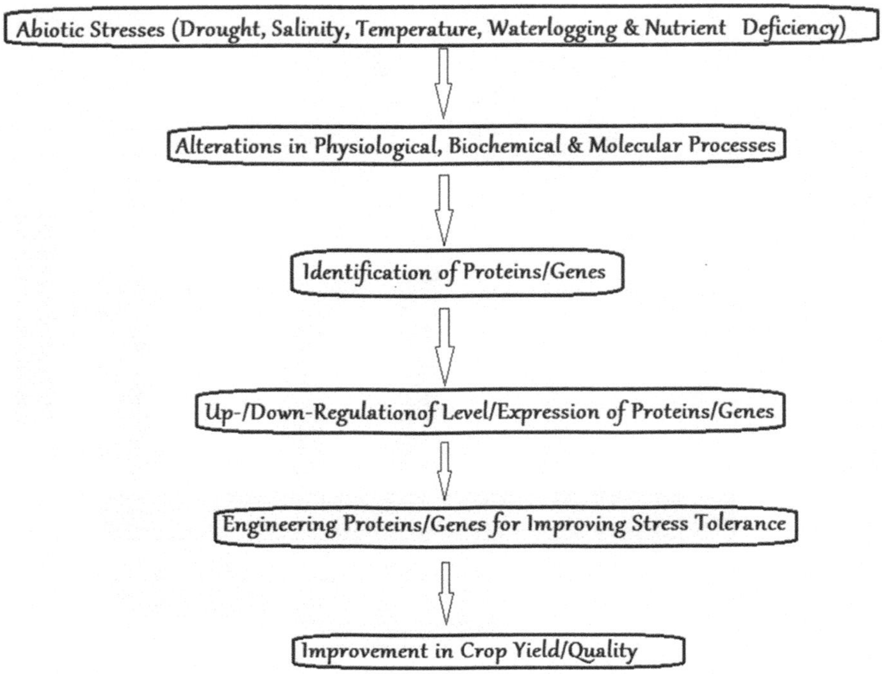

Fig. 5 Pathway for elucidation of mechanisms of heat tolerance and engineering heat tolerance. (Source: Ahmad et al. (2016) Reproduced with permission)

genomic, transcriptomic, proteomic and metabolomics data that is generated, to determine their interrelationships and compare them with the ever-expanding base of similar data. At present, Sri Lanka has only limited institutional and infrastructural capacity in all areas of 'omics' research.

7.2 Crop Management Options to Increase Climate Resilience of Cropping Systems in Sri Lanka

While heat-tolerant crop varieties contribute to increased climate resilience at the genetic level, crop management options aim to modify the crops' growing environment in a way which would mitigate the adverse conditions resulting from climate change (e.g. increased air temperatures) or to provide conditions/resources to the crops to counter the adverse conditions created by climate change (e.g. reduced water availability). A review of the research projects carried out in the national agricultural research institutes in Sri Lanka during the last two decades clearly highlights the fact that their research agenda is heavily biased towards development of new crop varieties through plant breeding. Improvement, optimization and increased precision of crop management have received much less attention than varietal improvement. Therefore, while efforts to develop heat-tolerant crop varieties could start from a relatively strong existing foundation of knowledge and expertise (at least in conventional breeding), efforts to increase climate resilience through modified crop management have to start from a much weaker level of existing knowledge and expertise. Furthermore, even after overcoming the institutional bias, the transfer of new or modified crop management technologies to the farmers/growers is much more difficult and complex than providing the seeds of a new heat-tolerant variety.

7.2.1 Increase of Tree Cover

Canopy cover of trees provides protection to shorter crops from the heat energy contained in incoming solar radiation and thereby reduces development of higher tissue temperatures, especially in the leaves. Also, by decreasing the radiation energy reaching the microenvironment above the foliage canopy of a short, annual crop, the tree canopy further decreases the possibility of developing higher leaf temperatures. By reducing the amount of radiation energy incident on both the foliage canopy and the soil, shade provided by a tree canopy overhead reduces evapotranspiration and thereby slows down the development of water stress, in the soil and plant tissues. Shade has also been shown to increase radiation use efficiency (i.e. amount of biomass produced per unit of radiation intercepted) of C_3 crops (Marshall and Willey 1983; Kasim and Dennett 1986), thus making the biomass production process more energy-efficient. Certain crops such as tea are routinely grown under shade. It has been shown that the reduction of irradiance on the tea canopy due to the presence of shade overhead decreases photo-inhibition (i.e. irreversible damage of chloroplasts due to accumulation of excess radiation energy) in tea leaf tissue (Mohotti and Lawlor 2002). Similar protection from photo-inhibitory damage could occur in other C_3 crops as well when tree cover is increased.

Most fields in which shorter annual crops (e.g. rice and other field crops) are grown in Sri Lanka have only scattered trees, especially in the dry and intermediate zones. Therefore, there is scope for increasing the tree density in these two zones, which have a higher vulnerability to climate change than the wet zone, where the tree cover in annual crop fields is higher. When increasing tree cover, the possibility of planting of indigenous fruit trees could be explored. Especially, the potential of indigenous tree species which are tolerant to drought (e.g. woodapple, *Limonia acidissima*, etc.) could be considered.

Despite the above-mentioned advantages, increasing tree cover in annual crop fields has to be done with caution, because of several potential negative feedbacks. Firstly, decreased radiation energy on the foliage canopy of the annual crop could potentially reduce its photosynthesis, biomass production and yield. However, in C_3 plants, which form the large majority of agricultural crops, photosynthesis of individual leaves is light-saturated at irradiances which are around 50% of full sunlight (Lawlor 1993; De Costa 2004a). Therefore, increasing shade up to a threshold is not likely to have a negative impact on photosynthesis and yields of annual crops, especially in environments of high irradiance levels (e.g. dry zone in the *yala* season). Secondly, when grown in a mixture with annual crops, tree species can compete with annual crops for essential resources (e.g. water and nutrients) causing yield reductions. Therefore, it is essential that tree species which absorb water and nutrients from deeper layers of the soil profile are selected to ensure adequate separation in the zones from which trees and annual crops obtain their resources. Thirdly, incorporation of trees, either as rows or individual trees, into annual crop fields in which trees were not present alters the air flow and aerodynamics of the entire landscape (Brenner 1996). This could have both positive and negative impacts. Reducing the velocity of wind flow above the foliage canopy of the annual crop could decrease evapotranspiration (Long and Persaud 1988) and conserve water in the soil and plant tissue, thus contributing to drought avoidance. On the other hand, reduced wind flow across the field could cause accumulation of heat energy within the field, thus causing a feedback increase in air temperature at the landscape level. Similarly, the vapour pressure deficit of the air above an annual crop could increase (Carr 1985) or decrease (Brown and Rosenberg 1971) depending on the magnitudes of changes in temperature and vapour pressure occurring as a result of introducing trees.

An attempt to incorporate trees into annual crop fields was made in the 1980s via the introduction of alley cropping (also known as hedgerow intercropping) to the dry zone of Sri Lanka. In alley cropping, annual crops were grown in alleys created by parallel rows of tree species. Leguminous tree species such as *Gliricidia sepium*, *Erythrina lithosperma* and *Calliandra calothyrsus* and non-leguminous species such as *Cassia spectabilis* were used in the alleys in which crop species such as maize and grain legumes were grown. The principal objective of introducing trees in the annual crop fields, particularly in the dry zone, was to improve soil fertility and ensure adequate crop yields in the uplands. This was expected via incorporation of tree loppings as green manure, by capturing additional nitrogen via biological nitrogen fixation and improved nutrient recycling via deeper rooting trees. Furthermore, the tree rows were expected to reduce soil erosion by acting as a physical barrier and reducing the length of the slope (Dharmasena 1994). Greater input of organic material in alley cropping was expected to improve soil physical properties such as the aggregate stability (Mapa and Gunasena 1995). In addition, loppings from tree rows were expected to provide a mulch (Van Noordwijk 1996) and thereby conserve soil moisture via reduced soil evaporation. Furthermore, tree hedges were expected to reduce weed growth during the fallow period between two cropping seasons by providing additional shade and also provide fuelwood and fodder. Accordingly, alley cropping was introduced as part of 'conservation farming' and 'sustainable agricultural practices' (Weerakoon and Seneviratne 1984; Weerakoon and Liyanage 1987; Somasiri et al. 1990). In the 1980s, alley cropping was promoted as a sustainable agroforestry practice in many other tropical countries in Asia and Africa (Kang et al. 1990; Kang 1993). A similar hedgerow intercropping system known as the 'Sloping Agricultural Land Technology (SALT)' was proposed and tested for the sloping lands in the mid- and high-elevation humid zones of Sri Lanka. Here, double hedgerows of the same tree species along with a few additions (e.g. *Flemingia congesta* and *Eupatorium inulifolium*) were grown along contours thus forming physical barriers against soil erosion and encouraging terrace formation in the long run. This technology was recommended and tested for both short-duration annual crops and perennial crops such as tea (Ekanayake 1994), which occupies a substantial land area in the sloping highlands of Sri Lanka.

However, despite the demonstration of many of the expected benefits in terms of reduced soil erosion and improved soil physical parameters (Keerthisena 1995), alley cropping did not progress beyond the research stations either in Sri Lanka or elsewhere, primarily because of poor farmer acceptance (Senaratne 2003). Similarly, SALT also has not been adopted widely either in the tea plantations or annual cropping systems of the sloping highlands of Sri Lanka. Both biophysical and socioeconomic factors have contributed to the poor farmer acceptance of alley cropping and SALT. Significant competition for resources (i.e. light, water and nutrients) by the tree hedges has been shown to cause resource limitation and yield reductions in both annual crops (De Costa and Chandrapala 2000a, b) and tea (De Costa and Surenthran 2005a, b). Increased labour requirement for hedgerow establishment and maintenance and occupation of 15–25% of the farm area by hedgerows and inadequate extension have been identified as the socioeconomic causes of poor farmer acceptance (Keerthisena 1995; Samarakoon and Abeygunawardena 1995; Senaratne 2003). Broadly, the same biophysical and socioeconomic causes have been identified as being responsible for poor farmer acceptance of alley cropping/hedgerow intercropping in other parts of the world as well (Francis and Atta-Krah 1989; Kang et al. 1990; Cannell et al. 1996). It has been shown that alley cropping is an agroforestry system where the tree-crop interface (i.e. the land area shared by both trees and crops) is longest, thus increasing the probability of resource competition. In comparison, scattered trees in farmlands are a system where the tree-crop interface is shorter, thus minimizing tree-crop competition.

7.2.2 Shifting of Crops

Each agricultural crop has a specific range of temperatures under which yields would be near maximum, if properly managed with adequate resources. Increasing temperatures could shift the temperatures of certain growing environments above the upper boundary of the optimum temperature range of certain crops. This could cause significant yield reductions in those crops in a future warmer climate unless new heat-tolerant varieties become available in the future. Therefore, the existing cropping systems may have to be modified by shifting certain heat-sensitive crops from the currently grown locations to locations where the future temperatures would still be within the optimum range. Likewise, certain locations which are too cold for growing certain crops could become suitable in the future due to increased temperatures. Therefore, a systematic modification of the composition of existing cropping systems via shifting of crops to environments within their optimum temperature ranges in a future climate could ensure climate resilience. As the magnitudes of historical temperature increases have been different in different agro-ecological regions of Sri Lanka (De Costa 2008), modification of existing cropping systems via crop shifting should be done based on adequately comprehensive predictions of future temperature and rainfall profiles.

Therefore, a substantial investment in climate modelling is required to generate predictions of future climate in Sri Lanka along with its spatial and temporal variation (De Costa 2012). At present, there is only limited capacity in terms of expertise in climate modelling in the Department of Meteorology of Sri Lanka and in the university faculties. Even though predictions of future climate are available from global-scale modelling work (Collins et al. 2013; Stocker et al. 2013a), they do not capture the spatial variation within Sri Lanka, which occupies only a few 'grid cells' in a global-scale Earth System Model or General Circulation Model. Accordingly, climate modelling which specifically focuses on downscaling predictions from global-scale models is an area in which Sri Lanka needs to develop its expertise urgently. Such downscaled predictions of future climate in Sri Lanka would provide an invaluable basis in designing climate change adaptation measures not only in the agriculture sector but also in all other sectors as well. A limited amount of results are available from such downscaling work done at the Department of Meteorology and elsewhere (Das et al. 2015; Patabendige et al. 2016; Dorji et al. 2017; Pattnayak et al. 2017).

7.2.3 Shifting of Planting Times

At present, planting times of rainfed annual crops in Sri Lanka are predominantly based on the availability of water as determined by the rainfall patterns. Based on the available evidence of historical changes in rainfall patterns in Sri Lanka (De Costa 2008), it is likely that planting times may have to be shifted to be in synchrony with future changes in the onset of rainfall seasons (e.g. Second Inter-monsoon and the North-East monsoon for *maha* planting and First Inter-monsoon and the South-West monsoon for *yala* planting). Here also, predictions with acceptable accuracy of the onset of rainfall seasons is required to shift planting times and improve climate resilience of future cropping systems in Sri Lanka. For cropping systems under major and minor irrigation schemes, planting time is preceded by the release of water from reservoirs and tanks for land preparation. Therefore, any shift in planting times has to be accompanied by changes in the scheduling of supporting services such as irrigation as well. Recently, the Department of Agriculture in Sri Lanka has recommended that land preparation be done with the rains received from the First and Second Inter-monsoons in *yala* and *maha* seasons, respectively. This would allow water from reservoirs and tanks to be released for planting, thus saving appreciable amounts of valuable irrigation water. Furthermore, earlier planting would reduce the probability of crops getting caught in adverse environmental conditions (e.g. drought, excessive rainfall) during the latter part of the seasons.

Shifting of planting times could also take into account the future increases in temperature, if they become available through improved climate modelling. While adjusting the planting time to be in synchrony with future changes in rainfall patterns, adjustments could also be possible so that heat-sensitive processes such as reproductive organ development, flowering and fertilization avoid periods of higher temperatures. This requires development of new varieties and/or careful selection of suitable existing varieties whose phenological development matches the temperature and rainfall regime of the growing environment. Therefore, a fine-tuning of the existing crop management practices and an effort to move towards 'precision agriculture' are needed to improve climate resilience of cropping systems in Sri Lanka.

Durations from planting to flowering and maturity are determined by developmental processes which are controlled by temperature, with increasing temperatures generally decreasing the durations (De Costa 2004a). Hence, temperature increases in a future climate could modify the timing of key developmental events such as panicle/cob initiation and flowering and alter the total crop duration. These alterations also will have to be taken into consideration when adjusting planting times, selecting crop varieties for cultivation in different agro-ecological regions and developing new crop varieties via plant breeding.

7.2.4 Introduction of a Third Season

A third season of crop cultivation has been introduced in selected agro-ecological regions in Sri Lanka during the last decade. This is a short cropping season which utilizes the residual soil moisture at the end of one cropping season to cultivate a short-duration crop prior to the commencement of the next major season. The third season often comes between the end of *yala* and the beginning of *maha*. Therefore, the residual moisture from the South-West monsoon is used to establish the crop, and water coming from the Second Inter-monsoon supplements the crop water requirement. Grain legumes such as mung bean are the most common among crops grown in the third season.

The third season could increase climate resilience of a cropping system by providing an option to compensate for crop losses in a major season due to any factor of climate change (e.g. supra-optimal temperatures) or increased climate variability (e.g. drought).

7.2.5 Conservation of Soil Moisture

Even before climate change and its possible impacts on water availability started being highlighted, measures for conservation of soil moisture had formed an integral component of cropping systems in environments where water was a limiting resource. Possible changes in precipitation patterns, increases in evapotranspiration rates and increased probability of drought have raised even further the importance of soil moisture conservation in a future warmer climate. Soil moisture conservation could be achieved through minimization of soil evaporation and optimization of irrigation by minimizing losses.

Mulching of crops is an age-old practice of conserving soil moisture. Traditionally, mulching has been done with plant-based material such as straw and prunings of hedgerows, shrubs or trees. Mulching with synthetic material such as polythene has come into practice in recent times, especially in crops grown in protected facilities such as polytunnels. While the benefits of mulching are well-known to farmers and extension workers, it is not practised extensively in Sri Lanka. This could be due to several reasons such as shortage of suitable mulching material, additional labour required and lack of proper guidance through extension. A series of multi-locational long-term field experiments conducted recently across a natural temperature and rainfall gradient in Sri Lanka involving several upland crops (i.e. maize, mung bean, chilli, tomato and potato) clearly showed that mulching combined with a 30% reduction of water input could achieve similar or greater yields than unmulched crops receiving 30% more water (Abhayapala et al. 2014; Malaviarachchi et al. 2014, 2016; Abhayapala 2017). These results were confirmed in the farmer fields in the Northern Province of Sri Lanka (Eeswaran 2018), which is identified as a region of high climate vulnerability (Punyawardena et al. 2013).

Apart from mulching, loss of soil moisture could be minimized by adopting an optimum density and planting arrangement and ensuring that the crop achieves canopy closure within the minimum possible time after planting. Here again, extension and technology transfer services in Sri Lanka can play a significant role in raising farmer awareness of optimum spacing/density requirements of their crops and of crop management practices required to achieve canopy closure quickly. These include optimum fertilizer and water management and increasing the precision of their application (e.g. split application of fertilizer).

Practices such as minimum tillage and zero tillage that minimize disturbance of soil during land preparation and crop establishment are being promoted in many countries as a means of increasing climate resilience of cropping systems by minimizing loss of soil moisture and arresting the decline of soil fertility (The Royal Society 2009). However, despite being aware of these practices during the last four decades, neither the extension workers nor the farmers in Sri Lanka have been keen on transferring and adopting these technologies. Some of the specific problems associated with the crops and soils in Sri Lanka probably contributed to low adoption of minimum and zero tillage. For example, soils belonging to the Great Group Rhodustalfs, known as 'Reddish Brown Earth' in the local soil classification (Panabokke 1996), which dominates the dry zone, are difficult to work with because they are sticky when wet and hard when dry. In many cropping systems, the working corridor is too narrow. Furthermore, minimum and zero tillage are heavily dependent on synthetic herbicides for weed control. In the prevailing socio-cultural environment in Sri Lanka, there is widespread scepticism against agrochemical-intensive agriculture. Therefore, it is unlikely that these practices would be adopted widely in the near- or medium-term future in Sri Lanka.

7.2.6 Modified Water Management

There is a high likelihood that the amount of water available for crops would decrease significantly in a future climate. While statistically significant declining trends in annual total precipitation have been shown only in a limited number of regions and locations, both globally (IPCC 2013) and locally (De Costa 2008), there has been clear evidence of increased rainfall variability (Hartmann et al. 2013; Stocker et al. 2013a). Such increased rainfall variability is likely to increase the frequency of both extreme precipitation events and the frequency and length of rain-free periods during successive precipitation events. Increased frequency of high-intensity precipitation causes soil erosion which in turn increases reservoir siltation. This results in reduced capacity for water storage and decreased volume of water available for irrigation. Furthermore, increased demand for water for other activities (e.g. human consumption, hydropower generation, industrial purposes, etc.) is likely to further reduce the amount of water for irrigation. All these changes that are taking place at present, and are likely to be accelerated with climate change in the future, necessitate a paradigm shift in irrigation and water management of crops. Because of the high dependence of Sri Lankan agriculture, especially rice cultivation, on irrigation, a major shift in the rationale and methodology of water management is required in Sri Lanka as well.

At present, irrigation of both lowland rice and upland crops in Sri Lanka is done predominantly using surface irrigation, which utilizes large volumes of water at relatively lower efficiency. In lowland rice, standing water is maintained for almost the entire crop duration until 2 weeks before harvesting, incurring substantial losses via direct evaporation, deep drainage and lateral seepage (Bouman et al. 2007). In upland crops, substantial losses of applied irrigation water occur via deep drainage and surface run-off. Furthermore, surface irrigation has very little control over the distribution of water within a field, often causing uneven irrigation. Therefore, in a scenario of limited water availability for irrigation, water-saving irrigation technologies have to be explored.

7.2.7 Water-Saving Irrigation Methods in Rice Cultivation

Rice is a crop which requires large quantities of water with irrigated lowland rice cultivation requiring 2500–3000 m^3 of water to produce 1 ton of rice grain (Bouman 2001). This is much higher than the corresponding value of 1000 m^3 of water per 1 ton of wheat grain. As 75% of global rice production comes from nearly 79 million ha of irrigated lowland rice, shortages of water in a future climate could have significant negative impacts on global rice production. A similar situation exists in Sri Lanka as well.

Water-saving irrigation technology in lowland rice cultivation is based on reducing the time period during which standing water is maintained and on reducing the downward hydrostatic pressure of standing water, thereby aiming to reduce evaporation and percolation (i.e. deep drainage) losses (Bouman et al. 2007). Three water-saving irrigation methods have been tested as alternatives to the current practice of maintaining standing continuously. These are (a) saturated soil culture (SSC), (b) alternative wetting and drying (AWD) and (c) aerobic rice culture. In SSC, the soil is maintained as close to saturation as possible without having standing water so that percolation and evaporation losses are minimized. Soil saturation is maintained by frequent shallow irrigations. In a compilation of 31 experiments, Bouman and Tuong (2001) estimated water savings ranging from 5% to 50% with an average of 23% saving relative to continuous standing water. Even though some yield reductions ranging from 0% to 12% (average 6%) were observed, the improvement of water productivity (WP) (i.e. amount of grain produced per unit of water input) meant that more grain could be produced with less water. However, SSC requires a high degree of control in the irrigation operation at the field level to enable frequent shallow irrigations (Tuong and Bouman 2003). This capability may not be present in most developing countries in Asia, including Sri Lanka, where lowland rice cultivation is predominantly practised. Furthermore, SSC is labour-intensive so that it may not be possible in situations where labour is scarce and expensive such as in Sri Lanka.

Alternative wetting and drying allows the soil water to decrease below saturation down to a certain level by withholding irrigation for several days after the disappearance of standing water from the previous irrigation. Irrigations, which are scheduled at longer time intervals than in SSC, are done to obtain a standing water depth of 2–5 cm. Substantial savings of irrigation water have been achieved in AWD via reductions in percolation, lateral seepage and evaporation (Cabangon et al. 2004; Belder et al. 2007). However, yield reductions have accompanied reductions in water input. Bouman and Tuong (2001) reported that yield reductions ranging from 0% to 70% were observed in 92% of their 31 experiments. However, they also reported that proportional yield reductions relative to maintaining continuous standing water were always less than proportional reductions in irrigation water input, thus increasing WP. The magnitude of increase in WP with AWD in rice varies with the number of days between successive irrigations and the soil and hydrological conditions of the growing location/region. For example, on heavy clay soils with shallow water tables, the magnitudes of water saving and increase in WP are higher than in lighter soils with deeper water tables.

The Rice Research and Development Institute (RRDI) of Sri Lanka has recommended AWD to farmers and has provided practical guidelines on how to determine the timing of irrigation. Because of the high sensitivity of the reproductive development of rice to water stress, the reproductive phase (i.e. from panicle initiation to heading) is excluded from AWD, with either standing water or SSC recommended during this phase. AWD in rice cultivation is widely practised in China and has been recommended for North-West India indicating that it is a mature technology which is gradually gaining farmer acceptance.

Despite its capability to save irrigation water and increase WP, AWD is not without drawbacks. Firstly, reductions in absolute yield, in spite of the increase in WP, could have a negative impact on total rice production at the national scale, thus reducing food security. Secondly, periods without standing water, especially during the vegetative phase of the rice crop, could promote incidence of weeds, which in turn could reduce yields. Therefore, AWD has to be combined with effective weed control, which will have to be predominantly via selective herbicides as the standard weed control method of maintaining continuous standing water is absent in AWD. Thirdly, periods without standing water could promote the emission of nitrous oxide, a potent greenhouse gas contributing to global warming and climate change. Nitrous oxide emissions could be higher in rice crops receiving high doses of nitrogen fertilizers. However, it should be noted that periods without standing water, while enhancing nitrous oxide emission, reduce emission of methane, another greenhouse gas, which has a higher global warming potential (GWP) than carbon dioxide, but a lower GWP than nitrous oxide.

Aerobic rice culture (ARC) involves growing lowland rice as an irrigated, upland crop enabling 30–50% reductions in water input while incurring 20–30% yield reductions in comparison to continuously flooded lowland rice (CFLR) culture and thereby achieving increases in WP (Bouman and Tuong 2001). Evaporation losses in

ARC are reported to be 50–75% lower than those in CFLR. Successful ARC requires development of varieties which combine the superior yielding ability and input responsiveness of newly improved lowland rice varieties with some of the typical characteristics of upland rice varieties such as the ability to grow under upland soil conditions and drought tolerance. Development of ARC for lowland rice environments is still at an early stage and requires overcoming some major challenges for its increased adoption (Tuong et al. 2005). These challenges include (1) breeding of suitable varieties, (2) development of effective weed control strategies, (3) measures to counter increased emission of nitrous oxide and (4) arresting the yield decline that has been observed after 3–4 years.

Germplasm screening to identify genotypes that have superior yield performance under different SSC and ARC was conducted at the RRDI, Batalagoda, Sri Lanka, during four consecutive seasons using germplasm from drought screening nurseries (DSN) and the International Network for Genetic Evaluation of Rice (INGER) of the International Rice Research Institute along with cultivated rice varieties of Sri Lanka. Genotype performance under different water management methods showed variation across seasons, indicating the presence of significant genotype x environment interaction. However, a few genotypes which showed consistently superior performance under SSC and ARC have been identified.

7.2.8 Water-Saving Irrigation Methods in Upland Crops

The current strategy of predominantly surface-based irrigation with little control over the amount and distribution of water applied would not be sustainable in a future climate of limited water availability for irrigation. Hence, irrigation technologies for upland crops have to move from surface irrigation to more water-efficient micro-irrigation technologies such as drip irrigation. During the last two decades in Sri Lanka, adoption of drip irrigation has gradually spread to large-scale commercial plantations of perennial crops such as fruits (e.g. mango) and to commercial vegetable production in protected culture facilities. However, adoption of drip irrigation by small-scale farmers, especially in the dry and intermediate zones, has been slow because of the high initial capital cost and certain field problems such as clogging of drip nozzles due to hardness of water.

Improved precision in land preparation by using laser-guided land levelling could improve uniformity and efficiency of surface irrigation. Similarly, irrigation scheduling based on real-time soil moisture measurements has the potential to provide the exact water requirement of a crop at a given time. However, such advanced technology is beyond the reach of subsistence upland farmers who form the majority of the upland farming community in Sri Lanka. On the other hand, there has been a slow increase of larger upland farms mainly focused on cultivating high-value fruit crops (e.g. mango and banana) in the sub-humid dry and intermediate zones. These farms and their growers represent a platform to promote advanced irrigation technology because of their greater capacity for investment. If successful adoption could be demonstrated in these larger farms, there is a possibility that at least a section of the smaller farming community would follow suit with adequate institutional support via financial incentives, extension and training.

7.2.9 Paradigm Shifts in Irrigation Strategy of Upland Crops

Deficit Irrigation (DI)

In addition to adoption of more efficient delivery methods, a paradigm shift is needed in determining the amount of water to be applied through irrigation, especially in a future climate of water scarcity. At present, determination of the irrigation requirement is based on bringing the soil up to field capacity by replacing the amount of water that is lost during the period from the last irrigation, with adjustments made to account for any natural water inputs via precipitation. However, in many parts of the world including Sri Lanka, there may not be enough water to provide this full requirement in the future, especially because of the projected water scarcity. Furthermore, for some crops at least, it has been shown that there is no need to provide the full irrigation requirement (Fereres and Soriano 2007). This offers the possibility of reducing the water input to certain crops without reducing the yield (Morison et al. 2008). This strategy is called 'deficit irrigation (DI)', which represents a paradigm shift for Sri Lankan farmers who are keen on flooding the soil with scant regard for large volumes of water required and substantial losses incurred before (i.e. losses during conveyance from source to field including evaporation) and after application (i.e. percolation, run-off, seepage and evaporation losses).

The concept of DI as shown in Fig. 6 seeks to apply water only up to the point (I_{def}) at which water productivity (WP) begins to decline. This could represent a substantial saving of water when compared with irrigation up to full saturation and beyond (I_{max}), which is currently the common practice among Sri Lankan farmers/growers. It can be noted from Fig. 6 that DI incurs a certain amount of yield loss. However, the yield obtained is produced at a higher WP, thus making the cropping system more climate-resilient, especially in a future climate of water scarcity. Furthermore, with the saved water, an additional area which is currently rainfed can be brought under DI. As the yields under DI are highly likely to be higher than rainfed yields, yield gains from the additional area brought under DI could compensate for the possible yield losses incurred when moving from irrigation up to full saturation to DI.

In addition to the above, DI can have multiple advantages. Firstly, irrigation up to full saturation would often induce a shallow root system as there is no need for the crop to divert additional assimilates to develop a deeper root system when water is readily available in the top soil. In contrast, DI induces a deeper root system, which is a clear advantage. A deeper root system not only extracts water that may be present in the deeper layers of the soil profile but also recaptures part of the water that percolates downwards from the top soil. Secondly, irrigation up to full saturation keeps the turgor pressure of leaf cells at maximum, thus allowing maximum leaf expansion rates leading to development of a large canopy with a high transpiration rate. In contrast, DI restricts cellular expansion rates, even if the leaf cells have maximum turgor pressure, by the partial inhibitory action of abscisic acid (ABA), which is translocated from root cells to the leaf via the xylem stream when roots experience water deficits in the soil (Davies and Zhang 1991). The ABA signalling from roots to shoots during DI also keeps the stomata partially closed (Wilkinson

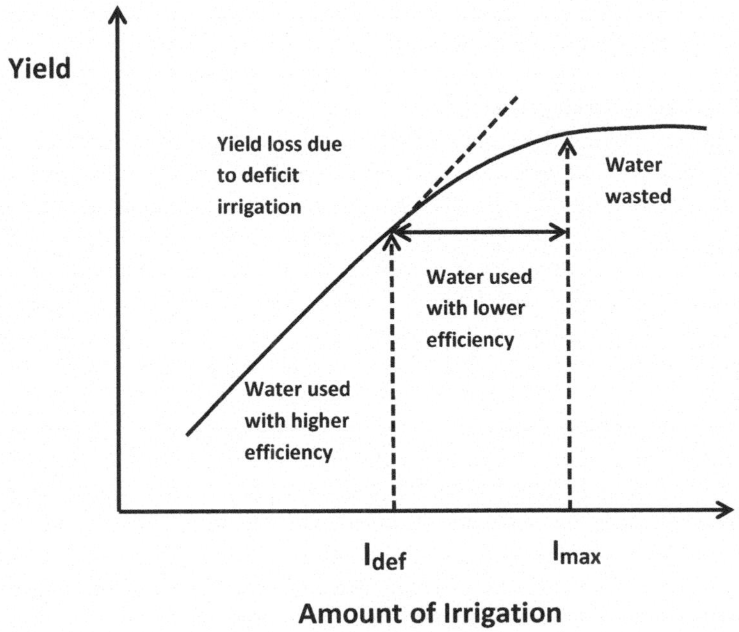

Fig. 6 Generalized relationships between applied irrigation water and crop grain yield. I_{def} indicates the point beyond which the productivity of irrigation water starts to decrease, and I_{max} indicates the point beyond which yield does not increase any further with additional water application. (Adapted from: Fereres and Soriano 2007)

and Davies 2002), which constitutes a third advantage of DI. The combination of restricted leaf expansion and stomatal opening decreases transpiration losses under DI in comparison to irrigation up to full saturation. This reduces the risk of water stress development within leaf tissue and allows regulated use of the limited pool of water in the soil. It has been shown that physiological water use efficiency (also called transpiration efficiency), defined as the amount of photosynthesis per unit of water transpired, is higher when stomata are partially closed (Jones 1976; Condon et al. 1990; Loomis and Connor 1992; De Costa 2004b). As stomata are partially closed in crops experiencing DI, photosynthesis occurs with a proportionately lower water loss, which could contribute to greater WP.

Deficit irrigation can take either of two forms: sustained deficit irrigation (SDI) practiced throughout the life cycle of a crop and, regulated deficit irrigation (RDI) practiced during selected periods of the crop's phenology, avoiding the critical periods (e.g. reproductive development phase) in which water deficits could have substantial negative impacts on yield. Therefore, while SDI would generally be more suitable for perennial crops and relatively drought-tolerant annual crops, RDI would be more appropriate for annual crops, especially those which are drought-susceptible. In addition to the nature of the crop, specific climatic and soil conditions in a given location/region determine which form of DI is most appropriate in a given situation. It is encouraging to note that increased water use efficiency has been reported by using RDI on maize in North China (Du et al. 2010).

Partial Root-Zone Drying (PRD)

Partial root-zone drying (PRD) is a novel strategy of irrigation where, at any given time, only one half of the root zone is irrigated, while the other half is kept un-irrigated (Kang and Zhang 2004). Irrigation of the two halves is alternated at 10–14-day intervals depending on the crop characteristics (e.g. extent of the root system), the prevailing atmospheric demand for water and the water holding capacity of the soil. By irrigating only half of the root system, the PRD achieves a 50% reduction of the irrigation water requirement. In comparison to the current practice of irrigation of the whole plant, PRD also represents a paradigm shift in irrigation strategy.

The rationale for PRD comes from a series of split-root experiments conducted by Davies and his co-workers (Blackman and Davies 1985; Zhang et al. 1987; Gowing et al. 1990; Davies and Zhang 1991). By growing a plant in a specially designed pot with two hydrologically separated compartments with the root system evenly distributed in both (i.e. a split-root experiment), Davies and his co-workers showed that keeping one compartment irrigated was sufficient to fulfil the entire water requirement of the plant and maintain maximum turgor pressure in its functional tissues (e.g. leaves). Interestingly, it was also shown that despite maintaining maximum turgor in its leaf tissue, gradual stomatal closure occurs when the un-irrigated compartment of the pot starts drying out gradually. It was shown that hormonal signals, specifically ABA, originating from roots of the un-irrigated compartment are responsible for inducing stomatal closure when they are translocated to the shoot via the xylem stream. These findings allowed the design of PRD which attempts to achieve a balance between reduced water input and maintenance of normal plant functioning to ensure minimum reduction in growth and yield (Davies et al. 2002). The hormonal signals induced from the un-irrigated half of the root system also exert a control on leaf expansion, thus restricting canopy growth and preventing the development of a plant/crop with a high water demand. Thus, when practised as an irrigation strategy, in crops grown under protected culture where compartmentation and separate irrigation of two halves of the root system are possible, PRD has the advantages of (1) regulating the water loss from a crop while ensuring an adequate water supply to essential physiological processes (e.g. photosynthesis, reproductive organ development, etc.); (2) reducing excessive vegetative growth; and (3) increasing physiological water use efficiency due to partial stomatal closure.

The advantages of PRD are partly dependent on the rate at which ABA is translocated in the xylem flow from root to shoot to induce partial stomatal closure and restrict leaf expansion (Stoll et al. 2000). It can be noted that xylem flow would be faster during periods of greater atmospheric demand for water, which occurs during warm, sunny periods with lower atmospheric relative humidity. These conditions are usually associated with periods of drought. Faster xylem flow could bring ABA to the shoots faster and induce greater stomatal closure, thus helping in conserving the limited water that would be available during a drought period (Davies et al. 2002). On the other hand, rate of xylem flow and root-to-shoot ABA transfer would be slower during periods of lower atmospheric demand characterized by cooler atmospheric conditions with higher relative humidity. Stomata respond directly to atmospheric humidity by increased opening when relative humidity is higher (Sheriff 1979; De Costa 2004b). Furthermore, slower ABA transfer would also allow greater stomatal opening during these periods, thus facilitating greater absorption of carbon dioxide and greater rates of photosynthesis. Accordingly, PRD achieves a balance between water conservation during stress periods and greater photosynthesis during milder periods. Therefore, PRD has considerable potential as a novel irrigation strategy to increase climate resilience of selected crops in a future climate of lower water availability. In North China, water use efficiency of maize (Du et al. 2010) and cotton (Du et al. 2006) has been increased with PRD practised by alternate furrow irrigation.

7.2.10 Home Gardens and Climate Resilience

Home gardens have been an integral part of the Sri Lankan society over its long history. Their structure and species diversity vary predominantly with the climate regime under which they have evolved (Hitinayake et al. 1996). Home gardens in the mid-elevation humid zone (mid-country wet zone, according to the local classification), which are also known as Kandyan Home Gardens, have high species diversity which includes both woody perennials and annuals, some of which are agricultural crops (Hitinayake et al. 1996; De Costa et al. 2006). The mixture of perennial and annual crops in Kandyan Home Gardens constitutes a complex multi-layered canopy structure, which is typical of multi-layered home gardens in the humid tropics elsewhere in the world (Kumar and Nair 2004, 2006). In contrast, home gardens in the lower-elevation, sub-humid dry zone of Sri Lanka are lower in plant density and species diversity and have lower vertical stratification (Pushpakumara et al. 2012).

Because of the presence of a limited amount of agricultural crops, home gardens have been touted as a possible means of ensuring food security in the face of climate change in Sri Lanka. However, a closer examination reveals that this claim is largely exaggerated and highly optimistic. At present, home gardens provide only a limited portion of the food supply of the majority of the Sri Lankan population, and that also is only on a seasonal basis. Furthermore, the scope for expanding home gardens is limited in areas where a majority of the Sri Lankan population lives. The presence of trees is expected to provide shade to the agricultural crops in a home garden and thereby reduce evapotranspiration and conserve soil moisture. De Costa et al. (2006) showed that trees in Kandyan Home Gardens have substantially high transpiration rates. Therefore, increasing productivity of home gardens of Sri Lanka in a warmer and drier climate in the future would be just as challenging as increasing productivity in large-scale agriculture in open fields.

7.2.11 Small Tanks, Cascade Systems and Climate Resilience

In Sri Lanka, crop production in the low-elevation sub-humid, dry and intermediate zones is supported by a large number of small and medium tanks, which are constructed according to a cascade system along elevation gradients (Udawattage 1985). These act as reservoirs of excess rainfall that is received predominantly from the North-East monsoon. As the South-West monsoon brings only limited rainfall to dry and intermediate zones, water stored in the tanks supports crop production in the *yala* season. As such, this network of tanks provides a buffer against rainfall variability and therefore contributes to climate resilience in the cropping systems.

In spite of their importance, a considerable number of small tanks and cascade systems have been destroyed during the establishment of large-scale, multipurpose irrigation schemes. Many of the remaining tanks have been neglected and reduced in capacity due to siltation (Dharmasena 1989), infestation of aquatic weeds and consequent eutrophication. Furthermore, increased frequency of high-intensity rainfall has caused increased soil erosion in almost all farming areas in Sri Lanka. This has had an adverse impact on tanks and reservoirs which have had to act as repositories of eroded soil particles leading to siltation and loss of capacity. Therefore, rehabilitation and restoration of the small and medium tanks need to be undertaken as an urgent priority to increase climate resilience of cropping systems in the dry and intermediate zones.

7.2.12 Management of Flooding and Water Logging in Crop Lands

Excess water in uplands during certain periods of the year is a regular occurrence even in the present climate. These periods include (a) the onset of *maha* season in low-elevation, sub-humid dry and intermediate zones where a high proportion of the annual precipitation is received within a short period of the Second Inter-monsoon and the North-East monsoon and (b) first half of the *yala* season in the low-elevation humid zone (i.e. 'wet zone') where substantial precipitation is received from the South-West monsoon. With increased frequency of extreme precipitation events in a future climate, incidence and duration of flooding and water logging in crop lands and consequent crop losses could increase. Therefore, management of water logging constitutes an important component of climate resilience in cropping systems in Sri Lanka.

Only limited options are available for combating the adverse impacts of water logging on crops. Development of flood-tolerant varieties in rice and varieties tolerant to limited periods of water logging in upland crops is the most promising option. Considerable progress has been made in this regard in rice where a flood-tolerant variety Bg455 has been released recently for the flood-prone regions in the wet zone. Molecular biological work to incorporate the quantitative trait locus (QTL) Submergence 1 (known as *Sub1*) (Perata and Voesenek 2007), which confers flood tolerance in rice, to Sri Lankan rice germplasm is in progress at the Rice Research and Development Institute, Batalagoda. Flood tolerance in rice is principally conferred by two mechanisms, i.e. tolerance of complete submergence and culm elongation with increasing depth of flood water to keep the foliage canopy above the water level (Sripongpangkul et al. 2000). The *Sub1* locus, which is located on chromosome 9 of rice, is responsible for submergence tolerance via an ethylene response factor (ERF), whose differential expression controls genotypic variation in flood tolerance (Fukao et al. 2006; Xu et al. 2006). However, apart from the work on rice, no research has been done in Sri Lanka to develop flood-tolerant varieties of other field crops or vegetables.

7.2.13 Crop Insurance Schemes

With increased climate variability and greater frequency of extreme climatic events, the importance of crop insurance schemes comes into focus as a means of protecting the farmer livelihoods, especially of the low-income subsistence farmers. In Sri Lanka, a government-funded indemnity-based crop insurance scheme is in place for rice and five other field crops (i.e. potato, big onion, chilli, maize and soybean). Farmers are paid up to Sri Lankan rupees 10,000.00 per acre for crop losses due to adverse climatic conditions such as droughts and floods. This insurance scheme is funded from a 1% levy on the profits of banks, financial institutions and insurance companies. However, farmer participation in crop insurance schemes was low in Sri Lanka until it was made mandatory for obtaining the government-sponsored fertilizer subsidy. Lack of awareness and trust on insurance schemes and reluctance to follow required procedures have been principal causes of low voluntary adoption of crop insurance schemes.

Crop insurance schemes based on climatic indices have been proposed as an alternative to schemes which compensate incurred crop losses (Biswas et al. 2009). Here, insured farmers in a defined area are compensated for their potential crop losses when predefined climatic thresholds are exceeded in a given season. Such index-based crop insurance schemes eliminate the need for crop loss assessment at the individual farmer level while removing 'insurance cheating' and lowered motivation on the part of the farmers to initiate adaptation measures to overcome climate-related crop losses. However, crop insurance linked to climate indices requires an adequately extensive network of climate monitoring stations covering the entire cropping area along with established relationships between climatic parameters and crop yields. None of the above two requirements are fulfilled currently in Sri Lanka. Therefore, it is likely that crop insurance schemes in Sri Lanka will continue to be indemnity-based for the foreseeable future.

7.2.14 A Holistic Approach to Management of Cropping Systems in a Future Climate

The foregoing discussion identified several options to increase climate resilience of crops and cropping systems in Sri Lanka along two broad parallel pathways, i.e. the 'soft science' option of initiating action via policies, strategies and action plans and the 'hard science' option of initiating technological advances via research. Technological advances could proceed along the parallel pathways of developing heat-tolerant crop varieties and modifying crop management. While each of these options has merit in their own right, their co-ordinated implementation with a holistic view of the entire farming system represents the best strategy of increasing climate resilience of cropping systems in Sri Lanka. Because of the diversity of climates and farming systems (Sect. 4) in different climatic zones of Sri Lanka (Sects. 2 and 3), a single model would not be universally applicable. Different holistic models which take into account the specific cropping systems, the natural landscape, its climate, the soil and the resource base along with socioeconomic and cultural factors have to be formulated to fit specific farming systems and natural environments in different agro-climatic zones.

Work on 'conservation farming' carried out by the DoA during the 1980s and 1990s represents an example of such a holistic approach to increase productivity and sustainability of small reservoir watersheds in the low-elevation, sub-humid dry zone of Sri Lanka. Conservation farming focused on the large number of minor watersheds in the dry zone which supported a farming system consisting of lowland rice and upland field crops via a small reservoir (called the 'village tank'), which collected its catchment run-off during the major rainfall season from the Second Inter-monsoon (SIM) and the North-East monsoon (NEM) in the *maha* season. Traditionally, shifting cultivation had been practised in the uplands for over 2000 years. However, with increasing populations and the consequently shorter rotation times and fallow periods, soil fertility and land productivity had declined. Furthermore, high soil erosion due to shifting cultivation had reduced the storage capacity of the village tanks due to siltation. Climate change could exacerbate many of the problems that the conservation farming approach was trying to address at a time when climate change had not yet become a topical issue. For example, increased frequency of high-intensity and highly erosive rainfall could accelerate soil erosion and siltation of small reservoirs. Increased air and soil temperatures increase decomposition rate of soil organic matter (SOM), which in combination with greater soil erosion could accelerate loss of soil fertility and land productivity. Therefore, agronomic and soil management practices that were promoted under conservation farming could have even greater relevance in the present times and the future when climate change is certain to impose substantial negative impacts on processes influencing soil fertility and land productivity in farming systems of small reservoir watersheds in the dry zone of Sri Lanka.

The conservation farming approach consisted of introducing practices (a) to reduce soil erosion, especially during fallow periods between cropping seasons; (b) to increase infiltration and soil water storage in the uplands so that the probability of upland crops experiencing water stress is minimized; (c) to increase catchment run-off during the high-precipitation period of the SIM and NEM so that water storage in the reservoir is maximized; and (d) to reduce the fallow period of uplands that had been used for shifting cultivation by incorporating practices which help regenerate their soil fertility. It can be noted that a delicate balance has to be achieved between increasing infiltration and soil water storage in uplands (which entails reduced run-off to the reservoir) and increasing catchment run-off to maximize water storage in the reservoir. In order to achieve the above goals, the following practices were evaluated in conservation farming: (1) zero tillage to reduce soil disturbance during crop establishment and thereby reduce soil erosion; (2) mulching with plant material to increase SOM content and thereby improve its water holding capacity and associated physical properties such as aggregate stability; and (3) alley cropping (discussed earlier in Sect. 7.2.1) to obtain a range of benefits such as reduced soil erosion and soil evaporation, improved SOM and greater overall land productivity.

Conservation farming experiments conducted in the dry zone demonstrated substantial benefits of mulching and alley cropping, while zero tillage appears to have received less attention. For example, maintaining a permanent live mulch as 0.75-m-wide strips spaced at 5 m apart along contours (i.e. 'strip-mulch farming') provided 80% protection against soil erosion in comparison to shifting cultivation, while alley cropping with *Gliricidia sepium* hedgerows at 4–5 m spacing provided 86% protection (Dharmasena 1994). Rice straw mulch applied at 4 t ha^{-1} increased cowpea yields by 36% (Weerakoon and Seneviratne 1984) and reduced soil loss and run-off in a maize field by 80% and 53%, respectively (Dharmasena 1992), in comparison to unmulched fields. Alley cropping increased maize and sesame yields by 168% and 87% above their respective mono-crops (Handawala and Kendaragama 1991). More importantly, Dharmasena (1994) demonstrated that strip-mulch farming and alley farming have specific rainfall-run-off relationships which are able to achieve the required hydrological balance by maximizing infiltration and soil water storage while also allowing greater catchment run-off during high rainfall periods. Climate change induces changes in precipitation patterns and in soil processes influencing soil fertility. Careful management of this hydrological balance becomes critically important to ensure sustainability and climate resilience of cropping/farming systems in small reservoir catchments in the dry zone of Sri Lanka. Therefore, revisiting the findings of conservation farming experiments carried out in the 1980s and 1990s could yield important guidelines for holistic approaches of soil, land and water management to ensure climate resilience of cropping systems of Sri Lanka in a future climate.

8 Constraints to Implementation of Measures to Increase Climate Resilience of Crops and Cropping Systems in Sri Lanka

It was mentioned earlier in this chapter that implementation of policy initiatives to increase climate resilience in Sri Lanka has been weak at best. Implementation of the different technological advances described above is also constrained by several factors, some of which have been described already in relation to Sri Lanka's current institutional and human resource capacity to develop heat-tolerant crop varieties. Many of these constraints apply equally in the development of crop management solutions to increase climate resilience through research and their subsequent introduction to the farmers via extension and technology transfer (ETT). Some of these constraints are described below.

8.1 Suboptimal Research Capacity and Inadequate Seed Production Capability

With a few exceptions, research at the crop research institutes of the public sector, especially those of the Department of Agriculture (DoA), proceeds at suboptimal levels because of severe shortages of manpower with expertise and an acute shortage of state-of-the-art research facilities. Even if the plant breeders were to be successful in developing relatively heat- and/or drought-tolerant crop varieties in the next few years, it would take several more years for the seeds of such varieties to reach the farmers. This is because seed production of new crop varieties that are coming through breeding programmes (towards which a major portion of institutional resources are channelled) is a major 'bottleneck' in Sri Lanka. The seed production capacity of the state sector seed farms has been scaled down substantially during the last two to three decades. Therefore, the DoA, the major organization which carries out development of new varieties, has to depend on the private sector organizations to multiply, produce and distribute adequate amounts of seed of the new varieties to the farmers.

8.2 Inadequacies in the Extension and Technology Transfer (ETT)

Farmers are often unaware of the new crop varieties that are being released for cultivation because of inadequacies in the ETT system of the DoA and the Provincial Directorates of Agriculture. This creates a vicious cycle, especially in a situation where the DoA has to depend on the private sector for adequate seed production. Lack of awareness and inadequate ETT do not create an adequate market demand for a particular variety for the private sector to produce adequate quantities of seed. Unavailability of seeds, in turn, prevents the variety being adopted by the farmers.

Farmer adoption of the crop management options to increase climate resilience is also constrained by weaknesses of the ETT system. There are several inherent weaknesses in the current ETT system in Sri Lanka. First and foremost, it is disjointed and lacks co-ordination because ETT is handled by several organizations within the state sector. In contrast to research which is under the purview of the Department of Agriculture, the responsibility of ETT is devolved to Provincial Councils, with their respective Provincial Departments of Agriculture as the responsible authority. There is only limited collaboration and co-ordination between the respective organizations of the Central Government and the Provincial Councils. While the respective research institutes of the DoA also have a few extension workers, they are severely inadequate to carry out effective ETT at the national scale or even at the regional scale. Similarly, the respective Provincial Departments of Agriculture are severely under-staffed in terms of trained and experienced ETT personnel. Therefore, the entire ETT system requires a major overhaul, both at the national and provincial levels. Recent initiatives of cyber extension through collaboration between the DoA and local mobile IT service providers have shown promise and indicate a pathway for ETT in the future.

9 Concluding Remarks

The foregoing description of the range of options available for increasing climate resilience in the crops and cropping systems of Sri Lanka and how they have progressed during the last two to three decades brings out the following key messages for future action: (a) Efforts to develop heat- and drought-tolerant crop varieties require substantial acceleration and upgrading with more precise screening methodology while harnessing the recent advances in molecular biology including the 'omics' approaches; (b) importance of crop management options to increase climate resilience has to be recognized and accorded the same level of priority as varietal improvement to be pursued through research and development; (c) a substantial investment is required to improve research capability of research institutes and universities via infrastructure and human resource development; (d) a major re-structuring of the extension and technology transfer network is needed for effective transfer of the technological advances generated through research to the farmers; and (e) capacity of seeds and planting material production in the state sector needs to be upgraded urgently to accelerate the transfer of seeds of improved crop varieties to farmers.

References

Abeysiriwardena DSDZ, Ohba K, Maruyama A (2002) Influence of temperature and relative humidity on grain sterility in rice. J Natl Sci Found Sri Lanka 30:33–41. https://doi.org/10.4038/jnsfsr.v30i1-2.2559

Abhayapala KMRD (2017) Assessment of climate resilience of tomato, chilli and potato in different agroecological zones of Sri Lanka and increasing their adaptation capacity to long-term climate change. PhD thesis, Postgraduate Institute of Agriculture, University of Peradeniya

Abhayapala KMRD, De Costa WAJM, Fonseka RM, Prasannath K (2014) Response of potato (Solanum tuberosum L.) to increasing growing season temperature under different soil management and crop protection regimes in the upcountry of Sri Lanka. Trop Agric Res 25:555–569. https://doi.org/10.4038/tar.v25i4.8061

Ahmad P, Abdel Latef AAH, Rasool S, Akram NA, Ashraf M, Gucel S (2016) Role of proteomics in crop stress tolerance. Front Plant Sci 7:1–14. https://doi.org/10.3389/fpls.2016.01336

Alexander LV, Zhang X, Peterson TC, Caesar J, Gleason B, Klein Tank AMG, Haylock M, Collins D, Trewin B, Rahimzadeh F, Tagipour A, Rupa Kumar K, Revadekar J, Griffiths G, Vincent L, Stephenson DB, Burn J, Aguilar E, Brunet M, Taylor M, New M, Zhai P, Rusticucci M, Vazquez-Aguirre JL (2006) Global observed changes in daily climate extremes of temperature and precipitation. J Geophys Res Atmos 111:1–22. https://doi.org/10.1029/2005JD006290

Allan RP, Soden BJ, John VO, Ingram W, Good P (2010) Current changes in tropical precipitation. Environ Res Lett 5:025205. https://doi.org/10.1088/1748-9326/5/2/025205

Anonymous (2010) National climate change adaptation strategy for Sri Lanka 2011 to 2016. Climate Change Secretariat, Ministry of Environment, Colombo

Anonymous (2012) The National Climate Change Policy of Sri Lanka. Sri Lanka

Anonymous (2016) National adaptation plan for climate change impacts in Sri Lanka 2016–2025. Climate Change Secretariat, Ministry of Mahaweli Development and Environment, Colombo

Anonymous (2018) Climate of Sri Lanka. In: Department of meteorology Sri Lanka. http://www.meteo.gov.lk/index.php?option=com_content&view=article&id=94&Itemid=310&lang=en. Accessed 18 July 2018

Beddington JR, Asaduzzaman M, Fernandez A, Clark ME, Guillou M, Jahn MM, Erda L, Mamo T, Bo N, Nobre CA, Scholes RJ, Sharma R, Wakhungu J (2012) Achieving food security in the face of climate change: final report from the Commission on Sustainable Agriculture and Climate Change. CGIAR research program on climate change. Agriculture and Food Security, Copenhagen

Belder P, Bouman BAM, Spiertz JHJ (2007) Exploring options for water savings in lowland rice using a modelling approach. Agric Syst 92:91–114. https://doi.org/10.1016/j.agsy.2006.03.001

Bindoff NL, Stott PA, KM AR, Allen MR, Gillett N, Gutzler D, Hansingo K, Hegerl G, Hu Y, Jain S, Mokhov II, Overland J, Perlwitz J, Sebbari R, Zhang X (2013) Detection and Attribution of Climate Change: from Global to Regional. In: Stocker TF, Qin D, Plattner G-K, Tignor M, Allen SK, Boschung J, Nauels A, Xia Y, Bex V, Midgley PM (eds) Climate change 2013: the physical science basis. Contribution of working Group I to the fifth assessment report of the Intergovernmental Panel on Climate Change. Cambridge University Press, New York, pp 867–952

Biswas B, Dhaliwal LK, Singh S, Chahal SK (2009) Weather based crop insurance in India: Present status and future possibilities. J Agrometeorol 11:238–241

Bita CE, Gerats T (2013) Plant tolerance to high temperature in a changing environment: scientific fundamentals and production of heat stress-tolerant crops. Front Plant Sci 4:1–18. https://doi.org/10.3389/fpls.2013.00273

Blackman PG, Davies WJ (1985) Root to shoot communication in maize plants of the effects of soil drying. J Exp Bot 36:39–48. https://doi.org/10.1093/jxb/36.1.39

Bouman BAM (2001) Water-efficient management strategies in rice production. Int Rice Res Notes 16:17–22

Bouman BAM, Tuong TP (2001) Field water management to save water and increase its productivity in irrigated lowland rice. Agric Water Manag 49:11–30. https://doi.org/10.1016/S0378-3774(00)00128-1

Bouman BAM, Humphreys E, Tuong TP, Barker R (2007) Rice and water. Adv Agron 92:187–237. https://doi.org/10.1016/S0065-2113(04)92004-4

Brenner AJ (1996) Microclimatic modifications in agroforestry. In: Ong CK, Huxley P (eds) Tree-crop interactions: a physiological approach. CABI/ICRAF, Oxford, pp 159–187

Brown KW, Rosenberg NJ (1971) Shelter-effects on microclimate, growth and water use by irrigated sugar beets in the Great Plains. Agric Meteorol 9:241–263. https://doi.org/10.1016/0002-1571(71)90025-2

Cabangon RJ, Tuong TP, Castillo EG, Bao LX, Lu G, Wang G, Cui Y, Bouman BAM, Li Y, Chen C, Wang J (2004) Effect of irrigation method and N-fertilizer management on rice yield, water productivity and nutrient-use efficiencies in typical lowland rice conditions in China. Paddy Water Environ 2:195–206. https://doi.org/10.1007/s10333-004-0062-3

Cannell MGR, Van Noordwijk M, Ong CK (1996) The central agroforestry hypothesis: The trees must acquire resources that the crop would not otherwise acquire. Agrofor Syst 34:27–31. https://doi.org/10.1007/BF00129630

Carpenter S, Walker B, Anderies JM, Abel N (2011) From metaphor to measurement: resilience of what to what? Ecosystems 4:765–781. https://doi.org/10.1007/si0021-001-0045-9

Carr MKV (1985) Some effects of shelter on the yield and water use of tea. In: Grace J (ed) Effects of shelter on the physiology of plants and animals. Swets and Zeitlinger B.V, Lisse, pp 127–144

Chou C, Lan CW (2012) Changes in the annual range of precipitation under global warming. J Clim 25:222–235. https://doi.org/10.1175/JCLI-D-11-00097.1

Chou C, Chiang JCH, Lan C-W, Chung C-H, Liao Y-C, Lee C-J (2013) Increase in the range between wet and dry season precipitation. Nat Geosci 6:263–267. https://doi.org/10.1038/ngeo1744

Collins M, Knutti R, Arblaster J, Dufresne J-L, Fichefet T, Friedlingstein P, Gao X, Gutowski W, Johns T, Krinner G, Shongwe M, Tebaldi C, Weaver A, Wehner M (2013) Long-term climate change: projections, commitments and irreversibility. In: Stocker TF, Qin D, Plattner G-K, Tignor M, Allen SK, Boschung J, Nauels A, Xia Y, Bex V, Midgley PM (eds) Climate Change 2013: the physical science basis. Contributions of the Working Group 1 to the fifth assessment of report of the Intergovernmental Panel on Climate Change. Cambridge University Press, New York, pp 1029–1136

Condon AG, Farquhar GD, Richards RA (1990) Genotypic variation in carbon isotope discrimination and transpiration efficiency in wheat. Leaf gas exchange and whole plant studies. Aust J Plant Physiol 17:9–22. https://doi.org/10.1071/PP9900009

Cossani CM, Reynolds MP (2012) Physiological traits for improving heat tolerance in wheat. Plant Physiol 160:1710–1718. https://doi.org/10.1104/pp.112.207753

Dai A (2011) Drought under global warming: A review. Wiley Interdiscip Rev Clim Chang 2:45–65. https://doi.org/10.1002/wcc.81

Dai A, Trenberth KE, Qian T (2004) A global dataset of Palmer Drought Severity Index for 1870–2002: relationship with soil moisture and effects of surface warming. J Hydrometeorol 5:1117–1130. https://doi.org/10.1175/JHM-386.1

Das L, Akhter J, Dutta M, Meher JK (2015) Ensemble-based CMIP5 simulations of monsoon rainfall and temperature changes over South Asia. National Institute for Agro-Environmental Sciences (NIAES), Tsukuba, pp 41–60

Davies WJ, Zhang J (1991) Root signals and the regulation of growth and development of plants in drying soil. Annu Rev Plant Physiol Plant Mol Biol 42:55–76. https://doi.org/10.1146/annurev.pp.42.060191.000415

Davies WJ, Wilkinson S, Loveys B (2002) Stomatal control by chemical signalling and the exploitation of this mechanism to increase water use efficiency in agriculture. New Phytol 153:449–460. https://doi.org/10.1046/j.0028-646X.2001.00345.x

De Costa WAJM (2004a) Principles of crop physiology: towards an understanding of crop yield determination and improvement, 2nd edn. University of Peradeniya, Peradeniya

De Costa WAJM (2004b) Plant water relations: principles and applications, 2nd edn. University of Peradeniya, Peradeniya

De Costa WAJM (2008) Climate change in Sri Lanka: myth or reality? Evidence from long-term meteorological data. J Natl Sci Found Sri Lanka 36:63–88. https://doi.org/10.4038/jnsfsr.v36i0.8048

De Costa WAJM (2009) Impacts of climate change on the long-term variation in the water balance of different rainfall seasons in selected locations of Sri Lanka. In: Nissanka S, Sangakkara U (eds) Global climate change and its impacts on agriculture, forestry and water in the tropics. CDM Study Centre, University of Peradeniya, Peradeniya, pp 335–350

De Costa WAJM (2010a) Adaptation of agricultural crop production to climate change: a policy framework for Sri Lanka. J Natl Sci Found Sri Lanka 38:79–89. https://doi.org/10.4038/jnsfsr.v38i2.2032

De Costa WAJM (2010b) Investing in research. J Natl Sci Found Sri Lanka 38:77–78. https://doi.org/10.4038/jnsfsr.v38i2.2031

De Costa WAJM (2012) Climate change research in Sri Lanka – are we investing enough? J Natl Sci Found Sri Lanka 40:281–282. https://doi.org/10.4038/jnsfsr.v40i4.5041

De Costa WAJM, Chandrapala AG (2000a) Competition between six hedgerow tree species and mung bean (vigna radiata (L.) wilczek) in the mid-country intermediate zone. J Natl Sci Found Sri Lanka 28:113–125. https://doi.org/10.4038/jnsfsr.v28i2.2680

De Costa WAJM, Chandrapala AG (2000b) Effects of different tree species on growth and yield of mung bean (Vigna radiata (L.) Wilczek) grown in hedgerow intercropping systems in Sri Lanka. J Agron Crop Sci 184:43–48. https://doi.org/10.1046/j.1439-037X.2000.00363.x

De Costa WAJM, Surenthran P (2005a) Resource competition in contour hedgerow intercropping systems involving different shrub species with mature and young tea on sloping highlands in Sri Lanka. J Agric Sci 143:395–405. https://doi.org/10.1017/S0021859605005563

De Costa WAJM, Surenthran P (2005b) Tree-crop interactions in hedgerow intercropping with different tree species and tea in Sri Lanka: 1. Production and resource competition. Agrofor Syst 63:199–209. https://doi.org/10.1007/s10457-005-1090-8

De Costa W, Amaratunga KSP, Udumullage RS (2006) Transpiration characteristics of some homegarden tree species in central Sri Lanka. In: Kumar BM, Nair PKR (eds) Tropical homegardens: a time-tested example of sustainable agroforestry. Springer, Dordrecht, pp 251–267

Dharmasena PB (1989) Optimum utilization of the storage in village tanks. Trop Agric 145:1–11

Dharmasena PB (1992) Soil erosion control measures for rainfed farming in the dry zone of Sri Lanka. PhD thesis, University of Peradeniya

Dharmasena P (1994) Conservation farming practices for small reservoir watersheds: a case study from Sri Lanka. Agrofor Syst 28:203–212

Donald CM (1968) The breeding of crop ideotypes. Euphytica 17:385–403. https://doi.org/10.1007/BF00056241

Donat MG, Alexander LV (2012) The shifting probability distribution of global daytime and night-time temperatures. Geophys Res Lett 39:1–5. https://doi.org/10.1029/2012GL052459

Dorji S, Herath S, Mishra B (2017) Future climate of Colombo downscaled with SDSM-Neural Network. Climate 5:24. https://doi.org/10.3390/cli5010024

Driedonks N, Rieu I, Vriezen WH (2016) Breeding for plant heat tolerance at vegetative and reproductive stages. Plant Reprod 29:67–79. https://doi.org/10.1007/s00497-016-0275-9

Du T, Kang S, Zhang J, Li F, Hu X (2006) Yield and physiological responses of cotton to partial root-zone irrigation in the oasis field of northwest China. Agric Water Manag 84:41–52. https://doi.org/10.1016/j.agwat.2006.01.010

Du T, Kang S, Sun J, Zhang X, Zhang J (2010) An improved water use efficiency of cereals under temporal and spatial deficit irrigation in north China. Agric Water Manag 97:66–74. https://doi.org/10.1016/j.agwat.2009.08.011

Easterling D, Evans J, Groisman PY, Karl T, Kunkel K, Ambenje P (2000) Observed variability and trends in extreme climate events: a brief review. Bull Am Meteorol Soc 81:417–425. https://doi.org/10.1175/1520-0477(2000)081<0417:OVATIE>2.3.CO;2

Eeswaran R (2018) Evaluation of a climate change-resilient agronomic package for selected upland annual crops under the farmer field conditions of the Northern Province of Sri Lanka. MPhil thesis, Postgraduate Institute of Agriculture, University of Peradeniya

Ekanayake PB (1994) Application of Sloping Agricultural Land Technology (SALT) in tea plantations. Tea Bull 14:3–17

Eriyagama N, Smakhtin V, Chandrapala L, Fernando K (2010) Impacts of climate change on water resources and agriculture in Sri Lanka: a review and preliminary vulnerability mapping. International Water Management Institute, Colombo

Fereres E, Soriano MA (2007) Deficit irrigation for reducing agricultural water use. J Exp Bot 58:147–159. https://doi.org/10.1093/jxb/erl165

Francis PA, Atta-Krah NA (1989) Sociological and ecological factors in technology adoption: fodder trees in southeast Nigeria. Exp Agric 25:1–10. https://doi.org/10.1017/S0014479700016380

Fukao T, Xu K, Ronald PC, Bailey-Serres J (2006) A variable cluster of ethylene response factor–like genes regulates metabolic and developmental acclimation responses to submergence in rice. Plant Cell 18:2021–2034. https://doi.org/10.1105/tpc.106.043000

Gowing DJG, Davies WJ, Jones HG (1990) A positive root-sourced signal as an indicator of soil drying in apple, Malus x domestica Borkh. J Exp Bot 41:1535–1540. https://doi.org/10.1093/jxb/41.12.1535

Groisman P, Knight R, Easterling D, Karl TR, Hegerl GC, Razuvaev VN (2005) Trends in precipitation intensity in the climate record. J Clim 18:1326–1350. https://doi.org/10.1175/JCLI3339.1

Handawala J, Kendaragama KMA (1991) Effect of Gliricidia sepium on upland soil fertility in the dry zone of Sri Lanka. In: Gunasena HPM (ed) Proceedings of the second regional workshop on multipurpose trees. University of Peradeniya, Kandy, pp 50–59

Hansen J, Sato M, Ruedy R (2012) Perception of climate change. Proc Natl Acad Sci 109:E2415–E2423. https://doi.org/10.1073/pnas.1205276109

Hartmann D, Klein Tank A, Rusticucci M, Alexander L, Brönnimann S, Charabi Y, Dentener F, Dlugokencky E, Easterling D, Kaplan A, Soden B, Thorne P, Wild M, Zhai P (2013) Observations: atmosphere and surface. In: Stocker TF, Qin D, Plattner G-K, Tignor M, Allen SK, Boschung J, Nauels A, Xia Y, Bex V, Midgley PM (eds) Climate Change 2013: the physical science basis. Contribution of Working Group I to the fifth assessment report of the Intergovernmental Panel on Climate Change. Cambridge University Press, New York, pp 159–254

Hasanuzzaman M, Nahar K, Alam MM, Roychowdhury R, Fujita M (2013) Physiological, biochemical, and molecular mechanisms of heat stress tolerance in plants. Int J Mol Sci 14:9643–9684. https://doi.org/10.3390/ijms14059643

Hegerl G, Black E, Allan R, Ingram W, Polson D, Trenberth K, Chadwick R, Arkin P, Sarojini B, Becker A, Dai A, Durack P, Easterling D, Fowler H, Kendon E, Huffman G, Liu C, Marsh R, New M, Osborn T, Skliris N, Stott P, Vidale P-L, Wijffels S, Wilcox L, Willett K, Zhang X (2015) Challenges in quantifying changes in the global water cycle. Bull Am Meteorol Soc 96 (7):1097–1116. https://doi.org/10.1175/BAMS-D-13-00212.1

Hitinayake HMGSB, De Costa WAJM, Jayaweera KGD (1996) Food trees in multi-layered homegardens in different agro-ecological regions of Kandy district. In: Gunasena HPM (ed) Multipurpose trees for food security, proceedings of the 7th regional workshop on multipurpose tree species. University of Peradeniya, Kandy, pp 252–264

IPCC (2013) Summary for policymakers. In: Stocker TF, Qin D, Plattner G-K, Tignor M, Allen SK, Boschung J, Nauels A, Xia Y, Bex V, Midgley PM (eds) Climate Change 2013: The physical science basis. Contribution of Working Group I to the fifth assessment report of the Intergovernmental Panel on Climate Change. Cambridge University Press, New York

Jagadish S, Craufurd P, Wheeler T (2007) High temperature stress and spikelet fertility in rice (Oryza sativa L.). J Exp Bot 58:1627–1635. https://doi.org/10.1093/jxb/erm003

Jagadish SVK, Septiningsih EM, Kohli A, Thomson MJ, Ye C, Redoña E, Kumar A, Gregorio GB, Wassmann R, Ismail AM, Singh RK (2012) Genetic advances in adapting rice to a rapidly changing climate. J Agron Crop Sci 198:360–373. https://doi.org/10.1111/j.1439-037X.2012.00525.x

Jones HG (1976) Crop characteristics and the ratio between assimilation and transpiration. J Appl Ecol 13:605–622. https://doi.org/10.2307/2401807

Kang BT (1993) Alley cropping : past achievements and future directions. Agrofor Syst 23:141–155. https://doi.org/10.1007/BF00704912

Kang S, Zhang J (2004) Controlled alternate partial root-zone irrigation: Its physiological consequences and impact on water use efficiency. J Exp Bot 55:2437–2446. https://doi.org/10.1093/jxb/erh249

Kang BT, Reynolds L, Atta-Krah AN (1990) Alley Farming. Adv Agron 43:315–359. https://doi.org/10.1016/S0065-2113(08)60481-2

Karl T, Melillo J, Pearson T (2009) Global climate change impacts in the United States. Cambridge University Press, New York

Kasim K, Dennett MD (1986) Radiation absorption and growth of Vicia faba under shade at two densities. Ann Appl Biol 109:639–650. https://doi.org/10.1111/j.1744-7348.1986.tb03221.x

Keerthisena RSK (1995) Fifty years of research in alley cropping at Maha Illuppallama: A Review. In: Gunasena HPM (ed) Proceedings of the sixth regional workshop on multipurpose trees. University of Peradeniya, Kandy, pp 20–29

Klein Tank AMG, Peterson TC, Quadir DA, Dorji S, Zou X, Tang H, Santhosh K, Joshi UR, Jaswal AK, Kolli RK, Sikder AB, Deshpande NR, Revadekar JV, Yeleuova K, Vandasheva S, Faleyeva M, Gomboluudev P, Budhathoki KP, Hussain A, Afzaal M, Chandrapala L,

Anvar H, Amanmurad D, Asanova VS, Jones PD, New MG, Spektorman T (2006) Changes in daily temperature and precipitation extremes in central and south Asia. J Geophys Res 111: D16105. https://doi.org/10.1029/2005JD006316
Kottek M, Grieser J, Beck C, Rudolf B, Rubel F (2006) World map of the Köppen-Geiger climate classification updated. Meteorol Zeitschrift 15:259–263. https://doi.org/10.1127/0941-2948/2006/0130
Kumar BM, Nair PKR (2004) The enigma of tropical homegardens. Agrofor Syst 61:135–152. https://doi.org/10.1023/B:AGFO.0000028995.13227.ca
Kumar BM, Nair PKR (2006) Tropical homegardens. Springer, Dordrecht
Lawlor DW (1993) Photosynthesis: molecular, physiological and environmental processes, 2nd edn. Longman, Harlow
Liu C, Allan RP (2013) Observed and simulated precipitation responses in wet and dry regions 1850–2100. Environ Res Lett 8:034002. https://doi.org/10.1088/1748-9326/8/3/034002
Lobell DB, Gourdji SM (2012) The influence of climate change on global crop productivity. Plant Physiol 160:1686–1697. https://doi.org/10.1104/pp.112.208298
Lobell DB, Bonfils C, Duffy PB (2007) Climate change uncertainty for daily minimum and maximum temperatures: a model inter-comparison. Geophys Res Lett 34:1–5. https://doi.org/10.1029/2006GL028726
Lobell DB, Schlenker W, Costa-Roberts J (2011) Climate trends and global crop production since 1980. Science 343:1017–1021. https://doi.org/10.5061/dryad.5t110.Supplementary
Long SP, Persaud D (1988) Influence of neem (Azardirachta indica) windbreaks on millet yield, microclimate, and water use in Niger, West Africa. In: Unger PW, Sneed TV, Jordan WR, Jensen R (eds) Challenges in dryland agriculture – a global perspective. Texas Agricultural Experimental Station, Texas, pp 313–314
Loomis RS, Connor DJ (1992) Crop ecology: productivity and management in agricultural systems. Cambridge University Press, Cambridge
Malaviarachchi MAPWK, De Costa WAJM, Fonseka RM, Kumara JBDAP, Abhayapala KMRD, Suriyagoda LDB (2014) Response of maize (Zea mays L.) to a temperature gradient representing long-term climate change under different soil management systems. Trop Agric Res 25:327–344. https://doi.org/10.4038/tar.v25i3.8043
Malaviarachchi MAPWK, De Costa WAJM, Kumara JBDAP, Suriyagoda LDB, Fonseka RM (2016) Response of mung bean (Vigna radiata (L.) R. Wilczek) to an increasing natural temperature gradient under different crop management systems. J Agron Crop Sci 202:51–68. https://doi.org/10.1111/jac.12131
Mapa RB, Gunasena HPM (1995) Effect of alley cropping on soil aggregate stability of a tropical Alfisol. Agrofor Syst 32:237–245. https://doi.org/10.1007/BF00711712
Marshall B, Willey RW (1983) Radiation interception and growth in an intercrop of pearl millet/groundnut. F Crop Res 7:141–160. https://doi.org/10.1016/0378-4290(83)90018-7
Meehl G, Stocker T, Collins W, Friedlingstein P, Gaye A, Gregory J, Kitoh A, Knutti R, Murphy J, Noda A, Raper S, Watterson I, Weaver A, Zhao Z-C (2007) Global climate projections. In: Solomon S, Qin D, Manning M, Chen Z, Marquis M, Averyt K, Tignor M, Miller H (eds) Climate Change 2007: The physical science basis. Contribution of Working Group I to the fourth assessment report of the Intergovernmental Panel on Climate Change. Cambridge University Press, New York, pp 746–845
Meehl GA, Tebaldi C, Walton G, Easterling D, McDaniel L (2009) Relative increase of record high maximum temperatures compared to record low minimum temperatures in the U.S. Geophys Res Lett 36:1–5. https://doi.org/10.1029/2009GL040736
Mohotti AJ, Lawlor DW (2002) Diurnal variation of photosynthesis and photoinhibition in tea: effects of irradiance and nitrogen supply during growth in the field. J Exp Bot 53:313–322. https://doi.org/10.1093/jexbot/53.367.313
Morison JIL, Baker NR, Mullineaux PM, Davies WJ (2008) Improving water use in crop production. Philos Trans R Soc B Biol Sci 363:639–658. https://doi.org/10.1098/rstb.2007.2175

Myhre G, Shindell D, Bréon F-M, Collins W, Fuglestvedt J, Huang J, Koch D, Lamarque J-F, Lee D, Mendoza B, Nakajima T, Robock A, Stephens G, Takemura T, Zhang H (2013) Anthropogenic and natural radiative forcing. In: Stocker TF, Qin D, Plattner G-K, Tignor M, Allen SK, Boschung J, Nauels A, Xia Y, Bex V, Midgley PM (eds) Climate Change 2013: the physical science basis. Contribution of Working Group I to the fifth assessment report of the Intergovernmental Panel of Climate Change. Cambridge University Press, New York, pp 658–740

Nelson DR (2011) Adaptation and resilience: responding to a changing climate. Wiley Interdiscip Rev Clim Chang 2:113–120. https://doi.org/10.1002/wcc.91

Palmer WC (1965) Meteorological drought. Research paper No. 45. US Department of Commerce Weather Bureau, Washington, DC

Panabokke CR (1996) Soils and agro-ecological environments of Sri Lanka. Natural Resources, Energy and Science Authority of Sri Lanka, Colombo

Patabendige CS, Kazama PS, Komori PD (2016) Near future climatic impact on seasonal runoff in Sri Lanka. https://www.researchgate.net/publication/307862797

Pattnayak KC, Kar SC, Dalal M, Pattnayak RK (2017) Projections of annual rainfall and surface temperature from CMIP5 models over the BIMSTEC countries. Glob Planet Change 152:152–166. https://doi.org/10.1016/j.gloplacha.2017.03.005

Perata P, Voesenek LACJ (2007) Submergence tolerance in rice requires Sub1A, an ethylene-response-factor-like gene. Trends Plant Sci 12:43–46. https://doi.org/10.1016/j.tplants.2006.12.005

Prasad PVV, Boote KJ, Allen LH, Sheehy JE, Thomas JMG (2006) Species, ecotype and cultivar differences in spikelet fertility and harvest index of rice in response to high temperature stress. F Crop Res 95:398–411. https://doi.org/10.1016/j.fcr.2005.04.008

Punyawardena BVR (2009) Climate of Sri Lanka. In: Nissanka S, Sangakkara U (eds) Global climate change and its impacts on agriculture, forestry and water in the tropics. CDM Study Centre, University of Peradeniya, Peradeniya, pp 7–20

Punyawardena BVR, Dissanayaka T, Mallawatantri A (2013) Spatial variation of climate change induced vulnerability in Sri Lanka. Natural Resources Management Centre, Department of Agriculture, Peradeniya

Pushpakumara DKNG, Marambe B, Silva GLLP, Weerahewa J, Punyawardena BVR (2012) A review of research on homegardens in Sri Lanka: the status, importance and future perspective. Trop Agric 160:55–125

Ranasinghe CS, Silva LRS, Premasiri RDN (2015) Major determinants of fruit set and yield fluctuation in coconut (Cocos nucifera L.). J Natl Sci Found Sri Lanka 43:253–264. https://doi.org/10.4038/jnsfsr.v43i3.7955

Ranasinghe CS, Kumarathunge MDP, Kiriwandeniya KGS (2018) Genotypic differences in cardinal temperatures for in vitro pollen germination and pollen tube growth of coconut hybrids. Exp Agric 54:731–743. https://doi.org/10.1017/S0014479717000357

Rosenzweig C, Karoly D, Vicarelli M, Neofotis P, Wu Q, Casassa G, Menzel A, Root TL, Estrella N, Seguin B (2008) Attributing physical and biological impacts to anthropogenic climate change. Nature 453:353–357. https://doi.org/10.1038/nature06937

Sacks WJ, Deryng D, Foley JA, Ramankutty N (2010) Crop planting dates: an analysis of global patterns. Glob Ecol Biogeogr 19:607–620. https://doi.org/10.1111/j.1466-8238.2010.00551.x

Samarakoon SMM, Abeygunawardena P (1995) An economic assessment of on-site effects of soil erosion in potato lands in Nuwara Eliya district of Sri Lanka. J Sustain Agric 6:81–92. https://doi.org/10.1300/J064v06n02_09

Senaratne DMAH (2003) Three decades of sustainable agricultural systems (SAS) in Sri Lanka: a review of institutional and policy issues. J Sustain Agric 21:61–84. https://doi.org/10.1300/J064v21n03

Sheffield J, Wood EF (2008) Global trends and variability in soil moisture and drought characteristics, 1950–2000, from observation-driven simulations of the terrestrial hydrologic cycle. J Clim 21:432–458. https://doi.org/10.1175/2007JCLI1822.1

Sheriff DW (1979) Stomatal aperture and the sensing of the environment by guard cells. Plant Cell Environ 2:15–22. https://doi.org/10.1111/j.1365-3040.1979.tb00769.x

Shindell DT, Faluvegi G, Koch DM, Schmidt GA, Bauer SE, Shindell DT, Faluvegi G, Koch DM, Schmidt GA, Unger N, Bauer SE (2009) Improved attribution of climate forcing to emissions. Science 326:716–718. https://doi.org/10.1126/science.1174760

Solomon S, Qin D, Manning M, Alley R, Berntsen T, Bindoff N, Chen Z, Chidthaisong A, Gregory J, Hegerl G, Heimann M, Hewitson B, Hoskins B, Joos F, Jouzel J, Kattsov V, Lohmann U, Matsuno T, Molina M, Nicholls N, Overpeck J, Raga G, Ramaswamy V, Ren J, Rusticucci M, Somerville R, Stocker T, Whetton P, Wood R, Wratt D (2007) Technical summary. In: Solomon S, Qin D, Manning M, Chen Z, Marquis M, Averyt K, Tignor M, Miller H (eds) Climate Change 2007: the physical science basis. Contribution of Working Group I to the fourth assessment report of the Intergovernmental Panel on Climate Change. Cambridge University Press, Cambridge

Somasiri S, Handawala J, Weerakoon WL, Dharmasena PB, Jayawardena SN (1990) Recommendations for rainfed upland farming in the dry zone. Agricultural Research Station, Maha Illuppallama

Sripongpangkul K, Posa GBT, Senadhira DW, Brar D, Huang N, Khush GS, Li ZK (2000) Genes/QTLs affecting flood tolerance in rice. Theor Appl Genet 101:1074–1081. https://doi.org/10.1007/s001220051582

Stocker T, Qin D, Plattner G-K, Alexander L, Allen S, Bindoff N, Bréon F-M, Church J, Cubasch U, Emori S, Forster P, Friedlingstein P, Gillett N, Gregory J, Hartmann D, Jansen E, Kirtman B, Knutti R, Krishna Kumar K, Lemke P, Marotzke J, Masson-Delmotte V, Meehl G, Mokhov I, Piao S, Ramaswamy V, Randall D, Rhein M, Rojas M, Sabine C, Shindell D, Talley L, Vaughan D, Xie S-P (2013a) Technical summary. In: Stocker TF, Qin D, Plattner G-K, Tignor M, Allen SK, Boschung J, Nauels A, Xia Y, Bex V, Midgley PM (eds) Climate Change 2013: the physical science basis. Contribution of Working Group I to the fifth assessment report of the Intergovernmental Panel on Climate Change. Cambridge University Press, New York, p 84

Stocker TF, Qin D, Plattner GK, Tignor M, Allen SK, Boschung J, Nauels A, Xia Y, Bex V, Midgley PM (2013b) Annex III: glossary. In: Planton S (ed) Climate Change 2013: the physical science basis. Contribution of Working Group I to the fifth assessment report of the Intergovernmental Panel on Climate Change. Cambridge University Press, New York, pp 1447–1466

Stoll M, Loveys B, Dry P (2000) Hormonal changes induced by partial root zone drying of irrigated grapevine. J Exp Bot 51:1627–1634. https://doi.org/10.1093/jexbot/51.350.1627

Stott PA, Jones GS, Christidis N, Zwiers FW, Hegerl G, Shiogama H (2011) Single-step attribution of increasing frequencies of very warm regional temperatures to human influence. Atmos Sci Lett 12:220–227. https://doi.org/10.1002/asl.315

The Royal Society (2009) Reaping the benefits: science and the sustainable intensification of global agriculture. The Royal Society, London

Trenberth KE (2011) Changes in precipitation with climate change. Clim Res 47:123–138. https://doi.org/10.3354/cr00953

Tuong TP, Bouman BAM (2003) Rice production in water-scarce environments. In: Kijne JW, Barker R, Molden D (eds) Water productivity in agriculture: limits and opportunities for improvement. CAB International/International Water Management Institute, Wallingford/Colombo, pp 53–67

Tuong P, Bouman BAM, Mortimer M (2005) More rice, less water—integrated approaches for increasing water productivity in irrigated rice-based systems in Asia. Plant Prod Sci 8:231–241. https://doi.org/10.1626/pps.8.231

Udawattage UDS (1985) The development of micro-catchments in Sri Lanka. J Hydrol 80:351–359. https://doi.org/10.1016/0022-1694(85)90127-1

Van Noordwijk M (1996) Mulch and shade model for optimum alley-cropping design depending on soil fertility. In: Ong CK, Huxley P (eds) Tree-crop interactions: a physiological approach. CABI/ICRAF, Oxford, pp 51–72

Wahid A, Gelani S, Ashraf M, Foolad MR (2007) Heat tolerance in plants: an overview. Environ Exp Bot 61:199–223. https://doi.org/10.1016/j.envexpbot.2007.05.011

Wassmann R, Jagadish SVK, Heuer S, Ismail A, Redona E, Serraj R, Singh RK, Howell G, Pathak H, Sumfleth K (2009) Climate change affecting rice production: the physiological and agronomic basis for possible adaptation strategies. Adv Agron 101:59–122. https://doi.org/10.1016/S0065-2113(08)00802-X

Weerakoon WL, Liyanage M d S (1987) Aspects of conservation farming. Agricultural Research Station, Maha illuppallama

Weerakoon WL, Seneviratne AM (1984) Managing a sustainable farming system in Sri Lanka. Trop Agric 140:41–50

Weerakoon WMW, Maruyama A, Ohba K (2008) Impact of humidity on temperature-induced grain sterility in rice (Oryza sativa L). J Agron Crop Sci 194:135–140. https://doi.org/10.1111/j.1439-037X.2008.00293.x

Wheeler T, Von Braun J (2013) Climate change impacts on global food security. Science 341:508–513. https://doi.org/10.1126/science.1239402

Wilkinson S, Davies WJ (2002) ABA-based chemical signalling: the co-ordination of responses to stress in plants. Plant Cell Environ 25:195–210. https://doi.org/10.1046/j.0016-8025.2001.00824.x

Xu K, Xu X, Fukao T, Canlas P, Maghirang-Rodriguez R, Heuer S, Ismail AM, Bailey-Serres J, Ronald PC, Mackill DJ (2006) Sub1A is an ethylene-response-factor-like gene that confers submergence tolerance to rice. Nature 442:705–708. https://doi.org/10.1038/nature04920

Zhang J, Schurr U, Davies WJ (1987) Control of stomatal behaviour by abscisic acid which apparently originates in the roots. J Exp Bot 38:1174–1181. https://doi.org/10.1093/jxb/38.7.1174

Microbial Pesticides Towards Eco-friendly Agriculture: Present Status and Future Prospects in Sri Lanka

D. M. De Costa

1 Pesticides as an Integral Part of the Present-Day Agriculture

Based on the current projections, the present world population of 7.6 billion has been predicted to reach 9.8 billion by 2050 (UN 2017). In order to meet the requirements of the growing population, the Food and Agriculture Organization (FAO) of the United Nations has forecasted the need of a 70% increase of food production by 2050 (Alexandratos and Bruinsma 2012). Scarcity of arable land and water, drastic changes of climatic conditions and emerging threats due to pest and diseases have become critical factors limiting the crop cultivation, worldwide. Nevertheless, agriculturists have to meet the global food requirement, which is a Herculean task, and, therefore, compelled to use intensive agriculture with high inputs such as high-yielding varieties, synthetic fertilizer and irrigation water which are the hallmarks of the Green Revolution. However, crops grown under high-input agriculture are prone to numerous types of biotic threats (i.e. pests, diseases and weeds) and reduce the potential crop yield. According to Oerke (2006), worldwide loss of potential yield due to pests at preharvest stage has been estimated to be 35%, and an additional 35% crop loss has been estimated to occur at the postharvest period. As reported by Oerke (2006), 70% of the crop yield could be lost if pesticides are not applied. Even with pesticide application, 30% of the theoretical yield is lost due to absence of effective pest management methods around the world in a uniform manner. Further, when the overall crop productivity is low and crop protection is limited to a certain extent of weed control, the losses due to pests account for more than 50% of the attainable production (Oerke 2006). These findings highlight the necessity of pesticides for achieving the targets of the agriculturists without sacrificing the crop yields. According to Popp et al. (2013), the crop yield per unit area can be increased by

D. M. De Costa (✉)
Department of Agricultural Biology, Faculty of Agriculture, University of Peradeniya, Peradeniya, Sri Lanka

R. P. De Silva et al. (eds.), *Agricultural Research for Sustainable Food Systems in Sri Lanka*, https://doi.org/10.1007/978-981-15-3673-1_7

efficient control of biotic stress factors rather than increasing the yield potentials. Hence, the use of pesticides has been increased dramatically since the early 1960s, and along with the increased use of pesticides, the average yield of major calorie crops (e.g. wheat, rice and maize) has been more than doubled.

2 Impact of Pesticide Usage

Worldwide consumption of pesticides has been reported to be 2 million tons per year (De et al. 2014), and of it, 47.5, 29.5, 17.5 and 5.5 percentages account for herbicides, insecticides, fungicides and the other types of pesticides, respectively. Even though the effect of pesticides in high-input intensive agriculture is impressive, use of pesticides is associated with numerous environmental and health issues. Accumulation and contamination of soil, water, air and food by pesticides have been reported as the harmful effects of abiotic components on the environment. Phytotoxicity, genetic/developmental disorders, non-infectious diseases, issues related to reproduction and physiological deformities in non-target animals are the negative impacts caused by pesticides to biotic components of the environment (Kumar 2015).

Direct impact of pesticides on humans and animals, contamination of food commodities, soil, surface and ground water and air, and direct/indirect effect on beneficial soil organisms and non-target organisms have been elaborated by Aktar et al. (2009) and Carvalho (2017) by providing examples from many corners of the world. Cancer risk of Vietnam war veterans and workers who were occupationally exposed to herbicides or dioxins, death of over 100 people in India due to consumption of wheat flour contaminated with parathion and drastic decline of the global bat population and near extinction of river dolphins in China and Pakistan due to exposure to organochlorine pesticides are to name a few of those examples (Aktar et al. 2009). In addition, examples on mass killings of nonhuman biota, such as bees, birds, amphibians, fish and small mammals, due to application of pesticides have been elaborated by Carvalho (2017) in the recent review on 'pesticides, environment and food security'. According to Carvalho (2017), approximately 25 million agricultural workers worldwide experience unintentional pesticide poisoning annually. Exposure to pesticides and many other agrochemicals has been related to cancers (especially prostate cancers), obesity, endocrine disruption and many other diseases in human (WHO 2017). In spite of food contaminations due to pesticides, the highest number of suicides by young Chinese women and Sri Lankan men and women has been reported due to self-poisoning of pesticides (Sri Lankan Ministry of Health 1995; WHO 2001). However, based on the recent statistics on suicides by WHO, Sri Lanka is presently ranked 22nd in the global league table of suicide incidence, which is a considerable improvement in comparison to previous decades (Knipe et al. 2015). Strict regulations related to availability of pesticides imposed by Sri Lanka have been considered as the major reason for the above reduction of pesticide-related suicides (Gunnell et al. 2007). Annually, a large wealth of information is being accumulated on contamination of the environment and toxic effects on all forms of biota due to pesticides, hence considerable efforts have been made to mitigate the problem by designing new chemicals, improving pesticide formulations and application devices and using degradable nanoparticles to deliver pesticides, which is a trendy area of research in pesticide technology (De et al. 2014). To this end, biopesticide is in the limelight.

3 Biopesticides and the Position of Microbial Pesticides

Even though biopesticides have been defined by different authors/organizations in different ways, according to US Environmental Protection Agency (USEPA), they are the pesticides derived from natural materials such as animals, plants, microorganisms and minerals. The definition of biopesticides proposed by USEPA is not used globally. For example, the International Biocontrol Manufacturer's Association (IBMA) and the International Organization for Biological Control (IOBC) encourage the use of the term 'biocontrol agents (BCA)' instead of biopesticides (Guillon 2003). Hence, BCA is an alternatively used term for biopesticides. According to USEPA, biopesticides are classified into three major groups based on the type of active ingredient it contained, namely, biochemicals, plant-incorporated protectants and microbial pesticides (USEPA 2008). Ingredient (s) in biochemical pesticides have been extracted from natural sources or synthesized to have the same structure and function as the naturally found chemical. Biochemical pesticides can interfere the mating behaviour of insect pests, attract the pest to trap plants and repel them from the crop plant. Plant-incorporated protectants or genetically modified crops express a deliberately introduced gene into the plant with a pesticidal action; hence the genetically modified plant possesses an in-built protection against the pest/pathogen attacks. Microbial pesticides contain the active ingredient of naturally occurring organisms or genetically modified organisms either in their live form or as their products or byproducts which have pesticidal activity. According to IBMA classification, BCA are grouped as macrobials, microbials, natural products and semiochemicals (chemicals modifying insect behaviours). Microbial pesticides suppress the target pests through a number of modes of action which have a direct or indirect impact on the pest concerned. The direct impacts by the microbial pesticides are competition, antibiosis, lysis and hyperparasitism, and induced host plant resistance is an indirect type of mode of action posed by microbial pesticides (Bonaterra et al. 2012).

4 Role of Microbial Biopesticides in Integrated Pest Management

All types of harmful agents of crop plants (i.e. insect pests, mites, plant pathogens, weeds) can be collectively considered as 'pest'. Integrated pest management (IPM) relies on combining of different crop protection practices, in a compatible manner, with careful monitoring of pests and their natural enemies or antagonists in general to overcome the limitations of any individual practice (Bajwa and Kogan 2002). The most significant feature of IPM is not to eradicate pest populations but to manage them below the levels that cause economic damage, except in the situations of production of propagative material such as seeds, seedlings and clones. Accordingly, the different pest management strategies that can be included in an IPM programme are, namely, selection of crop cultivars/varieties with complete or partial resistance, use of appropriate agronomic practices, use of physical methods, use of biological control agents, use of synthetic pesticides that have a high level of selectivity and are classified by the regulatory bodies as low-risk compounds whenever essential, use of decision supporting tools and use of biotechnological approaches to modify the genetic makeup of the host plant or biocontrol agent (Chandler et al. 2011).

According to a comprehensive study reported by Pretty (2008), 62 IPM research and development projects done in 26 countries, covering more than five million farm households, revealed the ability of IPM to reduce pesticide usage substantially. Over 60% of the studied projects resulted in an average of 70% reduction of pesticide usage while giving a 40% increase of the yield. In contrast, 20% of the farm projects resulted in a 60% reduction of pesticide usage but with a 5% yield loss. Even though some less successful results have been shown by the IPM systems practised for outdoor arable crop fields, use of IPM has been highly successful with protected crop cultivation (i.e. greenhouse crops). Therefore, over 90% of the greenhouse-grown tomato, cucumber and sweet pepper production in the Netherlands and almost all greenhouse-grown crops in the UK depend on IPM with a major contribution by biological control (van Lenteren 2000).

5 Natural Sources to Harness Microbial Antagonists

Microbial pesticides contain naturally occurring or genetically modified microbes, namely, bacteria, fungi, algae, virus, nematodes and protozoans (Mishra et al. 2015), either in their vegetative or reproductive stages (i.e. spores/cysts). These microbes suppress the growth or development of pest population or kill the population through their selective modes of action. Depending on the target pest, microbial pesticides can be of many forms: biofungicides, bioinsecticides, bioherbicides, bionematicides, bioalgaecides, biobactericides, bioacaricides, etc. Microbes dwell on or in many natural habitats either as epiphytic, endophytic and phoretic organisms. For example, rhizosphere, the narrow zone surrounding and influenced by plant roots, is a hotspot for a wide range of organisms, namely, bacteria, actinomycetes, fungi, oomycetes, nematodes, protozoans, algae, virus, archaea and arthropods (Mendes et al. 2013), due to the secretions of root exudates. In addition, microbial communities live on the aerial plant parts which are largely covered by the foliage and collectively referred to as the phyllosphere (Bulgarelli et al. 2013). Being one of the largest microbial habitats in the world, phyllosphere predominantly accommodates fungi, algae, bacteria, actinomycetes, archaea, yeast and bacteriophages and to a lesser extent of nematodes and protozoans. However, compared to the other types of microbiota, bacteria are the most prevalent group present on the phyllosphere which has been estimated to be approximately 10^6–10^7 microbial cells per cm^2 of leaf area (Lindow and Brandl 2003). Microbial inhabitants are prevalent in the endosphere of plants (inside the leaves, roots or stem tissues) and are termed as endophytes (Turner et al. 2013). As reported by Bulgarelli et al. (2013), in general, rhizosphere and root endosphere contain bacterial densities of 10^6–10^9 cfu/g and 10^4–10^8 cfu/g, respectively. Collectively the whole microbial community in or on the rhizosphere, phyllosphere or endosphere is termed as the plant microbiome or plant microbiota. In addition, microbes dwell in harsh environmental habitats, in aquatic bodies and in air samples. Virtually, there is no place free from microbes.

5.1 Plant Microbiome: A Gold Mine of Microbial Antagonists

Even though rhizosphere, phyllosphere, endosphere and many other phyto-habitats are heavily colonized with microbes, the microbial density and diversity of them have not been estimated accurately, especially by conventional culture-dependent methods. This is purely due to unculturable nature of the majority of microbes. With the introduction of culture-independent techniques including 16S rRNA gene sequencing, microarray, fluorescent in situ hybridization-confocal laser scanning microscopy and metagenomics, it has been estimated that 1 g of soil can harbour more than one million distinct bacterial genomes, which is a value exceeding several orders of magnitude than the previous estimates by culture-dependent methods (Gans et al. 2005).

Majority of the bacteria and fungi that dwell in the rhizosphere are capable of producing metabolites that have the ability to inhibit the growth or activity of competing microorganisms in the same habitat (Brakhage and Schroeckh 2011). *Trichoderma* sp. and *Agrobacterium radiobacter* are examples each for a fungus and a bacterium, respectively, which has the capability of producing antibiotics (antimicrobial compounds). Some rhizosphere microbes are capable of producing metabolites having low molecular weight and high vapour pressure. These are named as volatile organic compounds (VOCs). Various bacterial species, namely, *Stenotrophomonas maltophilia*, *Serratia plymuthica*, *Pseudomonas trivialis*, *Pseudomonas fluorescens*, *Bacillus subtilis* and *Burkholderia cepacia*, are reported to produce VOCs that can inhibit mycelial growth of fungal pathogens (Mendes et al. 2013). Rhizosphere microbiome has been reported to induce host plant resistance by modifying the plant's immune responses against pathogens. Phyllosphere-inhabiting bacteria such as *P. fluorescens* strain A506 and *Pantoea agglomerans* strains can inhibit colonization of *Erwinia amylovora*, the fire blight pathogen, once the phyllosphere is prior colonized by the bacterial inhabitants (Stockwell et al. 1999). Studies done on different grass species have identified *Azoarcus*, *Burkholderia*, *Gluconacetobacter*, *Herbaspirillum* and *Klebsiella* spp. as endophytic bacteria. It is also reported that inoculation of endophytic bacteria such as *G. diazotrophicus* and *Herbaspirillum* protects the inoculated plants from phytopathogenic fungi and bacteria through systemically acquired resistance (Monteiro et al. 2012).

6 Reasons to Introduce Microbial Pesticides as an Alternative to Synthetic Pesticides

Since the inception of Green Revolution, despite the benefits gained, negative impact caused by pesticides on human and other non-target organisms has been highlighted by many authors. Dating back to Carson (1962), through her book *Silent Spring*, a wealth of data is being accumulated on contamination of food and the environmental resources by pesticides (Aktar et al. 2009; Carvalho 2017; Nøstbakken et al. 2015), acute toxicity leading to death, potential health risks to human by chronic effects of pesticides (International Agency for Research on Cancer 2016, 2017), accumulation of pesticide residues and mass killing of beneficial non-target biota (WHO 2017; Paoli et al. 2015) caused by pesticides.

6.1 Sri Lankan Scenario

In parallel to the studies done globally, a reasonable number of pesticide-related studies have been done in Sri Lanka. To name a few, they include the assessment of pesticide usage in vegetable cultivation in the upcountry of the island (Jayetilleke and Bandara 1989; Padmajani et al. 2014), mobility patterns of pesticides in water resources (Aravinna et al. 2017), detection of pesticide residues in some selected vegetables collected from markets (Lakshani et al. 2017; Sharaniaya and Loganathan 2015), assessment of the residues of chlorpyrifos in selected vegetables (Marasinghe et al. 2011), health issues faced due to occupational exposure to acetylcholinesterase-inhibiting insecticides (Smit et al. 2003), pesticide poisoning in Sri Lanka (Jeyaratnam et al. 1982) and contamination of well water due to pesticides (Aponso et al. 2003).

Findings have revealed that majority of vegetable farmers in Sri Lanka are not practising the preharvest safety period (Sharaniaya and Loganathan 2015), use of pesticides containing active ingredients belonging to not permitted class (ib) and restricted use category (class (ii)), overdosing of the pesticides and use of cocktail mixtures of pesticides (Padmajani et al. 2014). Pesticide residue levels exceeding EU maximum residue levels (MRLs) have been reported for cabbage, capsicum and tomato available at the market or with wholesale dealers (Lakshani et al. 2017). According to a study done by Watawala et al. (2010), 45% of the upcountry farmers involved with intensive cultivation prefer to use more than the recommended amount of pesticides and also in higher frequencies, to avoid high risk of crop losses and to ensure a higher yield. Though lower than the MRLs, presence of residues of propineb, tebuconazole and chlorothalonil has been reported in milk samples collected from upcountry Sri Lanka and potential threat of having unidentified agrochemicals in those milk samples (Chaminda et al. 2012). Extremely high risk levels and high toxicity levels have been reported from ground and surface water of potato-cultivating areas of Sri Lanka due to heavy use of fungicides (Watawala et al. 2010). Hence, use of such water for human consumption poses health risks, and bioaccumulation of the chemicals in aquatic fauna is a concern.

Global and Sri Lankan scenario provide ample evidence of potential health and environmental threats posed due to complete dependence on pesticides in plant protection. Considering the alarming nature of those threats, it is a timely need to explore alternative options to synthetic pesticides, and microbial pesticides have become a prospective alternative.

7 Present Status of Microbial Pesticides

According to FAO (2017), the global pesticide production has been increased at a rate of about 11% per year (i.e. from 0.2 million tons in the 1950s to more than 5 million tons by year 2000). Sri Lanka has been ranked as the 4th pesticide user among the Asian countries with 114 active substances and nearly 440 agricultural pesticides (Lakshani et al. 2017). Therefore, public concern on usage of synthetic pesticides has been growing, and attention has been diverted to biological control of crop pests. Out of the total biopesticides used today in the world, 90% constitutes the microbial pesticides, the largest group of broad-spectrum biopesticides which are pest specific hence having no or less harmful impacts on non-target organisms (Koul 2011). Owing to the environmentally benign characteristics of microbial pesticides, worldwide microbial pesticide usage has been reported to be 60% bacterial biopesticides, 27% fungal biopesticides, 10% viral pesticides and 3% others such as nematode biopesticides (Mishra et al. 2015). Bacterial, fungal, viral, nematode and protozoan microbial pesticides available globally as commercial products are listed by Koul (2011), Chandler et al. (2011) and Mishra et al. (2015). As reported by Marrone (2007), there are about 1400 biopesticide products in the market, and 68 biopesticide-active substances are registered in the European Union (EU) and 202 in the USA. The EU biopesticides consist of 34 microbial pesticides, 11 biochemicals and 23 semiochemicals (EUPD 2010). Among the biopesticides registered in the USA, 102 are microbials, 52 are biochemicals, and 48 are semiochemicals. Even though there are registered microbial pesticides in Asia, the information is variable (Koul 2011). In general, the available biopesticide products represent 2.5% of the total pesticide market. As reported by Mishra et al. (2015), worldwide biopesticide use and market share percentage are dominated by the USA and Canada (44%), while the lowest use and the market share (3%) are reported from Africa. Europe, Asia, Australia and New Zealand and Latin America report 20, 13, 11 and 9 as the global biopesticide use and market share %, respectively.

8 Advantages of Microbial Pesticides in Comparison to Synthetic Pesticides

According to Koul (2011), in comparison to the growth of chemical pesticide market (which is about 1–2% per year), the growth of microbial pest control has been reported as 10% per year. Advantages of the microbial pesticides over the synthetic pesticides must be contributing to the increased growth rate of the use of microbial pesticides worldwide.

As reported by Koul (2011), the biggest strength of microbial pesticides is their non-toxic and nonpathogenic nature on non-target organisms. Since microbial pesticides are specific to a single group or species of pest, they have no or less negative effects on the beneficial organisms such as natural enemies of pests, namely, predators, parasitoids and pathogens. Further, the microbial pesticides can be used in many habitats such as aquatic bodies and indoor settings where chemical pesticide usage has limitations. Due to the non-hazardous nature of the residues of microbial pesticides, they can be applied without maintaining longer safety periods. As elaborated by Bonaterra et al. (2012), the fate of the microbial antagonists once they are applied to natural habitats such as rhizosphere, phyllosphere and fruit surfaces (for the management of postharvest pathogens) has been studied by many researchers. Declining of population density of the artificially introduced microbial antagonists more or less rapidly after the inoculation is a common observation, and dominating the introduced microbial population over the natural microbial community is a rare phenomenon. Therefore, imbalance of the microbial community and residual effects of the microbes are remote chances. Microbial pesticides can be used not only for the management of pathogens but also against the vectors responsible for transmitting the pathogens. Remote possibility to develop pest resistance, no detection of harmful residues and biodegradable nature are other desirable attributes possessed by the microbial pesticides (Sarwar 2015).

9 Limitations of Microbial Pesticides

Some of the advantages of microbial pesticides eventually become their own disadvantages. For example, due to the specific action of microbial pesticides, they are antagonistic against only a particular group of the pests present in the field. Therefore, a broad-spectrum control cannot be achieved. Environmental factors such as temperature, UV radiation, rain and wind can have negative influence on the efficiency of microbial pesticides, especially when they are applied to crops grown in open fields. Microbial pesticides which are applied to above-ground plant parts against foliar pathogens/insect pests are more susceptible to the effect of harsh environmental factors than those applied to soil for the control of soilborne pathogens/insect pests. Therefore, special care has to be paid on formulation, storage and delivery system to improve the efficiency of the microbial pesticides. Incorporation of water-soluble adjuvants, oils, stickers, emulsions and UV protectants with the antagonistic microbes is done at formulation stage to minimize the detrimental effects by the environmental factors (Kumar et al. 2014). Shelf life of the microbial pesticide is one of the major desirable attributes which determines successful commercialization of the product. Being organisms, it is the general belief and experience by the users that microbial pesticides are not long-lasting as the synthetic pesticides. Generally, the microbial antagonists which were multiplied in organic food base have longer shelf life than those who have been multiplied in inert or inorganic food bases (e.g. talc, vermiculite, kaolin). Shelf life of *Trichoderma* spp., a fungal antagonist against a wide range of plant pathogens, has been reported to be more than 18 months when multiplied in coffee husk (Kumar et al. 2014). Bulky nature and possible contaminations during the storage are some other limitations of the quality of microbial pesticides. Longer time taken to show a response and lack of consistence of the final effect of the microbials on target pests (due to the effect of crop, season, location, farming system, etc.) are some of the less attractive attributes of microbial pesticides in comparison to synthetic pesticides. Possible risks of clinical opportunistic infections have been reported for some of the microbial antagonists (e.g. *Burkholderia cepacia*, *Stenotrophomonas maltophilia*, *Pseudomonas aeruginosa*) effective against plant pathogens. However, phenotypic and genotypic studies conducted on such strains have not revealed differences between environmental and clinical isolates. Hence it demonstrates the ability of any given strain to grow freely in natural habitats and cause opportunistic infections which are encoded within them (Bonaterra et al. 2012). Moreover, most of the microorganisms that have been branded as potential opportunists in human are inhabitants of edible plant parts (e.g. *Stenotrophomonas maltophilia* in rice seeds, *Escherichia coli* in lettuce leaves). Accordingly, plants have been natural reservoirs of endophytic microorganisms; hence, they have been an integral part of our diet for a long time (Berg et al. 2015). Even though a microbial biological control agent that is nonpathogenic on non-target species is the ideal, it is a goal far from easy to achieve (Hayward et al. 2010). As such, the negative effects of such microbial endophytes should be insignificant. Less enthusiasm of the growers to use microbial pesticides and reluctance to use microbial pesticides by farmers as they contain live organisms are some other contributory factors to create a limited market for microbial pesticides. Therefore, investments on commercial products of microbial pesticides are limited.

10 Development of Microbial Pesticides

Development of a microbial pesticide from a plethora of microbes dwelling at numerous habitats is a complex task. To this end, a number of criteria, namely, antagonistic efficiency, ecological fitness, toxicological profiles, ability of mass production and legal and marketing aspects, have to be evaluated (Köhl et al. 2011). In general, the sequence of events essential to follow when developing a microbial pesticide has been summarized by Montesinos (2003), and the requirements of the current European Union regulation in developing microbial pesticides have been highlighted by Kiewnick (2007). The stepwise screening procedure of microorganisms for commercial biological control of plant pathogenic fungi and bacteria and evaluation criteria at various screening steps have been introduced more comprehensively and in a straightforward way by Köhl et al. (2011). Details related to each step of the screening procedure are discussed in Sect. 10.1.

10.1 Screening Steps

The procedure of development of a microbial pesticide starts with the screening of potential microbial antagonists, namely, bacteria, fungi, virus, protozoans and nematodes. To this end, sampling can be done from various sources such as environmental samples (rhizosphere, phyllosphere, anthosphere, endosphere, suppressive soils, water, etc.) and biological samples (e.g. insect and mite cadavers, infected live insects and mites, animal tissues, etc.). Choice of the sampling source and number of samples will give a better chance of hunting for a potential biological control agent with a broad spectrum of antagonism. Therefore, sampling from a broad range of geographical regions, crops managed under different management practices, different varieties of crops and crops subjected to seasonal variations would represent a wider range of microbial antagonists (Köhl et al. 2011).

Isolation of the biological control agent, from the sample, as a pure culture will be the next step. When isolating a biological control agent from environmental samples, use of selective or semi-selective culture media (e.g. Veens semi-selective medium for *Metarhizium* spp., selective media for *Beauveria* spp. and starch casein agar for *Streptomyces* spp.) and diagnostic media (e.g. to isolate fluorescent *Pseudomonas* spp.,) would be convenient for the researcher, especially with the samples rich with microbial density and diversity. Insect bait technique is a widely used tool for specific isolation of entomopathogenic fungi from environmental samples, where larvae of wax moth (*Galleria mellonella*), which is a highly susceptible and easy-to rare host, are used (Meyling 2007).

Ability of an organism to isolate onto and culture in synthetic medium will be a practical advantage; hence culturable microbes have a higher prospects towards commercialization than the unculturable microbes. Ability of *Bacillus thuringiensis* (Bt), *Pseudomonas fluorescens* and *Trichoderma* spp. to culture in simple media under in vitro conditions is one of the main desirable attributes of them to proceed successfully through the pipeline of development of biopesticides. In comparison, *Pasteuria penetrans*, a promising biological control agent (an obligate prokaryote) of root-knot nematodes (*Meloidogyne* spp.), and insecticidal virus have restrictions towards commercialization due to their unculturable nature in synthetic media.

Once isolated and grown in pure cultures, the culturable microorganisms can be identified and characterized and subjected to bioassays and pilot tests as parallel activities (Montesinos 2003; Kiewnick 2007). However, according to Köhl et al. (2011) procedure, identification of a potential microorganism would be complete when a microorganism had cleared the barriers of the preliminary assessments in rapid throughput screening systems. Spore production on agar, cell production in broth, germination at 37 °C, mycotoxin risks and survival on UV radiation are some of the criteria related to biomass production and safety features and ecological characteristics of the candidate microorganisms assessed by rapid throughput screening systems.

Identification of the candidate microorganism at species level needs to be done, and with the advanced molecular biological techniques and DNA sequencing and homology search available at present, this can be done more reliably than by depending on the morphology of the candidate microbe. Information gathered on the identity can be used to navigate the available databases to find in-depth details of the microorganism. For example, medical and microbiological databases such as the German Collection of Microorganisms and Cell Cultures (http://www.dsmz.de) and regulations (European Commission 2000) are searched for information to judge the safety and potential risks on human pathogenicity, allergenicity and toxicity (Brimner and Boland 2003). At this stage, it is worthwhile to check for the already patent-protected microbial antagonists, so that further attempts can be terminated.

In general, candidates satisfying the required expectations will be less than 1% of the originally isolated number of microbes from environmental samples (Montesinos 2003). The less number of microorganisms which has more promise needs to be tested under real field conditions and different combinations of conditions. The goal of this step will be to select a group of antagonists with moderate to high level of antagonism under a range of representative environmental conditions without high variation among batches of the microbial pesticide.

Pilot experiments on mass production have to be started for the microorganisms that have satisfied the above desirable features. Biomass production and spore production are measured under solid-state fermentation for fungal microbial antagonists, and liquid fermentation will be used to measure the production of biomass and spores for bacterial and yeast antagonists (Köhl et al. 2011). Despite the proper blend with industrial biotechnological techniques, if the candidates are not producing sufficient quantities of biomass and spores and if the production cost is very high, such candidates are not worth considering for the next steps of the development of microbial pesticides.

Once the microbial candidates that can be successfully mass-produced are identified, formulation has to be done. Vegetative cells and/or spores are formulated as dry powders, granules or wet formulations such as concentrated emulsions. Viability and shelf life of the microbial antagonist are the key parameters assessed at this stage. Köhl et al. (2011) have suggested more than 80% viability and 12 months of shelf life as reasonable criteria in decision-making. In parallel to the viability and shelf life studies, compatibility to synthetic pesticides and preliminary toxicological studies (using non-vertebrates) are to be conducted.

Based on the results of viability, shelf life and toxicological profiles of the pilot-produced formulations, field experiments will be conducted in environments which are conducive for pests. Then, the mass production of the microbial candidates showing promising results under field conditions will be upscaled, and cost will be assessed. As the final step of the screening, microbial pesticides will be integrated to existing cropping systems.

11 Sri Lankan Scenario on Microbial Pesticides: A Bibliometric Analysis on Research Related to Microbial Pesticides

Sri Lankan agriculture has an intimate relationship with the use of naturally available bioresources and timely use of cultural practice and religious rituals (Widanapathirana and Dassanayake 2013) for protection of crops from pests. However, such indigenous and nature-friendly practices have been continued by the farmers by tradition but without having a scientific basis on the underlying mechanisms. Even though some literature is available on the nature-friendly approaches practised by Sri Lankan farmers in different agricultural systems, use of microorganisms as an eco-friendly strategy has not been in the attention.

In this study, an attempt was done to capture the nature and the trend of scientific research conducted in Sri Lanka with a special emphasis on microbial pesticides. To this end, standard bibliometric methods practised elsewhere have been used (Sinha 2012). For a comprehensive coverage of the present topic, major keywords such as 'biopesticides', 'microbial pesticides', 'biological control', 'Sri Lanka', 'global trend', 'trend of usage of microbial pesticides', 'research on microbial pesticides', 'biopesticide market', etc. were used. All the words were connected with Boolean operators (e.g. 'OR', 'AND', 'NOT') to have two-word keywords. In addition, keywords related to sub-disciplines were also used (e.g. 'fungal', 'bacterial', 'viral', 'nematode', '*Trichoderma* spp.', '*Bacillus* spp.'). Biological control done on pathogens specific to different crops was also used as a search criterion. The web-based queries were run in Google search engine only. In addition, library material on undergraduate and postgraduate research conducted by national universities and postgraduate institutes and research conducted in major national research institutes of Sri Lanka were searched. This includes proceedings of research symposia, thesis, research reports/bulletins and newsletters. The information gathered was manually analysed on different criteria of interest. The information included research work related to various stages of the development procedure of microbial pesticides. Due to the dynamic nature of the discipline concerned and due to difficulties in accessing the information, there is a possibility to have some omissions of research work done in Sri Lanka on microbial pesticides or microbial biological control agents in general.

11.1 Research Towards Microbial Pesticides Against Insect Pests

Sri Lanka has a more than century-long history on biological control of insect pests using microorganisms. According to the review by Evans (1999), entomopathogenic fungi such as *Cordyceps barnesii* Thw., a pathogen of cockchafer or lamellicorn larva of coffee; *C. dipterigena* Thw., a widespread pathogen of dipterans; and *Verticillium lecanii* (Zimm.) (now renamed as *Lecanicillium lecani*) Viégas, a pathogenic fungus on green scale insects of coffee, have been reported more than one and half century ago from Sri Lanka. Since then, potential of several

Table 1 Research conducted in Sri Lanka on biological control of insect pests by entomopathogenic fungi

Crop	Pest	Entomopathogenic fungus	References
Coconut	*Oryctes rhinoceros*	*Metarhizium anisopliae* (Metschn.) Sorokin	Fernando et al. (1995)
Coconut	*Aceria guerreron is*	*Hirsutella thompsonii*	Edgington et al. (2008)
Coffee	*Hypothenemus hampei* (Ferrari)	*Beauveria bassiana* (Balsamo)	Yapa et al. (2007), Subasinghe et al. (2013)
Tea	*Oligonychus coffeae*	Fungal isolate (unidentified)	Amarasena et al. (2011)
Queen palm (*Livistona rotundifolia*)	Root borer (*Parapoynx stratiotata*)	*Nomuraea rileyi*	Svinningen et al. (2010)

entomopathogenic fungi in controlling major pests of plantation crops and ornamental crops has been investigated by several researchers (Table 1).

Limited number of research has been reported on the use of bacterial antagonists for the management of insect pests. Commercial preparations of *Bacillus thuringiensis* have been used against coconut caterpillar, *Opisina arenosella* Walker (Kanagaratnam et al. 1983), and tea tortrix, *Homona coffearia* (Danthanarayana 1967). In addition, an entomopathogenic nematode (*Heterorhabditis* spp.) has been used against the low-country live-wood termite (*Glyptotermes dilatatus*) of tea (Danthanarayana and Vitarana 1987) and upcountry live-wood termite, *Postelectrotermes militaris* (Amarasinghe and Hominick 1993; Amarasinghe et al. 1994; Amarasinghe 2008).

11.2 Research Towards Microbial Pesticides Against Plant Pathogens

Google search results revealed that the research on microbial antagonists against plant pathogens in Sri Lanka has been started in the early 1990s. From year 2000, the research on microbial antagonists against plant pathogens has shown an exponential growth, and it appears that still the research relevant to biological control of plant pathogens in Sri Lanka is exponentially-increasing (Fig. 1).

Percentage-wise breakdown of the research related to the type of microbial antagonist of plant pathogens is given in Fig. 2. The highest percentage of research has been done with fungal antagonists (35%) followed by the combination of fungal and bacterial antagonists (33%). *Trichoderma* spp. (e.g. *T. viride*, *T. polysporum*, *T. harzianum*, *T. viride*, *T. koningii*, *T. asperellum*) have dominated the research on fungal antagonists. *Bacillus subtilis*, *Pantoea agglomerans*, *Flavobacterium* sp., *Burkholderia spinosa*, *Bacillus megaterium* and *Pseudomonas syringae* pv. *phaseolicola* have been reported as the bacterial antagonists used for the management of plant pathogens. Research on the potential of bacteriophages in

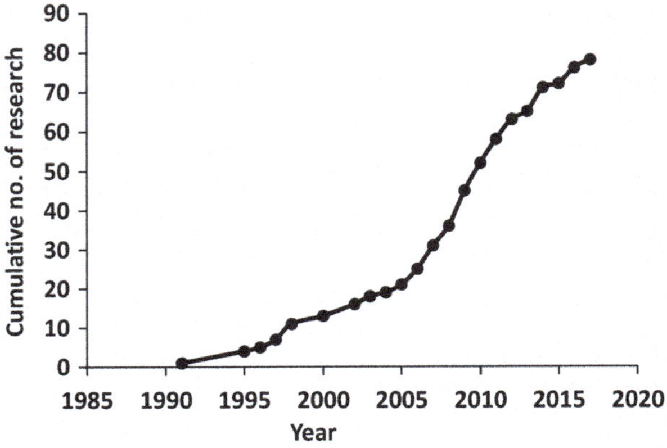

Fig. 1 Cumulative number of research on microbial pesticides/microbial biological control agents to control plant pathogens in Sri Lanka during 1990–2017. (Based on information compiled by the methodology described in Sect. 11)

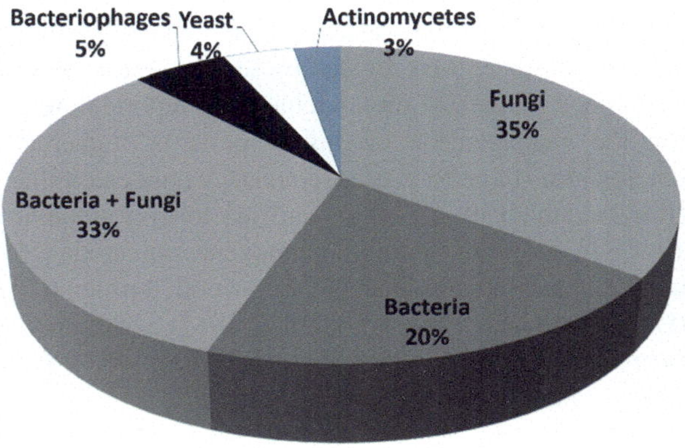

Fig. 2 Percentage-wise breakdown, based on the type of microbial antagonist in microbial pesticide-related research against plant pathogens conducted in Sri Lanka as reported during the period of 1990–2017

controlling the bacterial plant pathogens has only been focused on *Ralstonia solanacearum*, the bacterial wilt pathogen. Yeast spp. such as *Candida oleophila* and *Aureobasidium pullulans* have been attempted on management of postharvest pathogens of fruits. The potential of using an actinomycete (*Pasteuria penetrans*) has been investigated for the control of root-knot nematodes (*Meloidogyne* spp.).

The information collected in the present study on microbial biocontrol agents of plant pathogens was analysed with reference to diseases in different crops (Fig. 3). Accordingly, in Sri Lanka, 91% of the research on microbial biocontrol agents has been focused on plant pathogens of horticultural crops, namely, fruits, vegetables, root crops and leafy vegetables. The rest of the 9% of research has been focused on diseases of rice and plantation crops. Table 2 summarizes different types of diseases for which the research on microbial antagonists was conducted. Accordingly, it is evident that 67% of the research conducted so far has been focused on the control of soilborne diseases by microbial antagonists.

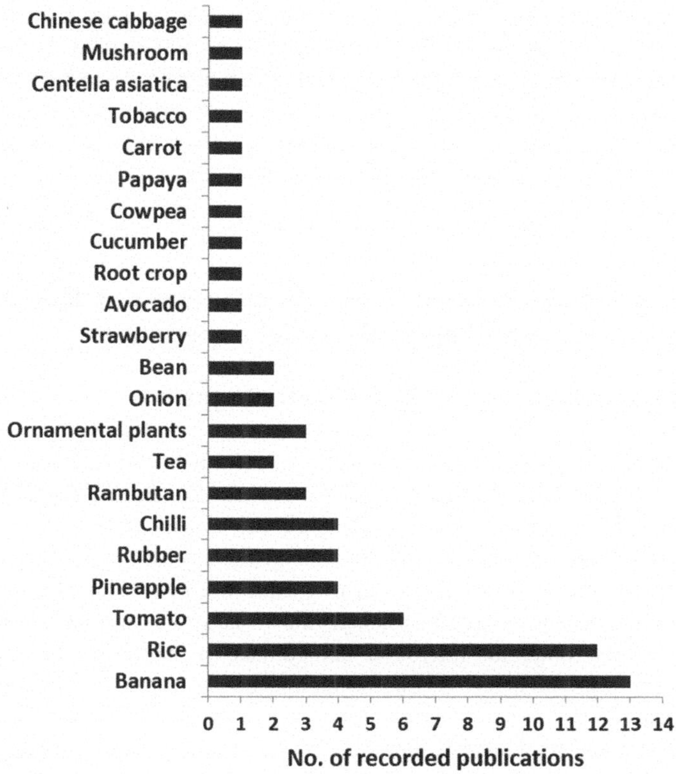

Fig. 3 Microbial pesticide-related research conducted in Sri Lanka during the period of 1990–2017 with reference to diseases in different crops

Table 2 Different types of plant diseases for which the research on microbial antagonists was conducted in Sri Lanka during the period of 1990–2017

Type of pathogen	Diseases
Fungi	Collar rot, rust, grey mould, white root rot, crown rot, sheath blight, black rot, stem end rot, corm rot, damping off, anthracnose, root rot, bulb rot, leaf twister disease, green mould, blister blight, fungal wilt, leaf spot
Bacteria	Bacterial wilt, bacterial soft rot
Nematodes	Root-knot nematodes

12 Research Towards Microbial Pesticides Against Weeds

A limited number of attempts have been reported on microbial control of weeds in Sri Lanka. Kelaniyangoda and Ekanayake (2010) have reported the potential use of *Puccinia melampodii* Diet. and Holow as a fungal biocontrol agent of *Parthenium hysterophorus*, a troublesome weed introduced to Sri Lanka in 1999. The efficiency of using fungal pathogens, namely, *Alternaria alternata*, *Cercospora rodmanii*, *Aspergillus* sp. and *Trichoderma* sp., as potential fungal antagonists for the control of water hyacinth (*Eichhornia crassipes*), a noxious invasive aquatic weed in Sri Lanka, has been reported by Queene et al. (2016). Further, Ekanayake et al. (2018) have identified fungal isolates showing pathogenic effects on water hyacinth and identified them as *Alternaria* sp., *Acremonium* sp., *Myrothecium* sp. and *Fusarium* sp.

13 Case Studies Related to the Development of Microbial Pesticides in Sri Lanka: Personal Experiences

13.1 Biological Control of Rice Sheath Blight

Sheath blight of rice, caused by *Rhizoctonia solani* Kühn AG1-1A, is a considerable biotic threat to many rice-growing countries worldwide, and Sri Lanka is not an exception. Yield losses due to rice sheath blight are considerable under high-input cropping systems, and cultural practices to manage the disease are not very effective. Further, rice cultivars resistant to sheath blight are not available. Therefore, apart from few fungicides, measures to manage the disease are limited. A study was initiated to screen potential microbial antagonists by exploring the microflora inhabiting on the phyllosphere of traditional and newly improved varieties of rice (De Costa et al. 2008a) as a non-chemical option to manage *R. solani.* Among a total of 196 bacterial and 91 fungal isolates, 12 bacterial and 2 fungal isolates showed more than 50% growth inhibition of *R. solani* under in vitro conditions; hence, they were tested for the in vivo antagonism under pot experiments. Out of the 14 antagonists, 6 bacterial and 1 fungal isolates reduced the disease incidence and severity more than 82% and 92%, respectively, for BG94-1 and IR8 varieties under pot experimental conditions (De Costa et al. 2008a). Field efficiency of the six bacterial and one fungal isolates was evaluated, and one of the bacterial isolates and the fungal isolate resulted in a 50% and 61% reduction of sheath blight severity, respectively, under field conditions for two consecutive seasons when applied as spore suspensions (De Costa et al. 2008a). The above bacterial isolate and the fungal isolate were identified as *Bacillus megaterium* and *Aspergillus niger*, respectively. Two bacterial antagonists (i.e. *Bacillus megaterium* and *B. subtilis*) and *A. niger* performed well under in vitro, in vivo and field conditions were subjected to mass culturing of spores (Soe and De Costa 2012). Spore-based formulations of the three antagonists (individual antagonist or combination of the antagonists) were prepared using talc as a formulating material, and the efficiency of the formulations in reducing rice sheath blight incidence was determined under pot experimental conditions. Soil applications and foliar applications of the talc-based formulations (as individuals or mixtures of antagonists) showed 45% reduction of the disease incidence (Soe and De Costa 2012). Spore viability and shelf life of the talc-based formulations were determined. It was revealed that spore containing talc-based formulations retained the spore viability over a period of 6 months when stored under refrigerated conditions or ambient temperature, and that was longer than the viability shown by spore suspensions (Soe and De Costa 2012). Treatment of a range of rice varieties with the microbial antagonists increased the level of defence enzymes such as β-1,3-glucanase and chitinase in rice sheath tissues indicating the ability of the microbial antagonists to induce host plant resistance (Soe and De Costa 2012).

13.2 Biological Control of Postharvest Diseases of Banana

Dessert-type banana (*Musa acuminata*) is highly susceptible to a range of postharvest diseases, namely, anthracnose, crown rot and blossom end rot. Some of these postharvest decays are latent infections, and some are wound infections. *Colletotrichum musae* is the causal fungal pathogen of banana anthracnose, and it is one of the causal agents of crown rot and blossom end rot of banana. Considering the consumer concerns on usage of fungicides, alternative non-pesticidal approaches were investigated to manage the postharvest diseases of banana. To this end, potential use of indigenous microbial antagonists for managing the postharvest pathogens was investigated. A bacterial antagonist was isolated from the fruit peel of banana (var. Seenikesel) (De Costa et al. 1997; De Costa and Subasinghe 1998), and it was identified initially as a non-fluorescent *Pseudomonas* spp. by conventional methods. Later metabolic fingerprinting methods such as fatty acid analysis identified the antagonist as *Burkholderia cepacia* GCB group (De Costa and Erabadupitiya 2005), and the BIOLOG identification system (MicroLog 1 4.2 version) gave the highest similarity index with *Burkholderia spinosa* (De Costa et al. 2008b). Molecular identification based on recA sequence analysis identified the isolate as a member of *Burkholderia cepacia* complex, but not an already identified species of the *B. cepacia* complex (De Costa and Erabadupitiya 2005; De Costa et al. 2008b). In vivo studies using different banana cultivars showed the ability of the bacterial isolate to control several postharvest diseases of banana either as a sole method (as a postharvest dip) or in combination with hot water treatment (at 50 °C for 3 min) (De Costa and Erabadupitiya 2005). A suspension of *B. spinosa* (10^5 cfu/ml) containing Tween 20 (0.02% v/v) was effective in controlling anthracnose and blossom end rot of a range of dessert banana varieties (87–95% and 81–82% disease reductions, respectively), while a concentration of 10^4 cfu/ml with Tween 20 (0.02% v/v) was sufficient to have a 86–98% control of crown rot (De Costa et al. 2008b). In vitro and in vivo studies have shown the broad-spectrum ability of *B. spinosa* to control several other postharvest pathogens, namely, *Colletotrichum gloeosporioides*, *Botryodiplodia theobromae*, *Thielaviopsis paradoxa*, *Ralstonia solanacearum* and *Erwinia carotovora* (De Costa et al. 2008b; Gunawardana and De Costa 2016; Subasinghe and De Costa 2017). Edible parts of banana and carrot which were treated with *B. spinosa* as a postharvest treatment were free from the antagonist (De Costa et al. 2008b; Samaranayaka et al. 2017). At a molecular level analysis of recA gene by PCR-RFLP (Polymerase Chain Reaction- Restriction Fragment Lenght Polymorphism) revealed a unique profile of DNA fragments for

B. spinosa which discriminated it from human pathogenic genomovar of *Burkholderia cepacia* complex (De Costa et al. 2008b).

14 Research on *Trichoderma* spp. Conducted in Sri Lanka

As per the Google search results and literature available from research symposia and undergraduate and postgraduate theses/dissertations, it is clear that majority of the research on biological control agents against plant pathogens has been done using *Trichoderma* spp.. Table 3 highlights the research conducted in Sri Lanka using different *Trichoderma* spp. for the management of a range of plant pathogens. It is evident that research has been conducted over a period of three decades on screening, identification, determination of in vitro and in vivo antagonism, mass production and formulation of *Trichoderma* spp. (Table 3).

15 Potential to Commercialize Microbial Pesticides in Sri Lanka

The rich diversity of climate, geology, topography and soils in Sri Lanka contributes to the species richness of its fauna and flora and to possess a wide array of ecosystems (Evans 1999). Therefore, Sri Lanka has been world renowned as one of the top biological hotspots in the globe. As each species and each environmental habitat harbour a guild of naturally dwelling microbes in the form of epiphytes, endophytes and saprophytes, harnessing of microorganisms having antagonistic ability against agricultural pests, among the natural inhabitants, is not a practical limitation. Hence, use of indigenous biological control agents for the management of crop pests is a possibility. For example, presence of a wide range of *Trichoderma* spp. (i.e. *T. asperellum*, *T. longibrachiatum*, *T. harzianum*, *T. viride*, *T. virens*, *T. polysporum*, *T. koningii*, *T. erinaceum*) highlights the species richness of the given biological control agent in different locations of the country (Table 3).

Increased public concern on the health and environmental hazards associated with synthetic pesticides has become a catalytic factor for developing alternative measures for the management of agricultural pests in Sri Lanka. Further, in comparison to the pesticide imports of 2013, an abrupt decline in pesticide imports (a 30% decline) has been reported in year 2014 (Lakshani et al. 2017). This is due to the stringent control over high-volume pesticides and banning of some weedicide formulations. Hence, 18% reduction of all pesticide formulations of imports to the country has been reported (Department of Agriculture 2015).

A growing interest on research related to biological control of agricultural pests in Sri Lankan researchers is clearly evident. Researchers of state universities, postgraduate institutes, Department of Agriculture, Sri Lanka and national research institutes have prioritized biological control research in their agenda. It is more interesting to note that the number of research involved with novel molecular biological techniques is being increasing and researchers are using these non-conventional

Table 3 Research conducted in Sri Lanka using different *Trichoderma* spp. for the management of a range of plant pathogens

Trichoderma spp.	Pathogen	Crop	References
T. harzianum RU01	*Uromyces appendiculatus*	*Phaseolus vulgaris*	Abeysinghe (2009a)
T. harzianum	*Rigidoporus microporus*	*Hevea brasiliensis*	Jayasuriya and Thennakoon (2007)
T. harzianum (TrH 40)	*Botryodiplodia theobromae*, *Colletotrichum gloeosporioides* and *Gliocephalotrichum microchlamydosporum*	*Nephelium lappaceum*	Sivakumar et al. (2002)
T. harzianum RU01	*Rhizoctonia solani*	*Solanum melongena* and *Capsicum annuum*	Abeysinghe (2009b)
T. harzianum and *T. viride*	*Pythium aphanidermatum*	*Solanum lycopersicum*	Jeyaseelan et al. (2012)
T. asperellum	*Thielaviopsis paradoxa* (*telemorph* = *Ceratocystis paradoxa*), *Colletotrichum gloeosporioides*	*Ananas comosus*, *Nephelium lappaceum*	Wijesinghe et al. (2010a, b, 2011)
T. asperellum, *T. longibrachiatum*	–	–	Rajapakse et al. (2016)
T. viride	*Meloidogyne* spp.	*Centella asiatica*	Shamalie et al. (2011)
T. viride	*Colletotrichum gloeosporioides*, *Fusarium oxysporum* f. sp. *cepae*	*Allium cepa*	Naguleswaran et al. (2014)
T. viride	*Sclerotium rolfsii*	–	Emerson and Mikunthan (2015)
T. viride and *T. koningii*	*Sclerotium rolfsii*	*Chlorophytum comosum 'Iaxum'*	Priyadarshani and Kelaniyangoda (2013)
T. viride and *T. harzianum*	*Fusarium oxysporum*	*Crossandra infundibuliformis* var. *Danica*	Jegathambigai et al. (2009a)
T. viride and *T. harzianum*	*Helminthosporium* sp.	*Chrysalidocarpus lutescens*	Jegathambigai et al. (2009b)
T. viride, *T. polysporum* and *T. harzianum*	*Ceratocystis paradoxa*	–	Kannangara et al. (2017)
T. erinaceum, *T. virens* and *T. asperellum*	*Fusarium oxysporum*, *Rhizoctonia solani*, *Colletotrichum gloeosporioides*, *Curvularia lunata*, *Corynespora cassiicola*, *Rigidoporus microporus* and *Phytophthora meadii*	–	Herath et al. (2015, 2017)

(continued)

Table 3 (continued)

Trichoderma spp.	Pathogen	Crop	References
Trichoderma spp.	*Rigidoporus lignosus*	*Hevea brasiliensis*	Wijesundara et al. (1991)
Trichoderma spp.	*Sclerotium rolfsii*	*Zamioculcas zamiifolia*	Jegathambigai et al. (2010)
Trichoderma spp.	*Fusarium* sp.	*Allium cepa*	Gunaratna et al. (2014)

techniques to reveal the in-depth information related to biological control at molecular level.

Even though there is no direct connection with the microbial biopesticides, Sri Lanka has previous experiences on successful biological control of agricultural pests, namely, *Salvinia molesta*, an aquatic weed; *Promecotheca cumingi*, the coconut leaf miner; and *Aceria guerreronis*, the coconut mite. These experiences will have a positive impact on the willingness to use microbial biopesticides by growers and invest on microbial pesticides by manufacturers.

16 Present Status of Development of Microbial Pesticides in Sri Lanka

Literature available shows that Sri Lanka has largely focused on microbial antagonists against plant pathogens in comparison to insect pests and weeds. Among the microbial antagonists, the biggest attention has been centred on *Trichoderma* spp., and potential of yeast, bacteriophage, nematode, virus and protozoan antagonists has not been explored yet. It is evident that research on microbial pesticides in Sri Lanka has been still limited to the initial steps of the development procedure, namely, isolation, screening, culture maintenance, identification and in vitro and in vivo antagonism. Very few microbial antagonists have passed the steps of mass production and formulation. No information is available on toxicological studies on non-target species and environmental impact of the potential microbial antagonists. Further, no microbial pesticide having an indigenous origin has been registered in the country yet. Until recently, the identification of the antagonists was solely based on conventional methods (morphology, biochemical tests). However, from the recent past, application of molecular methods for identification has been used. In addition to accurate identification of the antagonists, molecular methods can be used for evaluating the environmental impact, traceability and fate of the microbes in the environment and for improving the antagonistic efficiency of microbes against their hosts.

When the microbial pesticide market of the neighbouring countries in Asia is considered, China has 327 registered biopesticides as per the records till 2008. Out of them, 270 bacterial biopesticides have been obtained from 11 microbial species, mostly from *B. thuringiensis*. There are 22 registered fungal biopesticides in China originated from 6 fungal species. In addition, 35 registered viral biopesticides can be found, and 14 out of them have been developed from *Heliothis armigera* nuclear polyhedrosis virus (NPV) (ICAMA 2008). When the Indian biopesticide market is considered, most of the microbial pesticides have been developed using *Trichoderma* spp., *B. thuringiensis* and *Pseudomonas fluorescens.* Viral biopesticides contain NPV and granuloviruses (GV) (Rabindra 2005). By the year 2006, there were 12 registered microbial pesticides in India (Gupta 2006).

17 Future Directions

Based on the available literature, a substantial level of increase of the interest, knowledge and technology related to development of microbial pesticides has been observed over the past three decades in Sri Lanka. Despite these improvements, Sri Lanka has not been able to release a commercial product of microbial pesticide yet. To this end, a joint venture has to be undertaken by the researchers, manufacturers, government authorities and end users. Research should be continued till the commercialization without limiting to the research related to initial steps of the development procedure of microbial pesticides.

As microbial pesticides contain the whole organisms or their vegetative or reproductive units, maintaining the quality with respect to their antagonistic ability, viability (shelf life), purity and safety is of paramount importance for the physical and economic sustainability of the product. Similarly, mass culturing of the microbial antagonist is a must, to produce the final product in large quantities, to meet the demand. As microbes have diverse nutrient requirements, tailor-made mass culturing protocols with reference to the type of culture medium and equipment needed by them have to be determined. Further, mass culturing should be a cost-effective exercise to reach higher profit margins, from the manufacturer's point of view. Further, continuous monitoring of the product is needed to maintain the originality of the microbial pesticides, particularly due to the 'living nature' of it. Therefore, to release a successful final product of a microbial pesticide and even after the release, a heavy involvement of research studies is essential.

It is evident through the literature available that a reasonable number of laboratory studies on biological control of plant diseases have been done by Sri Lankan researchers with reference to isolation, identification, in vitro screening and a certain extent of in vivo screening. Considering the time, effort and resources spent on these findings, establishment of a national repository of the valuable microbial antagonists is a timely need.

Awareness of the end users on microbial pesticides needs to be promoted, and attitudinal changes have to be created in all strata of the society on the use of microbial pesticides for the management of crop pests in comparison to the use of synthetic pesticides. Though outcomes are not rapid and microbial pesticides do not result in a zero-level reduction of pest population, advantages of introducing microbial pesticides as an effective component in integrated management programmes need to be educated. Dissemination of such knowledge should be targeted from primary education to every stratum of the society; hence incorporating such information as appropriate to school curricula would be an effective modality. Most users are of the view that microbial pesticides can reach the targets as a sole measure of management. Considering the possible health and environmental threats of synthetic pesticides, government has to take the leading role in promoting and making available of microbial pesticides. Setting up an acceptable regulatory framework (while preserving the quality, safety and profit), encouraging the use of biological control through providing incentives/subsidies, establishing model farms and identifying model farmers/leader farmers are some of the options available to the government authorities. Considering the growing threat posed on health and environment due to complete dependence on synthetic pesticides, Sri Lanka should focus on a long-lasting country policy on the promotion of microbial pesticides/biopesticides, and it should not be faded away by political changes.

In Sri Lanka, the industrial biotechnology sector is in its infancy. As industrial biotechnology has an intimate relationship with the mass production of quality microbial pesticides, government support is needed in this regard. If private manufacturers are hesitant to manufacture microbial pesticides, government authority or a government-supported institute should get the responsibility of manufacturing while assuring the quality of the product. To this end, we can learn a lot from neighbouring countries. For example, even with less sophisticated technology, India has been successful in developing biopesticides at small scale to meet the local requirements by joint efforts of local sugar mills, cooperatives, state agricultural departments, integrated pest management centres and national universities (Alam 2000). Further, Shukla and Shukla (2012) elaborate the Indian scenario with reference to promoting research, production, registration and adoption of biopesticides in India in collaboration with the supporting organizations and schemes such as the Department of Biotechnology, Indian Council of Agricultural Research, National Agriculture Technology Project, National Centre for Integrated Pest Management and State Biocontrol Laboratories. Shukla and Shukla (2012) further highlight that amendments were done to Insecticide Act in 1968, for speeding up the development and production of biopesticides and their registration in India. Government should channel more funds on research related to microbial pesticides and human resource development. To this end, public-private partnership research grants have been promoted by some of the Sri Lankan research funding agencies. Farmers/growers should be patient enough to reap the advantages of the use of microbial pesticides, and consumers should be willing to accept cosmetic defects of food commodities treated with microbial pesticides if there are any. Such a collective effort definitely will be helpful to see the light at the end of the tunnel.

Acknowledgements The author wishes to acknowledge the efforts of all the Sri Lankan researchers who have contributed to the field of biological control of agricultural pests and the financial support provided by all the funding agencies for the research on biological control.

References

Abeysinghe S (2009a) Systemic resistance induced by *Trichoderma harzianum* RU01 against *Uromyces appendiculatus* on *Phaseolus vulgaris*. J Natl Sci Found 37(3):203–207

Abeysinghe S (2009b) Effect of combined use of *Bacillus subtilis* CA32 and *Trichoderma harzianum* RUOI on biological control of *Rhizoctonia solani* on *Solanum melongena* and *Capsicum annuum*. Plant Pathol J 8:9–16

Aktar W, Sengupta D, Chowdhury A (2009) Impact of pesticides use in agriculture: their benefits and hazards. Interdiscip Toxicol 2(1):1–12

Alam G (2000) A study of biopesticides and biofertilisers in Haryana, India. International Institute for Environment and Development

Alexandratos N, Bruinsma J (2012) World agriculture towards 2030/2050: the 2012 revision. ESA working paper no. 12-03. Rome, FAO

Amarasena PGDS, Mohotti KM, Ahangama D (2011) A locally isolated entomopathogenic fungus to control tea red spider mites (*Oligonychus coffeae Acarina*-Tetranychidae). Trop Agric Res 22 (4):384–391

Amarasinghe LD (2008) Entomopathogenic nematodes from the coastal belt of Sri Lanka and their efficiency in controlling termites. Pest Technol 2(2):125–129

Amarasinghe LD, Hominick WM (1993) Efficacy of entomopathogenic nematodes to control up-country live wood termite, *Postelectrotermes militaris*. Sri Lanka J Tea Sci 62(1):16–24

Amarasinghe LD, Hominick WM, Ried AP, Briscoe B (1994) Entomopathogenic nematodes for control of the tea termite, *Postelectrotermes militaris* in Sri Lanka. J Helminthol 68(4):277–286

Aponso GLM, Magamage C, Ekanayake WM, Manuweera GK (2003) Analysis of water for pesticides in two major agricultural areas of the dry zone. Ann Sri Lanka Dep Agric 5:7–22

Aravinna P, Priyantha N, Pitawala A, Yatigammana SK (2017) Use pattern of pesticides and their predicted mobility into shallow groundwater and surface water bodies of paddy lands in Mahaweli river basin in Sri Lanka. J Environ Sci Health B 52(1):37–47

Bajwa WI, Kogan M (2002) Compendium of IPM Definitions (CID). What is IPM and how is it defined in the worldwide literature? Integrated Plant Protection Center (IPPC), Oregon State University, Corvallis, OR, USA

Berg G, Erlacher A, Grube M (2015) The edible plant microbiome: importance and health issues. In: Principles of plant-microbe interactions. Springer, Cham, pp 419–426

Bonaterra A, Badosa E, Cabrefiga J, Francés J, Montesinos E (2012) Prospects and limitations of microbial pesticides for control of bacterial and fungal pomefruit tree diseases. Trees 26 (1):215–226. https://doi.org/10.1007/s00468-011-0626-y

Brakhage AA, Schroeckh V (2011) Fungal secondary metabolites–strategies to activate silent gene clusters. Fungal Genet Biol 48(1):15–22

Brimner TA, Boland GJ (2003) A review of the non-target effects of fungi used to biologically control plant diseases. Agric Ecosyst Environ 100(1):3–16

Bulgarelli D, Schlaeppi K, Spaepen S, van Themaat EVL, Schulze-Lefert P (2013) Structure and functions of the bacterial microbiota of plants. Annu Rev Plant Biol 64:807–838

Carson R (1962) The silent spring. Houghton Mifflin, Boston

Carvalho FP (2017) Pesticides, environment, and food safety. Food Energy Secur 6(2):48–60

Chaminda KGS, Marapana RAUJ, Serasinghe RT, Karunagoda RP (2012) Environmental impact and use of agrochemical in cattle feed and its effect on milk in Magastota, Nuwara Eliya, Sri Lanka. In: Centre for Environmental Justice/Friends of the Earth Sri Lanka. First National symposium proceedings, pp 27–30

Chandler D, Bailey AS, Tatchell GM, Davidson G, Greaves J, Grant WP (2011) The development, regulation and use of biopesticides for integrated pest management. Philos Trans R Soc Lond B Biol Sci 366(1573):1987–1998. https://doi.org/10.1098/rstb.2010.0390

Danthanarayana W (1967) Tea entomology in perspective. Tea Q 38(2):153–177

Danthanarayana W, Vitarana SI (1987) Control of the live-wood tea termite *Glyptotermes dilatatus* using *Heterorhabditis* sp. (Nemat.). Agric Ecosyst Environ 19(4):333–342

De Costa DM, Erabadupitiya HRUT (2005) An integrated method to control postharvest diseases of banana using a member of the *Burkholderia cepacia* complex. Postharvest Biol Technol 36 (1):31–39

De Costa DM, Subasinghe SSNS (1998) Antagonistic bacteria associated with the fruit skin of banana in controlling its postharvest diseases. Trop Sci 38(4):206–212

De Costa DM, Amaradasa BS, Wegiriya RNBPMRCL (1997) Antagonists of *Colletotrichum musae* associated with banana fruit skin. J Natl Sci Found 25(2)

De Costa DM, Samarasinghe SST, Dias HRD, Dissanayake DMN (2008a) Control of rice sheath blight by phyllosphere epiphytic microbial antagonists. Phytoparasitica 36(1):52–65

De Costa DM, Zahra ARF, Kalpage MD, Rajapakse EMG (2008b) Effectiveness and molecular characterization of *Burkholderia spinosa*, a prospective biocontrol agent for controlling post-harvest diseases of banana. Biol Control 47(3):257–267

De A, Bose R, Kumar A, Mozumdar S (2014) Targeted delivery of pesticides using biodegradable polymeric nanoparticles. Springer, New Delhi, pp 59–81

Department of Agriculture (2015) Pesticide statistics of the Office of the Registrar for the pesticides industry. Department of Agriculture, Peradeniya, Sri Lanka

Edgington S, Priyanthie Fernando LC, Jones K (2008) Natural incidence and environmental profiling of the mite-pathogenic fungus *Hirsutella thompsonii* fisher for control of the coconut mite in Sri Lanka. Int J Pest Manag 54(2):123–127

Ekanayake EMMS, Manage PM, Liyanage GY (2018) Isolation of fungi as biological control agents against water hyacinth (*Eichhornia crassipes*). In: Proceedings of international forestry and environment symposium, February, 22, 98p

Emerson FL, Mikunthan G (2015) Small scale production of *Trichoderma viride* on locally available liquid waste and other substrates. http://repo.lib.jfn.ac.lk/ujrr/handle/123456789/948. Accessed 19 June 2018

EUPD (2010) European union pesticides database. http://ec.europa.eu/food/plant/protection/evaluation/database_act_subs_en.htm

European Commission (2000) Directive 2000/54/EC on the protection of workers from risks related to exposure to biological agents at work. Off J Eur Community 262:21–45

Evans HC (1999) Biological control of weed and insect pests using fungal pathogens, with particular reference to Sri Lanka. Biocontrol News Inf 20(2):63N–68N

FAO (2017) http://www.fao.org/faostat/en/#home

Fernando LCP, Kanagaratnam P, Narangoda NK (1995) Some studies on the use of *Metarhizium anisopliae* (Metsch.) Sor. For the control of *Oryctes rhinoceros* in Sri Lanka. Cocos 10:46–52

Gans J, Wolinsky M, Dunbar J (2005) Computational improvements reveal great bacterial diversity and high metal toxicity in soil. Science 309(5739):1387–1390

Guillon ML (2003) Regulation of biological control agents in Europe. In: International symposium on biopesticides for developing countries. CATIE, Turrialba, pp 143–147

Gunaratna LNR, Deshappriya N, Jayaratna DL (2014) *Trichoderma* as a promising biological control agent against damping off disease in big onion (*A. cepa* L.). In: Proceedings of 15th annual research symposium, Faculty of Graduate Studies, University of Kelaniya, Sri Lanka, 94 p, 2009

Gunawardana DUM, De Costa DM (2016) Potential of using biocontrol agents as postharvest treatments to manage carrot soft rot. In: Proceedings of the 3rd symposium of Faculty of Agriculture Undergraduate Research Symposium, Faculty of Agriculture, University of Peradeniya, 16 December, p 130

Gunnell D, Fernando R, Hewagama M, Priyangika WDD, Konradsen F, Eddleston M (2007) The impact of pesticide regulations on suicide in Sri Lanka. Int J Epidemiol 36(6):1235–1242

Gupta PK (2006) Status of biopesticides – Indian scene. Toxicol Lett 164:S40

Hayward AC, Fegan N, Fegan M, Stirling GR (2010) *Stenotrophomonas* and *Lysobacter*: ubiquitous plant-associated gamma-proteobacteria of developing significance in applied microbiology. J Appl Microbiol 108(3):756–770

Herath HHMAU, Wijesundera RLC, Chandrasekharan NV, Wijesundera WSS, Kathriarachchi HS (2015) Isolation and characterization of *Trichoderma erinaceum* for antagonistic activity against plant pathogenic fungi. Curr Res Environ Appl Mycol 5(2):120–127

Herath HHMAU, Wijesundera RLC, Chandrasekharan NV, Wijesundera WSS (2017) Exploration of Sri Lankan soil fungi for biocontrol properties. Afr J Biotechnol 16(20):1168–1175

ICAMA (2008) Pesticide manual, the institute for the control of agrochemicals. Ministry of Agriculture, China. (in Chinese)

International Agency for Research on Cancer (2016) 2,4-Dichlorophenoxyacetic acid (2,4 D) and some organochlorine insecticides in IARC monograph on the evaluation of carcinogenic risk to humans. International Agency for Research on Cancer, Lyon, France

International Agency for Research on Cancer (2017) Some organophosphate insecticides and herbicides in IARC monograph on the evaluation of carcinogenic risk to humans. International Agency for Research on Cancer, Lyon, France

Jayasuriya KE, Thennakoon BI (2007) Biological control of *Rigidoporus microporus*, the cause of white root disease in rubber. J Rubber Res Inst Sri Lanka 75:61–70

Jayetilleke J, Bandara JMRS (1989) Pesticide management by the hill country vegetable farmers. Trop Agric Res 1:121–131

Jegathambigai V, Wijeratnam RSW, Wijesundera RLC (2009a) Control of *Fusarium oxysporum* wilts disease of *Crossandra infundibuliformis* var. *Danica* by *Trichoderma viride* and *Trichoderma harzianum*. Asian J Plant Pathol 3(3):50–60

Jegathambigai V, Wijeratnam RW, Wijesundera RLC (2009b) *Trichoderma* as a seed treatment to control *Helminthosporium* leaf spot disease of *Chrysalidocarpus lutescens*. World J Agric Sci 5:720–728

Jegathambigai V, Wijeratnam RW, Wijesundera RLC (2010) Effect of *Trichoderma* sp. on *Sclerotium rolfsii*, the causative agent of collar rot on *Zamioculcas zamiifolia* and an on farm method to mass produce *Trichoderma* species. Plant Pathol J 9(2):47–55

Jeyaratnam J, de Alwis Seneviratne RS, Copplestone JF (1982) Survey of pesticide poisoning in Sri Lanka. Bull World Health Organ 60(4):615

Jeyaseelan EC, Tharmila S, Niranjan K (2012) Antagonistic activity of *Trichoderma* spp. and *Bacillus* spp. against *Pythium aphanidermatum* isolated from tomato damping off. Arch Appl Sci Res 4(4):1623–1627

Kanagaratnam P, Pethiyagoda U, Velu MS (1983) Effect of four commercial preparations of *Bacillus thuringiensis* on *Opisina arenosella* walker. Cocos 1:7–10

Kannangara S, Dharmarathna RMGCS, Jayarathna DL (2017) Isolation, identification and characterization of *Trichoderma* species as a potential biocontrol agent against *Ceratocystis paradoxa*. J Agric Sci 12(1):51–62

Kelaniyangoda DB, Ekanayake HMRK (2010) *Puccinia melampodii* diet and Holow. as a biological control agent of *Parthenium hysterophorus*. J Food Agric 1(1):13–19

Kiewnick S (2007) Practicalities of developing and registering microbial biological control agents. CAB Rev 2:1–11

Knipe DW, Metcalfe C, Gunnell D (2015) WHO suicide statistics–a cautionary tale. Ceylon Med J 60(1):35. https://doi.org/10.4038/cmj.v60i1.7464

Köhl J, Postma J, Nicot P, Ruocco M, Blum B (2011) Stepwise screening of microorganisms for commercial use in biological control of plant-pathogenic fungi and bacteria. Biol Control 57 (1):1–12

Koul O (2011) Microbial biopesticides: opportunities and challenges. CAB Rev 6:1–26

Kumar S (2015) Biopesticide: an environment friendly pest management strategy. J Biofertil Biopestici 6(1):e127. https://doi.org/10.4172/2155-6202.1000e127

Kumar S, Thakur M, Rani A (2014) *Trichoderma*: mass production, formulation, quality control, delivery and its scope in commercialization in India for the management of plant diseases. Afr J Agric Res 9(53):3838–3852

Lakshani PWY, Rajapaksha MKLK, Sendthuran K (2017) Pesticide residues in selected vegetables in several growing areas by GC/MS using QuEChERS technique. Ann Sri Lanka Dep Agric 19 (2):188–208

Lindow SE, Brandl MT (2003) Microbiology of the phyllosphere. Appl Environ Microbiol 69 (4):1875–1883

Marasinghe JP, Magamage C, Shiromi MGD, Aravinna AGP (2011) Organophosphate pesticide residues in food commodities in Sri Lanka: a review. Ann Sri Lanka Dep Agric 13:81–94

Marrone PG (2007) Barriers to adoption of biological control agents and biological pesticides. In: CAB reviews: perspectives in agriculture, veterinary science, nutrition and natural resources 2, 51. CAB International, Wallingford

Mendes R, Garbeva P, Raaijmakers JM (2013) The rhizosphere microbiome: significance of plant beneficial, plant pathogenic, and human pathogenic microorganisms. FEMS Microbiol Rev 37 (5):634–663

Meyling NV (2007) Methods for isolation of entomopathogenic fungi from the soil environment-laboratory manual, 18 pp

Mishra J, Tewari S, Singh S, Arora NK (2015) Biopesticides: where we stand? In: Plant microbes symbiosis: applied facets. Springer, New Delhi, pp 37–75

Monteiro RA, Balsanelli E, Wassem R, Marin AM, Brusamarello-Santos LC, Schmidt MA, Tadra-Sfeir MZ, Pankievicz VC, Cruz LM, Chubatsu LS, Pedrosa FO (2012) *Herbaspirillum*-plant interactions: microscopical, histological and molecular aspects. Plant Soil 356(1–2):175–196

Montesinos E (2003) Development, registration and commercialization of microbial pesticides for plant protection. Int Microbiol 6(4):245–252

Naguleswaran V, Pakeerathan K, Mikunthan G (2014) Biological control: a promising tool for bulb-rot and leaf twisting fungal diseases in red onion (*Allium cepa* L.) in Jaffna district. World Appl Sci J 31(6):1090–1095

Nøstbakken OJ, Hove HT, Duinker A, Lundebye AK, Berntssen MH, Hannisdal R, Lunestad BT, Maage A, Madsen L, Torstensen BE, Julshamn K (2015) Contaminant levels in Norwegian farmed Atlantic salmon (*Salmo salar*) in the 13-year period from 1999 to 2011. Environ Int 74:274–280

Oerke EC (2006) Crop losses to pests. J Agric Sci 144(1):31–43

Padmajani MT, Bandara MACS, Aheeyar MMM (2014) Assessment of pesticide usage in up-country vegetable farming in Sri Lanka

Paoli D, Giannandrea F, Gallo M, Turci R, Cattaruzza MS, Lombardo F, Lenzi A, Gandini L (2015) Exposure to polychlorinated biphenyls and hexachlorobenzene, semen quality and testicular cancer risk. J Endocrinol Investig 38(7):745–752

Popp J, Pető K, Nagy J (2013) Pesticide productivity and food security. Agron Sustain Dev 33 (1):243–255

Pretty J (2008) Agricultural sustainability: concepts, principles and evidence. Philos Trans R Soc Lond B Biol Sci 363(1491):447–465

Priyadarshani KAL, Kelaniyangoda DB (2013) Effect of bio control agent *Trichoderma* (*T. viride* and *T. konnigiil* on basal rot of *Cloropytum comosum* 'Iaxum' caused by *Sclerotium rolfsii*. In: Proceedings of international forestry and environment symposium, September 2013

Queene CA, Safeena MIS, Zakeel MCM (2016) Identification of suitable potential pathogens for biocontrol of water hyacinth. In: 5th annual research sessions, South Eastern University of Sri Lanka, pp 231–236

Rabindra RJ (2005) Current status of production and use of microbial pesticides in India and the way forward. In: Microbial biopesticide formulations and applications. Project Directorate of Biological Control, technical document, 55, pp 1–12

Rajapakse RGAS, Fariz FS, Wickramarachchi WART, Dissanayake DMKK, Premarathne MPT, Kahawatte KJPK (2016) Morphological and molecular characterization of Trichoderma isolates used as bio-control agents in Sri Lanka. Trop Agric 164:17–27

Samaranayaka GRPM, De Costa DM, Priyadarshani TDC (2017) Potential of managing carrot soft rot caused by *Erwinia carotovora* using a bacterial antagonist, *Burkholderia spinosa*. In:

Proceedings of the 9th annual research symposium on sustainable agriculture for food security and poverty alleviation 2017, 21st September 2017, Faculty of Agriculture, Rajarata University of Sri Lanka, 74p

Sarwar M (2015) Biopesticides: an effective and environmental friendly insect-pests inhibitor line of action. Int J Eng Adv Res Technol 1(2):10–15

Shamalie BVT, Fonseka RM, Rajapaksha RGAS (2011) Effect of *Trichoderma viride* and Carbofuran (Curator) on management of root knot nematodes and growth parameters of gotukola (*Centella asiatica* L.). Trop Agric Res 23(1):61–69

Sharaniaya S, Loganathan P (2015) Vegetable growers perception of pesticide use practices and health effects in the Vavuniya District. Am Eurasian J Agric Environ Scie 15(7):1479–1485

Shukla R, Shukla A (2012) Market potential for biopesticides: a green product for agricultural applications. Int J Manag Res Rev 2(1):91

Sinha B (2012) Global biopesticide research trends: a bibliometric assessment. Indian J Agric Sci 82 (2):95–101

Sivakumar D, Wijeratnam RW, Wijesundera RLC, Abeyesekere M (2002) Control of postharvest diseases of rambutan using cinnamaldehyde. Crop Prot 21(9):847–852

Smit LA, van Wendel-de-Joode BN, Heederik D, Peiris-John RJ, van der Hoek W (2003) Neurological symptoms among Sri Lankan farmers occupationally exposed to acetylcholinesterase-inhibiting insecticides. Am J Ind Med 44(3):254–264

Soe KT, De Costa DM (2012) Development of a spore-based formulation of microbial pesticides for control of rice sheath blight. Biocontrol Sci Tech 22(6):633–657

Sri Lankan Ministry of Health (1995) Annual health bulletin. Ministry of Health, Colombo, 1997

Stockwell VO, McLaughlin RJ, Henkels MD, Loper JE, Sugar D, Roberts RG (1999) Epiphytic colonization of pear stigmas and hypanthia by bacteria during primary bloom. Phytopathology 89(12):1162–1168

Subasinghe SDSK, De Costa DM (2017) Potential of *Burkholderia spinosa* and *Bacillus megaterium* on plant growth promotion and suppression of selected soil-borne diseases of tomato (*Solanum lycopersicum* L.). In: Proceedings of the 4th symposium of Faculty of Agriculture Undergraduate Research Symposium, Faculty of Agriculture, University of Peradeniya, 12 January, 114

Subasinghe MRVN, Amarasinghe KGAPK, Dharmadasa M (2013) Possibility of use of agricultural Byproducts for mass production of *Beauveria bassiana* (Balsamo) Vuillemin to control coffee berry borer (*Hypothenemus hampi* (Ferrari)). J Food Agric 6:1–2

Svinningen AE, Jegathambigai V, Mikunthan G (2010) *Nomuraea rileyi*: a plausible fungi selectively controlling lepidopteron, *Parapoynx stratiotata* L. damaging queen palm (*Livistona rotundifolia* L.). Agric Appl Biol Sci 75:279–293

Turner TR, James EK, Poole PS (2013) The plant microbiome. Genome Biol 14(6):209

UN (2017) The world population prospects. United Nations, Department of Economic and Social Affairs, Population Division. 2017 Revision. https://esa.un.org/unpd/wpp/. Accessed 18 July 2018

USEPA (2008) What are biopesticides? http://www.epa.gov/pesticides/biopesticides/whatarebiopesticides.htm. Accessed 12 June 2018

van Lenteren JC (2000) A greenhouse without pesticides: fact or fantasy? Crop Prot 19(6):375–384

Watawala RC, Liyanage JA, Mallawatantri A (2010) Assessment of risks to water bodies due to residues of agricultural fungicide in intensive farming areas in the up-country of Sri Lanka using an indicator model. In: Proceedings of the National conference on water, food security, and climate change in Sri Lanka, BMICH, Colombo, June 9–11, 2009, vol 2. Water quality, environment, and climate change, p 69. IWMI

WHO (2017) Agrochemicals, health and environment: directory of resources. http://www.who.int/heli/risks/toxics/chemicalsdirectory/en/index1.html. Accessed 12 June 2018

WHO The World Health Report (2001) Mental health: new understanding, new hope. World Health Organization, Geneva

Widanapathirana CU, Dassanayake DLALA (2013) The use of plant parts in Pest control activities in traditional Sri Lankan agricultural systems. Int J Sci Technol Res 2(6):150–152

Wijesinghe CJ, Wijeratnam RW, Samarasekara JKRR, Wijesundera RLC (2010a) Biological control of *Thielaviopsis paradoxa* on pineapple by an isolate of *Trichoderma asperellum*. Biol Control 53(3):285–290

Wijesinghe CJ, Wijeratnam RW, Samarasekara JKRR, Wijesundera RLC (2010b) Identification of *Trichoderma asperellum* from selected fruit plantations of Sri Lanka. J Natl Sci Found 38(2)

Wijesinghe CJ, Wijeratnam RW, Samarasekara JKRR, Wijesundera RLC (2011) Development of a formulation of *Trichoderma asperellum* to control black rot disease on pineapple caused by (*Thielaviopsis paradoxa*). Crop Prot 30(3):300–306

Wijesundara RLC, Jeyanathan S, Liyanage NIS (1991) Some effects of isolates of *Trichoderma* on *Rigidipourus lignosus*

Yapa SWCRYMUSB, Dharmadasa M, Fernandopulle MND (2007) Possibility of use of *Beauveria bassiana* (Balsamo) for the control of coffee Berry Borer (*Hypothenemus hampei* (Ferrari)) (Coleoptra: Scolytidae). In: Proceeding of 7th agricultural research symposium, Wayamba University of Sri Lanka, pp 121–124

Statistics for Furtherance of Agricultural Research in Sri Lanka

S. Samita, T. Sivanathawerl, and L. D. B. Suriyagoda

1 Introduction

In almost all areas of research, use of statistics at the planning stage and analysis stage is essential to ensure that conclusions are valid. With recent advancements in information technology, access to use of statistics is quite high, and thus it is good to see an increasing trend of using statistics in research of all areas of work in general. Experimentation in agriculture and the use of statistics in agricultural research in Sri Lanka began with the establishment of the government organizations such as the Department of Agriculture (DOA), Tea Research Institute (TRI), Rubber Research Institute (RRI), and Coconut Research Institute (CRI) in the early twentieth century. Research results generated from those organizations have then been documented and published mainly through the journal of *Tropical Agriculturalist* (TA) published by the DOA from the early part of the twentieth century. Apart from that, the first Faculty of Agriculture (FOA) in the university system in Sri Lanka was established in 1948 at the University of Peradeniya. Researchers attached to the Faculty of Agriculture started contributing to experimentation in agriculture from 1948 onwards. Many of the graduates of the Faculty of Agriculture later became research officers at the DOA. Statistics became an integral part of the Faculty curriculum from 1974 onwards. In 1975, the Postgraduate Institute of Agriculture (PGIA) was established to offer taught and research-based higher degrees in agriculture, and PGIA also began publishing its research journal by the name *Tropical Agricultural Research* (TAR) from 1990 onwards. Thus TA and TAR together is the showcase of agricultural research that took place from the past to present. A good understanding of evolution of agricultural research in Sri Lanka can be obtained by analyzing articles in those two journals. This work was initiated to look into that aspect,

S. Samita (✉) · T. Sivanathawerl · L. D. B. Suriyagoda
Department of Crop Science, Faculty of Agriculture, University of Peradeniya, Peradeniya, Sri Lanka
e-mail: ssamita@pdn.ac.lk

R. P. De Silva et al. (eds.), *Agricultural Research for Sustainable Food Systems in Sri Lanka*, https://doi.org/10.1007/978-981-15-3673-1_8

especially with respect to use of statistics in agricultural research. The specific objectives of this article were to understand, (i) the type of research that the DOA, other key research institutes and PGIA have conducted, (ii) the use of statistics in their research, and (iii) identify the areas requiring improvements with respect to use of statistics, based on the published research work in those two journals. Specifically, the focus was on articles published during the period 1998–2017, and the number of articles published during this period by TAR and TA were 604 and 100, respectively.

2 Types of Research Conducted

Major areas of research conducted by the researchers from the DOA and PGIA are given in Table 1. Most of the research results published were on agronomy (22%) followed by genetics and breeding of crops, animal science, and social sciences in the descending order. Agronomy mostly consisted of crop management and eco-physiology-related research focusing toward ensuring sustainability of agricultural systems. Varietal screening, diversity/germplasm evaluation, and/or crop improvement-related research were the most common within the area of genetics and breeding. Experimental results published in the areas of floriculture, forestry and agroforestry, biostatistics, and climate science were relatively low (<2%). This may be due to the lack of interest that existed among the public on these themes in the past (Hall et al. 2009; Moss et al. 2010; Puri and Nair 2004; Spilsbury and Nasi 2006). However, the importance on these themes has recently been increased than ever (Luken et al. 2012; Moss et al. 2010; Turner et al. 2002). For instance, 11% articles (Table 1) related to pest and disease management has been published in the recent past.

Table 1 Major areas of research published in TA and TAR (1998–2017)

Area of research	Number	%
Agronomy	158	22.4
Genetics and breeding	111	15.8
Animal science	66	9.4
Social science	53	7.5
Food science	48	6.8
Environmental science	45	6.4
Economics	43	6.1
Pest management	43	6.1
Pathology	37	5.3
Postharvest management	34	4.8
Soil science	27	3.8
Biostatistics	11	1.6
Climate science	11	1.6
Forestry and agroforestry	10	1.4
Floriculture	7	1.0
Total	704	100

Table 2 Crops studied in research work of the articles listed in Table 1

Crop	Number	%
Rice	93	21.5
Vegetables	47	10.9
Fruits	32	7.4
Ornamentals	31	7.2
Tea	27	6.3
Tomato	19	4.4
Coconut	18	4.2
Onion	17	3.9
Chili	15	3.5
Trees	15	3.5
Maize	12	2.8
Sugarcane	12	2.8
Cowpea	8	1.9
Potato	8	1.9
Mung bean	7	1.6
Weeds	5	1.2
Rubber	4	0.9
Soybean	4	0.9
Cinnamon	4	0.9
Others[a]	54	12.5
	432	100

[a]Black gram, cassava, coffee, ginger, grasses, groundnut, home gardens, lime, medicinal plants, millets, mushroom, oil palm, pepper, pigeon pea, sorghum, sunflower, tobacco, yams

Key areas of interest in agronomic experiments included the evaluation of growth and development of different crops (or selected varieties of a crop) under different management systems such as types and/or rates of soil amendments including fertilizers and water management. In the sub-area of genetics and breeding, evaluation of germplasm of different crop species had been made aiming to improve the yield, pest and disease tolerance or resistance, and adaptability to environment using field and/or pot experiments. Similarly, molecular biological applications and varietal developmental results included in this sub-area were basically based on laboratory and pot trials. Animal science research consisted of productivity improvement through management, quality improvement of farm products, environmental impacts, behavioral studies, animal welfare, and disease prevention. Moreover, this included both terrestrial and aquatic animal farming systems.

Research findings are available for different crops ranging from annuals to perennials (Table 2). The highest number of research papers published was on rice, amounting over 21% of the total number of papers published in TA and TAR. Other crops which received higher attention were vegetables, fruits, and ornamentals (flowers and foliage) as groups of crops. Large range of vegetables, fruits, and ornamentals had been studied under each group, and thus statistics on individual crops are not presented. Tea, tomato, coconut, onion, and chili were among the single crops which received much attention after rice, vegetables, fruits,

Table 3 Conditions in which designed experiments were conducted

Condition	Number	%
Field	162	51.4
Lab	109	34.6
Greenhouse	44	14.0
	315	100.0

and ornamentals. Among the plantation crops, the number of articles published on rubber was fewer compared to tea and coconut. Despite the importance in sustainable cropping system management and contribution to national economy pulses such as mung bean, cowpea, soybean, and black gram, spices such as cinnamon and weeds have received considerably less attention (Table 2). Moreover, despite the adaptability to diverse range of environmental conditions, nutritional quality, and health benefits, millets, yams, and many other spices have gained less priority in their research programs. Crops in Table 2 represent a diverse range of species found in home gardens and forests with both agronomic and ecological importance. In summary, reasons for the observed variability in the number of publications presented for different crops in TA and TAR can be of diverse nature, such as priorities given to those individual crops, infrastructure available to conduct research, workload and directives given to research officers, life cycle of crops, availability of financial resources, and the relevance to publish those research findings in TAR and TA. Therefore, careful assessment and understanding is needed when interpreting this variability.

Most of the designed experiments were field experiments (59%) followed by laboratory and greenhouse (Table 3). As most of the experiments were field experiments, results generated from those experiments would have direct relevance and applicability to farmer field conditions. In addition, surveys for sociological aspects of agriculture; protocol, methodology, or product development research in sub-areas of animal science and food science; and computer-based research in sub-areas of statistics and climatology have also been conducted. The decision on the type of experiment/survey to be conducted is determined by the types of crop and/or aspects to be studied, i.e., objective of the experiment/survey.

3 Experimental Designs Used: Strengths and Weaknesses

In order to test the relevant hypotheses and draw valid conclusions, experiments should be conducted using the most appropriate experimental design, data should be collected on relevant variables, and data analysis should be done using appropriate statistical methods. The above aspects of the research reported in Table 1 were explored in this study.

Randomized complete block design (RCBD) was the most widely used experimental design for the studies (Table 4). Apart from that, completely randomized design (CRD) and split–plot design have also been used, but with a relatively less abundance. Most of the experiments were either single-factor experiments or

Table 4 Use of different experimental designs in published research work

Design used	Number	%
RCBD	100	49.8
CRD	84	41.8
Split–plot	17	8.5
Total	201	100.0

two-factor factorial experiments, and thus CRD or RCBD had been used. However, there were instances where a large number of treatments or treatment combinations were tested in agronomic experiments or in genetic material evaluation trials (Ranawake and Amarasinghe 2014; Aluwihare et al. 2016; Kekulandara et al. 2018). In such occasions, the appropriateness of using RCBD is questionable as the heterogeneity within a block can be increased due to the requirement of a large block size (Gomez and Gomez 1984; Cochran and Cox 1977). In such instances, the use of incomplete block designs (IBD) should be thought of, if it is a single-factor experiment, or the concept of confounding should have adopted if it is a factorial experiment (Cochran and Cox 1977). However, most researchers have ignored this necessity when selecting appropriate experimental designs. In fact, only one research was found with the use of IBD, and there was none using confounding. It is possible at least some experiments were modified by changing the objectives just to avoid the use of more sophisticated but appropriate design. With confounding, not only confounding of 2^n but 3^n and 4^n can also be easily thought with standard objectives of experiments. In fact, when all 2^n treatment combinations are not practically possible to be used, the experiment can be conducted using only a fraction of 2^n combinations using fractional factorial designs. A special case of fractional factorial design is a Taguchi design (Diamond 2001). In addition, when several continuous variables are studied with more than 2 levels considered for each variable, then it is not practicable to have all combinations used in the experiment, and response surface designs (Sreenivas et al. 2008) pave the way to conduct the experiment only with some combinations. It was surprising to see that the Latin square design had not been used at least in a single experiment although it is one of the effective designs that can control variability in field experiments. Another useful design recommended for field experiments, especially in genetic studies of varietal evaluation trials with a large number of varieties, is lattice design, and it has never been used in any of those research. Other common weaknesses observed with respect to design of experiments were as follows:

(i) Use of different locations as blocks in multi-locational experiments without replicating within locations. In the absence of replicates within locations, neither the location effect nor location × treatment interaction effect can be examined.
(ii) Consideration of interaction between location and other experimental factors when treatments are nested within location in multi-locational experiments. When factors are nested within locations, factor levels have to be compared by fixing locations, and nested designs have to be used to test the effects.

(iii) Use of RCBD in situations where split–plot and strip–plot designs are more appropriate in the presence of factors requiring large plots as well as small plots; using RCBD leads to poor estimation of all effects. By adopting split–plot design, high precision can be obtained at least for small plot factors. Similarly, by adopting strip–plot design, high precision can be obtained at least for the interaction effect.

(iv) Ignoring establishing blocks and arranging the experiment according to CRD when external heterogeneities (criteria to establish blocks) are available. Field experiments are often exposed to high heterogeneity, and thus use of CRD under such circumstances leads to imprecise estimates.

(v) Use of conventional designs to determine the optimum spacing for crops without using systematic designs (Thattil and De Costa 1988). Although there is evidence for the use of fan design, two-way and three-way parallel row designs for spacing trials of some field crops, use of systematic designs was very minimal due to their complexity of the analysis. Conventional designs have limited use for spacing trials because a large number of spacing levels cannot be tested using conventional designs.

(vi) Use of repeated measures as a factor. Since time such as 2 weeks and 3 weeks cannot be randomized, time cannot be used as a factor in experimental designs. Such situations should be handled by the analysis rather than by the design.

4 Statistical Methods Used: Strengths and Weaknesses

Even if the experimental design is appropriate for the study, unless the most relevant statistical method is used for data analysis, valid conclusions cannot be made. Different statistical methods require different assumptions to be satisfied. Thus if the data do not satisfy those assumptions, the interpretations from those analyses may not be valid. Apart from that, some researchers have used inappropriate designs when analyzing data such as (i) use of different sites as blocks from multi-locational experiments, (ii) consideration of time as an explanatory variable (i.e., factorial experiment) or as a blocking variable when repeated measurements were taken, (iii) ignoring the block effect and analyzing data as from a CRD, and (iv) ignoring covariate analysis when needed. Prior to 1996, recommendation of rice varieties to different locations was done using only the mean yield. However, Thattil et al. (1996) showed how stability analysis can be used for this purpose. Possible reasons for such incorrect decisions would be due to the lack of awareness on statistical methods and/or non-consultation of a statistician. Therefore, consultation of a statistician at the planning stage is essential to conduct a successful experiment.

Widely used statistical methods for data analyses by agricultural researchers are summarized in Table 5. Most of the researchers have used parametric and univariate statistical methods such as analysis of variance (ANOVA) and regression,

Table 5 Statistical methods used for data analyses

Analysis methods	Number	%
Univariate		
ANOVA/GLM	243	51.4
Regression	82	17.3
Chi-square analysis (two-way table)	40	8.5
Rank data analysis	19	4.0
Log-linear model	13	2.7
Logistic regression	13	2.7
Multivariate		
Cluster analysis	31	6.6
Principal component analysis	25	5.3
Discriminant function analysis	3	0.6
MANOVA	3	0.6
Factor analysis	1	0.2
Total	473	100

amounting 51% and 13%, respectively. The use of contingency tables (two-way tables) and chi-square test for testing associations was prominent among nonparametric approaches (8.5%). Moreover, 2.7% of the researchers have used log-linear models to find associations of multiway tables. Only 2.7% of the researchers have used logistic regression for binary data analysis, and 4% have used rank data analysis approaches. This reflects the type of data that researchers have generated from their experiments, i.e., mostly of interval and ratio data, and the least was ordinal-type data.

With two-way tables, chi-square test is appropriate only if the two variables are nominal. In case if the variables consist of ordered categories (ordinal), then the ordered nature of the variable should be taken into account. The ordinal model (log-linear model) is not saturated, while the corresponding nominal model is saturated. Thus, the association can be taken into account while retaining some degrees of freedom (*df*) for the goodness of fit test except for 2×2 table. For instance, if both variables are ordinal, linear-by-linear association model (Goodman 1979) is appropriate, and if only one variable is ordinal and the other is nominal, then row-effect or column-effect model is appropriate (Goodman 1979). With multiway nominal tables, if the objective is to visualize the observation in two-dimensional space, correspondence analysis can be performed (Greenacre 2007).

With a clear response variable, logit models are more appropriate compared to log-linear models. If the response variable consists of two categories, then the appropriate model is binomial logit model, and according to Table 5, it is good to see substantial use of that model in the research work of that period. However, no situation had been reported with response variable having more than two categories. The appropriate model for those situations is the multinomial logit model (Greene 2012). Within this too, if the response variable is nominal, baseline category model

(McCullagh and Nelder 1989) would be a good choice. If the categories of the response variable have some order, then cumulative logit such as proportional odds model, adjacent logit model, or continuation ratio logit can be thought of depending on the objective of the research (Agresti 2002). In case if different subjects have access only to certain choices, and explanatory variable values vary depending on the choice (called characteristics of the choices) compared to ordinary explanatory variables where explanatory variable values do not vary depending on the choice (called characteristic of the choosers), the appropriate model is discrete choice model (Lovreglio et al. 2014).

The methods listed above require observations to be independent. For a two-way table with dependent observations, McNemar test is an alternative to chi-square test. When modelling match pairs, conditional logistic models have to be fitted instead of logistic model.

Occurrence of clustered binary and multinomial responses is quite common in studies, and standard binomial and multinomial logistic models are often found to be not appropriate for such data. Random effect models (Collett 1991) or mixed models (generalized linear mixed models (GLMMs)) such as beta-binomial model (Williams 1975) and logistic normal binomial models (Pierce and Sands 1975; Samita 1995; Hughes et al. 1998) are possible alternatives for such data. Quasi-likelihood methods such as Williams method (Williams 1982) are also used as alternatives for clustered categorical responses. Item response models and Rasch models (Rasch 1961; Bock and Aitkin 1981; Holster and Lake 2016) are specific GLMMs that model situations such as with a battery of T questions, where the probability that ith subject makes the correct response to question t depends on overall ability of ith subject and easiness of question t.

Categorical repeated measures (longitudinal data) can also arise in studies, and standard logistic models are not appropriate for such situations either. A model that has been suggested for such situation is generalized estimation equations (GEE) (Liang and Zeger 1986; Nooraee et al. 2014).

It is important to note that more than 13% of the researchers have used multivariate methods for data analysis. Out of different multivariate methods, the use of cluster analysis for grouping individuals and principal component analysis (PCA) for grouping variables based on correlations were prominent. Despite the use of cluster analysis as the most common among multivariate methods, statistical significance of those identified clusters had not been tested, and it is useful in verifying such clusters using a follow-up analysis. Only in one instance researchers have used MANOVA for such purposes (Table 5). This indicates that researchers were not fully aware of the outcomes generated and interpretation of results, particularly from multivariate analysis. Another deficiency found with respect to use of multivariate methods was that sometimes the study is multivariate but separate univariate techniques have been used for the analysis. The biggest advantage of multivariate techniques is that inference is made by considering all response variables simultaneously, and thus possibility of identifying effects is higher with the use of multivariate techniques compared to use of separate univariate analysis. It is important to highlight that, if the awareness of researchers on multivariate statistical approaches

Table 6 Widely used statistical software for data analyses

Software used	Number	%
SAS	174	58.4
SPSS	47	15.8
Minitab	44	14.8
Excel	9	3.0
Others[a]	24	8.1
	298	100

[a]Stata, Polo Plus, MegaStat, Systat, PowerMarker, Amos, R, Genstat

can be improved, more precise and informative outcomes could be generated. Therefore, measures should be taken to improve the knowledge of researchers on data analysis and/or establish formal access to statisticians.

5 Statistical Software Used

Researchers have used different statistical software for data analyses. The most widely used statistical software was SAS (over 58%), while other software were Minitab and SPSS (Table 6). Therefore, more than 89% of the researchers have used widely accepted statistical software for data analyses, indicating that the researchers have the access to statistical software as well as knowledge of that software. Only 3% of the researchers have used MS Excel for their data analyses without using standard statistical software. Moreover, the use of an alternative software such as Stata, Polo Plus, MegaStat, Systat, PowerMarker, Amos, Genstat, and R was rare. Most of this alternate software are new and are tailor-made to be used by specific group of researchers. Therefore, unless proper training has been given, beginners may have difficulties in using such software. However, widely used statistical software such as SAS, SPSS, and Minitab have facilities to analyze a diverse range of data, and the algorithms they use are well accepted. Moreover, the sophisticated software such as SAS and R, which are open source, should encourage the use of comprehensive statistical analysis.

6 Misuse of Statistics

Misuse of data analysis in some instances was also detected as follows:

(i) Many researchers have used ANOVA procedures for count or proportion (percentage) data without testing for normality of those original variables. In situations where data transformations were made to reach normality, arcsine and square root transformations had commonly been used. However, none of the researchers have tested whether the transformed data have reached normality before performing ANOVA. In fact, transformation may not be necessary if appropriate nonparametric methods can be used depending on the type of data.

(ii) When measurements were made in successive time intervals, time was not treated as a repeated measure and instead was considered as another factor.
(iii) Initial differences in treatments were not recognized as covariates in situations where needed.
(iv) Even the percentages from binary responses were used with ANOVA to compare treatment effects.
(v) In situations where nonlinear relationships were suited, only linear relationships were generated in regression analysis.

Apart from the misuse of data analysis approaches, researchers also performed inadequate data analyses in some instances, e.g., though the use of log-linear models, logistic regressions, and multivariate methods was possible and appropriate, those attempts had not been made to generate meaningful results and reduce dimensionality. This information highlights that the awareness improvement programs on statistics and collaboration with statisticians are needed.

Widely used mean separation methods are presented in Table 7. Researchers abundantly used Duncan's new multiple range test (DNMRT) and least significant difference (LSD) representing 59% and 33%, respectively. The use of Dunnett's and Tukey's methods was not common among the researchers. This may be due to the lack of awareness about these methods. Though these mean separation techniques have their own advantages (strengths) and disadvantages (limitations), it was apparent that the researchers are not aware of it. Therefore, the most appropriate mean separation method for the study had not been used in some instances. Despite its importance, the use of least square means (LSmeans) procedure was rare (2.4%). It is important to highlight that most of the agricultural experiments ended up with incomplete datasets (missing data) due to the occurrence of unexpected events by nature. In such situations, though the use of LSmeans procedure for the mean separation is advisable, researchers have used it occasionally. Researchers had used typical, stereotype mean separation techniques without knowing the strengths and weaknesses of those approaches. Similarly, there were hardly any group comparisons. Perhaps such analysis had not been performed due to lack of knowledge on implementing them through software. Orthogonal contrasts can be performed under such situations, and thereby useful and meaningful comparisons can be done.

Table 7 Mean separation methods used

Mean separation methods	Number	%
DNMRT	99	59.3
LSD	55	32.9
Tukey	6	3.6
LSmeans	4	2.4
Dunnett's	3	1.8
Total	167	100

7 Concluding Remarks

It was good to see statistics have been used in almost all research. This indicates that researchers are aware of the importance of the use of statistics as well as aware that the use of them is essential. However, according to the findings, there is a huge potential to improve the use of statistics in research work. Use of most appropriate design and statistical technique are key areas to be improved. Huge improvement of these publications could have been achieved with improvements in the use of statistics. In fact, some very useful findings may have been concealed due to improper use of statistics. Researchers have to be more informed about this fact, and a culture of researchers having an effective dialogue with a statistician from planning to publishing the research work should be established. Authorities should also seriously address this issue and make attempts to provide and facilitate these requirements. For instance, having a qualified statistician at every research station should be made compulsory, and thereby access to statisticians can effectively be enhanced. Altogether a collective attention of researchers and authorities on these aspects is essential to assure furtherance and sustainability of agricultural research in Sri Lanka.

References

Agresti A (2002) Categorical data analysis, 2nd edn. Wiley Interscience, New York

Aluwihare YC, Ishan M, Chamikara MDM, Weebadde CK, Sirisena DN, Samarasinghe WLG, Sooriyapathirana SDSS (2016) Characterization and selection of phosphorus deficiency tolerant rice genotypes in Sri Lanka. Rice Sci 23:184–195

Bock RD, Aitkin M (1981) Marginal maximum likelihood estimation of item parameters: application of an EM algorithm. Psychmetrika 46:443–459

Cochran G, Cox GM (1977) Experimental designs. Wiley, New York

Collett D (1991) Modelling binary data. Chapman and Hall, London

Diamond WJ (2001) Practical experiment designs: for engineers and scientists, 3rd edn. Wiley, New York

Gomez KA, Gomez AA (1984) Statistical procedures for agricultural research, 2nd edn. Wiley, Hoboken

Goodman LA (1979) Simple models for the analysis of associations in cross classifications having ordered categories. J Am Stat Assoc 74:537–552

Greenacre M (2007) Correspondence analysis in practice, 2nd edn. Chapman & Hall/CRC, London

Greene WH (2012) Econometric analysis, 7th edn. Pearson Education, Boston

Hall TJ, Dennis JH, Lopez RG, Marshall MI (2009) Factors affecting growers' willingness to adopt sustainable floriculture practices. HortScience 44:1346–1351

Holster TA, Lake JW (2016) Guessing and the Rasch model. Lang Assess Q 13:124–141

Hughes G, Munkvold G, Samita S (1998) Application of the logistic–normal–binomial distribution to the analysis of eutypa dieback disease incidence. Int J Pest Manag 44:35–42

Kekulandara DS, Sirisena DN, Bandaranayake PCG, Samarasinghe G, Wissuwa M, Suriyagoda LDB (2018) Variation in grain yield, and nitrogen, phosphorus and potassium nutrition of irrigated rice cultivars grown at fertile and low-fertile soils. Plant Soil. https://doi.org/10.1007/s11104-018-3663-0

Liang KY, Zeger SL (1986) Longitudinal data analysis using generalized linear models. Biometrika 73:13–22

Lovreglio R, Borri D, dell' Olio L, Ibeas A (2014) A discrete choice model based on random utilities for exit choice in emergency evacuations. Saf Sci 62:418–426

Luken H, Clarke JL, Muller R (2012) Genetic engineering and sustainable production of ornamentals: current status and future directions. Plant Cell Rep. https://doi.org/10.1007/s00299-012-1265-5

McCullagh P, Nelder JA (1989) Generalized linear models, 2nd edn. Chapman and Hall, London

Moss RH, Edmonds JA, Hibbard KA, Manning MR, Rose SK, van Vuuren DP, Carter TR, Emori S, Kainuma M, Kram T, Meehl GA, Mitchell JFB, Nakicennovic N, Riahi K, Smith SJ, Stouffer RJ, Thomson AM, Weyant JP, Wilbanks TJ (2010) The next generation of scenarios for climate change research and assessment. Nature 463:747–756

Nooraee N, Molenberghs G, van den Heuvel ER (2014) GEE for longitudinal ordinal data: comparing R-geepack, R-multgee, R-repolr, SAS-GENMOD, SPSS-GENLIN. Comput Stat Data Anal 77:70–83

Pierce DA, Sands BR (1975) Extra Bernoulli variation in binary data, Technical report. Department of Statistics, Oregon State University, USA

Puri S, Nair PKR (2004) Agroforestry research for development in India: 25 years of experiences of a national program. Agrofor Syst 61:437–452

Ranawake AL, Amarasinghe UGS (2014) Relationship of yield and yield related traits of some traditional rice cultivars in Sri Lanka as described by correlation analysis. J Sci Res Rep 3:2395–2403

Rasch G (1961) On general laws and the meaning of measurement in psychology. In: Proceedings of the fourth Berkeley symposium on mathematical statistics and probability, vol 4: contributions to biology and problems of medicine. University of California Press, Berkeley, CA, pp 321–333. https://projecteuclid.org/euclid.bsmsp/1200512895

Samita S (1995) Analysis of aggregated plant disease incidence data, PhD thesis. University of Edinburgh, UK

Spilsbury MJ, Nasi R (2006) The interface of policy research and the policy development process: challenges posed to the forestry community. Forest Policy Econ 8:193–205

Sreenivas RR, Kumar CG, Prakasham RS, Hobbs PJ (2008) The Taguchi methodology as a statistical tool for biotechnological applications: a critical appraisal. Biotechnol J 3:510–523

Thattil RO, De Costa WAJM (1988) Spacing experiment on maize X mungbean intercropping system using a "3 way" systematic design. Trop Agric 144:109–122

Thattil RO, Marambe B, Abeynayake NR, Abeysiriwardena S De Z (1996) Recommendation of rice cultivars based on stability analysis. In: Proceedings of the annual research seminars, University of Peradeniya, pp 1–9

Turner RK, Paavola J, Cooper P, Farber S, Jessamy V, Georgiou S (2002) Valuing nature: lessons learned and future research directions. CSERGE working paper EDM, no. 02-05. University of East Anglia, The Centre for Social and Economic Research on the Global Environment (CSERGE), Norwich

Williams DA (1975) The analysis of binary responses from toxicological experiments involving reproduction and teratogenicity. Biometrics 31:949–952

Williams DA (1982) Extra binomial variation in logistic linear models. Appl Stat 31:144–148

Paddy Field and Constructed Wetland: The Equivalencies

M. I. M. Mowjood, K. B. S. N. Jinadasa, and B. F. A. Basnayake

1 Introduction

Wetlands are among the most productive bio-system on the Earth (US-EPA 1995). It provides services such as food production, flood control, drought mitigation, groundwater recharge, biodiversity, pollution control, and nutrient recycling. Paddy fields (PF) and constructed wetlands (CW) are man-made ecosystems (Plate 1) that mimic natural wetlands used for rice production and pollution control, respectively. In spite of their different purposes, both systems have been originated from the same concept of ecosystem that is submerged. These systems have been extensively and inclusively studied and improved for decades with the main objectives of increasing paddy yield and improvement of water quality, respectively. However, these man-made ecosystems are yet to reach its potential to deliver more than what we can achieve now, thus requiring an out-of-box thinking.

Paddy fields and CW are characterized by unique biotic and abiotic components and processes. Agronomic practices in PF and design and operation of CW decide the components and processes of each systems. The present practices in both bio-systems on water/effluent management, paddy varieties/plant species, chemicals usage, and other agronomic practices may positively or negatively contribute toward the basic approach in natural wetland management. For example, weeds and insects are not considered as unwanted in the natural wetland system but in PF or CW management. Therefore, this paper reviews the components and functional

M. I. M. Mowjood (✉) · B. F. A. Basnayake
Department of Agricultural Engineering, Faculty of Agriculture, University of Peradeniya, Peradeniya, Sri Lanka
e-mail: mmowjood@pdn.ac.lk

K. B. S. N. Jinadasa
Department of Civil Engineering, Faculty of Engineering, University of Peradeniya, Peradeniya, Sri Lanka

R. P. De Silva et al. (eds.), *Agricultural Research for Sustainable Food Systems in Sri Lanka*, https://doi.org/10.1007/978-981-15-3673-1_9

Plate 1 Paddy field (left) and constructed wetlands (right)

equivalencies of PF and CW with natural wetland so that each ecosystem can be synergized for its higher potential.

1.1 Natural Wetlands and Its Importance

Wetlands are identified as areas that are inundated or saturated by surface or groundwater at a frequency and duration sufficient to support a prevalence of vegetation typically adapted for life in saturated soil conditions. Stormwater collected in wetlands is temporarily retained for a period and slowly released into downstream or permeates underneath to groundwater. Therefore, it attenuates and provides lag time for peak flow, thus playing a major role in mitigation of flood and drought and climate change scenarios.

Water is a universal solvent and thus dissolves almost everything and gets contaminated easily. The contaminants are removed through biotic and abiotic processes when they flow through natural wetlands. Therefore, wetlands are considered as the "kidney" of the Earth.

Wetlands are also called as "biological supermarkets" as it is a meeting place for all biological components and bio-functions. It provides habitat for various fauna and flora and enhances biomass production and nutrient recycling. The average biomass production in wetland is comparable to the rain forest and coral reefs. *Ramsar* convention identified six wetland sites in Sri Lanka out of 2309 locations in the world as protected considering their importance in the biodiversity (https://www.ramsar.org/).

1.2 Paddy Field

Paddy fields are man-managed ecosystems (anthropogenic habitats) with variable degree of intensity (Bambaradeniya 2000; Bambaradeniya and Amerasinghe 2003) and as man-managed temporary wetlands (Lupi et al. 2013).

Paddy fields are characterized by the presence of standing water that is seasonal. However, it can be operated under three water regimes:

- Continuous flood (free surface water)
- Intermittent flood
- Non-flooded (subsurface water or aerobic)

Continuous flooding is a common practice in rice cultivation for various reasons such as weed control, reduction of the loss of nutrients, and warming the soil. PF are irrigated in rotation that leads intermittent flooding in field to meet the crop water requirement which varies with the growth stage and vegetative, reproductive, and ripening phase. Accordingly, the fauna, flora, and biological niches vary within the season (Kirinde et al. 2017).

1.3 Constructed Wetland

Constructed wetlands are engineered systems (anthropogenic habitats) which are designed and constructed for treating wastewater, utilizing the same processes as in PF involving vegetation, soils, and associated microbial assemblages. In the early stage, the application of CW was mainly for treating domestic and municipal wastewater. At present, the use has significantly expanded to purify agricultural effluents, industrial effluents, landfill leachates, polluted rivers & lakes, and urban highway & stormwater runoff.

CWs are designed to take advantage of many of the same processes that occur in natural wetlands within a more controlled environment. Some of these systems have been designed and operated with the sole purpose of treating wastewater, while others have been implemented with multiple uses such as aesthetic and restoration of wetland habitat for wildlife use and environmental enhancement.

Several types of CWs with varying hydrology and hydraulics have been used (Table 1). Free water surface flow systems are with the water flowing over the soil surface at shallow depths. On the other hand, subsurface flow systems are designed to create subsurface flow through a permeable medium, keeping the water being treated below the surface, thereby avoiding the appearance of water. The hydraulic in CW can be horizontal (HF) or vertical (VF) flow depending on the inflow and outflow arrangement. Several combinations of HF, VF, and its combination in sequence as hybrids have been used for different purposes and type of wastewaters (Tanaka et al. 2006). All types maintain ponded or saturated substrate as seen in PF. The pollutant removal efficiency varies with the type of hydraulic regime (Truong et al. 2011). These efforts basically bring both systems closer to the natural wetland which undergoes varying combination of water regime for the betterment.

Free water surface CW closely resembles to PF with continuous flooding. Subsurface vertical flow hybrid CW has shown high pollutant removal efficiency comparable to that in alternative wetting and drying cycle in PF with higher nutrient uptake (Sasikala et al. 2010; Weerakoon et al. 2013, 2016; Sellathurai 2015).

Table 1 Types of CWs

Conventional CWs		Hybrid CWs		Enhanced CWs
Free water surface flow Subsurface water flow	Horizontal flow (HF) Vertical flow (VF)	Multistage	HF + VF VF + VF + HF VF + HF + VF	Artificial aerated Baffled flow Variable loading Circular flow

Nutrient removal by plants plays a major role in both PF and CW systems under aerobic and anaerobic conditions which facilitate the mineralization.

2 Components of Paddy Fields and CW

Figure 1 shows the components of PF and CW. PF can be designed on existing land surface where the CW can be built at, below, or even above the existing land surface depending on the site topography.

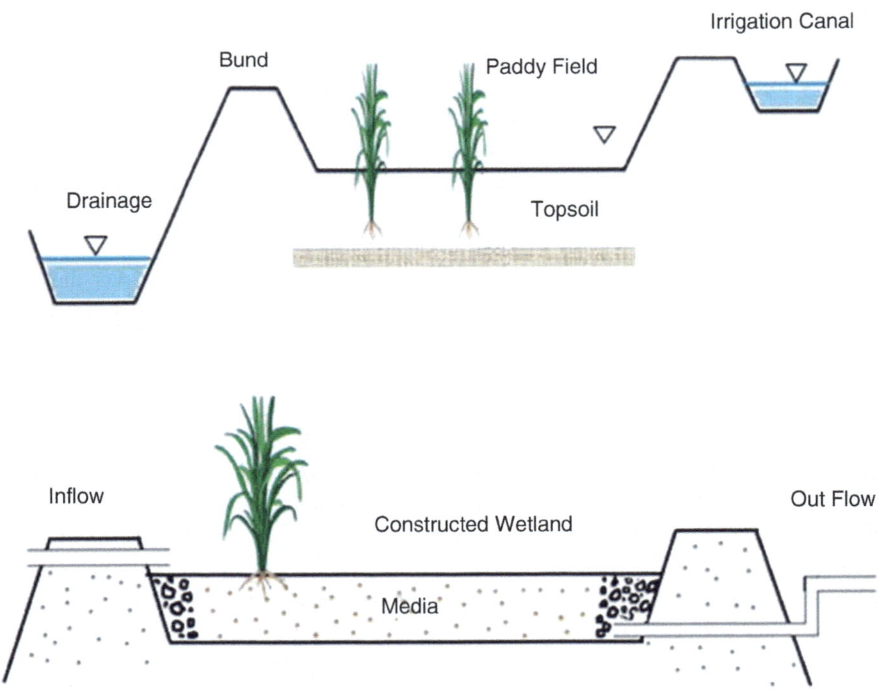

Fig. 1 Schematic diagrams of PF (above) and CW (below)

Boundaries Considering the large extent of area, PF are made into plots by earthen bunds with the height of 30–40 cm to hold the water up to 10–20 cm depth. Smearing is practiced to the bunds by clays to seal the cracks and reduce seepage. In the case of CW, the side boundaries are earthen, if it is placed in natural depression. Cemented masonry walls also have been used to reduce the seepage. The bunds play a major role for providing place for several fauna and flora. The bottom of the PF is sealed by plow layer formed by compaction due to the action of plow bottom and machine use in land preparation. In most cases, the bottom of the CW is cemented or lined to hydrological isolation of the unit from local groundwater.

Media

The existing soil is used as media in PF. Because of the plowing, puddling, and leveling by primary and secondary tillage, the larger soil particles settle at the bottom. Higher silt and clay content can be found in top layer in PF. Natural soil, aggregates, and other filter materials are used as media in CW (Weragoda et al. 2010). A good horizontal or vertical flow is expected through media filled in CW. Thus, larger aggregates are arranged at the bottom, while soil is refilled at the top. This media is very important for microbial community in PF and CW.

Plants

Rice plants occupy the entire plot in mono-cropping. Other field crops are not incorporated in PF due to their susceptibility to submerged condition. However, weeds (defined as unwanted plants) which grow along the rice plant are reduced considerably using weedicide in early stage as they compete with rice and reduce yield. On the other hand, selected plants have been used in CW depending on the tolerance and potential to remove the target contaminants. Three types of plants have been used depending on the growth position in relation to water level as shown in Fig. 2 (Tanaka et al. 2006, 2011; Jinadasa et al. 2008; Sewwandi et al. 2010):

- Floating plants: *Pistia stratiotes* (water lettuce), *Lemna minor* (duckweed), and *Eichchornia crassipes* (water hyacinth)
- Submerged plants: *Hydrilla verticillata*
- Emergent: *Typha latifolia* L., *Typha angustifolia* L., and *Phragmites australis*

Microbial Community

Both ecosystems comprise of microbial communities which are attached to the surfaces of the aquatic plants and aggregates or soils. Decomposers such as bacteria, fungi, and actinomycetes are active in any wetland. A portion of the microbial community is adapted to the aerobic surface water and uppermost soil layer, while some microbes are active in anaerobic condition in subsoils. The facultative microbes are also found because of the diurnal or seasonal changes in the water table. This varying oxygen regime makes the wetland highly vibrant system for nutrient dynamics.

Fig. 2 Types of plants used in CW

3 Functions/Processes

Paddy fields and CW consist of complex processes involving physical chemical and biological means. It can also be categorized as abiotic (physical and chemical) and biotic processes.

3.1 Abiotic Processes

3.1.1 Hydrological Processes

Figure 3 shows the hydrological processes identified in PF same as in CW. Precipitation, infiltration, surface runoff, seepage, evaporation, and

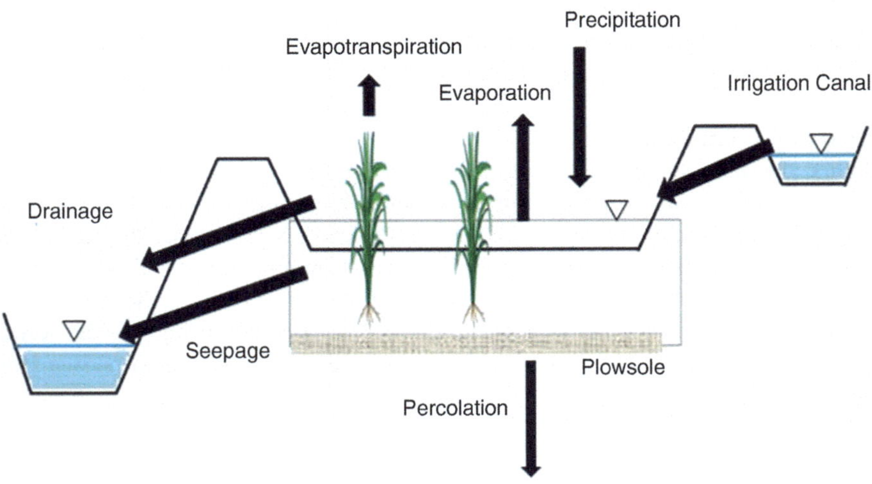

Fig. 3 Hydrological processes in PF

evapotranspiration are considered as major components of the hydrological process. Water flows into the wetland through irrigation/influent, precipitation, surface water, and also the groundwater. Water losses from the wetland are due to evapotranspiration, surface runoff, subsurface water outflow, and percolation. The water movement through and from the wetland affects the temporal fluctuations in water levels and controls the water balance and the water storage within a wetland. The changes in water regime determine the hydraulic retention time, one of the important parameter that influences the nutrient removal.

3.1.2 Chemical Processes

Various chemical transformations occur in wetlands. Decomposition, transformation, and adsorptions are the main chemical processes in PF and CW. Oxidation and reduction occur depending on the availability of oxygen in submerged condition (Reddy and Patrick 1974). Rice and reed plants have shown the ability to transfer atmospheric oxygen to root zone (Brix 1993; Iijima et al. 2017). Several researches have been conducted on media selection for CW particularly for its ability to adsorb heavy metals (Sewwandi et al. 2012, 2013). Types of clay mineral and organic matter increase Cation Exchange Capacity in PF and CW.

3.2 Biotic Processes

3.2.1 Mineralization

Conversion of complex substances into simple components through microbial processes is called as mineralization. Dissolved biodegradable material is decomposed by microorganisms.

Organic matters are the main “skeleton” for biological system. Both PF and CW are able to accumulate higher organic carbon due to fixing capacity of atmospheric CO_2 and release O_2 by photosynthesis (Mowjood and Kasubuchi 1998, 2002; Usui et al. 2003). On other hand, both systems usually show a low Biochemical Oxygen Demand (BOD) due to higher rate of decomposition by microbes.

Nitrogen is a major nutrient to be added in PF and major element to be removed from CW. Biochemical cycling of nitrogen is a complex process involving interconversions between different nitrogen species and transfer between various stores. Ammonification, volatilization, nitrification, and denitrification are major processes in nitrogen dynamics (Kaldec and Knight 1996) as shown in Fig. 4. This transformation depends on oxic and anoxic conditions, C:N ratio, and availability of supplementary nutrients in the system for microbial growth.

Aquatic plants play an important role in supporting the mineralization processes. Certain aquatic plant species transport atmospheric oxygen into their submerged stems, roots, and tubers (Brix 1993; Iijima et al. 2017). Oxygen is then utilized by the microbial decomposers.

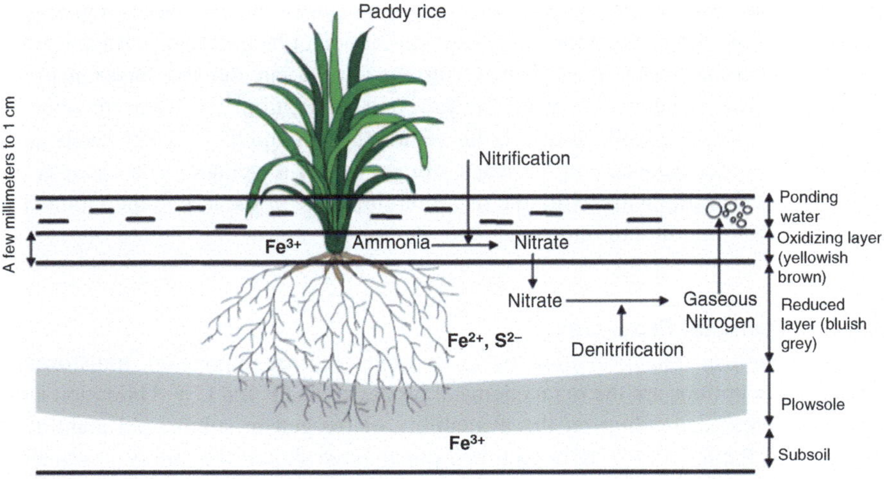

Fig. 4 Nitrogen transformation in PF

3.2.2 Ecological Processes

Successful performance of PF and CW depends on ecological functions that are similar to those of natural wetlands, which are based largely on interactions within plant communities. PF and CW provide high-quality habitat for fauna and flora. Fishes, amphibians, reptiles, mammals, monotremes, insects, and invertebrates are the major components in fauna in PF and CW. The flora includes variety of aquatic and macrophytes. Other major species that can be seen in the wetlands are algae which can vary in size and color.

Benthic invertebrates play an important role in the ecology of both ecosystems. They release dissolved nutrients by their feeding activities, excretion, and burrowing into sediments and increase the rate of decomposition of particulate matter (Covinch et al. 1999). Benthic invertebrates are considered key components of rice field fertility due to its significant roles in nutrient translocation and organic matter decomposition (Roger et al. 1987).

The composition of benthic macrobenthos showed that insect larvae were the most abundant in PF. The predominance of insects in PF was reported from Sri Lanka (Bambaradeniya and Amerasinghe 2003; Bambaradeniya and Edirisinghe 2008), Ethiopia (Desta et al. 2014), and Japan (Yosihiro 2010). The availability of more protected habitable niche by paddy plants would have favored the insect to thrive. The vegetative and reproducing growth stages of the rice plant such as tillering, booting, and flowering stages attract a variety of insects (Bambaradeniya and Edirisinghe 2008).

As paddy plants grow, sunlight penetration to the bottom of the paddy field reduces, thus reducing decomposition rates and resulting in decrease in detritus quantity. Further, the growth of paddy plants leads to the increase in paddy root structures; thus, gradually decreasing habitable area for benthic fauna results in decreased abundance as the crop season progressed. Similar trends were reported in CW also (US-EPA 2000).

Growth of the plant and the mineral composition of plant parts have been studied in PF as well as in CW. Sashikala et al. (2005) reported the suitable harvesting time for bulrush (*Scirpus grossus*) and regrowth responses to leachate in CW with coir-based media. Somaweera et al. (2016) reported the nutrient uptake and its partitioning in different portion of rice plant under continuous flooding, alternate wetting, and drying conditions. A similarity in partitioning of nutrient or mineral is observed in both PF and CW.

4 Engineering Options

4.1 Design

The design of PF and CW should mimic the natural wetland in general. PF are designed largely to target for higher efficiency of inputs such as water, plants, chemicals, machine, and labor. Plot size and orientation toward irrigation and drainage channel, land level, height of the bund, and water control structures are the main components to be designed (Mizutani et al. 1999). In the case of CW, the designs are largely governed by wastewater characterization and effluent quality requirements. Both designs are driven for better plant growth through various processes and easy operations.

4.2 Operation and Maintenance

All engineered systems have to be effective throughout the growth stages of the plants. The management of water, media, biota, and other basic components affects the effectiveness of ecosystems. Irrigation and drainage schedule, application of agro-chemicals, and harvesting are the main operations in PF, while wastewater inflow and outflow and harvesting of biomass are the major operations in CW. Land/media preparations, planting, and other infrastructure maintenance are carried out in between the two harvestings in both systems.

Crop water requirement has been used to estimate the water duty in irrigation schedule in PF, while required hydraulic retention time decides the inflow rate in CW.

5 Issues and Concerns

5.1 Food Security

The increase in population creates demands for food production as well as wastewater treatment. With the green revolution, paddy production has been increased with the compromise to the environment over the time. Application of synthetic fertilizer, especially P-contained fertilizer, may increase the accumulation of Cd in paddy soil (Chandrajith et al. 2012). There is a tendency of plants to uptake heavy metals that may subsequently transfer into the food chain. The ability of heavy metals to persist in nature, bioaccumulation, and biomagnifications led to eco-toxicity to plants, animals, and human beings. Therefore, the application of nature-based approaches becomes popular in most parts of the world. However, there is a debate on how the world population can be catered without intensive agriculture.

Toxic chemicals should be trapped before it reaches the food chain. The CW with nonedible plant species plays a greater role in translocation of toxic chemicals such as heavy metals.

5.2 Biodiversity

The rice fields, being agroecosystems, are managed with a variable degree of intensity through different agronomic measures and practices that affect the abundance and composition of the biotic community (Halvart and Kaule 1996). Use of modern high-yielding rice varieties responsive to fertilizers and pesticides and the increase in the number of crops grown per year by planting short-duration varieties are the features in modern agriculture. The use of pesticides has been an important contributor to improve rice productivity. Farmers and policy makers consider pesticides as a guarantee against crop failure and a necessary input for rice production. The widespread promotion and indiscriminate use of insecticides and the introduction of a limited number of rice varieties on a very large scale to replace the diverse array of varieties grown previously have been the major factors responsible for the rapid multiplication of rice pests and diseases (Bambarandeniya et al. 2004).

Although insecticides are known to have rapid curative action in preventing economic damage, indiscriminate use of insecticides has led to the destruction of natural enemies, causing the resurgence of new pest species and the development of insecticide-resistant pest populations. Composition of the ecosystem mainly depends on the inputs used, management practices, and variety (Duru et al. 2015). Therefore, it is very important to evaluate ecosystems including their biotic (flora and fauna) and abiotic (soil and water) components under different farming systems.

5.3 Urbanization

Wetlands are among the most productive as well as most threatened ecosystem on the Earth. Worldwide, the extent of natural wetlands continues to decline as land is converted to accommodate the demand for increasing populations and development. Continuous conversion of PF into nonagricultural activities is having an increasing trend. This creates an imbalance in natural drainage system and causes drought and floods. PF in the urban landscape serve as substitute for wetlands for its services.

5.4 Greenhouse Gas Emission

The carbon dioxide and methane are the two most potent greenhouse gases, which contribute largely to global warming (IPCC 2007). Submerged condition as in PF and CW are potential for methane emission. However, it is also reported that PF and CW are fixing carbon. Mowjood and Kasubuchi (1998) have shown that dissolved oxygen increases in the daytime due to photosynthesis in PF. Usui et al. (2003) have shown that dissolved CO_2 decreases in the water in paddy field in the daytime. Bhattacharyya et al. (2013) have reported that the lowland flooded rice ecosystem has the capacity to store carbon in soil and can behave as net carbon sink. Carbon distribution pattern in lowland flooded rice ecology revealed that carbon entered into the system mainly through net ecosystem production of rhizome deposition, algal biomass, lost from the system through crop harvest, dissolved organic carbon leaching, and methane emissions. Further studies are needed to elucidate the carbon fixing in PF and CW.

6 Lessons Learned and the Way Forward

Design Wetland design has been based on design equations used for wastewater treatment systems. Constructed wetlands are complex systems in terms of biology, hydraulics, and nutrient recycling. Decisions are made on design parameters by aggregating performance data from a variety of wetlands, which leads to uncertainties about the validity of the parameters. On other hand, design parameters of PF are more or less on agronomic perspective than engineering perspective.

Oxygen Dynamics Oxygen is critical in submerged condition. Constructed wetlands can be designed to remove nitrogen, if sufficient aerobic (open water) and anaerobic (vegetated) zones are provided. Constructed wetlands have been used in conjunction with aerobic treatment processes that can nitrify to remove nitrogen. Open windows (open water areas) increase the oxygen exchange from atmosphere and also facilitate the penetration of solar radiation for photosynthesis of submerged plants and algae as reported by Mowjood and Kasubuchi (1998) and Sewwandi et al. (2012).

Water Regime The water regime is critical for plant growth and pollutant removal in both PF and CW. The impact of continuous flooding and alternative wetting and drying (AWD) in terms of mineralization and nutrient uptake by plant is reported (Junzeng Xu et al. 2013). Somaweera et al. (2016) have studied uptake, accumulation, and allocation of nitrogen, phosphorus, and potassium in different tissues of a rice plant under different soil moisture management conditions. Alternative wetting and drying reduced root growth in the early growth stage and enhanced after flowering. Alternative wetting and drying after flowering resulted in better uptake of N, P, and K in comparison to continuous flooding. Sellathurai (2015) identified the impact of AWD cycle in paddy field on soil organic matter and soil nitrogen.

Shorter dry spells (4 days) enhanced the organic matter accumulation compared to longer dry spells (12 days). The nitrogen in pore water was higher in 12 days compared to 4-day dry spell. This shows that 12-day dry spell is adequate to create aerobic condition for nitrification followed by denitrification.

Habitat Manipulation/Ecological Intensification In most instances, the species richness and abundance of the predator populations may be greater than those of the pest populations when little or no insecticides are used (Way and Heong 1994). In Sri Lanka, more than 50% of the terrestrial arthropod species in rice fields have been shown to consist of predators, with spiders as the dominant predatory group comprising 60 species (Bambaradeniya and Edirisinghe 2008). The use of CW also has a concern on plant growth without pest damage. Therefore, an approach on habitat manipulation or ecological intensification may decrease the pest damage in both cases without use of pesticides. Food production and environmental quality can be improved with ecological friendly practices in both PF and CW.

7 Conclusions

Almost all studies in PF and CW for improvement of the systems can be considered as efforts toward the natural wetland system. Hybrid system, variable hydraulic loading, alternative wetting and drying, and heterogeneous biological composition show better performance compared to conventional practices in both PF and CW. More studies are required for engineered ecosystem with the cyclic pattern of wet and dry condition. The lesson learned in design, water regime, oxygen dynamics, and ecological intensification in both systems warranted focus for further capitalization of compromised benefits as in wetlands.

References

Bambaradeniya CNB (2000) Ecology and biodiversity in an irrigated rice field ecosystem in Sri Lanka. PhD thesis. University of Peradeniya, Sri Lanka, p 525

Bambaradeniya CNB, Amerasinghe FP (2003) Biodiversity associated with the rice field agro ecosystem in Asian countries in Asian countries: a brief review. Working paper 63. International water management Institute, Colombo, Sri Lanka

Bambaradeniya CNB, Edirisinghe JP (2008) Composition, structure and dynamics of arthropod communities in a rice agro-ecosystem. Cey J Sci Biol Sci 37(1):23–48

Bambarandeniya CNB, Edirisinghe JP, Silva DE, Gunathilake CVS (2004) Biodiversity associated with an irrigated rice agro ecosystem in Sri Lanka. Biodivers Concerv 13:1715–1753

Bhattacharyya P, Neogi S, Roy KS, Dash PK, Neogi S, Tripathi R, Rao KS (2013) Net ecosystem CO2 exchange and carbon cycling in tropical lowland flooded rice ecosystem. Nutr Cycl Agroecosyst 95:133–144. https://doi.org/10.1007/s10705-013-9553-1

Brix H (1993) Macrophyte-mediated oxygen transfer in wetlands: transport mechanisms and rates. In: Mishiri GA (ed) Constructed wetland for water quality improvement. Lewis Publishers, Boca Raton

Chandrajith R, Dissanayake N, Dissanayake CB (2012) Arsenic and other heavy metals in rice from Sri Lanka- preliminary results with ICP-MS. In: Symposium proceedings: international symposium on water quality and human health: challenges ahead

Covinch AP, Palmer MA, Crowl TA (1999) The role of benthic invertebrate species in freshwater ecosystems. Bioscience 49(2):119–127

Desta L, Prabha Devi L, Sreenivasa V, Amede T (2014) Studies on the ecology of the paddy and fish co – culture system at Dembi Gobu microwater shed at Bako, Ethiopia. Int J Fish Aquat Stud 1(3):49–53

Duru M, Olivier T, Guillaume M, Roger MC, Marie-Angélina M, Eric J, Etienne-Pascal J, Jean-Noël A, Serge S, Jacques-Eric B (2015) How to implement biodiversity bases agriculture to enhance ecosystem services: a review. Agron Sustain Dev 2015:1259–1281

Halvart MM, Kaule BG (1996) Activity pattern of fish in rice field. Hydrobiologies 145:159–170

Iijima M, Hirooka Y, Kawato Y, Watanabe Y, Wada KC, Shinohara N, Nanhapo PI, Wanga MA, Yamane K (2017) Short term evaluation of oxygen transfer from rice (Oryza sativa) to mixed planted drought-adapted upland crops under hydrophonic culture. Plant Prod Sci 20(4):434–440

IPCC (2007) Climate Change 2007: Synthesis report. Contribution of working groups I, II and III to the Fourth Assessment Report of the Intergovernmental Panel on Climate Change (Core Writing Team, eds. Pachauri RK, Reisinger A). IPCC, Geneva, Switzerland, 104 pp

Jinadasa KBSN, Tanaka N, Sasikala S, Werellagama DRIB, Mowjood MIM, Ng WJ (2008) Impact of harvesting on constructed wetlands performance – a comparison between Scirpus grossus and Typha angustifolia. J Environ Sci Health 43:664–671

Junzeng Xu, Qi Wei, Yanmei Yu, Shizhang Peng, Shihong Yang (2013) Influence of water management on the mobility and fate of copper in rice field soil. J Soils Sediments 13 (7):1180–1188

Kadlec RH, Knight RL (1996) Treatment wetlands, 1st edn. CRC Press, Boca Raton

Kirinde GWRWMRMWK, Dayawansa NDK, Mowjood MIM (2017) Water regimes and ecosystem in lowland Paddy field: a case study in Awlegama, Kurunegala, Sri Lanka. Trop Agric Res 28(4):334–346

Lupi D, Anna R, Bruno R (2013) Benthic macro invertebrates in Italian rice fields. J Limnol 72 (1):184–200

Mizutani M, Hasegawa S, Koga K, Goto A, Murty VVN (1999) Advanced paddy field engineering. Japanese Society of Irrigation, Drainage and Reclamation Engineering. Shinzan-Sha Sci & Tech Publishing, Tokyo

Mowjood MIM, Kasubuchi T (1998) Dynamics of Dissolved Oxygen (DO) in the ponded water of a paddy field. Soil Sci Plant Nutri 44(3):405–413

Mowjood MIM, Kasubuchi T (2002) The effect of convection on the exchange coefficient of oxygen and estimation of net production rate of oxygen in ponded water of a paddy field. Soil Sci Plant Nutri 48(5):673–678

Reddy KR, Patrick WH (1974) Effect of alternate aerobic and anaerobic conditions on redox potential, organic matter decomposition and nitrogen loss in a flooded soil. Soil Biol BioChem 7(2):87–94

Roger PA, Grant IF, Reddy PM, Watanabe I (1987) The photosynthetic aquatic biomass in wetland rice fields and its effect on nitrogen dynamics. In: Efficiency of nitrogen fertilizers for rice. International Rice Research Institute, Los Banos, pp 43–68

Sashikala S, Mowjood MIM, Basnayake BFA (2005) Formulation of harvesting regimes from bulrush (*Scirpus grossus*) growth responses to MSW leachate in constructed wetlands with coir based media. Trop Agric Res 17:114–124

Sasikala S, Tanaka N, Jinadasa KBSN, Mowjood MIM (2010) Comparison study of pulsing and continuous flow for improving effluent water quality and plant growth of a constructed wetland to treat domestic wastewater. Trop Agric Res 21(2):147–156

Sellathurai T (2015) Postgraduate Institute of Agriculture. MPhill thesis. University of Peradeniya, Sri Lanka

Sewwandi BGN, Weragoda SK, Mowjood MIM, Tanaka N, Sasikala S (2010) Effect of submerged and floating plants on dissolved oxygen dynamics and nitrogen removal in constructed wetlands. Trop Agric Res 21(4):353–360

Sewwandi BGN, Vithanage M, Wijesekara SSRMDHR, Rajapaksha AU, Jayaratna DGLM, Mowjood MIM (2012) Characterization of aqueous Pb(II) and Cd(II) biosorption on native and chemically modified Alstonia macrophylla saw dust. Biorem J 16(2):113–124

Sewwandi BGN, Vithanage M, Wijesekara SSRMDHR, Mowjood MIM, Hamamoto S, Kawamoto K (2013) Adsorption of Cd(II) and Pb(II) onto humic acid treated coconut (Cocos nucifera) husk, ASCE's. J Hazard Toxic Radioact Waste 18(2):1–10. https://doi.org/10.1061/(ASCE)

Somaweera KATN, Suriyagoda LDB, Sirisena DN, De Costa WAJM (2016) Accumulation and partitioning of biomass, nitrogen, phosphorus and potassium among different tissues during the life cycle of rice grown under different water management regimes. Plant Soil 401 (1–2):169–183

Tanaka N, Jinadasa KBSN, Werellagama DRIB, Mowjood MIM, Ng WJ (2006) Constructed tropical wetlands with integrated submergent – emergent plants for sustainable water quality management. J Environ Sci Health A A41(10):2221–2236

Tanaka N, Ng WJ, Jinadasa KBSN (2011) Wastewater treatment by constructed wetlands. World Scientific Publishing/Imperial College Press, London. isbn:13978-1-84816-297-6, 10 1-84816-297-9

Truong HD, Quang LN, Nguyen HC, Brix H (2011) Treatment of high-strength wastewater in tropical constructed wetlands planted with Sesbanta Sesban: horizontal subsurface flow versus vertical down flow. Ecol Eng 37:711–720

U.S Environmental Protection Agency (2000) Manual constructed wetlands treatment of municipal wastewaters EPA625-R-99-010

U.S. Environmental Protection Agency (1995) Wetlands fact sheets. EPA843-F-95-001. Office of Water, Office of Wetlands, Oceans and Watersheds

Usui Y, Mowjood MIM, Kasubuchi T (2003) Absorption and emission of CO2 by ponded water of a paddy field. Soil Sci Plant Nutri 49(6):853–857

Way MJ, Heong KL (1994) The role of biodiversity in the dynamic and management of tropical irrigated rice. Bull Ecol Res 84:567–587

Weerakoon GMPR, Jinadasa KBSN, Herath GBB, Mowjood MIM, Van Bruggen JJA (2013) Impact of the hydraulic loading rate on pollutants removal in tropical horizontal subsurface flow constructed wetlands. Ecol Eng 61(2013):154–160

Weerakoon GMPR, Jinadasa KBSN, Herath GBB, Mowjood MIM, Zhang D, Tan SK, Jern NW (2016) Performance of tropical vertical subsurface flow constructed wetlands at different hydraulic loading rates. J Clean Soil Air Water 44(9999):1–11

Weragoda SK, Tanaka N, Sewwandi BGN, Mowjood MIM (2010) Efficiency of coconut coir-pith as an alternative substrate in the treatment of submerged macrophyte wetland systems in tropical conditions. Chem Ecol 26(6):445–452

Yosihiro N (2010) Ecosystem services by paddy fields as substitutes of natural wetlands in Japan. Ecol Eng 56:97–106. https://www.ramsar.org/

Integrated Plant Nutrient Management in Major Agricultural Soils of Sri Lanka: A Review of the Current Status and the Way Forward

W. S. Dandeniya and R. S. Dharmakeerthi

1 Introduction

Being exposed to the tropical hot and humid climate, majority of the soils in Sri Lanka is highly weathered. Soil erosion, high rates of organic matter decomposition, leaching of basic cations, and removal of nutrient at a high rate through biomass produced under moist environmental conditions are common natural processes in such climatic regions resulting in low inherent soil fertility. Thus, supplying inputs externally to replenish soil nutrient pools is crucial to maintain or enhance crop productivity on these soils. Low productivity in agricultural lands usually leads to less agricultural inputs by farmers, which in turn aggravated the soil fertility depletion. To minimize land degradation through nutrient impoverishment and increase the productivity, fertilizers are applied since 1940s, under numerous subsidy schemes, as many farmers are not in a financially sound position to afford the cost of fertilizers. The mineral fertilizers, such as rock phosphates and dolomite, and chemical fertilizers, such as urea, muriate of potash, and triple superphosphate, which are collectively referred to as synthetic fertilizers hereinafter, are the most popular external nutrient inputs used for cropping in Sri Lanka. Fertilizer recommendations in the country have evolved, integrating synthetic and organic fertilizers in nutrient management programs over the years (Nagarajah 1986; Palm and Sandell 1989). Yet, overapplication of fertilizers polluting the environment and underapplication causing nutrient mining are often observed in farmers' fields (Kendaragama 2006). Although self-sufficiency in some crop sectors has been achieved in Sri Lanka due to the introduction of improved varieties periodically, subsidized fertilizers, and related technological advances in the agriculture sector, sustenance of the achieved productivity levels is becoming increasingly challenging.

W. S. Dandeniya (✉) · R. S. Dharmakeerthi
Department of Soil Science, Faculty of Agriculture, University of Peradeniya, Peradeniya, Sri Lanka
e-mail: warshisd@pdn.ac.lk

R. P. De Silva et al. (eds.), *Agricultural Research for Sustainable Food Systems in Sri Lanka*, https://doi.org/10.1007/978-981-15-3673-1_10

It is estimated that more than 50% of the arable land in the country are subjected to soil fertility deterioration due to poor nutrient management (Nayakekorale 1998). Decrease in area for food crop production with changes in land use, population increase, and rapid urbanization creating sinks for agricultural produce are continuing, forcing the agriculture sector in Sri Lanka to increase productivity of land while being conscious on environmental quality. At present, the per capita arable land extent in Sri Lanka is only 0.16 ha, and it is one of the lowest in Asia (FAO 2017). To bring marginal lands back to successful crop production, it is vital to address soil fertility issues hindering crop productivity. In this chapter, the history of soil fertility management in Sri Lanka is examined along with currently adopted management practices, lessons learned, and the way forward for environmentally sustainable soil fertility and plant nutrition management.

2 Management of Soil Fertility and Plant Nutrients: Paradigm Shift

Low productivity of agricultural lands is a major contributing factor to inadequate food production and poor nutritive value of food resulting in undernourishment and malnutrition among the people in developing countries, whom accounts to nearly two-third of the world's population. Soil fertility decline is one of the key reasons for reducing the productivity of arable lands in many countries including Sri Lanka (Sanchez 1997; Nayakekorale 1998). Traditionally, soil fertility is viewed as the ability of a soil to support plant growth. This concept has a long history as proven by the writings dating back to 2500 B.C. of ancient Mesopotamians. Historical evidences indicate Chinese, Greek, Indian, and Egyptian civilizations, among others, practiced manure application to increase productivity of agricultural lands. Some of them used green manure crops, wood ash-like materials, and closely available mineral salts. However, the precise role of nutrients on productivity of land was not established until Justus Von Liebig (1803–1873), the father of modern fertilizer industry, has scientifically proven the value of mineral elements found in soils for plant growth.

With the dawn of fertilizer industry, nutrient availability in soil was identified as the major factor deciding plant growth, which may have led to overlook the other fertility parameters. It was expected that a fertile soil should be able to provide all essential mineral nutrients for plants in available forms, in quantities, and in a suitable balance. Hence, soil fertility management mainly focused on management of plant nutrient supply from the soils. Different types of fertilizers started being manufactured and applied in Europe since late 1800s, and manufacturing processes improved commencing from early 1900s. A number of fertilizers were available for the research work that initiated for "green revolution" by the 1930s. Thus, after World War II, in postwar period (1950s) with "green revolution," the application of external inputs, supplying nutrients mainly via synthetic fertilizers, emerged as the first soil fertility management paradigm in the modern history (Sanchez 1994). Synthetic fertilizers together with improved germplasm during "green revolution" increased the crop production in the world as never seen before. By the 1970s, synthetic fertilizer usage has become popular in many parts of the world. The significant impact of the fertilizers had on crop production made it the silver bullet people seek for soil fertility improvement for any crop for a few decades to come. However, failure of this concept in improving yields under resource poor settings such as in sub-Saharan Africa and increased notion toward sustainable agriculture and conscience on environmental health urged a paradigm shift in soil fertility management (Sanchez 1997; Bationo 2009).

With advancements in knowledge on soil nutrient dynamics and the interactive role of soil physical, chemical, and biological properties on plant performances, the scope of soil fertility has expanded resulting in a second paradigm shift in soil fertility management in the late 1980s (Sanchez 1997). Accordingly, an integrated soil fertility management (ISFM) approach that encourages the use of both synthetic and organic fertilizers/amendments was introduced to minimize the total reliance on synthetic fertilizers. In addition, social, cultural, and economic challenges like availability of inputs, cost, technical feasibilities, etc. in adopting soil fertility management under different settings were also considered in ISFM (Sanchez 1997; Bationo 2009). Although still an external input paradigm, the second paradigm is viewed as more appropriate due to numerous benefits the approach had on physical, chemical, and biological aspects of soil fertility and socioeconomic feasibility. In parallel to ISFM, site-specific soil fertility management addressing the spatial and temporal heterogeneity in soil fertility status also started to evolve.

At present, the soil fertility management research is at the verge of a third paradigm shift, in which enhancing the nutrient-supplying power of soil with renewable soil fertility replenishment strategies is being looked into (Sanchez 1997; Ajayi et al. 2007). These approaches include relying more on biological processes (adapting germplasm to adverse soil conditions, enhancing soil biological activity, optimizing nutrient cycling to minimize external inputs and maximize the efficiency of their use, crop rotations with legume crops, etc.) and improving soil organic carbon pool through C sequestration with deep-rooted crops, use of carbonaceous compost, biochar application, etc. to improve overall soil fertility. The paradigm shift in soil fertility management is illustrated in Fig. 1. Considering these developments, conceptualizing soil fertility as the capacity of a soil to receive, store, and transmit energy to sustain plant growth as indicated by Food and Agriculture Organization of the United Nations (FAO) is more meaningful.

2.1 A Historical Account on Plant Nutrient Management Approaches in Sri Lanka

The present land use pattern in Sri Lanka is a legacy of land policy of the colonial past where export-based commercial plantation agriculture was superimposed on a traditional subsistence farming system (Mapa 2003). Before being colonized by the British in late seventeenth century, Sri Lankan farmers used organic inputs, crop rotations, shifting cultivation, and crop diversification like practices and biodynamic farming techniques for maintaining or improving crop productivity. Such organic

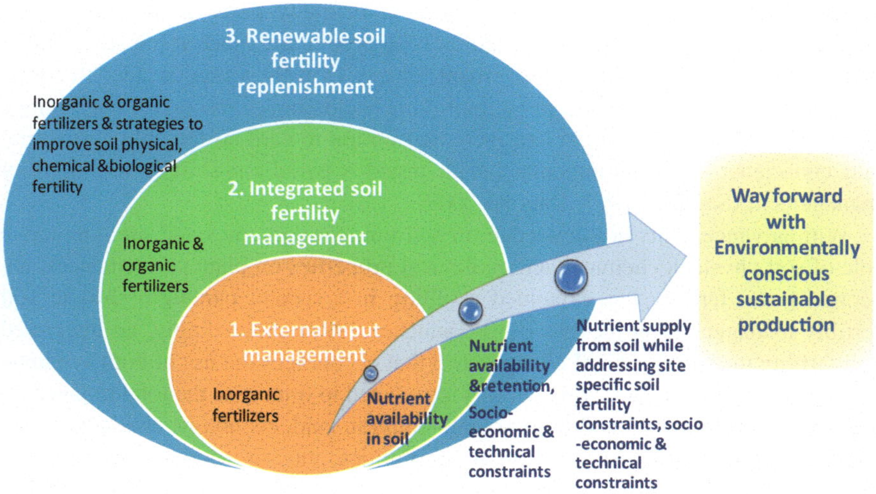

Fig. 1 Paradigm shift in the approaches of soil fertility and plant nutrient management highlighting the main factors considered and inputs used at different stages

and biodynamic means of farming were continued until the 1940s for all the crops except for large-scale tea and rubber plantations established under the British governance. With the postwar recession in economy in the 1950s, food shortage in the world limiting importations, and increase in population, there was a clear need to increase crop yields, thus making synthetic fertilizers a popular choice for soil fertility management. During the 1940s to 1960s, the traditional crop breeds/varieties that performed well under biodynamic means of farming at subsistence scale were replaced with nutrient-responsive high-yielding varieties cultivated at commercial scale in many parts of the country, demanding a change in the means of plant nutrients management. This change came as countrywide adoption of synthetic fertilizer application for nutrient management, and under this context, Sri Lanka reaped the benefits of green revolution on one hand (Nagarajah 1986; Palm and Sandell 1989). On the other hand, some farmers used integrated approach in nutrient management by combining biodynamic methods and organic farming techniques they have used for generations with synthetic fertilizer application.

Government institutions in Sri Lanka such as Department of Agriculture and Plantation Crop Research Institutes and universities are conducting research related to soil fertility and plant nutrition in parallel to other parts of the tropics since long. However, changes in fertilizer recommendations to reflect the paradigm shift in soil fertility and plant nutrient management happened at a very slow pace in Sri Lanka due to various socioeconomic challenges. The integrated soil fertility management (ISFM) approach appeared in nutrient management recommendations in late 1990s, nearly a decade later from its introduction in other parts of the world. The nutrient recommendations being practiced for many crops, including rice, tea, rubber, and coconut, are still based on yield targets defined by agroclimatic conditions or

Table 1 A historical account on soil fertility and plant nutrient management approaches in Sri Lanka

Soil fertility management approach	Introduction	Remarks
Natural means for replenish soil fertility	Precolonial era	Natural methods were used to replenish soil fertility Used organic inputs, crop rotations, shifting cultivation, and crop diversification like practices and biodynamic farming methods
External input paradigm	1940s	Bone meal, a natural fertilizer, first introduced as an external nutrient input for rice, tea, and rubber
	1950s	Synthetic fertilizers were introduced for rice farming and plantation crops as blanket recommendations Countrywide adoption of synthetic fertilizers for nutrient management in annual crops
	1970s	Blanket fertilizer recommendations changed into more specific recommendations; Rice: Based on soil types, climatic regions, agroclimatic regions and yield targets Plantation crops: Soil types/region Other crops: Blanket recommendations
Integrated soil fertility management (ISFM)	1990s	Site-specific fertilizer recommendations containing both synthetic and organic sources were introduced; Rice: Agroclimatic region/water management method/yield target and soil test based Plantation crops: Soil series level and field level using soil and foliar analysis (rubber) and later for coconut and tea
	2018	Other crops: Blanket recommendations with organic and synthetic fertilizers as sources of nutrients Site-specific nutrient requirements decided based on soil test kits are under investigation
Renewable soil fertility replenishment	2010	Related technologies are mostly at research level and still not appearing as recommendations

Sources of information: Amarasiri (1986), Nagarajah (1986), Yogaratnam and Silva (1987), TRI (2000), Bandara et al. (2003), CRI (2008), DOA (2013), Sirisena et al. (2015), Sirisena and Suriyagoda (2018)

irrigation management than site-specific soil fertility constraints. Table 1 summarizes the evolution of soil fertility and plant nutrient management approaches in Sri Lanka. Details of the recommendations for nutrient management are discussed later in this chapter.

3 Fertility Status of Soils of Sri Lanka

Agricultural landscapes are often very diverse in terms of cropping patterns, geophysical environments, input usage, resource availabilities, etc. Plate 1 depicts a specific situation related to the hill country where cropping systems and their management vary within a very small area. Further, soil resource also shows high

Plate 1 Diversity of an agricultural landscape under intensive cultivation in hilly Nuwara Eliya District, Sri Lanka, showing high spatial heterogeneity. (Photo credit: Dr. R.S. Dharmakeerthi)

spatial heterogeneity. Thus, soil fertility constraints are spatially variable, and the ISFM programs should acknowledge and address this heterogeneity by identifying site-specific constraints in order to be successful. As in other parts of the tropics, Sri Lanka has problem soils such as acid sulfate and saline soils, which are not suitable for cropping under their natural state. Problem soils present a unique set of soil fertility constraints, hence requiring special management approaches. The focus of this chapter is limited to soil fertility constraints in agricultural lands that do not encompass problem soils.

Many tropical soils are impoverished in mineral nutrients (Tiessen et al. 1994), and agricultural soils of Sri Lanka are no exception. Analysis of soil from different parts of Sri Lanka under different crops revealed deficient status of macro-, secondary, and micronutrients (Indraratne and Thilakarathne 2009; Kumaragamage and Indraratne 2011). In a study conducted using soil samples from 52 locations of Sri Lanka, Indraratne and Thilakarathne (2009) reported that most of the tested soils are deficient in available N, P, K, Ca, Mg, Cu, Zn, S, Mn, and B for plant growth. Low availability of N is largely due to poor organic matter content, and high losses of N due to leaching, volatilization, and denitrification (Kumaragamage 2010). High fixation reduces the availability of P in soil (Withana and Kumaragamage 1995). Alfisols, the dominant agricultural soils in the dry and intermediate zones, may display deficiencies of sulfur and cationic micronutrients such as copper (Cu) and zinc (Zn) (Kumaragamage 2010), whereas Ultisols, the main agricultural soils in the wet zone, generally have low cation exchange capacity and base saturation, are acidic in nature, and often exhibit deficiency of potassium (K) and magnesium (Mg) (Kumaragamage et al. 1999) and Al toxicity. The use of high inputs in intensive cultivation of soils having high infiltration rates such as sandy regosols in Kalpitiya and red latosols in Jaffna Peninsula, and soils with shallow depth to groundwater, leads to pollution of water with nutrients that leach through soil, which has serious implications on human and environmental health (Kuruppuarachchi 2010; Young et al. 2010; Jayasingha et al. 2011). Studies conducted in coastal sand aquifers under intensively cultivated vegetable cropping systems have revealed that the buildup of nitrate is quite dramatic and indicated that intensive vegetable cultivation is a major source of nitrate to groundwater (Kuruppuarachchi 2010; Jayasingha et al. 2011). According to a study conducted over a 2-year period using 58 sampling locations, Jayasingha et al. (2011) reported that nitrate-N in well water in Kalpitiya ranged from 0.60 to 212.40 mg/L in the dry seasons and 0.20 to 148.50 mg/L in rainy seasons with 50% of the samples in each season exceeding WHO standards for nitrate-N in drinking water (10 mg/L).

In tropical soils, the cation exchange capacity (CEC) is very low due to many reasons, which includes, but not limited to, the dominance of iron and aluminum oxides and kaolinite minerals (Indraratne 2006), low soil organic matter (OM) contents, and low pH values. Because of the low CEC, applied nutrients are rapidly leached below the root zone of annual crops. To increase nutrient use efficiency in annual crop production, techniques must be developed to retain applied nutrients in the root zone of the crop. Restoring and maintaining soil OM content in the long term is essential to enhance soil physical, chemical, and biological fertility aspects and thereby improve nutrient-supplying power.

Intensive cultivation without adopting any soil fertility restoration mechanisms exploits soil resource resulting in decline in soil fertility and depletion of the nutrient-supplying power of soil (Kendaragama et al. 2001). Continuous cropping and removal of harvest deplete soil nutrient pool, and this is more of a concern for annual crops than perennials, since most annual crops produce high amount of biomass within a short period of time, thus removing large quantities of plant nutrients from the soil. For instance, the removal of P from soil with the harvest of some popular crops such as capsicum, cabbage, tomato, carrot, and potato cultivated in the wet zone of the country is recorded as 5, 18, 18, 24, and 27 kg/ha, respectively (Wijesundara 1990). Nearly 20 kg of P per season is estimated to be removed with the harvest of lowland rice at an average yield of 6 t/ha (Suriyagoda et al. 2014). The soils are not always replenished with nutrients via fertilizer application since the decision on nature and amount of fertilizer applied with the next cropping is affected by number of factors like availability of material, willingness of the farmer to pay, market value of the crop, risks taken, etc. (Wjewardana and Amarasiri 1990; Sirisena and Suriyagoda 2018). When applied at recommended dosages, the nutrient input to soil via fertilizers for vegetable crops is higher than the nutrient removal with the harvest of these crops since the recommendations allow a compensation for the nutrient unavailable for plant uptake due to fixation in soil and losses due to runoff and leaching. For example, P input with synthetic fertilizer recommendation for vegetable crops ranges from 50 to 70 kg/ha (Wijewardena 2005), whereas removal of P with the harvest is only 5 to 30 kg/ha (Sirisena and Suriyagoda 2018). In one hand, many commercial-scale vegetable producers in up-country wet zone are known to apply synthetic fertilizers, foliar fertilizers, and animal manures, at or above the recommended dosages, when they cultivate high-value crops, sometimes leading to building up of nutrients such as P (Kendaragama et al. 2001; Sirisena and Suriyagoda 2018) and heavy metals like Cd, Cu, Zn, Ni, and Pb in soils (Premarathna et al. 2005). Wijesundara (1990) reported 68% of the intensively vegetable-cultivated fields had soils with available P (Olsen P) contents greater than 100 mg/kg, whereas in general, the optimum value of Olsen P ranges from 10 to 24 mg/kg depending on the crop. On the other hand, most of the low-value vegetable crops (pumpkin, radish, lettuce, etc.) and other field crops (maize, millets, sorghum, etc.) receive only partial amount of the recommended fertilizer dosage (Wjewardana and Amarasiri 1990; Nayakekorale 1998) leading to nutrient mining from soil. Depletion of nutrients negatively affects biological activity in soils, which in turn lowers physical fertility aspects such as aggregate stability, moisture holding capacity, etc., thus deteriorating overall soil fertility.

4 Inputs for Plant Nutrient Management: Use of Synthetic Fertilizers

The use of synthetic fertilizers has become an essential crop management practice in present-day agriculture. Synthetic fertilizers are applied with the sole objective of increasing plant available nutrient levels in soil. Even though 14 elements are identified as essential plant nutrients supplied by soil (N, P, K, Ca, Mg, S, Fe, Mn, Zn, Mo, Cu, B, Cl, and Ni), the major plant nutrients that are being supplied as chemical fertilizers in Sri Lanka are N, P, K, and Mg, while Zn, Cu, and B are also supplied under specific conditions for tea (TRI 2000), paddy (DOA 2013), or fruits and other annual crops (Wijewardena 2005).

Marking the history of introduction of fertilizers for crop nutrient management, the government of Sri Lanka freely issued bone meal, an organic fertilizer containing 3% N and 22% P_2O_5, to rice growers in the 1940s (Nagarajah 1986). Since then, synthetic fertilizers as straight fertilizers (a source targeting to supply one macronutrient) and different fertilizer mixtures (a source targeting to supply more than one nutrient at a time) have been used in the country depending on the evolution of fertilizer recommendations. For example, the fertilizer recommendation for rice changed 11 times from 1950 to 2013, fertilizer mixtures were recommended in three instances (1950, 1971, and 1980), and straight fertilizers were recommended in other eight events (Nagarajah 1986; Sirisena and Suriyagoda 2018). An account on evolution of fertilizer recommendations and the types of synthetic fertilizers used for plantation crops, rice, and other annual crops until 1985 is presented in Volume 4 of Soil Science Society of Sri Lanka Journal published in 1986 and discussed later in this chapter.

4.1 Sources of Phosphorous (P)

Prior to 1970, use of "low-analysis fertilizer" such as rock phosphate (RP) and bone meal, which in general has less than 30% of the total available primary nutrient, was a common feature not only in perennial crop sectors but also for annual crops such as paddy (Wijewardena 2005). Bone meal was recommended for rice farming in 1940 as the P source. Fertilizer recommendations for rice in 1950, 1956, 1967, and 1971 included rock phosphate and triple superphosphate as P sources, and the recommendations in 1959 and 1964 included only rock phosphate as the P source. With the increased adoption of new improved varieties, which are highly responsive to fertilizers, superphosphates and fused phosphates became the sole source of P since 1980. At present, the most popular P source in annual cropping systems is TSP. In the plantation crop sector and other perennial cropping systems, P is supplied as phosphate rock powder. The imported rock phosphates (IRP) were replaced gradually by locally produced Eppawala rock phosphate (ERP) since the discovery of this ore at Eppawala (8.1456° N, 80.4048° E), Sri Lanka, in 1971. Being an igneous and/or metaigneous rock, apatite in this ore shows low solubility in water and in citric acid (Zoysa et al. 2001). Much research has been conducted over the last few decades to increase the P availability of ERP for annual crops. Selectively mined and powdered Eppawala apatite crystals are marketed as high-grade ERP (HERP) which has about 35–42% P_2O_5 content (Dahanayake et al. 1995). Mixing ERP or HERP with peat (Dahanayake et al. 1991; Rathnayake et al. 1993), composts (Mawalagedera et al. 2012), P-solubilizing microorganisms (Tennakoon et al. 2016; Hettiarachchi et al. 2017), organic wastes, or acid-forming N fertilizers (Zoysa et al. 2001) to increase P availability has been tested and gained varying degrees of success at laboratory or greenhouse studies. However, the effectiveness of such technologies under field conditions with different ecophysiological situations needs confirmation through further research. Single superphosphate had been produced by acidulating ERP (ESP) by using nitric acid (Tennakone and Weragama 1992) or using sulfuric acid by the Lanka Phosphate (Pvt) Ltd. Field evaluations have been carried out on this product during 2010/2011 by the Department of Agriculture and have confirmed that ESP can be used as an effective P fertilizer for paddy-growing soils in Sri Lanka (D.N. Sirisena, Personal Communications). However, it has still not been produced at commercial scale.

4.2 Sources of Nitrogen (N)

In the early 1960s, ammonium sulfate was used as the N fertilizer; however, during the period of 1967–1978, urea became the major source of N, because of its low cost per unit of N and also due to severe soil acidification observed under application of ammonium sulfate in wet zone soils (Wickremasinghe et al. 1981; Dharmakeerthi and Thenabadu 1996). Since the 1980s, the fertilizer research in Sri Lanka focused more on improving fertilizer use efficiency and reducing groundwater and surface water pollution. In order to increase the efficiency of urea in paddy-growing soils, prilled urea was introduced in 2013. Slow-releasing N fertilizers like sulfur-coated urea and IBDU are available in international market but have not been introduced to Sri Lanka yet. At present, research is being conducted on producing slow-released fertilizers using nanotechnology to improve the N fertilizer use efficiency in agricultural systems (Kottegoda et al. 2011, 2016).

4.3 Sources of Potassium (K), Magnesium (Mg), and Calcium (Ca)

Muriate of potash (MOP) is the most commonly used K source in the country since the late 1960s. In nursery formulations in the plantation crop sector, and for some Cl-sensitive crops such as tobacco, sulfate of potash (SOP) is being used. Dolomite is another fertilizer that is locally produced and used as both Mg supplement and ameliorant of soil acidity. Dolomite also supplies Ca. This is mostly used as a liming material in tea plantations and vegetable production systems in the up-country of Sri Lanka and also known to provide benefits in upland cropping systems in low country dry zone of Sri Lanka (Eeswaran et al. 2016). The use of kieserite and Epsom salt as Mg supplement can be observed under nursery conditions or to correct Mg deficiency in plants.

4.4 Fertilizer Formulations

Both urea and ammonium sulfate (SA) readily release N, increasing plant available N contents in soil, and since most of the arable lands have N-deficient situations for crop growth, plants respond quickly when N fertilizers are added. This sometimes led to mismanagement of fertilizers by farmers (Nagarajah 1986). Overapplication of N fertilizers beyond recommended dosage and disregard of application of P and K sources due to lack of visible response from the crop for these two nutrients resulted in introduction of fertilizer mixtures (formulations) as recommendations for most crops (Amarasiri 1986; Nagarajah 1986). However, when fertilizer formulations are used, it is difficult to address site-specific issues related to nutrient imbalances, and that results in risk of overloading or underapplying a nutrient to soil environment and difficulty in matching nutrient supply with crop growth requirement. Thus, fertilizer formulations could lead to poor fertilizer use efficiency (Nagarajah 1986). The formulations were removed since the 1990s from fertilizer recommendations of annual crops.

In the plantation crop sector, fertilizer mixtures have been formulated predominantly using urea and/or SA, ERP, and MOP. Even when the nutrient requirement for a specific field is identified through soil and foliar analysis, a fertilizer mixture for that NPK ratio is often formulated in plantation crop sector to minimize application cost, and it ensures correct proportions and quantities of NPK that is required for the optimum growth of the plant in that field. Fertilizer formulations are also popular in floriculture and home gardening, which operate at comparatively less diverse agroecological conditions compared to annual cropping systems. The formulations

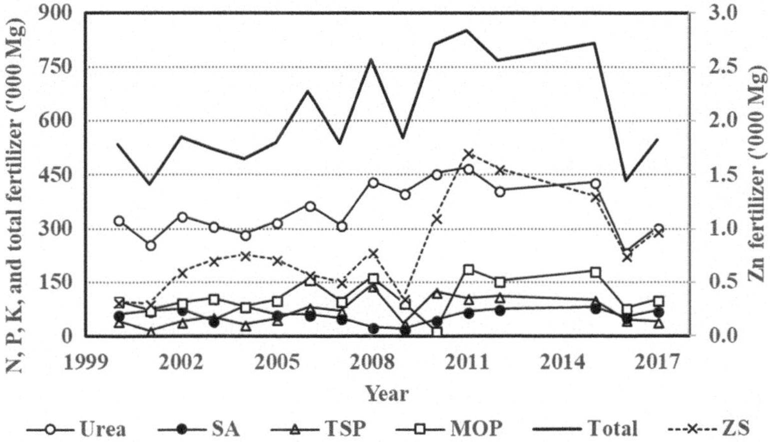

Fig. 2 Imported quantities of NPK fertilizers [urea, ammonium sulfate (SA), triple superphosphate (TSP), muriate of potash (MOP)] and Zn sulfate (ZS) during 2000 to 2017. (Source: National Fertilizer Secretariat, Sri Lanka 2017)

prepared at present for floriculture and home gardening purposes contain number of nutrients in addition to N, P, and K, and the composition depends on the targeted growth stage of the crop and mode of application (to soil or foliar).

4.5 Present Situation of Synthetic Fertilizer Usage in Sri Lanka

Although a number of plant nutrients are deficient in Sri Lankan soils (Kumaragamage and Indraratne 2011), most commonly used chemical fertilizers are urea, SA, TSP, ERP, MOP, and dolomite, supplying only N, P, K, Ca and Mg. The use of chemical fertilizers containing micronutrients is often less than the recommended levels. Quantities of major fertilizers imported into the country during the last two decades are given in Fig. 2. The quantities imported have a direct influence from changing government policies on fertilizer subsidy scheme (Weerahewa et al. 2010; Wijetunga and Saito 2017). In 2015, Sri Lanka had imported nearly 816,000 tons of synthetic fertilizer which dropped by about 33% in 2017 (Fig. 4.1) mainly due to a policy change from a material subsidy to a cash subsidy. From the imported synthetic fertilizers, 54% was urea and about 20% was MOP. The proportion of TSP importation dropped continuously from 14% in 2012 to about 7% by 2017.

From the total chemical fertilizer usage in the country in 2017 (0.6 million tons), 30% was consumed by rice cultivation (0.8 million ha) and 24% by the tea plantations (0.2 million ha) followed by 12% by vegetable crops (0.1 million ha) (Table 2). Similar proportions could be observed in other years as well. Hence, fertilizer usage in Sri Lanka vary between 80 and 700 kg/ha depending on the crop sector, with a national average of about 270 kg/ha.

Table 2 Distribution of some important chemical fertilizers in different crop sectors in Sri Lanka in 2017

	Urea	SA	TSP	ERP	MOP	DOL	ZS	Other	Total
Crop	‘000 Mg								
Paddy	142.9	2.6	14.1	<0.1	18.8	<0.1	0.03	2.27	180.7
Tea	73.5	27.1	0.4	21.6	32.1	3.0	0.40	10.7	168.8
Rubber	4.4	0.9	<0.1	3.0	2.3	0.3	0.00	0.8	11.7
Coconut	17.4	3.3	2.4	7.8	15.1	2.8	<0.01	1.2	50.0
OFC	2.3	2.4	2	<0.1	2.2	0.3	<0.01	1.5	10.7
Vegetable	17.6	22.7	10.5	1.4	13.9	2.8	0.11	8.29	77.3
Other	46.2	12.3	11.1	5.4	16.8	3.0	0.44	5.16	100.4
Total	304.3	71.3	40.5	39.2	101.2	12.2	0.98	29.92	599.6

SA ammonium sulfate, *TSP* triple superphosphate, *ERP* Eppawala rock phosphate, *MOP* muriate of potash, *DOL* dolomite, *ZS* zinc sulfate, *OFC* other field crops
Source: National Fertilizer Secretariat, Sri Lanka. Sector-wise Fertilizer Distribution Report (2017)

4.6 Regulation of Synthetic Fertilizer Usage in Sri Lanka

Out of many synthetic fertilizers required to improve crop growth and yield in Sri Lankan soils, only rock phosphates and dolomite are locally produced, which account for only 9% of the total fertilizer usage (Table 2). All others are imported to the country by state-owned and private companies under the strict regulation set by the government. Fertilizer formulations are prepared locally from the imported raw materials by private sector and state-owned companies. The importation, formulation, and distribution of chemical fertilizers in the country are regulated by the Regulation of Fertilizer Act No. 68 of 1988. The Fertilizer Advisory Committee, which comes under the purview of the National Fertilizer Secretariat according to this law, has the sole authority to recommend the fertilizers to be imported to Sri Lanka. Currently, the importation of bio- and organic fertilizers is not allowed except for research purposes, and this move aims to protect the biodiversity in the country. Any chemical fertilizer imported to Sri Lanka should abide by quality standards set forth by the Sri Lanka Standards Institute (SLSI), and this is implemented by the National Fertilizer Secretariat. A comparison of quality standards for potentially toxic trace elements adopted by some countries is given in Table 3.

5 Inputs for Plant Nutrient Management: Use of Organic Fertilizers

Organic materials have an important place in soil fertility management due to their short-term effect on nutrient supply and long-term contribution to soil organic matter replenishment (Lal 2004). Increase of soil organic carbon pool in degraded soils increases crop yields through beneficial effects on nutrient supply, biological activities, water holding capacity, and soil structure (Chathurika et al. 2015; Mariaselvam et al. 2016; Chathurika et al. 2016). According to Lal (2004), an increase of 1 ton of soil carbon pool of degraded cropland soils may increase

Table 3 Maximum permissible levels (total content mg/kg) of the five potentially toxic trace elements for synthetic fertilizer in some countries

Country	Cd	As	Pb	Cr	Hg	References
Brazil (For simple mineral fertilizers and compound fertilizers containing macro and micro nutrients)	4–57	2–500	20–1000	40–500	0.05–10	Goncalves Jr. et al. (2014) DOI: https://doi.org/10.5772/57268
USA (Texas) (Any type of fertilizer)	39	41	300	17	NA	Goncalves Jr. et al. (2014) DOI: https://doi.org/10.5772/57268
EU countries (Phosphate fertilizers)	50–275	NA	NA	NA	NA	
New Zealand (Phosphate fertilizers)	280	NA	NA	NA	NA	
Canada (Any type of fertilizers)	20	75	500	NA	NA	
Australia (Any type of fertilizers)	10–300	NA	100	NA	5	National Code of Practice for Fertilizer Description and Labelling – Australia (2011)
Japan (Phosphate fertilizer)	343	50	100	5	NA	Goncalves Jr. et al. (2014) DOI: https://doi.org/10.5772/57268; Journal of Scientific Research and Reports (2014): Vol 3(4): 610–620, JSRR.2014.007
China (Any type of fertilizers)	8	50	100	500	5	
Sri Lanka (Any type of fertilizers)	0.2–3.0	0.3–25	0.2–30	3–50	0.1–1.0	Sri Lanka Standards Institute, 2014 (SLS 644, 748, 812, 1104)

NA not available, a range of values indicate different standard limits for different types/categories of fertilizers

maize yield by 10–20 kg/ha. An increase in soil organic C (SOC) stock leads to increased crop yield even in high-input agriculture (Bauer and Black 1994) but especially in SOC-depleted soils (Johnston 1986).

5.1 Conventional Organic Fertilizers

Animal manures and green manures are conventional organic inputs utilized by farmers. These amendments, generally of high quality, contain low amounts of lignin- and phenol-like compounds, release nutrients quickly, and are known to improve soil fertility and crop yield. The decomposition rates of animal and crop residues are high in soil and, thus, should be added repeatedly and in high quantities to achieve significant improvement in soil fertility and carbon sequestration (Sommerfeldt et al. 1988; Sukartono et al. 2011). *Crotalaria juncea*, *Tithonia diversifolia*, *Calliandra calothyrsus*, *Gliricidia sepium*, *and Vigna radiata* are examples of commonly used green manure crops in Sri Lanka (Palm and Sandell 1989). In addition, in the plantation crop sector of Sri Lanka, particularly in rubber-, coconut-, and oil palm-grown soils, the cultivation of leguminous cover crop species such as *Pueraria phaseoloides*, *Desmodium ovalifolium*, and *Mucuna bracteata* is a common practice to conserve soil from erosion. An additional advantage of this practice is the addition of organic matter by cover-crops into soils.

Cattle manure is popular among Sri Lankan farmers even before the introduction of synthetic fertilizers. Poultry farming is more organized and happens at a large scale in the country under intensive management practices compared to cattle farming. Therefore, poultry manure (i.e., litter mixed with feces and urine of chickens) is more available and cheaper than cattle manure; thus, poultry manure is the most commonly used animal manure in the country. Application of untreated poultry manure could lead to spread of antibiotic resistance in the environment, which is an emerging threat to human health management (Martinez 2009; Herath et al. 2016). High rates of manure application can cause water pollution due to nutrients and greenhouse gas emissions.

Compost is another popular organic fertilizer in Sri Lanka. With the government interventions in the late 1990s to recycle solid waste at municipal council level, several composting stations were established in the country. As quality control measures, standards for compost produced from municipal solid waste and agricultural waste were introduced in 2003 by Sri Lanka Standards Institution (SLSI 1246:2003). A few companies are producing compost at large scale in the country, and several small-scale producers are also contributing to the market. Compost is a popular organic fertilizer applied in floriculture industry, nursery management, and home gardening but not widely used in commercial-scale cultivation of annual and perennial crops.

5.2 Carbon-Rich Organic Soil Amendments

Carbonaceous organic soil amendments having high C/N ratio contain high level of recalcitrant C, release C to the soil slowly over time, and retain in the soil for a long time. Such materials are added with the objective of increasing overall soil fertility than as sources of nutrients; hence, it is better to refer to as organic soil amendments. Further, these materials give the best results when incorporated in IPNM along with synthetic fertilizers (Mariaselvam et al. 2014).

Common high C/N ratio amendments such as rice straw, sawdust, and rice husks can be used as soil amendments to retain C for a long period. However, due to low N content in these high C/N amendments, N immobilization may occur soon after the application to soils. Rao and Mikkelsen (1976) found that when soil and rice straw were not incubated prior to planting rice seedlings, applied N was immobilized, causing inhibition of plant growth and low N content in plants. Mariaselvam et al. (2014) studied different C-rich mixtures prepared using cattle manure, rice straw, wood shavings, and sawdust in fresh and incubated forms on nutrient release from an alluvial soil (Entisol) and a red-yellow podzolic soil (Ultisol) to identify their suitability as soil amendments. They found that some organic materials mineralize faster (rice straw-based materials) than the others (wood-based materials). Further, the contribution of wood-based materials to long-lasting C pools in soil is higher than rice straw-based materials. Therefore, all organic materials are not equally suitable as organic soil amendments to improve soil carbon sequestration. The results further indicate that the organic materials affect nutrient cycling in soil, and the effect of the material on short-term nutrient cycling should be considered when selecting a material as an amendment. Otherwise, crop growth may impede due to temporary nutrient immobilization in soil when C-rich material is added.

Recently, application of biochar is being promoted to arrest fertility degradation, and even to revert degraded agricultural soils in the tropics, because of number of positive characteristics in biochar (Lehmann 2007; Dharmakeerthi et al. 2015). High in porosity, specific surface area, charge density, and pH as well as being a source of plant nutrients together with high recalcitrant organic C content are the main features of biochar that enhance soil physical, chemical, and biological fertility in many soils (Enders et al. 2012; Vasujini et al. 2014). But the quality of biochar varies with the pyrolysis conditions and feedstock used (Enders et al. 2012; Vasujini et al. 2014). Slow-release fertilizers are being developed using biochar and N fertilizers to increase the fertilizer use efficiency in Sri Lankan agricultural soils (Gamage et al. 2013; Kottegoda et al. 2011, 2016). However, lack of effective technologies to produce good-quality biochar in large scale has delayed the adaptation of this technology in Sri Lankan agriculture.

It has been suggested that amending soils with biochar could arrest, or sometimes even revert, degradation of tropical agricultural lands (Kimetu et al. 2008). Different application rates of biochar tested in alluvial (Entisol), reddish brown earth (Alfisol), and reddish brown latosolic (Alfisol) soils from Sri Lanka indicated that soil fertility status could be improved by adding biochar at the rate of 1 Mg/ha. Cattle manure and sawdust mixed and incubated for 2 months (CSi) and applied at 2 Mg/ha rate has given comparable results to biochar at 1 Mg/ha with respect to soil fertility and productivity (Chathurika et al. 2015). In nursery rubber plants, Dharmakeerthi et al. (2012) observed a significant growth improvement (26–61%) when an acidic Ultisol is amended with 1% (w/w) rubber wood biochar in combination with N and Mg fertilizers.

A major constraint for commonly used or presently recommended organic soil amendments like compost, green manure, and animal manure is their unavailability to apply at the recommended rate (10 Mg/ha) in each season. Further, the availability of materials shows high variability in regional scale. Thus, requirement of material at a lower rate (nearly ten times) with long-lasting effects on soil are positive traits related to biochar and CSi. Moreover, sawdust, which is the base material in both organic soil amendments, is a waste product in timber mills with little economic value. Use of sawdust to produce these organic amendments will be a solution to manage waste of one industry. The concern on contaminants, pests, and diseases that could be introduced through these two organic amendments is minimal in comparison to other alternatives like compost.

5.3 Biofertilizers

Another important group of organic fertilizers that is being researched in Sri Lanka is biofertilizers (Palm and Sandell 1989; Seneviratne et al. 2011). Nitrogen-fixing inoculants were introduced in the late 1980s but did not become popular due to restrictions of supply of the material to farmers and need of repeated application, among other reasons (Granhall et al. 1987; Palm and Sandell 1989; Catroux et al. 2001). Further, the inoculant did not show significant advantage over the use of synthetic fertilizers, and the effects on crop performance were inconsistent over field and with time (Granhall et al. 1987). The nature of indigenous microbial population of soil, the inoculation rate, the adaptability of the inoculants to environmental stresses, the purity of the inoculants, and the shelf life of the formulations affect the significance of the impact of using biofertilizers in agriculture (Catroux et al. 2001; Herrmann and Lesueur 2013). Thus, the features of biofertilizers such as successful inoculation and maintenance of quality in processing and storage have to be improved in order to promote adoption of biofertilizers for nutrient management (Catroux et al. 2001; Herrmann and Lesueur 2013). A number of success stories exist from different countries on the use of plant growth-promoting bacteria inoculants and mycorrhizal fungi as biofertilizers to improve crop productivity, soil fertility, and plant tolerance for stresses in sustainable agriculture (Adesemoye et al. 2009; Mahdi et al. 2010; Bhardwaj et al. 2014; Gracia-Fraile et al. 2015). In a study conducted in Auburn, USA, supplementing 75% of the recommended fertilizer rate with inoculants to a tomato crop resulted in yield and nutrient uptake statistically equivalent to the full fertilizer rate without inoculants (Adesemoye et al. 2009). Ability to reduce synthetic fertilizer usage when applied with biofertilizers without sacrificing crop productivity under tropical conditions has been observed in many other studies as summarized in reviews by Bhardwaj et al. (2014) and Garcia-Fraile et al. (2015).

In Sri Lanka, since 2003, several products to increase decomposition of straw, solubilize ERP, and fix P forms in soil, and N_2 fixation, were researched and tested at different scales for annual crops (Seneviratne et al. 2009; Kumari et al. 2010; Rajapaksha 2010; Rajapaksha et al. 2011; Katulanda and Rajapaksha 2012) and plantation crops (Seneviratne et al. 2011; Hettiarachchi et al. 2014, 2017; Tennakoon et al. 2016) with the participation of Department of Agriculture and various research institutes. Based on greenhouse-scale and field-scale experiments with rice cultivation, Rajapaksha et al. (2011) reported that 50% of TSP could be substituted with ERP along with seed inoculants formulated with *Enterobacter gergoviae*, *Bacillus mycoides,* and *Bacillus pumilus*. These products, if successful, are along the line of third paradigm of soil fertility and plant nutrient management. Improvement of soil fertility under the application of biofertilizers has been observed (Rajapaksha 2010; Seneviratne et al. 2011). The diversity with respect to soil conditions creating diverse ecological conditions for microbial communities and the diversity in farming systems (input usage, agronomic practices, crop rotations, etc.) stand as challenges in developing effective biofertilizers because the microorganisms used show niche specificity (Granhall et al. 1987; Catroux et al. 2001; Rajapaksha 2010; Herrmann and Lesueur 2013). A few government-funded projects to formulate biofertilizers for annual crops (rice, vegetables) and plantation crops (tea and rubber) are in progress at present.

6 Fertilizer Recommendations in Sri Lanka

Fertilizer application for different crops was initially based on experiences of the growers on ad hoc fertilizer applications. Therefore, initial fertilizer recommendations were based solely on crop demand and yield goals. Systematic fertilizer trials conducted later for different crop-soil systems revealed variable responses of plant growth and yields to added fertilizers (e.g., Constable and Hodnett 1953 for rubber). Although initial fertilizer recommendations comprised of synthetic fertilizers, only with the adoption of ISFM approach organic fertilizer inputs were introduced to fertilizer recommendations of most crops (Yogaratnam and Silva 1987; Bandara et al. 2003). The rates of organic fertilizers to be applied in integrated nutrient management approach vary across crops, and this is discussed under the section on integrated plant nutrient management in this chapter.

Even though fertilizer recommendations for different crop sectors initially started as blanket application irrespective of soil variability, most of the synthetic fertilizer recommendations developed at present encourage soil test-based nutrient management (Bandara et al. 2003; Weerasinghe 2017). However, detailed information generated through soil survey and classification efforts (Moorman and Panabokke 1961; Mapa 2005; Mapa et al. 1999, 2009) have not been effectively utilized in fertilizer recommendations for any crop in Sri Lanka.

6.1 Plantation Crops

From the plantation crops, rubber was the first to incorporate soil series-based fertilizer recommendations. By the late 1970s, the Rubber Research Institute of Sri Lanka introduced a site-specific fertilizer recommendation based on soil and foliar analysis for mature rubber plants (Yogaratnam and Silva 1977). For example, fertilizer recommendation for rubber (*Hevea brasiliensis* L.) varies according to the K status of the soil (Yogaratnam et al. 1984). Similarly, tea and coconut research institutes in Sri Lanka later introduced soil- and foliar-based fertilizer recommendations for mature plantations (CRI 2008; TRI 2000).

The second highest chemical fertilizer consumer in Sri Lanka is tea plantations, consuming about 24% of the imported fertilizer. Fertilizer usage in rubber plantations is low, with about 240–675 kg of NPK fertilizer/ha/year depending on the soil type and the age (Samarappuli 2001), while fertilizer use in tea plantations is the highest among traditional plantation crops, with 350–1750 kg of NPK fertilizer/ha/year depending on the age and yield target (TRI 2000). The very high rates of N fertilizer use, initially as sulfate of ammonia and currently as urea, have contributed to soil acidification in tea-growing areas. Hence, these soil require special management practices such as the application of 1–2 Mg of dolomite/ha once in every 3 years. Fertilizer requirement in coconut plantations in Sri Lanka varies from 230 to 600 kg of NPK fertilizers /ha/year depending on the age of the palm, soil type, and cycle of replanting. Low productivity in coconut lands of the acidic wet zone soils of Sri Lanka has been due to Mg deficiency, and as a prophylactic measure, application of 240–320 kg of ground dolomite is usually recommended for wet zone soils (CRI 2008). However, the application of chemical fertilizers by coconut growers in Sri Lanka is very low. Compared to tea, rubber, and coconut, the extent under oil palm is low and only 8000 ha in Sri Lanka, but because of the labor scarcity and very high profit margin, there is a tremendous pressure on converting rubber plantations in the wet zone of Sri Lanka to oil palm cultivation. The government aims to increase the total extent under oil palm up to 20,000 ha by year 2020. The fertilizer requirement of oil palm is very high due to high nutrient removal from the fields (Goh 2004). Even though there is no fertilizer recommendation as such for oil palm in Sri Lanka, commercial planters apply between 420 and 2100 kg of NPK fertilizers per ha per year (Prof. A Nugawela, personal communications). Residual effects and environmental impacts when high doses of chemical fertilizers are applied to oil palm plantations need to be assessed critically.

6.2 Paddy

Paddy consumes a large proportion of the imported fertilizer, i.e., 45–55% of urea, 35–55% of TSP, and 15–35% of MOP, and the recommended rate is around 225 to 340 kg of NPK fertilizers per hectare per season (Weerasinghe 2017). In comparison to fertilizer usage in other countries, this is a moderate amount. However, fertilizer for paddy cultivation in Sri Lanka is provided free or at a very high subsidy since the 1940s (Nagarajah 1986; Weerahewa et al. 2010); thus, actual chemical fertilizer usage in some farmer fields could be much higher than the recommended rates. Kendaragama (2006) observed that in rice-based cropping systems, the actual fertilizer usage has ranged from 50 to 300% of the recommended rates. The low rates were observed in rice-rice cropping systems in the coastal areas and in some up-country rice-vegetable cropping systems.

Fertilizer recommendation for rice cultivation has been revised from time to time since its first tentative recommendation in 1950 (Nagarajah 1986), considering research findings, farmer experience, and/or government policies. Because of very low yield response to added P fertilizers, application of TSP has now been recommended once in two seasons. This could be partly due to P buildup in soil due to continuous application of P fertilizers (phosphate rocks until the 1990s and concentrated superphosphate thereafter). However, measurement of available P may not be helpful to explain the low responses to added P in soils with very low Olsen P levels and/or positive response in soils with high Olsen P. Sirisena and Suriyagoda (2018) speculated that the degree of P saturation may be a better indicator to predict crop response to added P. Moreover, the basal application of N fertilizer has now been shifted to 2 weeks after transplanting in order to increase the fertilizer use efficiency (DOA 2013). Since the burning of rice straw in paddy fields is discouraged, the amount of K fertilizer requirement could be decreased up to 50% (Sirisena et al. 2015) without affecting the crop yield. Zn is deficient in most rice-growing soils, and therefore application of 5 kg of $ZnSO_4$ has been recommended to all soils as a precaution, but only a few farmers have adopted this recommendation.

6.3 Other Annual Crops

Considering the nutrient requirement of different crops, various fertilizer mixtures have been introduced in the first fertilizer recommendations released for cereals, vegetables, and fruit crops in 1980 (Nagarajah 1986). Basal application from the relevant NPK mixture varied depending on the crop species from 125 kg/ha (soybean) to 625 kg/ha (vegetables) to 1500 kg/ha (up-country potato). Generally, vegetables and potato, which consume the highest amount of chemical fertilizers in this category of crops, are cultivated in steeply dissected hilly landscapes in the up-country wet zone of Sri Lanka and/or sandy soils in the dry zone that has a shallow ground water table with quick recharge. Vegetable and potato cultivation is a year-round practice in these areas, and cropping at least three times a year has become a common practice. In order to minimize the rate of soil fertility degradation due to intensive cultivation of high nutrient-demanding crops without adopting proper soil conservation practices, farmers in these areas apply large quantities of organic manures in addition to the recommended doses of chemical fertilizers. Application of around 10 Mg of poultry manure/ha/season that is rich in P and readily available N or composts made using such manure or solid wastes is popular among these farmers. For example, Kendaragama et al. (2001) observed that chemical fertilizer usage in vegetable cropping systems in these areas is about six to nine times higher than the recommended rates. High levels of available P (Mehlich-3 extractable P ranging from 17 to 298 mg/kg with 95% of analyzed soils having values greater than 45 mg/kg) and P sorption capacity (as indicated by single-point adsorption capacity, P_{150}, of soils ranging from 185 to 1167 mg/kg with a mean value of 665 mg/kg) are reported for up-country intensively cultivated vegetable fields (Amarawansha and Indraratne 2010). High levels of NO_3-N (300 ppm) in the ground waters of the vegetable-cultivated areas with sandy soils (Kuruppuarachchi 2010) have been observed. For food crop sector, a soil test-based fertilizer recommendation was first introduced in 1993 in order to minimize the excessive building up of nutrients and increased productivity (Weerasinghe 2017).

7 Integrated Plant Nutrient Management

Effective and efficient management of plant nutrients can be achieved only when the nutrient management plan addresses the physical, chemical, and biological fertility aspects of soils simultaneously (Chathurika et al. 2015; Mariaselvam et al. 2016). Identification of soil fertility constraints and selection of soil amendments and/or corrective measures to improve soil fertility should be essential in a nutrient management plan (Chathurika et al. 2015; Mariaselvam et al. 2016). Improving soil organic matter pool is essential in most of the degraded tropical soils. Lal (2015) suggested that restoration of soil organic C to threshold levels of at least 11 to 15 g kg − 1 (1.1–1.5% by weight) within the root zone is critical to reducing soil and environmental degradation risks.

Understanding the importance of soil organic matter pool on sustainability of Sri Lankan agriculture, application of organic amendments is highly encouraged for all cropping systems. From plantation crops, rubber and coconut adopt the establishment of leguminous cover throughout its life span, in addition to recommending the application of composts and organic residues as mulches around tree (Samarappuli 2001; Tennakoon and Bandara 2003). Cover crops minimize the erosion of organic matter containing top soil while helping to provide organic matter through its litter. In tea cultivations, rehabilitation of tea lands by growing Mana or Guatemala grasses for 2 years prior to replanting, burying of prunings, and application of compost at replanting has been recommended in addition to the synthetic fertilizer application (TRI 2000, 2016). For rice fields, returning paddy straw back into the field and application of 6 Mg of compost/ha/year have been recommended on top of the synthetic fertilizer inputs. Recently, the application of 0.6 Mg of partially burnt paddy husk/ha/year has been encouraged among paddy farmers. In most of the annual cropping systems managed with synthetic fertilizer inputs, application of 10 Mg of organic manure has been recommended (Wijewardena 2005). Most of these recommendations rather help to arrest further degradation of soil organic matter reserves than increasing the reserves up to the sustainable limit suggested by Lal (2015) for developing countries in the tropics.

Biochar is being considered as an organic soil amendment that can be effectively used in ISFM approach. Under Sri Lankan conditions, application of synthetic fertilizers along with biochar has decreased soil acidity, increased nutrient availability (Dharmakeerthi et al. 2012), improved soil physical properties (Gamage et al. 2016), and enhanced soil carbon pool and crop yields (Mariaselvem et al. 2014; Chathurika et al. 2016). In addition to organic and synthetic fertilizers, the use of rock powder as soil amendment to improve soil fertility status has been reported in the tropics (Silva et al. 2013; Chathurika et al. 2015). Improvement in maize yield with the inclusion of rock powder in ISFM was observed by Chathurika et al. (2015).

8 Way Forward in Soil Fertility Management

More than 75 years have passed since the introduction of synthetic fertilizers for plant nutrient management. Yet the use efficiency of the synthetic fertilizers remains around 10–50% under farm field condition. Strategies to improve N fertilizer use efficiency up to 40% was introduced by the Rice Research and Development Institute (DOA 2013), but it is not being practiced by majority of rice farmers yet due to number of environmental, social, political, and technical reasons. Therefore, soil fertility and plant nutrition management plans should be researched and developed to maximize the resource use efficiency in agricultural systems, which will provide additional benefits such as reducing the indirect costs on recovering from environmental and health issues created due to agricultural pollutants.

Cultivation of same crop(s) in a plot of land and application of synthetic fertilizers over several decades have changed the soil fertility in most agricultural lands. Therefore, man-made spatial heterogeneity is becoming more prominent among farmer fields than the inherent heterogeneity. Soil test-based fertilizer recommendation is effective under these conditions. However, lack of effective soil testing mechanism for farmers, together with very high fertilizer subsidy, hinders the adoptability of this site-specific technology and harnessing the positive impacts on soil and the environment. Soil fertility evaluations and land suitability for cropping are conducted by only a few large-scale agro-industries. Research should be carried out to develop a farmer-friendly soil test kit and quick and easy way to obtain the fertilizer recommendations by a farmer to his field. The Department of Agriculture is developing a farmer-friendly fertilizer management protocol using soil test kits and plant sap test kits to determine nutrient status of plants and available nutrients in soils (mainly N, P, and K) based on colorimetric assays. This research program was initiated in 2017, and now the protocol is being tested at pilot scale.

The soil testing is currently carried out by the Department of Agriculture, or respective crop research institutes with the mandate, or by some private sector companies that are still limiting their analysis for available nutrients, pH, and electrical conductivity and occasionally for soil texture. These analyses would reveal only the potentially available level of nutrients to support plant growth in the short run. Therefore, a package of soil analyses that warrant the understanding of overall soil fertility (chemical, biological, and physical fertility) at an affordable price to the farmers should be developed, and farmers should be encouraged to keep records on soil health for crop production. This is being effectively practiced in some European countries and in North America, which encourages the farmers to be accountable and environmentally conscious.

Development of soil fertility management apps aiming for e-agriculture could be more effective in the present-day farming. Moreover, further research is required to improve the currently adopted algorithms and/or to develop new algorithms where required, to determine the optimum fertilizer rates for a given soil-crop system based on the soil test values. A better understanding on soil fertility constraints and their spatiotemporal variability is required to support the development of effective site-specific nutrient management strategies.

Applying fertilizer inputs to a degraded soil or to a soil having restrictions for crop growth such as compaction, poor drainage, poor biological activity, and nutrient toxicities is not cost-effective. Soil physical and biological fertility are relatively new concepts to the stakeholders involve in management of soil fertility and plant nutrition in Sri Lanka. Thus, a consorted effort should be taken to educate farmers, agricultural officers, and other stakeholders about the environmental consequences related to improper soil fertility management and to encourage conscious adoption of best management practices to enhance productivity in the long run with environmental consciousness.

Incorporation of organic fertilizers or organic soil amendments is beneficial in soil fertility and plant nutrition management (Sirisena et al. 2015). Development of soil-crop system-specific technologies to increase soil organic matter pool above a critical level in Sri Lankan agricultural lands should be encouraged. Moreover, technologies to produce organic soil amendments in required quantities by recycling agricultural and urban wastes need to be developed or made available to the farmers. Effective N-fixing, P-solubilizing, and/or plant growth-promoting soil microorganisms have been already isolated from various soil-crop systems; however, their commercialization needs to be promoted. Regulations with respect to quality of external inputs should be developed to safeguard the environment and enhance the food safety, especially when animal manures and compost from municipal solid wastes are being used. These regulations should not be limited to nutritive composition but extend to assessing the contamination levels by potentially toxic trace elements (Cd, Pb, Co, As, Hg, etc.) and antibiotics and biocides. Moreover, policies on fertilizer usage in the country should be focused not only to increase the productivity of the land but also to safeguard the environment. Therefore, the material fertilizer subsidy given to farmers needs to be critically evaluated, and a better environmental-friendly policy should be adopted.

References

Adesemoye AO, Torbert HA, Kloepper JW (2009) Plant growth-promoting rhizobacteria allow reduced application rates of chemical fertilizers. Microb Ecol 58(4):921–929

Ajayi OC, Akinnifesi FK, Sileshi G, Chakeredza S (2007) Adoption of renewable soil fertility replenishment technologies in the southern African region: lessons learnt and the way forward. Nat Res Forum 31(4):306–317. https://doi.org/10.1111/j.1477-8947.2007.00163

Amarasiri SL (1986) Some considerations in arriving at fertilizer recommendations for annual crops. J Soil Sci Soc Sri Lanka 4:30–34

Amarawansha EAGS, Indraratne SP (2010) Degree of phosphorous saturation in intensively cultivated soils in Sri Lanka. Trop Agric Res 22(1):113–119

Bandara WMJ, Wickramasinghe WMADB, Sirisena DN (2003) Potassium fertilization of paddy in Sri Lanka. In: Importance of potash fertilizers for sustainable production of plantation and food crops in Sri Lanka, pp 143.11

Bationo A (2009) Soil fertility-paradigm shift through collective action. CTA, Wageningen, The Netherlands. http://knowledge.cta.int/en/Dossiers/Demanding-Innovation/Soil-health/Articles/Soil-FertilityParadigm-shift-through-collective-action. Accessed 14 May 2018

Bauer A, Black AL (1994) Quantification of the effect of soil organic matter content on soil productivity. Soil Sci Soc Am J 58(1):185–193

Bhardwaj D, Ansari MW, Sahoo RK, Tuteja N (2014) Biofertilizers function as key player in sustainable agriculture by improving soil fertility, plant tolerance and crop productivity. Microb Cell Factories 13(1):66–76

Catroux G, Hartmann A, Revellin C (2001) Trends in rhizobial inoculant production and use. Plant Soil 230:21–30

Chathurika JAS, Indraratne SP, Dandeniya WS, Kumaragamage D (2015) Beneficial management practices on growth and yield parameters of maize (*Zea mays*) and soil fertility improvement. Trop Agric Res 27(1):59–74

Chathurika JAS, Kumaragamage D, Zvomuya F, Akinremi OO, Flaten DN, Indraratne SP, Dandeniya WS (2016) Woodchip biochar effects on selected soil fertility parameters in two Chernozemic soils. Can J Soil Sci 96:472–484

Constable DH, Hodnett GE (1953) The manuring of *Hevea brasiliensis* at Dartonfield. Ceylon Empire J Exp Agric 21:131–136

CRI (2008) Use of fertilizer for coconut and land suitability (in Sinhala). Advisory circular no. 01. Coconut Research Institute, Sri Lanka

Dahanayake K, Senaratne A, Subasinghe SM, Liyanaarachchi A (1991) Potential use of naturally occurring sulphuric acid to beneficiate poorly soluble phosphate from Eppawala, Sri Lanka. Fertil Res 29(2):197–201

Dahanayake K, Ratnayake MP, Sunil PA (1995) Potential of Eppawala apatite as a directly applied low-cost fertilizer for rice production in Sri Lanka. Fertil Res 41(2):145–150

DOA (2013) Department of Agriculture Fertilizer Recommendation for Rice, Sri Lanka: Department of Agriculture

Dharmakeerthi RS, Thenabadu MW (1996) Urease activity in soils: a review. J Natl Sci Found Sri 24(3):159–195

Dharmakeerthi RS, Chandrasiri JA, Edirimanne VU (2012) Effect of rubber-wood biochar on nutrition and growth of nursery plants of *Hevea brasiliensis* established in an Ultisol. Springer Plus 1:84

Dharmakeerthi RS, Hanley K, Whitman T, Woolf D, Lehmann J (2015) Organic carbon dynamics in soils with pyrogenic organic matter that received plant residue additions over seven years. Soil Biol Biochem 88:268–274

Eeswaran R, De Costa WAJM, De Costa DM, Dandeniya WS, Sivakumar S, Suriyagoda LDB (2016) Evaluation of a climate change-adaptive, eco-friendly agronomic package for potato (*Solanum tuberosum* L.) cultivation in the farmer fields of the Jaffna district of Sri Lanka. Trop Agric Res 27(2):190–202

Enders A, Hanley K, Whitman T, Joseph S, Lehmann J (2012) Characterization of biochar to evaluate recalcitrance and agronomic performance. Bio Resour Technol 114:664–653

FAO (2017) Special report FAO/WFP crop and food security assessment mission to Sri Lanka. 22 June 2017, Food and Agriculture Organization of the United Nations

Gamage DV, Mapa RB, Dharmakeerthi RS, Biswas A (2016) Effect of rice-husk biochar on selected soil properties in tropical Alfisols. Soil Res 54(3):302–310

Gamage A, Basnayake B, Costa J, Vidanagamage K (2013) Evaluation of total N, P, K and organic matter contents of soil amended with paddy husk charcoal coated urea and comparison of the yield of paddy. http://dl.lib.mrt.ac.lk/handle/123/8892. Accessed 23 June 2018

Garcia-Fraile P, Menendez E, Rivas R (2015) Role of bacterial biofertilizers in agriculture and forestry. Bioengineering 2(3):183–205. https://doi.org/10.3934/bioeng.2015.3.183

Goh KJ (2004) Fertilizer recommendation systems for oil palm: estimating the fertilizer rates. In: Soon CP, Pau TY (eds) Proceedings of MOSTA best practices workshops: agronomy and crop management. Malaysian Oil Scientists and Technologies Association, Kuala Lumpur, pp 235–268

Granhall U, Kulassoriya SA, Hirimburegama WK, De Silva RSY, Lindberg T (1987) Nitrogen fixation in some rice soils in Sri Lanka. MIRCEN J Appl Microbiol Biotechnol 3(4):367–388

Herath EM, Palansooriya AGKN, Dandeniya WS, Jinadasa RN (2016) An assessment of antibiotic resistant bacteria in poultry litter and agricultural soils in Kandy District, Sri Lanka. Trop Agric Res 27(4):389–398

Herrmann L, Lesueur D (2013) Challenges of formulation and quality of biofertilizers for successful inoculation. Appl Microbiol Biotechnol 97(20):8859–8873

Hettiarachchi RP, Dharmakeerthi RS, Jayakody AN, Seneviratne G, De Silva E, Gunathilake T, Thewarapperuma A (2014) Effectiveness of fungal bacterial interactions as biofilmed biofertilizers on enhancement of root growth of *Hevea* seedlings. J Environ Prof Sri Lanka 3(2)

Hettiarachchi RP, Dharmakeerthi RS, Seneviratne G, Jayakody AN, De Silva E, Gunathilake T, Thewarapperuma A, Maheepala CK (2017) Availability and leaching of nutrients after biofilm biofertilizer applications into a red yellow Podsolic soil. J Rubber Res Inst Sri Lanka 94:43–53

Indraratne SP (2006) Occurrence of organo-mineral complexes in relation to clay mineralogy of some Sri Lankan soils. J Natl Sci Found Sri 34(1):29–35. https://doi.org/10.4038/jnsfsr.v34i1.2073

Indraratne SP, Thilakarathne JKDAK (2009) Sorption capacity and deficiency of nutrients in some soils of Sri Lanka. J Soil Sci Soc Sri Lanka 29:17–26

Jayasingha P, Pitawala A, Dharmagunawardhane HA (2011) Vulnerability of coastal aquifers due to nutrient pollution from agriculture: Kalpitiya, Sri Lanka. Water Air Soil Pollut 219 (1–4):563–577. https://doi.org/10.1007/s11270-010-0728-y

Johnston AE (1986) Soil organic matter, effects on soils and crops. Soil Use Manag 2(3):97–105

Katulanda P, Rajapaksha CP (2012) Response of maize grown in an alfisol of Sri Lanka to inoculants of plant growth promoting rhizobacteria. J Plant Nutr 35(13):1984–1996

Kendaragama KMA (2006) Fertilizer use efficiency in farming systems in Sri Lanka. J Soil Sci Soc Sri Lanka 8:1–18

Kendaragama KMA, Lathiff MA, Chandrapala AG (2001) Impact of vegetable cultivation on fertility status of soils in the Nuwara Eliya area. Ann Sri Lanka Dep Agric 3:95–100

Kimetu JM, Lehmann J, Ngoze SO, Mugendi DN, Kinyangi JM, Riha S, Verchot L, Recha JW, Pell AN (2008) Reversibility of soil productivity decline with organic matter of differing quality along a degradation gradient. Ecosystems 11(5):726–739

Kottegoda N, Munaweera I, Madusanka N, Karunaratne V (2011) A green slow-release fertilizer composition based on urea-modified hydroxyapatite nanoparticles encapsulated wood. Curr Sci 10:73–78

Kottegoda N, Madusanka N, Sandaruwan C (2016) Two new plant nutrient nanocomposites based on urea coated hydroxyapatite: efficacy and plant uptake. Indian J Agric Sci 86(4):494–499

Kumaragamage D (2010) Site specific nutrient management in soils of Sri Lanka: a review of recent work using a systematic approach to formulate fertilizer recommendations for annual crops. J Soil Sci Soc Sri Lanka 22:1–16

Kumaragamage D, Indraratne SP (2011) Systematic approach to diagnosing fertility problems in soils of Sri Lanka. Commun Soil Plant Anal 42:2699–2715

Kumaragamage D, Nayakekorala HB, VidhanaArachchi LP (1999) Risks and limitations of wet zone soils. In: Mapa RB, Somasiri S, Nagarajah S (eds) Soils of the wet zone of Sri Lanka. Soil Science Society of Sri Lanka/Survodaya Publishers, Peradeniya, pp 139–159

Kumari PDSU, Nanayakkara CM, Bandara JMAU (2010) Development of a fungal inoculum for efficient phosphate utilization in agriculture. In: Proceedings of the 15th international forestry and environment symposium, 26–27 November 2010, Nugegoda, Sri Lanka

Kuruppuarachchi DSP (2010) A review on the leaching of nitrate from agricultural soils and pollution of ground water in Sri Lanka. J Soil Sci Soc Sri Lanka 22:37–49

Lal R (2004) Soil carbon sequestration impacts on global climate change and food security. Science 304:1623–1627

Lal R (2015) Restoring soil quality to mitigate soil degradation. Sustainability 7(5):5875–5895. https://doi.org/10.3390/su7055875

Lehmann J (2007) Bio-energy in the black. Front Ecol Environ 5:381–387

Mahdi SS, Hassan GI, Samoon SA, Rather HA, Dar SA, Zehra B (2010) Bio-fertilizers in organic agriculture. J Phytol 2(10):42–54

Mapa RB (2003) Sustainable soil management in the 21st century. Trop Agric Res Ext 6:44–48

Mapa RB (ed) (2005) Soils of the intermediate zone of Sri Lanka. Soil Science Society of Sri Lanka/ Survodaya Publishers, Peradeniya

Mapa RB, Somasiri S, Nagarajah S (eds) (1999) Soils of the wet zone of Sri Lanka. Soil Science Society of Sri Lanka/Survodaya Publishers, Peradeniya

Mapa RB, Somasiri S, Dassanayake AR (eds) (2009) Soils of the dry zone of Sri Lanka. Soil Science Society of Sri Lanka/Survodaya Publishers, Peradeniya

Mariaselvam AA, Dandeniya WS, Indraratne SP, Dharmakeerthi RS (2014) High C/N materials mixed with cattle manure as organic amendments to improve soil productivity and nutrient availability. Trop Agric Res 25(2):201–213

Mariaselvam AA, Dandeniya WS, Indraratne SP, Dharmakeerthi RS (2016) Beneficial nutrient management practice for improving maize (*Zea mays*) yield in a tropical Entisol. Trop Agric Res 27(2):147–158

Martinez JL (2009) Environmental pollution by antibiotics and by antibiotic resistance determinants. Environ Pollut 157(11):2893–2902

Mawalagedera SM, Weerakkody WA, Premaratne KP (2012) Circulation culture of tomato for efficient nutrient uptake and high yield in tropical greenhouses. Trop Agric Res 23(3):204–217

Moorman FR, Panabokke CR (1961) Soils of Ceylon. Trop Agric 117:5–69

Nagarajah S (1986) Fertilizer recommendations for rice in Sri Lanka: a historical review. J Soil Sci Soc Sri Lanka 4:4–14

Nayakekorale HB (1998) Human induced soil degradation status in Sri Lanka. J Soil Sci Soc Sri Lanka 10:1–35

Palm O, Sandell K (1989) Sustainable agriculture and nitrogen supply in Sri Lanka: farmers' and scientists' perspective. Ambio 18:442–448

Premarathna HMPL, Hettiarachchi GM, Indraratne SP (2005) Accumulation of cadmium in intensive growing soils in the up country. Trop Agric Res 17:93–103

Rajapaksha RMCP (2010) Exploring unseen: a review on the diversity and functions of soil microorganisms of selective ecosystems in Sri Lanka. J Soil Sci Soc Sri Lanka 22:51–70

Rajapaksha RMCP, Herath D, Senanayake AP, Senevirathne MGTL (2011) Mobilization of rock phosphate phosphorus through bacterial inoculants to enhance growth and yield of wetland rice. Commun Soil Sci Plant Anal 42(3):301–314

Rao DN, Mikkelsen DS (1976) Effect of rice straw incorporation on rice plant growth and nutrition. Agron J 68:752–756

Rathnayake MP, Kumaragamage D, Dahanayake K (1993) A comparison of citric acid solubilities of commercially available Eppawala rock phosphate and selectively mined primary apatite crystals after mixing with peat. Trop Agric Res 5:69–76

Samarappuli L (2001) Nutrition. In: LMK T, Nugawela A (eds) Handbook of rubber, vol 1: Agronomy. Rubber Research Institute of Sri Lanka, Agalawatta, pp 156–175

Sanchez PA (1994) Tropical soil fertility research: towards the second paradigm. In: Transactions of the 15th world congress of soil science, vol 1. Mexican Society of Soil Science, Chapingo, pp 65–88

Sanchez PA (1997) Changing tropical soil fertility paradigms: from Brazil to Africa and back. In: Moniz AC et al (eds) Plant-soil interactions at low pH. Brazilian Society of Soil Science, Piracicaba, pp 19–28

Seneviratne G, Thilakaratne RMMS, Jayasekara APDA, Seneviratne KACN, Padmathilake KRE, De Silva MSDL (2009) Developing beneficial microbial biofilms on roots of non legumes: a novel biofertilizing technique. In: Microbial strategies for crop improvement. Springer, Berlin/Heidelberg, pp 51–62

Seneviratne G, Jayasekara APDA, De Silva MSDL, Abeysekera UP (2011) Developed microbial biofilms can restore deteriorated conventional agricultural soils. Soil Biol Biochem 43 (5):1059–1062

Silva B, Paradelo R, Vazquez N, Garcia-Rodeja E, Barral MT (2013) Effect of the addition of granitic powder to an acidic soil from Galicia (NW Spain) in comparison with lime. Environ Earth Sci 68:429–437

Sirisena D, Suriyagoda LD (2018) Toward sustainable phosphorus management in Sri Lankan rice and vegetable-based cropping systems: a review. Agric Nat Resour 52:9–15

Sirisena D, Wanninayake W, De Silva AGS (2015) Potassium dynamics in irrigated low land paddy soils and its relation to plant K and grain yield of rice. Trop Agric 163:89–99

Sommerfeldt TG, Chang C, Entz T (1988) Long-term annual manure applications increase soil organic matter and nitrogen, and decrease carbon to nitrogen ratio. Soil Sci Soc Am J 52 (6):1668–1616

Sukartono WH, Utomo KZ, Nugroho WH (2011) Soil fertility status, nutrient uptake, and maize (*Zea mays* L.) yield following biochar and cattle manure application on sandy soils of Lombok, Indonesia. J Trop Agric 49:47–52

Suriyagoda L, De Costa WAJM, Lambers H (2014) Growth and phosphorus nutrition of rice when inorganic fertilizer application is partly replaced by straw under varying moisture availability in sandy and clay soils. Plant Soil 384(1–2):53–68

Tennakone K, Weragama UT (1992) Nitric acid treated phosphate fertilizer from Eppawala apatite in Sri Lanka. Fertil Res 32(1):115–119

Tennakoon NA, Bandara SH (2003) Nutrient content of some locally available organic materials and their potential as alternative sources of nutrients for coconut. Cocos 15:23–30

Tennakoon PL, Rajapaksha RM, Hettiarachchi LS (2016) Potentials of plant growth promoting Rhizobacteria based bio-fertilizers for tea. In: Proceedings of the 232nd meeting of the experiments and extension forum. The Tea Research Institute, Sri Lanka, pp 26–39

Tiessen H, Cuevas E, Chacon P (1994) The role of soil organic matter in sustaining soil fertility. Nature 371:783–785

TRI Advisory Circular (2000) Fertilizer recommendations for mature tea. Circular no. SP3, serial no. 00/3. Tea Research Institute, Sri Lanka

TRI Advisory Circular (2016) Fertilizer recommendations for mature tea in small holdings. Interim circular no. SP10, serial no. 01/16. Tea Research Institute, Sri Lanka

Vasujini P, Dandeniya WS, Dharmakeerthi RS (2014) Assessing the quality of biochar produced from coconut husk waste. J Soil Sci Soc Sri Lanka 24:21–28

Weerahewa J, Kodithuwakku SS, Ariyawardana A (2010) The fertilizer subsidy program in Sri Lanka. In: Pinstrup-Andersen P, Cheng F (eds) Food policy for developing countries: case studies. Cornell University, Ithaca

Weerasinghe P (2017) Best practices of integrated plant nutrition system in Sri Lanka. Best practices of integrated plant nutrition system in SAARC Countries, 135

Wickremasinghe KN, Sivasubramaniam S, Nalliah P (1981) Urea hydrolysis in some tea soils. Plant Soil 62(3):473–477

Wijesundara SM (1990) Removal of nutrients by vegetable crops cultivated in the mid-and upcountry wet zone. Trop Agric 146:79–85

Wijetunga CS, Saito K (2017) Evaluating the fertilizer subsidy reforms in the rice production sector in Sri Lanka: a simulation analysis. Adv Manag Appl Econ 7(1):31

Wijewardena JDH (2005) Improvement of plant nutrient management for better farmer livelihood, food security and environment in Sri Lanka. In: Improving plant nutrient management for better farmer livelihoods. Food Security and Environmental Sustainability, Beijing, pp 73–93

Withana AK, Kumaragamage D (1995) Evaluation of phosphorous extraction methods for assessing phosphorous availability of some Sri Lankan soils. Trop Agric Res 7:143–151

Wjewardana JDH, Amarasiri SL (1990) Comparison of phosphate on growth of vegetable crops on acid soil. Trop Agric 6:57–68

Yogaratnam N, Silva P (1977) Use of leaf analysis as a guide to manuring of rubber. Bull Rubber Res Inst Sri Lanka 12:46–50

Yogaratnam N, Silva FPW (1987) Use of organic materials as fertilizers. Bull Rubber Res Inst Sri Lanka 24:17–20

Yogaratnam N, Silva FPW, Weerasuriya SM (1984) Recent developments in the nutrition of Hevea in Sri Lanka. In: Proceedings of the international rubber conference, vol 1, Part (1) September 17–19. Rubber Research Institute of Sri Lanka, Colombo, Sri Lanka, pp 207–247

Young SM, Pitawala A, Gunatilake J (2010) Fate of phosphate and nitrate in waters of an intensive agricultural area in the dry zone of Sri Lanka. Paddy Water Environ 8(1):71–79

Zoysa AK, Loganathan P, Hedley MJ (2001) Comparison of the agronomic effectiveness of a phosphate rock and triple superphosphate as phosphate fertilisers for tea (*Camellia sinensis* L.) on a strongly acidic Ultisol. Nutr Cycl Agro ecosyst 59(2):95–105

Geo-informatics: Contribution from Spatial Sciences for Agricultural Development and Food Security

N. D. K. Dayawansa and Ranjith Premalal De Silva

1 Introduction to Geo-informatics and Allied Technologies

Geo-informatics is a science and technology which utilizes spatial data to find answers to the real-world problems. It integrates different disciplines which are dealing with spatial information (Srivastawa 2014). Karimi (2009) defines geo-informatics as a science and technology in gathering, analyzing, interpreting, and using geo-referenced or spatial data. The Encyclopedia of Information Science and Technology (2015) defines geo-informatics as an academic discipline or career of working with geographical data for better understanding and interpretation of human interactions with the earth. It further defines the term broadly as a number of different technologies, approaches, processes, and methods to interpret issues and controversies relating to the earth's surface for collaborative decision-making. Development of remote sensing technology for the collection of spatial data followed by GIS and GPS made significant changes in surveying and cartography. Subsequent developments in this technology pushed the academia and industry to bring these disciplines under one umbrella of geo-informatics (Srivastawa 2014). Since spatial data or geographical data are heavily involved in finding answers to simple to very complex real-world questions, geo-informatics and allied disciplines are flourishing in recent times. The main components of geo-informatics are remote sensing, Geographical Information Systems (GIS), and Global Positioning Systems (GPS). In addition, surveying, cartography, and photogrammetry also play a major role. Remote sensing is the oldest out of these technologies developed with the objective of collecting geographical or spatial data remotely. Aerial photography and satellite remote sensing technologies have achieved a vast development during the past 100 years. A large number of satellite systems with varying capabilities are orbiting around the earth collecting very useful spatial information. The latest

N. D. K. Dayawansa (✉) · R. P. De Silva
Department of Agricultural Engineering, University of Peradeniya, Peradeniya, Sri Lanka
e-mail: dammid@pdn.ac.lk

R. P. De Silva et al. (eds.), *Agricultural Research for Sustainable Food Systems in Sri Lanka*, https://doi.org/10.1007/978-981-15-3673-1_11

addition to remote sensing is drone technology which can be easily used to acquire spatial information in a relatively small area at low altitudes. There are many definitions to remote sensing. According to Lillesand et al. (2014), remote sensing is the science and art of obtaining information about an object, area, or phenomenon through the analysis of data acquired by a device that is not in contact with the object, area, or phenomenon under investigation. According to Sabins (1996), remote sensing is the science of acquiring, processing, and interpreting images that records the interaction between energy and matter. According to Campbell and Wynne (2011), remote sensing is the practice of delivering information about the earth's land and water surfaces using images acquired from an overhead perspective, using electromagnetic radiation in one or more regions of the electromagnetic spectrum, reflected or emitted from the earth's surface. In this definition, they identified that it is not universal and stresses the omission of meteorological and extraterrestrial remote sensing in the definition.

Geographical Information Systems (GIS) is the most popular among all and is widely used in analyzing spatial data and producing maps. Different scientists have defined GIS based on its technological capabilities as a spatial data analysis tool, database approach, or usefulness in various applications. Burrough (1996) and Burrough and McDonnel (1998) define GIS as a powerful set of tools for collecting, storing, retrieving at will, transforming, and displaying spatial data from the real world for a particular set of purposes and can be considered as the most widely used definition of GIS. Cowen (1988) defined GIS as a decision support system which involves integration of spatially referenced data in a problem-solving environment. According to Devine and Field (1986), GIS is a Management Information System (MIS) that allows map display of general information. Maguire (1991) provides an overview of GIS and describes the difficulty of providing a proper definition to GIS. He has analyzed different definitions of GIS to comment on that all of them are general definitions with a common feature of GIS as a system which deals with spatial information. He suggests three views of GIS: as map view which focuses on cartographic aspects of GIS, database view which emphasizes the usefulness of well-designed and well-implemented database, and the spatial analysis view which focuses on the usefulness of spatial analysis. In addition, the application view of GIS identifies the technological capability of GIS to deal with global scientific problems.

Global Positioning Systems is a satellite-based navigation system originally designed by the United States for military navigation purposes. The Global Positioning System (GPS) is the only fully functional Global Navigation Satellite System (GNSS). Utilizing a constellation of at least 24 medium earth orbit satellites that transmit precise microwave signals, the system enables a GPS receiver to determine its location, speed, direction, and time. GPS has enabled the collection of spatial information on the dynamics of the earth surface as they occur. At present, GPS is a very essential component of everyday life.

Integration of remote sensing, GIS, and GPS with other allied disciplines such as cartography, photogrammetry, and surveying has made a significant progress during past few decades. The availability of high-resolution spatial and spatiotemporal data provides immense opportunities to gather new knowledge and to understand the complex geographical problems effectively (Guo and Mennis 2009). Mapping has become an integral part of day-to-day work with the integration of multiple sensors and multiple platforms for fast acquisition and processing of spatial data. Tao and Li (2007) explain that the availability of Internet and wireless communication networks and advances in sensors has made mapping process mobile and dynamic. Development of big data analytical capabilities also has positively impacted on working with remotely sensed data. According to Chi et al. (2016), a large number of space and airborne sensors provide a massive amount of remotely sensed data every day, and they discuss the opportunities and challenges that big data analytics bring in the context of remote sensing.

2 Usefulness of Location-Enabled Data in Agriculture

Spatial data can be identified as data collected at a specific location with one or more attributes. Spatial data are special due to their characteristics, namely, all of them have a specific location on the earth, carry attributes to describe them, and create spatial relationships with the surrounding objects. In addition, spatial data are subjected to change with time and also consist of metadata. In day-to-day life, majority of problems that come across are spatial in nature. Knowingly or unknowingly, people use a huge amount of spatial data in their routine activities. As examples, spatial data are essential in natural resources management, transportation, retail site selection and customer analysis, vehicle tracking and routing, agricultural activities, epidemiology, criminology and defense activities, etc.

Spatial data play a major role in agriculture from site selection for suitable crops, crop establishment, management, harvesting, and postharvest handling to food and nutritional security assessment. All geo-informatics technologies deal with location-enabled data, and they provide capabilities for improved decision-making through location-specific data analysis. Real-world applications provide evidences of the role of remote sensing, GPS, and GIS in agriculture. Pakistan has developed a crop forecasting and estimation system using GIS and remote sensing for wheat, rice, cotton, sugarcane, maize, and potato (Ahmad et al. undated). Winter wheat yield has been estimated using remote sensing in the United States (Salazar et al. 2007). To support improved irrigation scheduling in the Syr Darya basin, Uzbekistan, the ISAREG model integrated with GIS has been used (Fortes et al. 2005). Agricultural land use suitability analysis using GIS has been widely applied in many countries (Akinci et al. 2013; Zolekar and Bhagat 2015; Joshua et al. 2013).

3 Methodology Adopted for Review

This paper presents an overview of the use of geo-informatics technology in agriculture and food security in Sri Lanka based on a review of published articles during a limited period. The review is based on relevant research articles found in journals, proceedings, World Wide Web, and other documents and has been limited to the published literature on Sri Lankan studies after 2000. Within the selected period, the review was restricted to some selected articles which have a considerable contribution from geo-informatics technology to agriculture directly or indirectly.

4 Natural Resource Management and Geo-informatics

Agriculture and natural resources have a strong nexus, and soil and water resources are important components of agricultural systems. Natural resources management is a vast subject area which is heavily involved with spatial data. Natural resources are subjected to change with time due to natural causes and human influences. Hence, most up-to-date data is required for making decisions correctly. Remote sensing, GIS, and GPS technologies definitely help in managing natural resources successfully since they help to capture frequent changes in natural resources and analyze the data in multidimensional space considering diverse factors which may influence on resources and resource management. Management of watersheds helps to protect land and water resources which are vital for the success of agriculture. Geo-informatics technologies are very much helpful to identify critical issues in watershed scale such as soil erosion, land use changes, hydrological properties, water pollution, etc. (Panday et al. 2007; Chowdary et al. 2009; Basanyat et al. 2000). A number of studies have been carried out in Sri Lanka at watershed scale to estimate hydrological processes, soil erosion, land use changes, etc. De Silva (2005) has estimated spatial distribution of groundwater recharge in Deduru Oya catchment using remote sensing and GIS. Rainfall and acute evapotranspiration data have been derived from NOAA satellite data at 10-day interval for a year, and a simple water balance equation was used to estimate recharge. Since satellite data helps to derive rainfall and evapotranspiration basin wide, it was possible to estimate the spatial distribution of recharge in the basin. The author identified a number of limitations in the study and highlighted the usefulness of using remotely sensed data for better spatial representation of recharge compared to point measurements. The use of GIS is not highlighted in the methodology. Water sampling and spatial interpolation techniques were used in groundwater quality assessment and mapping in selected river basins (Piyadasa et al. 2006).

Soil erosion and land degradation (Manawadu and Karunarathne 2004; Jayarathne et al. 2010; Jayasekera et al. 2018) and land use changes and their impacts (Gunawardena et al. 2004; Adikari 2004; Kumarihamy and Dayawansa 2009) were key studies carried out under natural resources. Kumarihamy and Dayawansa (2009) explained a participatory GIS approach for sustainable land use planning at catchment scale. Majority of soil erosion studies have followed a similar methodology by using Universal Soil Loss Equation (USLE) with the help of GIS to derive required data layers and to integrate them to find soil loss at watershed scale. Jayarathne et al. (2010) coupled USLE-derived soil erosion map with socioeconomic factors such as population density, land/man ratio, agro land/man ratio, and beneficiaries of social benefits to improve the land degradation assessment. Wijesundara et al. (2018) used

Revised USLE to estimate soil erosion in Kirindi Oya basin. Overlay analysis in GIS is common across all these studies in integrating diverse spatial data layers. Aerial photography and satellite remote sensing have been used in assessing land use changes. Lack of accuracy assessment in land use identification can be highlighted as a limitation in majority of these studies. Also it is obvious that the advanced spatial data analysis functionality has not been used in most of the methodologies and the entire spatial analyses have been confined to overlay analysis.

Integration of GIS into distributed hydrological models helps input of spatially varying data into these models. A number of hydrological modeling attempts using GIS were identified. A study conducted in a small watershed in Walawe river basin utilized remote sensing and GIS technology for hydrological modeling with Soil Water Assessment Tool (SWAT) model and HEC-HMS lumped model (Weragala 2004). The SWAT was used to simulate flow at the watershed outlet, and since it is a distributed parameter model, GIS has played a major role compared to HEC HMS. SRTM Digital Elevation Model has provided topographic information to the model, while Landsat 7 satellite images were used to derive land use information. GIS has been used to derive other required spatial data layers. In a study by Jayasekera et al. (2005) of hydrological modeling in Upper Uma Oya catchment using SWAT model, GIS was used to generate basin spatial data layers required as input data to the model. IRS satellite data was used to derive land use information. Muthuwatta and Jayakody (2006) also used SWAT in mesoscale hydrological modeling with GIS.

In hydro-geological modeling of Samanalawewa watershed, Jayawardena and Gunatilake (2005) used surface analysis and 3D analysis capabilities of GIS together with remotely sensed data to identify and locate the fractures, joints, and fault patterns. A number of map layers related to geological properties and leakage problems in the areas were developed and analyzed using GIS. Delineating groundwater recharge potential sites in Ambalantota, Senanayake et al. (2015) also adopted a similar approach. Land use, soils, geology, geomorphology, drainage density, topography, and lineament density maps were developed using surface and 3D analysis techniques in GIS. Weighted overlay analysis was employed to develop an index to identify and map recharge potential zones. Rainfall map for the study area was developed using surface interpolation without mentioning the technique used. Kumar et al. (2016) also used overlay analysis in GIS to identify groundwater potential zones in Kilinochchi.

Few studies have been conducted in Sri Lanka on the use of GIS and remote sensing in mapping climatic parameters such as rainfall and water resources management including agriculture-related pollution. Ten-day estimates of actual evapotranspiration in Sri Lanka were derived using NOAA AVHRR satellite images supplemented with ground-based estimations (Chandrapala and Wimalasuriya 2003). Those estimates together with soil moisture status estimated with satellite images and rainfall data were incorporated in a GIS. The authors stated that this is a solid basis for understanding water consumed by environment and agriculture and also helpful in understanding the efficiency in major irrigation systems. Bastiaanssen and Chandrapala (2003) described a new procedure for hydrological data collection and assessment for agricultural and environmental water use. The variability of annual water balance in Sri Lanka was derived with measured rainfall and remotely sensed actual evaporation rates. In the study, the annual water balance of 103 river basins in Sri Lanka was presented. The results identified that the total agricultural water use and net environmental system use are 15% and 51% of the net water inflow, and the consumptive use of rain-fed and irrigated agriculture is approximately equal. A study conducted by Jayasinghe et al. (2011) used GIS to map the spatial and temporal variation of nutrient pollution of groundwater in the unconfined sandy aquifers of Kalpitiya peninsula where intensive agriculture is practiced. Vegetation growth zonation for improved water resources planning was carried out by Muthuwatta and Chemin (2013) using NOAA-AVHRR satellite data. Satellite-derived vegetation indices are useful tools to monitor vegetation dynamics. In this study, they identified the usefulness of actual observations from satellite imagery on vegetation growth over estimated vegetation growth derived using multiple data layers such as soil, slope, climate, etc. in conventional agroecological zoning. They highlighted the usefulness of these vegetation growth zones in land use and water resources planning.

Ecosystem health is a key factor for successful agriculture. Degradation of ecosystems can adversely affect the soil and water resources to bring negative impacts on agriculture. Also, the vital ecosystems can be adversely affected due to poor agricultural practices. In addition, urbanization and development activities also lead to damage to the important ecosystems such as wetlands. Spatial sciences can be used to monitor, analyze, and map the vulnerability of these ecosystems and identify the risk factors.

Dahdouh-guebas et al. (2002) studied the land use changes in Pambala-Chilaw lagoon complex to identify the impact of shrimp farming on mangrove ecosystem. The study used aerial photography and GIS to identify the land uses and to assess the land use change. Samarasinghe and Dayawansa (2013) used remote sensing and a GIS-based approach to study the degradation risk of Kolonnawa marsh. Risk mapping was carried out using an index, and it was helpful to prioritize the areas for conservation. Gunawardena et al. (2014a) used satellite images from ALOS (AVNIR2) and Landsat ETM+ sensors for identification, evaluation, and change detection of highly sensitive wetlands in South Eastern Sri Lanka. In a study conducted in Horton Plains to map the plant and animal habitats, IRS LISS III satellite images were used to identify the land use. Other spatial data layers such as slope, aspect, soil, elevation (using Digital Elevation Model), and agroecological regions were derived using GIS. GIS helped to integrate the data layers to identify the land forms which are helpful in identifying habitats of plants and animals. The study reveals the usefulness of high spatial resolution satellite images to identify the presence of some animals in open areas such as grasslands. It highlights the usefulness of integrating remote sensing and GIS in this type of studies (Jayarathne et al. 2005) where an in-depth analysis is required combining ecosystem conditions and requirements for wildlife. A considerable number of studies in forestry sector also have been carried out. The studies include mapping of forest cover, assessment of biomass content (Gunawardena et al. 2008; Gunawardena et al. 2014b), Gunawardena et al. 2016), etc.

Soil plays a major role in agriculture, and spatial and attribute information on soils are useful in land management in agricultural systems. Remote sensing and GIS coupled with soil surveys are very helpful approaches in collecting, storing, and presenting spatial and temporal variations of soil properties. Use of geo-informatics technology in developing soil databases, mapping soils, and identifying soil suitability in agriculture is well documented (Zhu et al. 2001; Sculla et al. 2003; Liengsakul et al. 1993). GIS technology coupled with extensive fieldwork was used to develop a digital soil database for dry, wet, and intermediate zones of Sri Lanka (Mapa et al. 2004, Mapa et al. 2006). This database has been used to classify and map the soil in these climatic zones. The potential uses of this database were identified as suitability mapping, management of problem soils, erosion hazard assessment, agroecological zoning, land use planning, and mathematical simulation of processes as solute or pesticide movement in groundwater. Soil salinity can adversely affect crop growth. To identify the spatial variation of salinity levels, salt-affected soils were mapped using interpolation of field sampled data taken at different soil depths in Hambantota district (Gunasena et al. 2006). Nayanaka et al. (2010) have mapped the agronomically important soil properties in paddy growing Alfisols in Pulasthigama in Polonnaruwa district of Sri Lanka. Rathnayake et al. (2016) have mapped the organic carbon concentration of paddy growing soils in Northern Sri Lanka (2016). Geostatistical mapping of chemical properties of red latosols in Puttalam district was done by Eranga et al. (2015). Spatial variability of soil properties in an Alfisol soil catena was studied by Rosemary et al. (2017). In all these studies, the authors have identified the usefulness of mapping spatial variability of soil parameters to guide site-specific fertilizer applications, intercropping and other management practices, irrigation and tillage operations, land use planning, developing a national carbon accounting system, etc.

Climate change can bring positive and negative impacts on agriculture depending on the nature of change. Hence, knowledge on possible changes in future climate will help the decision-makers in planning and management of land and water resources. Geo-informatics coupled with mathematical models is widely used all over the world in predicting and mapping climate change-related scenarios. A limited number of studies have been carried out in Sri Lanka on changes of climate using geo-informatics. De Silva et al. (2007) used HadCM3 model to assess the impact of climate change on irrigation water requirements of paddy in Sri Lanka. Rainfall and paddy irrigation requirement were mapped using GIS to identify the spatial variation of these parameters.

5 Land Suitability Analysis for Agriculture

Agricultural land suitability assessment is the process of assessing the suitability of land for different crops. This involves a large amount and variety of spatial data. Selecting an appropriate location is related to multicriteria decision-making since it involves a large number of attributes and various criteria (Joerin et al. 2001; Reshmidevi et al. 2009). GIS is a very useful tool in handling this vast amount of spatial data and integrating them together to identify suitable land parcels. Studies on land suitability assessment for agriculture in Sri Lanka using GIS are limited. Jayasinghe and Machada (2008) conducted a study to develop a web-based online consultative system with crop land suitability analysis for Kandy, Nuwara Eliya, Badulla, and Matale districts of Sri Lanka. Requirement factors for crops such as rainfall, soil, topography, temperature, and land use were developed using GIS, and weighted overlay was performed to identify the crop suitability classes. This study developed a flexible and user-friendly online crop consultation system using web GIS tools. Senanayake et al. (2017) also developed a geodatabase to recommend suitable crops at village level in Sri Lanka. Authors mentioned that the "CROPREC" desktop software helps offline access of the database, while web portal is for online access to provide location-based agriculture information. Suthakar (2015) used GIS and multicriteria analysis to assess the land suitability for some selected crops in Jaffna peninsula in Sri Lanka. Author mentioned that the integration of multicriteria decision-making (MCDM) techniques with GIS considerably improves decision-making through conventional map overlay technique, and as a result, much attention has been paid to utilize MCDM techniques in GIS to assess land suitability for different purposes.

6 On-Farm Resource Management Through Geo-informatics

Optimal levels of water and nutrients are essential for a satisfactory crop growth. In this regard, timely and frequent availability of data is of paramount importance. Collection of data using field surveys is costly and time-consuming. Hence, remote sensing, GIS, and GPS technologies are very much helpful in collecting data at farm or system level. According to Goswami et al. (2012), precision agriculture is the application of technology to improve agricultural productivity as compared to conventional methods and to reduce the impact on environment. Goswami et al. (2012) carried out a review on the use of geo-informatics technology on precision agriculture and commented on the usefulness of GIS, remote sensing, and GPS in collecting and analyzing spatial data in the field. Few studies have been carried out in Sri Lanka with respect to assessment of agricultural systems and agricultural resource management with geo-informatics tools. Bandara (2003) used NOAA satellite data to assess three large irrigation systems in Sri Lanka. The study used satellite images to compute land surface parameters for actual evapotranspiration, potential evapotranspiration, and biomass for the command area of three irrigation systems. The crop in the systems is different varieties of paddy. The author mentioned that the examination of spatial variation of irrigation system parameters is difficult with the 1.1 km spatial resolution of NOAA AVHRR data due to small irrigation extents.

Drought is a complex and recurrent phenomena which bring adverse impacts on human and nature (Burcheld and Gilligan 2016). There are different classifications to drought. Meteorological drought occurs due to reduced rainfall, agricultural drought is a result of soil moisture stress, reduced canal flow and reservoir storage brings hydrological drought, and restricted water access caused by economic or political power results in socioeconomic drought (Heim 2002). Burcheld and Gilligan (2016) used remote sensing and GIS to assess the agricultural adaptation to drought in dry zone of Sri Lanka. To identify double-cropping agricultural communities, the study used MODIS satellite data by deriving enhanced vegetation index. In this study, remotely sensed data helped to identify large-scale pattern of cultivation. Remotely sensed data analysis was then coupled with a GIS utilizing the characteristics of agricultural communities to identify communities with similar structural characteristics. Priyadarshanee et al. (2013) assessed the performance of Mahaweli System C in GIS environment using socioeconomic indicators. The education attainment, quality of housing, access to household consumer durables, access to vehicles in households, and income level of household were selected as indicators for performance analysis. A hamlet was taken as the smallest spatial unit to collect data and for mapping. The overall performance index was calculated by summing up the set of indicators by providing relative weights. Indicator values were used to map the spatial variability of the performance of farming communities. A GIS-based decision support system was developed in managing tea plantations in St. Coombs Estate in Talawakelle, Sri Lanka (Wellala et al. 2012). Digital spatial data layers on fields, land use, drainage, buildings, soils, and socioeconomic data were integrated into the database. Proximity and overlay analysis tools available in GIS were used to identify the areas that need protection.

7 Harvesting and Postharvesting Operations Through Geo-enabled Data

Remote sensing is widely used in assessing crop health and forecasting yields. In most studies, yield forecasting is based on statistical-empirical relationships between yield and vegetation indices. Normalized Difference Vegetation Index (NDVI) is the most commonly used for this purpose (Lopresti et al. 2015; Prasad et al. 2006; Wall et al. 2008). Delecolle et al. (1992) explain that it is questionable to use remote sensing for direct estimation of crop production due to indirect link between remote sensing and crop state variables. Authors suggested that crop models can be used together with remotely sensed data. Spatial modeling of crop production is important since soil conditions, weather, and management interventions can be spatially varying (Priya and Ryosuke undated; Nagamani and Mariappan 2017). In many crop production models, GIS is used to input required data and to present results. A limited number of studies have been conducted in Sri Lanka related to crop yield estimations and forecasting using remote sensing technologies. Sirisena and Dammalage (undated) used MODIS 8-day composite images to identify paddy cultivated areas and yield forecasting using NDVI and Enhanced Vegetation Index (EVI2) in Kurunegala district of Sri Lanka. They have identified the paddy cultivated land with an accuracy of 77% and commented that it can be improved with the use of high-resolution satellite data. The authors identified that both NDVI- and EVI2-based models provide more accurate yield estimations at 80 days after transplanting.

Extraction of phonological parameters such as transplanting, panicle initiation, and flowering is useful in the purpose of irrigation water and nutrient management, yield prediction, and crop yield mapping (Jayawardhana and Chathurange 2016). Jayawardhana and Chathurange (2016) have used MODIS 16-year time series data to extract phonological parameters of paddy in Sri Lanka. A GIS-based methodology and field-collected data were used to assess the productivity of potato in Nuwara Eliya and Badulla districts in Sri Lanka (Fernando and Premasiri 2006). Available natural resources and conditions for potato production such as soil properties, climatic parameters together with yield, cost of production, cropping pattern, etc. collected at field level were integrated in GIS to map the potato production characteristics in the study area. However, the paper has not detailed how the decision support system was developed and functioned.

8 Transportation, Distribution, and Marketing of Agricultural Produce Using Geo-informatics Tools

Identification of most cost-effective routes and suitable market centers is important in marketing agricultural produce. Geo-informatics can play a significant role in this subject area. Network analysis capability in GIS is used to assess the transport routes to identify travel time and distances of haulage routes (Devlin et al. 2008). Identification of suitable market centers and management of transportation networks can be carried out with geo-informatics tools more effectively. However, there is no published related literature in Sri Lanka found during the period considered for this review.

9 National Food and Nutritional Security with Geo-informatics Resource Base

Food and nutritional security assessment in a county is important to identify vulnerable areas which need government interventions. Huge quantity of spatial data is required for this purpose. Hence, geo-informatics tools can play a significant role in assessing food and nutritional security. A study conducted in Anuradhapura district in Sri Lanka developed Knowledge-Based Scoring (KBS) approach to classify Grama Niladhari Divisions (smallest administrative unit in Sri Lanka) based on food insecurity (Satharasinghe and Sheinkman 2005). A vulnerability index was calculated for each GN division to map the spatial variation of food insecurity. The authors highlighted the usefulness of these maps for prioritization of GN divisions for development and poverty eradication programs without conducting baseline studies. Wijerathne (2005) assessed the spatial dimensions of poverty in Udawalawe left bank irrigation development project. Poverty can make a significant impact on food security of people. In this study also, GN division was used as the smallest spatial unit for analysis. Study collected poverty-related data through a household survey to estimate the poverty levels. Poverty maps at GN level were created by clustering households. This study has used GIS only for the mapping purpose. The study identified the usefulness of mapping spatial variation of poverty and identified the access to water as a key factor which makes these changes. Ranasinghe et al. (2005) demonstrate that basic GIS applications can be effectively used to identify vulnerable populations/areas using secondary data available with health authorities to develop and optimize nutrient surveillance system in Sri Lanka. This study used nutrition indices to map the spatial variability of nutrient levels of pregnant mothers and children. Dassanayake et al. (2007) discussed the use of GIS for spatially targeted interventions to control child malnutrition at district level. This study focused on mapping pockets of child malnutrition which is important in targeting interventions. The study highlighted the usefulness of GIS in evidence-based decision-making. All the above four studies have used GIS only for the mapping purposes. De Silva (2007) has conducted a detailed food security and vulnerability assessment for Sri Lanka using a large number of spatially distributed variables. However, the spatial details of this study have been confined to district level, and factor analysis has been used to reduce the dimensionality of the data within the spatial boundaries. In general, these studies demonstrate the power of GIS in displaying spatial variability of important parameters.

10 Conclusions and the Way Forward

A review was conducted to identify the application of geo-informatics tools in agriculture and allied disciplines in Sri Lanka based on some published literature after 2000. It was evident that the published literature on application of geo-informatics in agriculture is limited in Sri Lanka. The published literature is mostly based on isolated studies conducted as academic exercises or on personal interest. A wide range of applications are published on natural resources management with the use of remote sensing and GIS technologies. Out of these studies, most relevant applications related to agriculture are soil erosion assessment, land use/land cover change assessment at watershed scale, development of soil databases and maps, and hydrological studies. Application of USLE with overlay functions in GIS is the most common method applied in estimating soil erosion. Hydrological studies in watershed scale have attempted to use models coupled with GIS. Majority of studies have used GIS for spatial analysis and mapping, while use of remote sensing is limited. Few attempts have been made to derive some climatic parameters to estimate water budget in agricultural systems using remotely sensed data.

Though there is a vast potential of using geo-information technology in agriculture from planting of crops up to the consumer, available literature suggests that Sri Lanka has so far not utilized the potential. Also, there is a possibility of not publishing studies or projects conducted by some organizations due to various reasons. Land suitability assessment for different crops for crop diversification, irrigation and fertilizer management at system level, identification of market facilities, optimum transportation root selection, food security, and nutrition management are some of the most potential and useful applications of geo-informatics technology in agriculture. Since spatial data plays a key role in decision-making in agriculture, it is a timely need to enhance the use of geo-informatics in government and corporate sector institutions in Sri Lanka. The capacity on this subject area has been developed during the past few decades through short-term and postgraduate level training obtained within and outside the country by government and nongovernment officers. With this background, the country is equipped with required knowledge, expertise, opportunities, and resources to work with this advancing technology especially in the agriculture sector development. As a vision for the future, use of this technology should not only be limited to basic mapping exercises with GIS but should include the use of remote sensing and other allied disciplines if needed with much improved analytical capabilities.

References

Adikari SB (2004) Land use changes in Kandy lake catchment and water pollution in Kandy lake. In: Proceedings of the first national symposium on geo-informatics, geo-informatics society of Sri Lanka, pp 163–173

Ahmad I, Ghafoor A, Bhati MI, Akthar IH, Ibrahim M, Rehman O (undated) Satellite remote sensing and GIS based crops forecasting & estimation system in Pakistan, Available at http://www.fao.org/fileadmin/templates/rap/files/Project/Expert_Meeting__17Feb2014_/P2-2_Satellite_Remote_Sensing_and_GIS_based_Crops_Forecasting___Estimation_System_in_Pakistan.pdf

Akinci H, Özalp AY, Turgut B (2013) Agricultural land use suitability analysis using GIS and AHP technique. Comput Electron Agric 97:71–82

Bandara KMPS (2003) Monitoring irrigation performance in Sri Lanka with high frequency satellite measurements during the dry season. Agric Water Manag 58:159–170

Basanyat P, Teeter LD, Lockaby KB, Flynn GM (2000) The use of remote sensing and GIS in watershed level analyses of non-point source pollution problems. For Ecol Manag 128 (1–2):65–73

Bastiaanssen WGM, Chandrapala L (2003) Water balance variability across Sri Lanka for assessing agricultural and environmental water use. Agric Water Manag 58:171–192

Burcheld E, Gilligan J (2016) Agricultural adaptation to drought in the Sri Lankan dry zone. Appl Geogr 77:92–100

Burrough PA (1996) Principles of geographical information systems for land resources management. Clarendon Press, 196p

Burrough PA, McDonnel RA (1998) Principles of geographical information systems. Oxford University Press, Oxford, 352p

Campbell GB, Wynne RH (2011) Introduction to remote sensing, 5th edn. The Guilford Press, New York

Chandrapala L, Wimalasuriya M (2003) Satellite measurements supplemented with meteorological data to operationally estimate evaporation in Sri Lanka. Agric Water Manag 58:89–107

Chi M, Benediktsson JA, Sun Z, Shen J, Zhu Y (2016) Big data for remote sensing: challenges and opportunities. Proc IEEE 104:2207–2219

Chowdary VM, Ramakrishnan D SYK, VinuChandran JA (2009) Integrated water resource development plan for sustainable management of Mayurakshi watershed, India using remote sensing and GIS. Water Resour Manag 23(8):1581–1602

Cowen DJ (1988) GIS verses CAD verses DBMS: What are the differences? Photogramm Eng Remote Sens 54:1551–1554

Dahdouh-guebas F, Zetterstro T, Ronnback P, Troell M, Wickramasinghe A, Koedam N (2002) Recent changes in land use in the Pambala-Chilaw lagoon complex (Sri Lanka) investigated using remote sensing and GIS: conservation of mangroves vs development of shrimp farming. Environ Dev Sustain 4:185–200

Dassanayake DMUAJK, Mahawithanage STC, Chandrasekera GAP (2007) Use of GIS for spatially targeted interventions to control child malnutrition at district level, geospatial innovations serving humanity. In: Proceedings of the fourth national symposium on geo-informatics, geo-informatics society of Sri Lanka, pp 87–93

De Silva CS, Weatherhead EK, Knox JW, Rodriguez-Diaz JA (2007) Predicting the impacts of climate change—A case study of paddy irrigation water requirements in Sri Lanka. Agric Water Manag 93:19–29

De Silva RP (2005) Estimating groundwater recharge in the Deduru Oya catchment, Sri Lanka using RS and GIS. Decadal Proceedings 1990–2004, Sweden International Training Course on Remote Sensing Education for Educators, De Silva RP (ed), Geo-Informatics Society of Sri Lanka, pp 57–63

De Silva RP (2007) Food insecurity and vulnerability assessment for Sri Lanka, Geo-Informatics Society of Sri Lanka

Delecolle R, Maas SJ, Guerif M, Frederic B (1992) Remote sensing and crop production models: present trends. ISPRS J Photogramm Remote Sens 47(2):145–161

Devine HA, Field RC (1986) The gist of GIS. J For:17–22

Devlin GJ, McDonnell K, Ward S (2008) A study of travel times and distances for haulage routes in Ireland using GPS and GIS. J Transp Geogr 16(1):63–72. https://doi.org/10.1088/1755-1315/20/1/012050

Encyclopedia of Information Science and Technology (2015) Information science reference, 2015

Eranga RLAA, Karunaratne SB, Abeysinghe DC, Gajanayake B (2015) Geostatistical mapping of selected soil chemical properties in Red Latosols in Sri Lanka, climate, environment and earth sciences. In: Proceedings of the iPURSE, p 71

Fernando KMEP, Premasiri HMR (2006) Evaluation of productivity of potatoes in Nuwara Eliya and Badulla districts with the aid of GIS techniques. Vidyodaya J Sci 13:49–64

Goswami SB, Matin S, Saxena A, Bairag GD (2012) A review: the application of remote sensing, GIS and GPS in precision agriculture. Int J Adv Technol Eng Res (IJATER) 2(1):50–54

Fortes PS, Platonov AE, Pereira LS (2005) GISAREG – A GIS based irrigation scheduling simulation model to support improved water use. Agric Water Manag 77:159–179

Gunasena CP, Weerasinghe KDN, Subasinghe S, Sumith P (2006) Mapping salt affected lands in Hambantota district using GIS, geo-informatics for environmental conservation and management. In: Proceedings of the third national symposium on geo-informatics, geo-informatics society of Sri Lanka, pp 11–21

Gunawardena A, Fernando T, Takeuchi W, Wickramasinghe CH, Samarakoon L (2014a) Identification, evaluation and change detection of highly sensitive wetlands in South-Eastern Sri Lanka using ALOS (AVNIR2, PALSAR) and Landsat ETM+ data. IOP Conf Ser Earth Environ Sci 20 (1):012050

Gunawardena AR, Fernando TT, Nissanka SP, Dayawansa NDK (2014b) Assessment of spatial distribution and estimation of biomass of *Prosopis juliflora* (Sw.) DC. in Puttalam to Mannar Region of Sri Lanka using remote sensing and GIS. Trop Agric Res 25(2):228–239

Gunawardena AR, Nissanka SP, Dayawansa NDK (2008) Development of merchantable timber volume estimation of Pinus caribaea plantations using multi-spectral remotely sensed data. Engineer XXXXI(05):68–73

Gunawardena AR, Nissanka SP, Dayawansa NDK, Fernando TT (2016) Above ground biomass estimation of mangroves located in Negombo-Muthurajawela wetland in Sri Lanka using ALOS PALSAR data. Trop Agric Res 27(2):137–146

Gunawardena GMWL, Dayawansa NDK, De Silva RP (2004) Estimation of land use change and reservoir sedimentation using remote sensing data. In: Proceedings of the SLAAS, Published by Sri Lanka Association for the Advancement of Science, December, 2004

Guo D, Mennis G (2009) Spatial data mining and geographic knowledge discovery—an introduction. Comput Environ Urban Syst 33:403–408

Heim RR (2002) A review of twentieth century drought indices used in the United States. Bull Am Meteorol Soc 83(8):1149–1165
Jayarathne KDBL, Dayawansa NDK, De Silva RP (2010) GIS based analysis of biophysical and socio-economic factors for land degradation in Kandaketiya DS division. Trop Agric Res 21 (4):361–367
Jayarathne KDBL, De Silva RP, Dayawansa NDK (2005) Habitat mapping at Horton Plains using IRS satellite data. In: De Silva RP (ed) Decadal proceedings 1990–2004, Sweden International Training Course on Remote Sensing Education for Educators, Geo-Informatics Society of Sri Lanka, pp 13–30
Jayasekera AS, Dayawansa NDK, De Silva RP, Mutuwatte LP (2005) Spatially distributed hydrological modeling of upper Uma Oya catchment using soil and water assessment tool (SWAT). In: Decadal proceedings of the United Nations/Sweden international training course on remote sensing education for educators 2005, pp 66–85
Jayasekera MJPTM, Kadupitiya HK, Vitharana UWA (2018) Mapping of soil erosion hazard zones of Sri Lanka. Trop Agric Res 29(2):135–146
Jayasinghe P, Pitawala A, Dharmagunawardhane HA (2011) Vulnerability of coastal aquifers due to nutrient pollution from agriculture: Kalpitiya, Sri Lanka. Water Air Soil Pollut 219:563–577
Jayasinghe PKSC, Machada T (2008) Web based online GIS system with crop land suitability identification. Agric Inf Res 17(1):13–19
Jayawardena DT, Gunatilake AAJK (2005) Application of GIS in surface geological and three dimensional hydro-geological modeling of the Samanalawewa hydropower project in Sri Lanka. In: Geo-informatics for future of Sri Lanka proceedings of the second national symposium on geo-informatics, geo-informatics society of Sri Lanka, pp 95–108
Jayawardhana WGNN, Chathurange VMI (2016) Extraction of agricultural phenological parameters of Sri Lanka using MODIS, NDVI time series data. Procedia Food Sci 6:235–241
Joerin F, Thériault M, Musy A (2001) Using GIS and outranking multicriteria analysis for land-use suitability assessment. Int J Geogr Inf Sci 15(2):153–174
Joshua JK, Anyanwu NC, Ahmed AJ (2013) Land suitability analysis for agricultural planning using GIS and multi criteria decision analysis approach in Greater Karu urban area, Nasarawa State, Nigeria. Afr J Agric Sci Technol (AJAST) 1(1):14–23
Karimi HA (2009) Handbook of research on geo-informatics. https://doi.org/10.4018/978-1-59140, 518p. ISBN: 1591409969
Kumar P, Herath S, Avtar R, Takeuchi K (2016) Mapping of groundwater potential zones in Killinochi area, Sri Lanka, using GIS and remote sensing techniques. Sustainable Water Res Manage 2(4):419–430
Kumarihamy RMK, Dayawansa NDK (2009) Sustainable land use planning for Kurundu Oya catchment: a participatory GIS approach for integrated water resources management. In: Proceedings of the fourth South Asia water conference, interfacing poverty, livelihood and climate change in water resources development: lessons in South Asia, May 4–6, 2009, Kathmandu, Nepal, pp 64–77
Liengsakul M, Mekpaiboonwatana S, Pramojanee P, KeesBronsveld K, Huizing H (1993) se of GIS and remote sensing for soil mapping and for locating new sites for permanent cropland — a case study in the "highlands" of northern Thailand. Geoderma 60(1–4):293–307
Lillesand T, Kiefer RW, Chipman J (2014) Remote sensing and image interpretation. Wiley, Hoboken, 704p
Lopresti MF, Di Bella CM, Degioanni AJ (2015) Relationship between MODIS-NDVI data and wheat yield: a case study in Northern Buenos Aires province, Argentina. Inf Process Agric 2 (2):73–84
Maguire DJ (1991) An overview and definition of GIS, Available at: www.ciat.cgiar.org
Manawadu L, Karunarathne YA (2004) Identification of erosion prone areas in Kukule watershed using geographic information systems. In: Proceedings of the first national symposium on geo-informatics, geo-informatics society of Sri Lanka, pp 121–132

Mapa RB, Munasinghe MK, Dassanayake AR, Kendaragama KMA, Kadupitiya HK (2006) Development of a soil database for Sri Lanka and its applications for environmental concerns, geo-informatics for environmental conservation and management. In: Proceedings of the third national symposium on geo-informatics, geo-informatics society of Sri Lanka, pp 1–9

Mapa RB, Munasinghe MK, Kendaragama KMA, Dassanayake AR (2004) Development of a digital soil database for Sri Lanka and its applications, geo-informatics research and applications. In: Proceedings of the first national symposium on geo-informatics, geo-informatics society of Sri Lanka, pp 59–65

Muthuwatta L, Chemin Y (2013) Vegetation growth zonation in Sri Lanka for improved water resources planning. Agric Water Manag 58:123–143

Muthuwatta L, Jayakody P (2006) Development and testing a spatially distributed hydrological model for a meso scale watershed, a case study from Hulanda Oya watershed, Sri Lanka, geo-informatics for future of Sri Lanka. In: Proceedings of the second national symposium on geo-informatics, geo-informatics society of Sri Lanka, pp 59–71

Nagamani K, Nethaji Mariappan VE (2017) Remote sensing, GIS and crop simulation models – a review, Int J Curr Res Biosci Plant Biol 4: 8, 80–92

Nayanaka VGD, Vitharana WAU, Mapa RB (2010) Geostatistical analysis of soil properties to support spatial sampling in a paddy growing Alfisol. Trop Agric Res 22(1):34–44

Panday A, Chowdary VM, Mal BC (2007) Identification of critical erosion prone areas in the small agricultural watershed using USLE, GIS and remote sensing. Water Resour Manag 21 (4):729–746

Piyadasa RUK, Weerasinghe KDN, Harsha Kumara PW, Maier D (2006) Groundwater distribution and quality characteristics in the right bank of Nilwala ganga (Badulla Oya and Kirama Oya basins) Sri Lanka, geo-informatics for environmental conservation and management. In: Proceedings of the third national symposium on geo-informatics, geo-informatics society of Sri Lanka, pp 73–80

Prasad AK, Chai L, Singh RP, Kafatos M (2006) Crop yield estimation model for Iowa using remote sensing and surface parameters. Int J Appl Earth Obs Geoinf 8:26–33

Priya S, Ryosuke S (undated) National spatial crop yield simulation using GIS based crop production model, Available at http://citeseerx.ist.psu.edu/viewdoc/download?doi=10.1.1.110.2857&rep=rep1&type=pdf

Priyadarshanee MML, Dayawansa NDK, De Silva RP (2013) Assessment of irrigation system performance with socio-economic indicators: a GIS-based study in the Medagama Block of Mahaweli system C. Trop Agric Res 24(3):215–227

Ranasinghe PHTD, Mahawithanage STC, Madatuwa TMJC (2005) Geographic information systems (GIS): an option to optimize nutrient surveillance system, geo-informatics for future of Sri Lanka. In: Proceedings of the second national symposium on geo-informatics, geo-informatics society of Sri Lanka, pp 39–45

Rathnayake RR, Karunaratne SB, Lessels JS, Yogenthiran N, Rajapaksha RPSK, Gnanavelrajah N (2016) Digital soil mapping of organic carbon concentration in paddy growing soils of Northern Sri Lanka. Geoderma Reg 7(2):167–176

Reshmidevi TV, Eldho TI, Jana R (2009) A GIS-integrated fuzzy rule-based inference system for land suitability evaluation in agricultural watersheds. Agric Syst 101:101–109

Rosemary F, Vitharana UWA, Indraratne SP, Weerasooriya R, Mishra U (2017) Exploring the spatial variability of soil properties in an Alfisol soil catena. Catena 150:53–61

Sabins FF (1996) Remote sensing: principles and interpretation, 3rd edn. Freeman and Co., New York, 512p

Salazar L, Kogan F, Roytman L (2007) Use of remote sensing data for estimation of winter wheat yield in the United States. Int J Remote Sens 28(17):3795–3811

Samarasinghe YMP, Dayawansa NDK (2013) A remote sensing and GIS based study in assessment of the degradation risk of the Kolonnawa marsh. J Natl Sci Found 41(4):327–335

Satharasinghe A, Sheinkman M (2005) Use of local knowledge for mapping food insecurity. In: Geo-Informatics for Future of Sri Lanka Proceedings of the Second National Symposium on Geo-Informatics, Geo-Informatics Society of Sri Lanka, pp 1–12

Sculla P, Franklina J, Chadwick OA, McArthur D (2003) Predictive soil mapping: a review. Prog Phys Geogr 27(2):171–197

Senanayake IP, Dissanayake DMDOK, Mayadunna BB, Weerasekera WL (2015) An approach to delineate groundwater recharge potential sites in Ambalantota, Sri Lanka using GIS techniques. Geosci Front. Available at http://dx.doi.org/10.1016/j.gsf.2015.03.002

Senanayake SS, Munasinghe MAK, Wickramasinghe WMADB (2017) Development of geodatabase to recommend suitable crops at village level in Sri Lanka. Annal Sri Lanka Dep Agric 19:209–228

Sirisena PMTS, Dammalage TL (undated) Cultivated paddy area identification and rice yield estimation using free satellite images. Available at: http://a-a-r-s.org/acrs/administrator/components/com_jresearch/files/publications/Ab%200127.pdf

Srivastawa GS (2014) Introduction to geo-informatics. McGraw Hill Education (India) Private Limited, New Delhi, 300p. ISBN: 9781259058462

Suthakar K (2015) Assessment of land suitability potential for selected field crops using GIS-based Multi-Criteria Analysis (MCA): evaluating the case for Jaffna Peninsula, Sri Lanka Available at SSRN: https://ssrn.com/abstract=3152293

Tao CV, Li J (2007) Advances in mobile mapping technology. Taylor & Francis Group, London, p xi. ISBN 978-0-415-42723-4

Wall L, Larocque D, Léger P-M (2008) The early explanatory power of NDVI in crop yield modeling. Int J Remote Sens 29(8):2211–2225

Wellala NNK, Gunatilake J, Shyamalie HW (2012) Use of geographic information system in tea plantation management: a case study at St. Coombs Estate, Talawakelle, Sri Lanka. J Tea Sci 77 (Part (1/2)):70–82

Weragala DKN (2004) Use of GIS/RS technologies in distributed hydrological modeling of the Walawe river basin, geo-informatics research and applications. In: Proceedings of the first national symposium on geo-informatics, geo-informatics society of Sri Lanka, pp 31–42

Wijerathne D (2005) Spatial dimension of poverty within an irrigated agricultural setting, the case of Uda Walawe irrigation agricultural setting. In: Geo-informatics for future of Sri Lanka proceedings of the second national symposium on geo-informatics, geo-informatics society of Sri Lanka, pp 13–25

Wijesundara NC, Abeysingha NS, Dissanayake DMSLB (2018) GIS-based soil loss estimation using RUSLE model: a case of Kirindi Oya river basin, Sri Lanka. Model Earth Syst Environ 4 (1):251–262

Zhu AX, Hudson B, Burt J, Lubich K, Simonson D (2001) Soil mapping using GIS, expert knowledge, and fuzzy logic. Soil Sci Soc Am J 65:1463–1472

Zolekar RB, Bhagat VS (2015) Multi-criteria land suitability analysis for agriculture in hilly zone: remote sensing and GIS approach. Comput Electron Agric 118:300–321

Experiences of Biochar Applications for Sustainable Agriculture in Sri Lanka

B. F. A. Basnayake, R. T. K. Ariyawansha, D. A. S. Gamage, and A. K. Karunarathna

1 Introduction

We were learning the art of cultivating a "*chena*," when we stumbled on an abandoned one, way back in 1979. It was a revelation. We called it the "*Kala Pola*" (jungle market). It had flourishing and lushes growth of many types of peppers, tomato, pumpkin, and aubergine in clusters. They were from seeds that had sprouted from the previous cultivations. The immediate response was to examine the soil, and we were astonished to find the soil rich in very small particles of char and nuggets of char sparsely spread across these dense clusters. The Center for Agricultural Machinery, Land, Water, and Forestry Management (CEMAGREF) in France welcomed the idea of making char from agricultural wastes. They were interested in using it for fueling gasifiers to run tractors, while our intention was for replacing fertilizer for increasing agriculture productions.

At that time, commercial use of charcoal was not known, but there had been the use of it throughout history. The well-known discovery in the Amazon basin of Terra Preta de Indio (Black Soils of the Indians) by the Spaniards came to light only in the recent past in 2001 (Bezerra 2015), because that civilization got wiped out contracting diseases that the Spaniards had brought with them from Europe. It is believed that the ancient Indians engineered the poor tropical soils with the additions

B. F. A. Basnayake (✉) · A. K. Karunarathna
Department of Agricultural Engineering, Faculty of Agriculture, University of Peradeniya, Peradeniya, Sri Lanka
e-mail: benb@pdn.ac.lk

R. T. K. Ariyawansha
Solid Waste Management Research Unit, Department of Agricultural Engineering, Faculty of Agriculture, University of Peradeniya, Peradeniya, Sri Lanka

D. A. S. Gamage
Department of Chemical and Process Engineering, Faculty of Engineering, University of Peradeniya, Peradeniya, Sri Lanka

R. P. De Silva et al. (eds.), *Agricultural Research for Sustainable Food Systems in Sri Lanka*, https://doi.org/10.1007/978-981-15-3673-1_12

Fig. 1 Oxisol and Terra Preta. (Source: B. Glaser cited by https://www.geozentrum-hannover.de/DE/Gemeinsames/Nachrichten/Veranstaltungen/2009/Hauskolloquium_2009_2010/2009_10_06_abstracts.html)

of biochar, or they crafted the properties of char into biochar. These soils have biochar, pot pieces, and animal skeletons; thus, they were formed under the trash piles of the indigenous people 2 m deep, 7000 years old, much more fertile than the surrounding soil (Fig. 1). There are reports of replenishing back soils of the Indians. These fertile soils are mined for commercial use now, and they find that these soils are especially helpful as a natural growing medium. In other words, the soils expand. Perhaps the microcosms in those soils have evolved to recycle and also adsorb CO_2 from the atmosphere as the substrate, perhaps undergoing at some later stage anaerobic conditions in this Amazon basin to utilize CO_2 as an energy source as well as for cell growth, a mechanism for their survival.

Although we do not have historical evidence of char use in Sri Lankan agriculture, the influence of Portuguese in Sri Lanka was similar to the report of Francisco de Orellana, the Spanish conquistador in 1542. He had ventured along the Rio Negro, one of the Amazon basin's great rivers, hunting for a hidden city of gold, and his expedition found a network of farms, villages, and even huge-walled cities (https://www.youtube.com/watch?v=0Os-ujelkgw). According to Sanderatne and de Alwis (2018), in Sri Lanka, the high death rate in the nineteenth century and early twentieth century was primarily due to several infective and parasitic diseases, unhygienic living conditions, and inadequate medical services. Incidences of dysentery, tuberculosis, cholera, typhoid, and malaria were exceedingly high. One should wonder, what happened in the fifteenth century? We would have lost the actual craft of engineering very similar ferruginous soils like in the Amazon basin and the know-how of making charcoal. Instead, they were driven to clearing jungles, and they cultivated a chena in isolation, preventing diseases from spreading. One of the farmers, whom we met in Kalawewa many years ago, recounted his experiences of settling in Kalawewa from a village in Kandy after the restoration of the reservoir during the time of British rule. When they found many of the settlers succumbing to numerous diseases, including malaria, they fled and survived in the forest, cultivating *chenas*. He was not surprised when we narrated the observations of "*Kala Pola,*" and we still find that most farmers are aware of the importance of char, but they are unable to quantify its exact value to cultivation. So are most scientists, researchers, students, and administrators or some are not willing to accept alternatives to inorganic fertilizers because of self-interests in exploiting farmers for monetary gains.

This review is a testimony in quantifying the importance of biochar for sustainable agricultural productions. It entails laboratory investigations to determine kinetic properties, testing of an "engineering model" to determine the gas flow characteristics of straw bales with and without pyrolytic conditions. Mathematical expressions from these results were developed for simulating the pyrolysis front in the design of a 1 Mg/h pyrolyzer. Test runs were done to optimize the thermal process conditions and the products. We describe the potential use of the technology by arresting the polluting practice of straw burning and the twin benefits of biochar and power generations from paddy straw. We describe the development of pyrolyzers in Sri Lanka and the issues related to investments for promoting the developed technologies. From a scientific point of view, it was necessary to examine the quality of biochar in relation to the behavior of nutrients, particularly nitrogen and phosphorus. We describe field experimentations that were conducted firstly in order to reduce the use of inorganic fertilizer and secondly to demonstrate the role of biochar in enhancing phosphorus movements in rice cultivation. We then conclude convincingly that biochar is essential for sustainable agriculture productions in poor tropical soils of Sri Lanka.

2 Development of a Mobile Pyrolyzer (Carbonization)

Pyrolysis is a thermal decomposition process, requiring a supply of heat, preferable without oxygen or any excess oxidizing agent in the media of supplying heat until the biomass is converted to char, pyrolysis gases, pyrolysis products, and water. The product yield during the thermochemical conversion of biomass depends on temperature, pressure, time, reaction conditions, rate of heating, and added reactants or catalysts (Paul 1982; Demirbas and Kucuk 1997). The degree of equilibrium attained between these products will depend on the rate of heat supplied. Fast thermal transfer at higher temperatures maximizes the production of bio-oil and/or gases, while slow thermal transfer at lower temperatures maximizes the yield of carbon/charcoal. The rate of heating of 10 °C/s, <500 °C, is ideal for biochar productions. It is an endothermic reaction requiring the supply of heat at a specific rate of heating or variable, which will depend on the reactor configuration. In the thermal decomposition process, depending on the temperature regime as given in Table 1, the biomass

Table 1 The mean post-pyrolysis feedstock residues resulting from different temperatures and residence times (IEA Bioenergy 2007)

Mode	Conditions	Products (%)		
		Liquid	Biochar	Syngas
Fast pyrolysis	Moderate temperature, around 500 °C, short hot vapor residence time ~ 1 S	75	12	13
Intermediate pyrolysis	Moderate temperature, around 500 °C, moderate hot vapor residence time ~ 10–20 S	50	20	30
Slow pyrolysis (carbonization)	Low temperature, around 400 °C, very long solid residence time	30	35	35
Gasification	High temperature, around 800 °C, long vapor residence time	5	10	85

undergoes drying, torrefaction, carbonization, and gasification. We find that the structure of the biomass, consisting of hemicelluloses, cellulose, lignin, and other substances, including various inorganic matter, vastly influences the quality of the char and the pyrolysis products. In the design of equipment required for any of these processes, it is important to determine the kinetics of pyrolysis, including the rate of pyrolysis at a given supply of heat, and it will depend on the size of particles, isothermal or non-isothermal reactor configurations, and the type of biomass. We can observe the mass reductions in an isothermal thermobalance (Fig. 2) at a given temperature of different particle sizes as shown in Fig. 3 (Basnayake 1986).

Each of these particles has undergone a first-order rate reaction, and we can express the unit mass reduction for a single run as:

$$m = \frac{M_t - M_f}{M_i - M_f} \tag{1}$$

where M_i = initial mass, M_f = final mass, and M_t = mass reduction at time t.

The first-order rate reaction is expressed as

$$\frac{dm}{dt} = -k[m] \tag{2}$$

where k = the rate constant.

Integrating Eq. 2 gives

$$m = m_0 e^{-kt} \tag{3}$$

We can substitute Eq. 3 in Eq. 1 to give

$$\frac{dm}{dt} - k.m_o e^{-kt} \tag{4}$$

We can then plot a graph of $ln(m)$ vs time. The slope of the straight line gives $-k$ for that temperature, and we could denote it as k_1. If we then do the same for the same

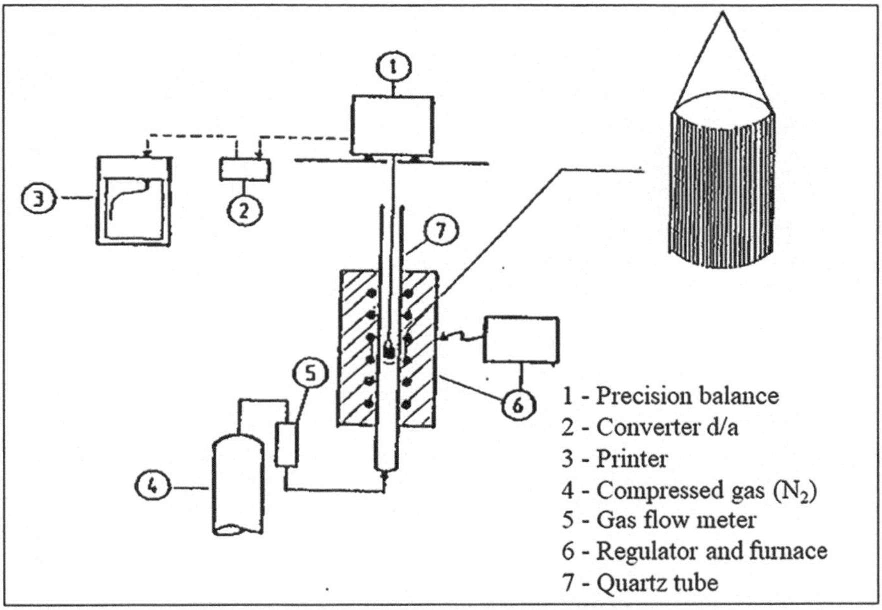

Fig. 2 An isothermal thermobalance (Basnayake 1986)

mass of particle for different temperatures, k_2, k_3, k_4, ...k_n can be obtained. The energy activation value can be obtained if *ln(k)* is plotted against 1/*T*, where T_1, T_2, T_3, T_4.....T_n are the temperatures corresponding to k_1, k_2, k_3...k_n values. The slope of the straight line is equal to $\frac{-\Delta E}{RT}$, and the factor of frequency is the intercept, thus giving

$$k = k_0 e^{-\frac{\Delta E}{RT}} \tag{5}$$

Therefore, Eq. 5 can now be inserted in Eq. 4 to replace *k*. We found that the energy of activation is about 1050 kJ/kg for most materials. But it will depend on heat transfer characteristics of the materials under investigation. The pyrolysis products can be used for supplying the required energy. Therefore, the products leaving the reaction zone can be written as

$$m_p = m_o - m_o e^{-kt} = m_0\left(1 - e^{-kt}\right) \tag{6}$$

If the required quantity of the products is recycled and combusted to provide adequate energy for the endothermic reactions of pyrolysis, a system can be designed to continuously produce char from biomass. We wanted to develop a carbonization reactor to pyrolyze bales of straw, and we needed to find out the pressure losses encountered to send high-temperature flow of fumes using a high-temperature centrifugal fan. At high temperatures, there could be preferential passages through

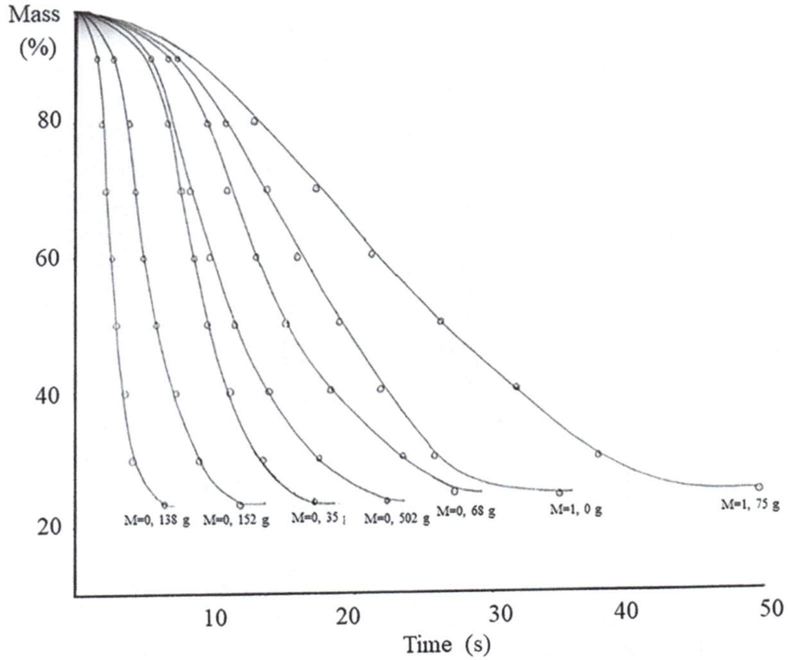

Fig. 3 The mass reductions in an isothermal thermobalance at a given temperature of different particle sizes (mass % vs time (s) of straw at 600 °C) (Basnayake 1986)

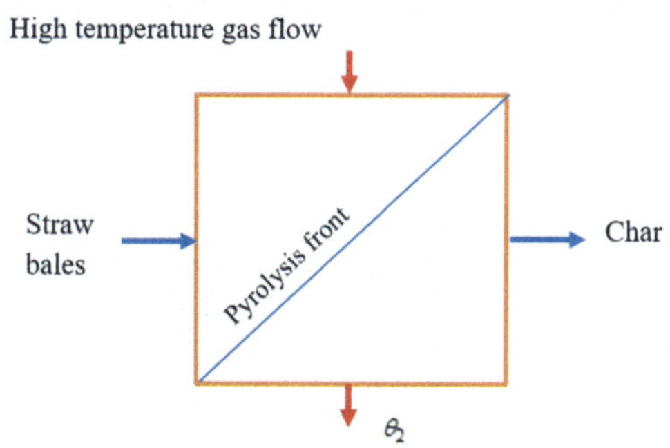

Fig. 4 Conceptual pyrolysis front (Basnayake, 1986)

a bale of straw, but if the bales are replenished on a continuous basis, the preferential paths get filled with char or fresh straw. It was conceptually perceived to have a pyrolysis front as shown in Fig. 4.

Therefore, we built an engineering model to determine the characteristics of the gas flow. In the first instance, ambient air was used to determine the pressure loss, and it amounted to

$$\Delta P = \lambda_1 v + \lambda_2 v^2 = 510v + 470v^2 \tag{7}$$

where ΔP = pressure loss in N/m^2 and v = velocity of air flow m/s and λ_1 and λ_2 are coefficients describing porosity of the material.

The first term depends on the potential energy, thus along the length of the passage and average diameter of the passages of the straw bales. The second term is the kinetic energy of entry and exit losses. λ_1 and λ_2 get modified, if the temperature increases due to changes in viscosity, the average diameter of passages through the bales, Reynolds number, and the changes in density ρ. We verified the predicted pressure losses with combusted gasses being sent through the bales. The pyrolysis front should be a diagonal, if there is a cross flow as shown in Fig. 5. The energy contents as given in Table 2 were used to determine the mass balances of the reactor, and the predicted performances were found to be similar to the experimental results of the mobile pyrolyzer. Test runs were done to optimize the thermal process conditions and the products by observing the pyrolysis front for automation of feeding. The granulation techniques of small particles were one of the aspects of product development.

3 Potential Use of Technology in Sri Lanka

We were very keen in transferring these developed technologies and know-how to interested parties in Sri Lanka. Unfortunately, none of the companies in Sri Lanka were interested in the production and conversion of char into fertilizer. A concerted effort was made to popularize and industrialize the processes by conducting trials and even obtaining the first patent titled "Discovery of a process to retard the release of nitrogen fertilizer by using charcoal and manioc" (Basnayake 1994). Although the Department of Agriculture advocates use of straw for composting, the necessary technology transfer of straw baling and know-how to store and preserve it have not been forthcoming. Therefore, farmers still continue to burn the straw by spreading the cut straw over the stubble. This practice seems to have some major effects, namely, weed suppression, pest control (John 2013), and the stubble getting converted to char. It is estimated that 50% of the paddy straw is burnt in Sri Lanka. The remaining 50% is now converted to compost, animal feed, and other uses with 20% decomposing in the fields contributing to methane productions. Therefore, residue burning is still widely practiced although officially banned in most countries, including Sri Lanka. However, most of the straw is burnt and blown away, and the amount of char remaining in the field is less than 10%, considering the high ash content in paddy (Ariyawansha et al. 2014). The burning is contributing to air pollution (Putun et al. 2004), human health problems (Iranzo et al. 2004; Kadam et al. 2000), and substantial nutrient losses (Haefele et al. 2011). There is a valid case for using straw for biochar productions and the excess heat used for power generation. It was estimated that just over 100 MW of power generation from small power plants within a radius of a collection area of 5 km can be a feasible project for Sri Lanka. The environmental costs of 108.5 million USD annually can be avoided in

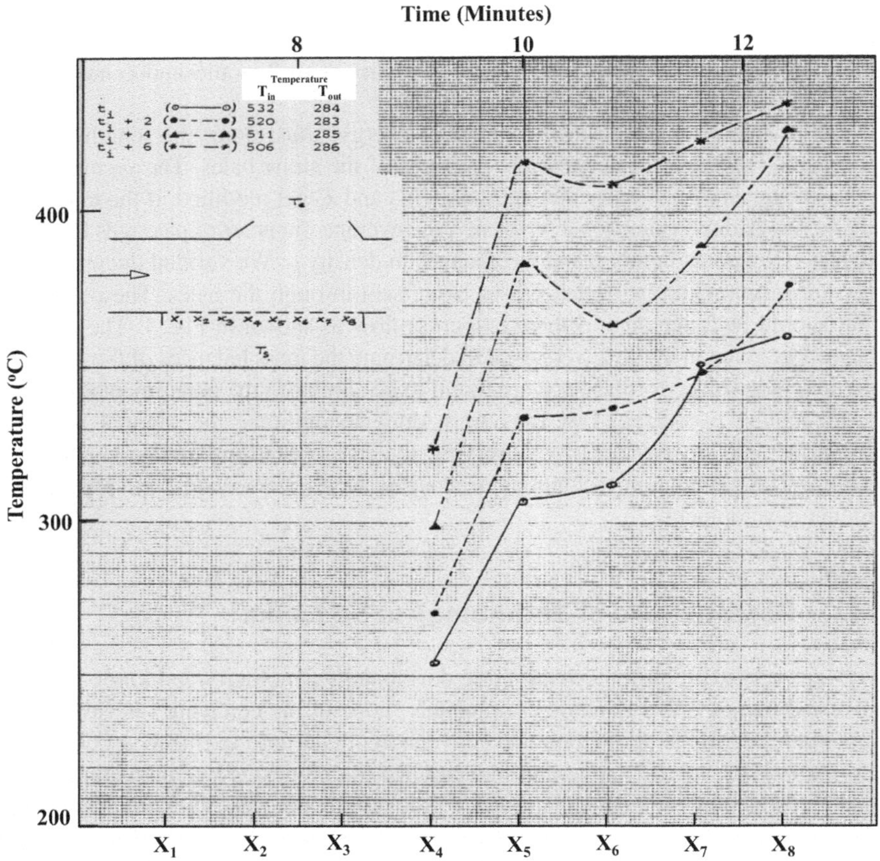

Temperature gradient for the points X_4, X_5, X_6, X_7, X_8 for time (minutes) t_{i+2}, t_{i+4}, t_{i+6}

Fig. 5 Experimental pyrolysis front (Basnayake 1986). (Note: *ti* initial time, *Tin* inflow temperature, *Tout* outflow temperature)

this twin system. Most of all, the saving on fertilizer applications is 153 million USD annually (Ariyawansha et al. 2014).

We were hoping to introduce straw balers to Sri Lanka and launch the long-awaited project of making char with a similar but more advanced reactor than an earlier one at CEMAGREF. We even experimented on several smaller units to convert paddy husk to char. It was an inclined reactor without recycling of pyrolysis gases and tar/oil by giving just about enough air for combusting part of those products as illustrated in Fig. 6 (Manel 2010; Gamage et al. 2012). A much larger unit was designed and was made operational with selected materials from municipal solid waste (MSW). It had a capacity of 100 kg/hr (Fig. 7). A gasifier was used to supply energy for drying and pyrolysis, such that the pyrolysis gases were to be used

Table 2 Enthalpy values of specific heat capacity and sensible heat of pyrolysis, fumes of the combustible (Basnayake 1986)

Material	HHV (kJ/kg)	LHV (kJ/kg)	Specific heat capacity (kJ /kg/K)			Sensible heat (kJ/kg at 25 °C)			ET (kJ/kg)
Straw	19,560	18,015	300 °C	400 °C	500 °C				380
Gas	9610	8958	1.34	1.4	1.47	368.5	525	694	
Pyrolysis products	14,640	12,350	1.75	1.82	1.89	481	682.5	898	
Oil/tar	32,160	30,240	1.055	1.09	1.12	288	408	534	
Carbon	28,890	27,860	1.095	1.147	1.2	287	430	570	
Water			1.997	2	2.12	550	750	1007	

Note: *HHV* high heating value, *LHV* low heating value, *ET* endo-thermocity

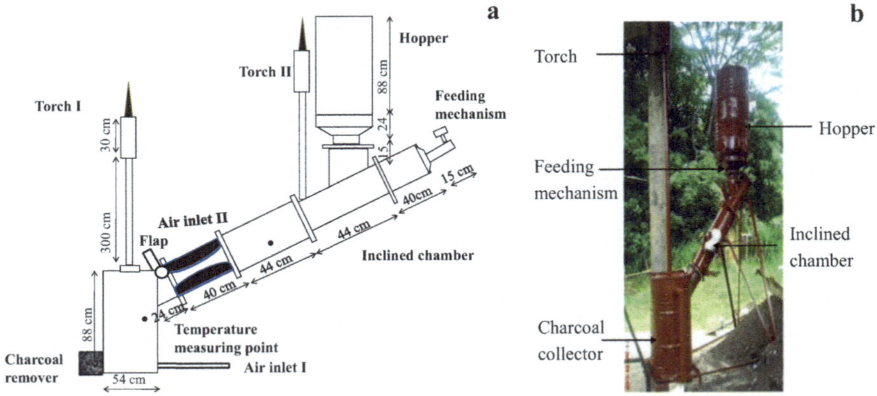

Fig. 6 Pyrolyzer. (**a**) A schematic diagram of the front view of the pyrolyzer. (**b**) Constructed pyrolyzer (Manel 2010; Gamage et al. 2012)

Fig. 7 Fabricated biochar plant at Gohagoda MSW disposal site (**a**) biochar plant, (**b**) biochar plant at operation (**c**), (**d**) producer gas flame from gasifier

for even a larger reactor, thus having a series of them (Fig. 8). In the unit that was built, recycling of gases was tried by cooling and filtering the pyrolysis products. The tars/oils that condensed within input materials were pyrolyzed once more in the

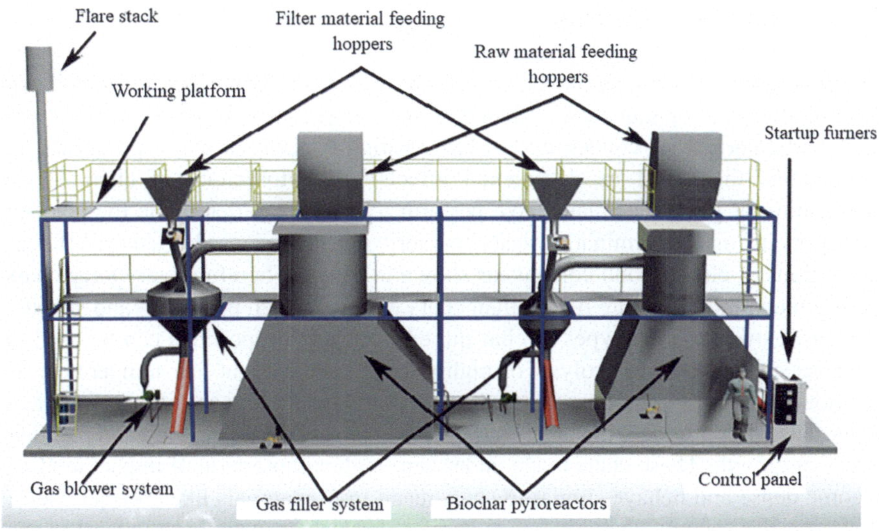

Fig. 8 Conceptual design of series of pyrolyzers

tall reactor, thus encountering fewer problems in blockages of pipes and valves. It was necessary to maintain temperatures >150 °C inside the reactor to continuously drive off moisture out of the reactor so that water will not accumulate but condense in the heat exchanger pipes. A steam-releasing system was incorporated to ensure safety. We eventually used a very old cast iron-cased centrifugal fan, discarded from a tea estate because the un-condensable gases still had some considerable amount of sensible heat, causing problems for most low-temperature fans readily available in the market.

Unfortunately, the government agencies and private companies turned down biochar proposals. It became very apparent that industries, agricultural officials, and academics preferred chemical applications rather than promoting effective cultivation techniques with biochar applications for sustainable development. It was naive in their part because throughout history, charcoal has been used in agriculture. At the same time of the work at CEMAGREF, a handful of scientists recognized the value of charcoal and had published few papers on the subject (Kishimoto and Sugiura 1985; Santiago and Santiago 1989). However, after the advent of discovering biochar use in ancient civilizations of the Amazon region in 2001, there have been considerable research and development to industrialize the processes and apply biochar (Lehmann and Joseph 2009). In most of the research studies, particularly in the tropical belt, benefits of biochar applications are well documented. Nevertheless, there are pessimistic points of view on biochar use cited in one or two review papers (Mukherjee and Lal 2014).

4 Quality of Biochar

Different points of view on the use of biochar arise from fundamental principles of defining the quality parameters for biochar. In our experience, if we use newly made char, particularly with condensed tar, having high concentrations of polyphenolic compounds, there will be detrimental effects of phytotoxicity of charcoal on establishing or growth of plants. We perform a bioassay to determine the level of phytotoxicity using germination tests for very sensitive seeds by using different proportions of char to sand contents as shown in Fig. 9. Therefore, very tarry chars require much more mixture of soil than chars with very high porosities and specific surfaces. However, such types of char mixed in correct proportions can be used to deter insects and pests. Pyrolysis of highly dense feedstock at low temperature as mentioned before will produce char with less porosity and low specific surface than straw. The latter produces unstable char but is very useful in the seeding and planting stages of growth. These chars made under very high compression of impact loading become dense and behave similar to chars made from materials like coconut shells. However, the coconut shell chars can last for decades or even thousands of years. Also curing of stable (dense) char with soil over a period of time reduces phytotoxicity and increases the nitrogen content. Such chars can then be termed biochar. The microbial actions in the mixture of char and soil are enhanced with a carbon source like compost. If we use green manure mixed with char and soil, and then cured, it will have very different properties. Whenever a phase of curing has taken place or by allowing the char to remain in the soil for a period without causing phytotoxicity, we could use the term "biochar." However, it is best to define the terms as accepted by the international community. According to Lehmann and Joseph (2009), biochar is a charcoal carbon product derived from biomass that can enhance soils, sequester or store carbon, and provide useable energy. Biochar is produced by thermal decomposition of organic material under limited supply of oxygen (O) and at relatively low temperatures (<700 °C) (Lehmann and Joseph 2009). Therefore, it is defined as charcoal made for the application in soils.

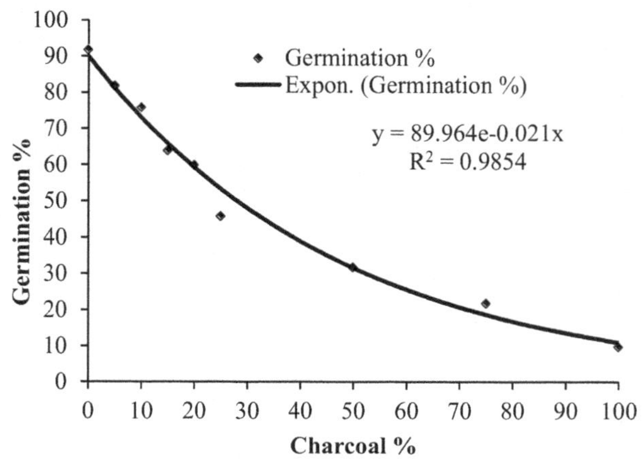

Fig. 9 Relationship between charcoal percentage and germination of lettuce (*Lactuca sativa L.*) seeds (Manel 2010)

Table 3 Properties of MSW compost and biochar (Manel 2010)

Parameter	MSW compost	Charcoal
MC (%) (w.b)	13.37	0.33
Volatile matter (%) (d.b)	32.62	24.91
Ash (%) (d.b)	65.60	72.61
Fixed carbon (%)(d.b)	1.78	2.48
pH	7.5	9.2
Temperature (°C)	22.9	19.5
Conductivity (μs/cm)	6560	1087
Salinity (%o)	3.6	0.5
Bulk density (kg/m^3)	1090	620

Rice husk charcoal-coated urea can potentially be used as a slow-releasing nitrogen fertilizer which reduces leaching losses of urea. It also helps to reduce the phosphate and potassium from leaching. The cyclic effect of phosphate release is an important finding, which could be the central issue in defining microbial behavior in soils. The fluctuations of phosphate may have cyclic effects of 28 days. Charcoal-coated urea also increased the pH and redox potential value of the soil up to the desired levels (Gamage 2015). Sometimes, municipal solid waste compost cannot be used directly as a fertilizer, due to the presence of pathogens, unwanted materials, and toxic compounds such as heavy metals (Zucconi et al. 1985; Gajdos 1997), but it can be pyrolyzed to char reducing the mobility of heavy metals (Paz-Ferreiro et al. 2014) and getting rid of pathogens (Bond et al. 2017). The proximate analysis for this study is given in Table 3. The inorganic content in this charcoal showed a higher pH value of 9.1 due to a high ash content.

We did an interesting experiment to develop a biocatalyst char (BCC) activated with market wastes and Eppawala rock phosphate (ERP) in aerobic and anaerobic reactors mixed in water and then applied to garden wastes to make compost. The addition of charcoal when needed was the method used to control the pH; thus, the activation level of the digesting slurry was controlled, before adding the slurry into garden wastes. The catalyst-added compost was relatively high in nutrients compared to the control compost. In addition, aerobic catalyst-added compost had high phosphorus content than anaerobic catalyst-added compost, but the ammonium nitrogen content was high in anaerobic catalyst-added compost than others. The level of decomposition was higher in the compost with the aerobic catalyst (Gunasekara 2014). Biochar and biochar-biocatalyzed fertilizer notably having high pH can be used to neutralize acidic soils in most parts of the country and world, thus eliminating the use of lime which is an abiotic resource or transportation of alkaline soils that have less nutrient value than biochar-based fertilizers. In other words, such fertilizer neutralizes acid soils while providing nutrients in one application.

5 Field Experimentations

Apart from the differences in biochar quality, it was essential to determine the effect of char as a replacement for urea. Therefore, we undertook a study knowing the fact that the rice cultivars in Sri Lanka require very high nitrogen contents and thus respond best to inorganic fertilizers. It certainly did for the plots with recommended inorganic fertilizers (1) giving highest yields in the *Yala* season (Table 4), whereas the plots with inorganic fertilizers mixed with compost (3) gave a better yield than inorganic alone in the *Maha* season (Table 5). In that season, the plots with charcoal and 30% w/w recommended inorganic fertilizers (2) performed better than inorganic alone but with much less nitrogen. The yields were not significantly different in the *Yala* season between (1), (3), and (4). Namely, (4) were the plots with charcoal, paddy straw compost, and 30% (w/w) recommended inorganic fertilizers in the *Yala* season (Gamage 2015). The paddy straw compost was processed in a simulation lysimeter of a landfill bioreactor (Thivyatharsan et al. 2012). The study aimed at maximizing methane emission from a mixture of rice straw, cow dung, and green materials and evaluating the quality of the resulting compost. The color of the compost was very dark brown, and it did comply with SLS guidelines for quality of compost. The study shows that there is a high potential of producing compost from landfill bioreactors using rice straw (Gamage et al. 2010).

We wanted to promote vegetable cultivation with biochar that can be grown under coconut in the Hambantota district. We selected *Amaranthus oleraceus* for the field experiment with six treatments including soil as control (Ariyawansha et al. 2013). The charcoal was prepared using the traditional pit method; thus, lighter materials oxidize to form ash that get mixed with made char. We observed that the quality of biochar had an influence on plant growth and yield. There were considerable differences between treatments. The best performance including the highest yields of plant growth was observed from treatment 3, biochar made from coconut shells, and it has followed an ideal growth curve as shown in Fig. 10. Such high yields obtained from inter-cultivation under coconut have symbiotic benefits, such as it increases the productivity of coconut lands and consumption of *Amaranthus oleraceus*, thus reducing the extent of cultivation required in a small country like Sri Lanka. Both the statistical analysis and kinetic studies indicate a significant difference of biochar application with organic and inorganic fertilizer applications. The biochar which was produced with coconut husk and *Gliricidia sepium* sticks showed growth retardations due to inhibitions caused by high ash content. Total fresh weight recorded per plot can be ranked as highest from being from coconut shell biochar (CSB) (4.36 kg) followed by coconut husk biochar (CHB) (2.78 kg), then biochar made from *Gliricidia sepium* sticks (2.4 kg), recommended fertilizer (2.08 kg), compost (1.82 kg), and lowest the control (1.27 kg). This study proved that a localized production of nutrient supply by biochar applications reduces the cost of production.

We conducted another study to investigate the possibility to use coconut-based biochar as soil amendments in integrated *Zea mays* cultivation in a coconut land (Waduge 2011). There were five treatments. The best mean yield was obtained from

Table 4 Effect of different treatments on the growth and yield parameters of rice-Yala (Gamage 2015)

Treatments	Number of tillers	Number of panicles	Root dry weight (g)	Shoot dry weight (g)	Yield per plot (kg)	Yield kg/ha
1	$8.60^{b} \pm 0.26$	$6.66^{b} \pm 0.14$	$3.1^{b} \pm 0.1$	$5.0^{b} \pm 0.10$	$12.59^{a} \pm 0.08$	$5036^{a} \pm 32.0$
2	$8.00^{b} \pm 0.40$	$7.00^{b} \pm 0.40$	$3.0^{b} \pm 0.0$	$4.9^{b} \pm 0.26$	$10.18^{d} \pm 0.02$	$4072^{d} \pm 8.0$
3	$10.66^{a} \pm 0.23$	$8.00^{a} \pm 0.00$	$3.9^{a} \pm 0.4$	$6.0^{a} \pm 0.00$	$10.74^{c} \pm 0.01$	$4296^{a} \pm 4.0$
4	$11.33^{a} \pm 0.64$	$8.33^{a} \pm 0.18$	$4.1^{a} \pm 0.1$	$6.2^{a} \pm 0.26$	$11.48^{b} \pm 0.59$	$4592^{a} \pm 20.0$

Note: Means followed by the same letter at each column are not significantly different ($p < 0.05$), and each value represents the mean of ten hills. Treatment 1, inorganic fertilizer only (urea, TSP, and MOP); treatment 2, rice husk charcoal-coated urea, TSP, and MOP; treatment 3, inorganic fertilizer (urea, TSP, and MOP) with rice straw compost only; treatment 4, rice husk charcoal urea, TSP, and MOP with rice straw compost

Table 5 Effect of different treatments on the growth and yield parameters-Maha (Gamage 2015)

Treatments	Number of tillers	Number of panicles	Root dry weight (g)	Shoot dry weight (g)	Yield per plot (kg)	Yield (kg/ha)
1	$9.33^{b} \pm 0.39$	$6.1^{c} \pm 0.17$	$3.1^{b} \pm 0.08$	$4.7^{b} \pm 0.17$	$6.66^{b} \pm 0.36$	$2664^{b} \pm 12$
2	$8.00^{c} \pm 0.00$	$7.3^{b} \pm 0.26$	$3.0^{b} \pm 0.00$	$4.8^{b} \pm 0.08$	$6.94^{b} \pm 0.18$	$2776^{b} \pm 4$
3	$9.40^{b} \pm 0.45$	$7.6^{b} \pm 0.10$	$3.8^{a} \pm 0.07$	$5.7^{a} \pm 0.09$	$7.77^{a} \pm 0.06$	$3108^{a} \pm 24$
4	$10.1^{a} \pm 0.10$	$8.6^{a} \pm 0.10$	$3.9^{a} \pm 0.1$	$6.0^{a} \pm 0.00$	$6.11^{b} \pm 0.01$	$2444^{b} \pm 4$

Note: Means followed by the same letter at each column are not significantly different ($p < 0.05$), and each value represents the mean of ten hills. Treatment 1, inorganic fertilizer only (urea, TSP, and MOP); treatment 2, rice husk charcoal-coated urea, TSP, and MOP; treatment 3, inorganic fertilizer (urea, TSP, and MOP) with rice straw compost only; treatment 4, rice husk charcoal urea, TSP, and MOP with rice straw compost

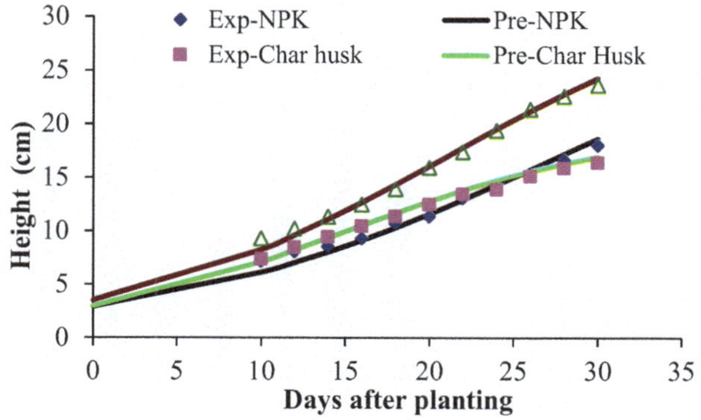

Fig. 10 Relationship between average plant height and days after planting (Ariyawansha et al. 2013)

Table 6 Summary of *Zea mays* yield (Waduge 2011)

Treatment	Average seed weight/cob(g)	Total seed weight (kg)
T1 – Recommended fertilizer (RF)	88.24	3.00
T2 – Coconut Shell biochar (CSB)	24.41	0.83
T3 – CSB + RF	99.26	3.37
T4 – Coconut husk biochar (CHB)	41.91	1.42
T5 – CHB + RF	76.03	2.58

CSB with recommended fertilizer (RF), and it reduces in the order of RF, CHB with RF, CHB, and least of all CSB (Table 6). The results show that the quantity of fertilizer can be reduced. This study also revealed that we can optimally use the applied chemical fertilizers by adding a stable form of biochar to the soil (Ariyawansha et al. 2013).

We conducted a field experiment with ten treatments to evaluate the effect of biochar applications with 30% of recommended inorganic fertilizer on *Cucumis sativus* growth (Weerasekara et al. 2013). It was a collaborative research project with HJS Condiments Limited. The combined effect of biochar and inorganic fertilizer gave the best yield (Fig. 11) in comparison to other treatments as a result of reduction of N fertilizer leaching from the soil, similar to reported by Chan et al. (2009). Several mechanisms contribute to the decrease in nutrient leaching which are related to increased nutrient use efficiency by increased water and nutrient retention (residence time in the root zone) and availability. It is likely to be related to an increase in internal reactive surface area of the soil-biochar matrix, thus decreasing water percolation below the root zone. Notably, the conditions were favorable for increased plant water use (increased evaporative surface) and increased plant nutrient use through enhanced crop growth (Basnayake et al. 2011). In relation to pest and disease attacks, the biochar treatments withstood well to aphid attack and downy mildew in comparison to the use of inorganic fertilizers. The least of the attacks was observed in the biochar with inorganic recommended fertilizers. Biochar may have a

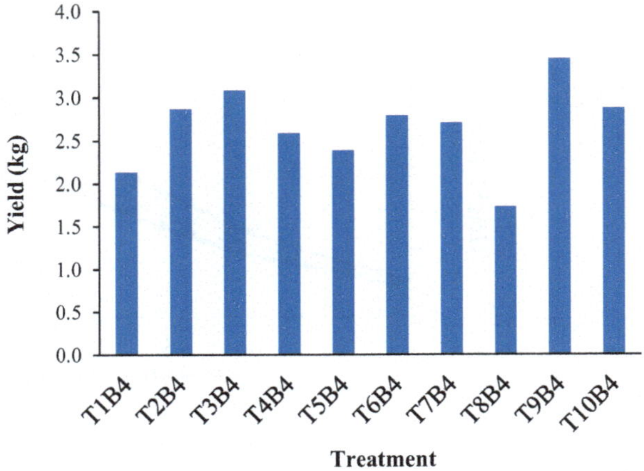

Fig. 11 Grade 1 (export quality) yields of the treatments in Block 4 (Weerasekara et al. 2013). Note: T1 = control (growing on soil), T2 = standard practices recommended by HJS Condiments Ltd., T3 = soil + inorganic fertilizer (IF) recommendation, T4 = soil + BC (BC replacing all recommended inputs on dry weight basis (db)), T5 = soil + BC + compost (BC replacing all IF on db), T6 = soil + BC + IF (replace compost with BC on db), T7 = soil + BC + compost + IF – urea (replace urea with BC), T8 = soil + BC + IF – urea (replace compost + urea with BC), T9 = soil +70% BC+ 30% IF and compost, and T10 = soil + BC + compost + IF

positive impact on plant resistance to disease due to its suppressive effect on soil pathogens (Matsubara et al. 2002).

The replacement of inorganic nitrogen with nitrogen-rich plants or even trapping available nitrogen in composting processes was the focus of recent research. The field experiments conducted with developed fertilizer of granules made with active phase composting with biochar show very promising results. In a pot experiment, we grew *Solanum lycopersicum* (tomato) (Chularathna 2015). The plant heights were best with the developed fertilizer (1) and in order of lesser heights were 70% developed fertilizer with 30% recommended fertilizer (2), recommended fertilizer (3), and lastly the control with only soil (4). However, the mixed treatment gave highest yields followed by developed organic fertilizers and then recommended fertilizers and finally the control. The high vegetative growth could be attributed to available nitrogen, but the phosphorus and potassium would have been limiting for producing more fruits. As we have reported, the cyclic effect of phosphorus could be the central issue in defining microbial behavior in soils. Soil microorganisms act as sink and source of phosphorus (P) and mediate key processes in the soil P cycle, e.g., P mineralization and immobilization (Oberson and Joner 2005). The genetic and biochemical mechanisms of these transformations are not yet completely understood (Ohtake et al. 1996).

6 Sustainable Agriculture Productions with Biochar Applications

The sustainability of agriculture is very much dependent on the availability of phosphorus. In fact, the cumulative annual yields of paddy plotted against cumulative annual phosphorus and potassium applications follow a hyperbolic function than a straight line indicating exhaustion and degrading soils and poor cultivation practices (Basnayake et al. 2011). In applying biochar, we can overcome reductions in yields and increase productions, because our finding of cyclic effect of phosphorus is the key factor in differentiating soils ameliorated with biochar being better and more effective media for plant growth as opposed to only soil. The results of the studies throughout the world indicate that the role of biochar can be defined as materials with varying degrees of promoting optimum soil microcosms to provide nutrients while functioning as a "receptor," which makes it a "nutrient regulator." It has many diverse properties unlike the effects of one-way nutrient supply in inorganic fertilizer applications. Therefore, it requires in-depth research to find the mechanisms of fixing both carbon and nitrogen in biochar at required levels of nutrients like phosphorus availability to the plant while mobilizing it from the soil. In response to enhancing the property of "nutrient regulator," the development of new biochar-based fertilizers should focus on finding types of substrates needed for microcosms to anchor phosphorus in the matrix of biochar structures and release it when needed by the plant, thus functioning as supplier and receptors. It is then necessary to introduce new cultural practices rather than rely on continuous fertilizer applications, since the buildup of receptors every season will eventually fix adequate nitrogen and carbon from the atmosphere, while phosphorus remains available to microcosms, like Terra Preta de Indio.

In view of disseminating the findings, it will depend on biomass generation, gathering and processing systems, making available enough feedstocks for advanced thermal conversions in different ecological regions in the country. Increasing the biomass generations of intercropping, including so-called weeds between cultivation seasons, is useful in fixing large quantities of nitrogen, which are mostly lost in current cultivation practices. We hope that the academics, scientists, researchers, engineers, administrators, and politicians understand the usefulness of land consolidation in sustainable development and formulate a policy with a sound strategy and a planning process to implement large-scale use of paddy straw for making fertilizer. Therefore, the use of paddy straw bales, cow dung, and green manure is a "need of the hour" for firstly producing energy and refrigeration requirements for preserving seeds and processing of food and secondly extracting valuable compost from landfill bioreactors for making biochar-based fertilizer that can contribute toward self-sustainability of the farming communities. Moreover, the findings in 1994 are still valid and proven beyond doubt with enhanced and matured knowledge bases in the country to assist small and medium enterprises in partnership with the farming communities to generate electricity that can be supplied to the national grid while producing fertilizer. It is better than relying on large trading companies to drive the miserable and ailing third world economy of this country.

Acknowledgments The authors thank CEMAGREF and France for the financial and technical support given for R&D activities and Kandy Municipal Council, Coconut Cultivation Board, HJS Condiments Ltd., and academic and nonacademic staff of the Department of Agricultural Engineering and Postgraduate Institute of Agriculture, University of Peradeniya for supporting the research and development efforts. We wish to thank Aurora Basnayake for English corrections and valuable comments on comprehension.

References

Ariyawansha RTK, Basnayake BFA, Vijenayaka DMGP, Waduge NS, Gamage DAS (2013) Biochar applications for enhancing nutrient availability in tropical soils. In: Proceedings of the international conference on solid waste 2013 – Innovation in Technology and Management, Hong Kong SAR, P.R. China, 5 – 9 May 2013, pp 676–679

Ariyawansha RTK, Senevirathne SADN, Basnayake BFA (2014) Potential of converting paddy straw to bio-char and electricity in Sri Lanka. In: Proceedings of 11th Asia Pacific Roundtable on Sustainable Consumption and Production (APRSCP). 19–20 May 2014 in Plaza Athénée, Bangkok, Thailand, pp 207–222

Basnayake BFA (1986) Conception and functioning of a multi-product pyrolyzer (Conception et fonctionnement D'un carbonisateurmultiproduits). DEng Thesis. Université de Pierre et Marie Curie, Paris, France

Basnayake BFA(1994) Patent of discovery of a process to retard the release of nitrogen fertilizer by using charcoal and manioc. No: 10665

Basnayake BFA, Ariyawansha RTK, Senevirathne SADN (2011) A mathematical predictions on poor response to fertilizer applications for increased paddy and tea productions and remedial measures with biochar and integrated pest management programmes in Sri Lanka. In: International symposium on agriculture and environment, On November 9th 2011, at Faculty of Agriculture, University of Ruhuna, Sri Lanka

Bezerra J (2015) Chapter 2: Terra Preta de Índio and Amazonian History. In: The Brazilian Amazon world forests 21. Springer, Cham

Bond T, Tse Q, Chambon CL, Fennell P, Fowler GD, Templeton MR (2017) The feasibility of char and bio-oil production from pyrolysis of pit latrine sludge. Environ Sci Water Res Technol 3 (6):1171

Chan KY, Van Zwieten L, Meszaros I, Downie A, Joseph S (2009) Agronomic values of green waste biochar as a soil amendment. Aust J Soil Res 45:629–634

Chularathna WGGS (2015) Evaluation and characterization of organic biochar fertilizer produced through active phase composting. Undergraduate thesis. Department of Natural Resources, Faculty of Applied Sciences, Sabaragamuwa University of Sri Lanka

Demirbas A, Kucuk M (1997) Biomass conversion processes. Energy Convers Manage 38 (2):151–161

Gajdos R (1997) Effects of two composts and seven commercial cultivation media on germination and yield. Compost Sci Util 5:16–37

Gamage DAS (2015) Development of nutrient management technologies for sustainable rice farming for mitigating water and atmospheric pollution. PhD thesis, Postgraduate Institute of Agriculture, University of Peradeniya, Peradeniya, Sri Lanka

Gamage DAS, Sarachchandra NL,Basnayake BFA, Costa WAJM (2010) Lysimeter simulation of paddy straw landfill bioreactor to generate biogas and minimize ground water pollution. In: 9th Asia Pacific Roundtable for Sustainable Consumption and production 10th – 12th June 2010. Colombo Sri Lanka

Gamage DAS, Basnayake BFA, Costa J, Vidanagamage K (2012) Evaluation of total N, P, K and organic matter contents of soil amended with paddy husk charcoal coated urea and comparison of the yield of paddy. In: Proceedings of international conference on sustainable built environment, Sri Lanka

Gunasekara NS (2014) Development of an effective organic fertilizer through a novel approach of biocatalyst derived from biochar. B.Sc. thesis, Department of Agricultural Engineering, Faculty of Agriculture, University of Peradeniya, Sri Lanka
Haefele SM, Konboon Y, Wongboon W, Amarante S, Maarifat AA, Pfeiffer EM, Knoblauch C (2011) Effects and fate of bio-char from paddy residues in paddy-based systems. Field Crop Res 121:430–440
https://www.geozentrum-hannover.de/DE/Gemeinsames/Nachrichten/Veranstaltungen/2009/Hauskolloquium_2009_2010/2009_10_06_abstracts.html. Accessed on 28 Apr 2018
https://www.youtube.com/watch?v=0Os-ujelkgw. Accessed on 01 May 2018
IEA Bioenergy (2007) Biomass pyrolysis. http://www.ieabioenergy.com/wp-content/uploads/2013/10/Task-34-Booklet.pdf. Accessed on 01 May 2018
Iranzo M, Canizares JV, Roca-Perez L, Sainz-Pardo I, Mormeneo S, Boluda R (2004) Short communication: characteristics of paddy straw and sewage sludge as composting materials in Valencia (Spain). Bioresour Technol 95:107–112
John A (2013) Alternatives to open-field burning on paddy farms. Options 18
Kadam KL, Forrest LH, Jacobson WA (2000) Paddy straw as a lignocellulosic resource: collection, processing, transportation, and environmental aspects. Biomass Bioenergy 18:369–389
Kishimoto S, Sugiura G (1985) Charcoal as a soil conditioner. In: Symposium on forest products research, international achievements for the future, vol 5, pp 12–23
Lehmann J, Joseph S (2009) Biochar for Environmental Management, Earthscan, Dunstan House, 14a St Cross Street, London, ECIN 8XA, UK
Manel UGE (2010) Development of a municipal solid waste pyrolyzer for producing high quality charcoal. B.Sc. thesis, Department of Agricultural Engineering, Faculty of Agriculture, University of Peradeniya, Sri Lanka
Matsubara Y, Hasegawa N, Fukui H (2002) Incidence of Fusarium root rot in asparagus seedlings infected with arbuscular mycorrhizal fungus as affected by several soil amendments. J Jpn Soc Hortic Sci 71:370–374
Mukherjee A, Lal R (2014) The biochar dilemma. Soil Res 52:217–230
Oberson A, Joner EJ (2005) Microbial turnover of phosphorus in soil. In: Turner BL, Frossard E, Baldwin DS (eds) Organic phosphorus in the environment. CABI, Wallingford, pp 133–164
Ohtake H, Kato J, Kuroda A, Taguchi K, Sakai Y (1996) Chemolactic signal transduction in pseudomonas aeruginosa in pseudomonas. In: Nakazawa T, Furukawa K, Hass D, Silver S (eds) Molecular biology and biotechnology. American Society for Microbiology, Washington DC, pp 188–194
Paul S (1982) Bio-energy re-news. J Energy Biomass Recycl 1:1–48
Paz-Ferreiro J, Lu H, Fu S, Méndez A, Gascó G (2014) Use of phytoremediation and biochar to remediate heavy metal polluted soils: a review. Solid Earth 5:65–75
Putun EA, Apayd E, Putun E (2004) Paddy straw as a bio-oil source via pyrolysis and steam pyrolysis. Energy 29:2171–2180
Sanderatne N, de Alwis S (2018) The demographic transition in Sri Lanka: A socio-economic analysis. Marga Institute, 941/1, Jayanthi Mawatha, Kotte Road, Ethul Kotte
Santiago A, Santiago L (1989) Charcoal chips as a practical substrate for container horticulture in the humid tropics. Acta Hortic 238:141–147
Thivyatharsan R, Gunarathna HAYR, Basnayake BFA, Kuruparan P (2012) Performance and suitability of a landfill bioreactor with low cost biofilm contained clay-waste polyethylene-clay composite liner system for tropical climates of Asian countries. J Mater Cycles Waste Manage. https://doi.org/10.1007/s10163-011-0039
Waduge NS (2011) Use of coconut based biochar as soil amendment for integrated maize cultivation in coconut lands. Directed study report. Postgraduate Institute of Agriculture, University of Peradeniya, Sri Lanka
Weerasekara IACS, Ariyawansha RTK, Basnayake BFA, Karunarathne AK, Pathirage AC, Basnayake CA (2013) Evaluation of biochar influence on nutrient regulation for growth and yield improvements of gherkin (*Cucumis sativus* L.). In: Proceedings of the international

conference on solid waste 2013 – Innovation in Technology and Management, Hong Kong SAR, P.R. China, 5–9 May 2013, pp 684–686

Zucconi F, Monaco A, Forte M, Bertoldi MD (1985) Phytotoxins during the stabilization of organic matter. In: Gasser JKR (ed) Composting of agricultural and other wastes. London/New York, Elsevier Applied Science, pp 73–88

Taking *Trans* Fats Out of the Food Supply

Terrence Madhujith and Subajiny Sivakanthan

1 Introduction

Lipids are a group of chemically diverse compounds which are soluble in organic solvents and insoluble in water (Fahy et al. 2011). Food lipids are classified based on their physical state at ambient temperature. Food lipids in solid forms are termed fats, while liquid forms are named oils. Food lipids also can be classified based on their polarity as nonpolar lipids (e.g. triacylglycerols and cholesterol) and polar lipids (e.g. phosphoacylglycerols) indicating the differences in their solubility and functional properties. Foods vary in their quantity and the composition of lipids (Damodaran and Parkin 2017).

Lipids are made up of fatty acids and glycerol. Fatty acids are carboxylic acids carrying mostly an even number of carbon atoms in aliphatic chain, a carboxylic acid, and a methyl group at two ends (Fig. 1). Fatty acids with odd number of carbon atoms can be found in minute quantities in some microorganisms and dairy fats. Aliphatic chain of the fatty acids may contain double bonds; thus, fatty acids are classified as saturated and unsaturated (monounsaturated and polyunsaturated) (Lannes and Ignácio 2013; Chatgilialoglu et al. 2014). Majority of the naturally occurring unsaturated fatty acids occur in *cis* forms (Valenzuela and Valenzuela 2013). Most dietary *trans* fats are generated by the partial hydrogenation of unsaturated vegetable oils. Partial hydrogenation is carried out in order to increase the oxidative stability and thereby to extend the shelf life and to enhance the physicochemical properties of foods. The lipids modified through partial hydrogenation can potentially replace animal fats such as lard, tallow, and butter (Puligundla et al. 2012;

T. Madhujith (✉)
Department of Food Science and Technology, Faculty of Agriculture, University of Peradeniya, Peradeniya, Sri Lanka

S. Sivakanthan
Department of Agricultural Chemistry, Faculty of Agriculture, University of Jaffna, Kilinochchi, Sri Lanka

R. P. De Silva et al. (eds.), *Agricultural Research for Sustainable Food Systems in Sri Lanka*, https://doi.org/10.1007/978-981-15-3673-1_13

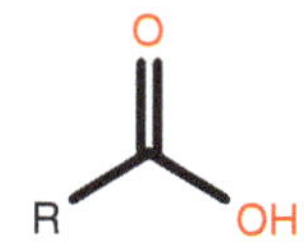

Fig. 1 Structure of fatty acid (R denotes aliphatic chain)

Lannes and Ignácio 2013). Epidemiological studies have reported that there is a strong positive correlation between the intake of industrially produced *trans* fatty acids and cardiovascular diseases (de Souza et al. 2015). However, the conjugated linoleic acid (CLA), a type of *trans* fats with conjugated double bonds naturally present in the ruminant meat and milk, is known to possess a myriad of health benefits (Wannamethee et al. 2018).

As a result of scientific findings indicating the adverse health effects, *trans* fat becomes one of the centers of focus in the food industry. Several high-income countries commenced restricting *trans* fat contents of foods; however, low- and middle-income countries failed to mitigate *trans* fats in their food supplies. Studies on *trans* fat consumption in Sri Lanka is lacking; however, it is suspected that a considerable number of food items commonly consumed in the country contain high levels of *trans* fat. Therefore, there is a timely need to urge the regulatory authorities to take necessary actions to mitigate the *trans* fat levels in the Sri Lankan diet. In this backdrop, this chapter reviews the chemistry and health effects of *trans* fats, global scenario on regulations on *trans* fat levels, and the possible measures to mitigate *trans* fats in Sri Lankan food supply.

2 Chemistry of *Trans*-Fatty Acids

Trans-fatty acids are unsaturated fatty acids with at least one ethylenic double bond in the *trans* geometrical configuration. However, for labelling purposes, restricted definitions have been proposed for *trans* fats which exclude fatty acids containing conjugated double bonds which are known to possess health benefits (Ledoux et al. 2007; Silva et al. 2014). Fatty acids with *cis* configuration carry the hydrogen atoms on the same sides of the molecule providing a "V" shape, and this is considered the usual form of fatty acids naturally found in foods. On the other hand, the fatty acids with *trans* configuration bearing the hydrogen atoms on the opposite sides of the molecule leading to a near linear configuration (Fig. 2) (Żbikowska 2010), are more rigid and straight molecules with a higher melting point than their *cis* counterparts (Remig et al. 2010).

3 Classification of *Trans* Fats

Trans fatty acids can be classified as naturally occurring and industrially produced (artificial). *Trans* fatty acids are produced by the gut flora of ruminant animals via the process known as biohydrogenation; thus, *trans* fatty acids occur naturally in minute quantities in meat and milk derived from ruminants such as cattle, sheep, and goats.

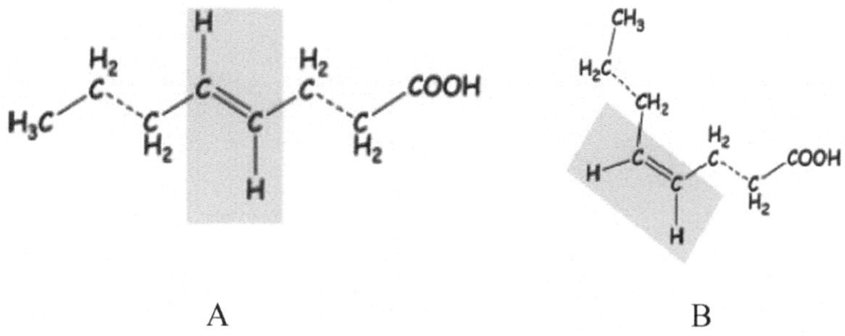

Fig. 2 Configuration of *trans*-fatty acid (**a**) and configuration of *cis* fatty acid (**b**)

However, the origin of most of the *trans* fats in the human diet is not ruminant products. The main sources of *trans* fatty acids are partially hydrogenated vegetable oils, margarines, shortenings, other fat spreads produced through partial hydrogenation, baked goods, and frying fats and fried products. In addition, refining of edible oils, frying, and food irradiation also lead to formation of considerable amounts of *trans* fats in foods (Martin et al. 2007).

3.1 Natural *Trans* Fats

3.1.1 The Origin

The microflora found in rumen of ruminants carry isomerases, which isomerize naturally existing form of fatty acids (*cis)* into *trans* form. Biohydrogenation of 18-carbon polyunsaturated fatty acids such as linoleic acid forms primarily vaccenic acid (11-*trans*-octadecenoic acid) and conjugated linoleic acid (CLA) (9-*trans*, 11-*cis*-octadecadienoic acid). *Trans*-fatty acids generated in the rumen of ruminants are transported to milk and muscles where they are deposited in fat (Żbikowska 2010; Remig et al. 2010). As these *trans* fats occur naturally, it is impossible to eliminate *trans* fats completely from these sources. However, the amount of *trans* fats arising from the ruminants is less likely to cause adverse health effects (Uauy et al. 2009; Downs et al. 2013b). In fact, to the contrary, CLA is known to provide a myriad of health benefits (Silva et al. 2014); thus, it is encouraged to have CLA in the diet (Fig. 3).

3.1.2 Health Benefits of Natural *Trans* Fats

Even though there are sufficient evidences to suggest the adverse health effects of *trans* fats, recently, a paradox arises that the natural *trans* fats produced by ruminants such as vaccenic acid and CLA do not harm the human health. In the past decade, CLA gained much attention due to their potential health benefits (Wannamethee et al. 2018). It is proven that CLA possesses therapeutic potential such as antiatherogenic, antidiabetic, anti-inflammatory, and anti-oncogenic effects and immune-modulating properties (Silva et al. 2014; Wannamethee et al. 2018). Thus, the definition of *trans* fat in the Codex Alimentarius standard, as well as

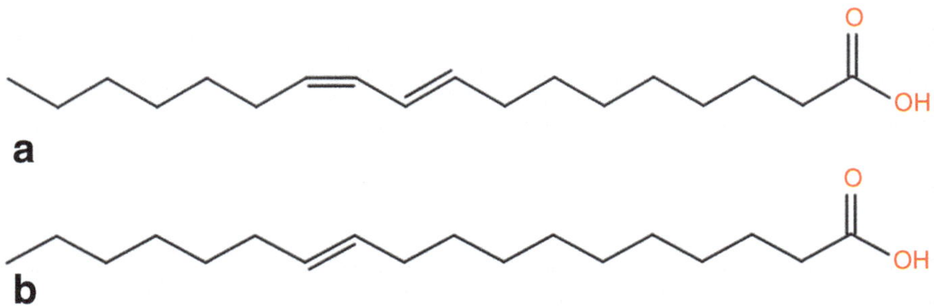

Fig. 3 Structure of conjugated linoleic acid (9-*trans*, 11-*cis*-octadecadienoic acid) (**a**) and vaccenic acid (11-*trans*-octadecenoic acid) (**b**)

definition by some other countries such as the United States, Canada, and Denmark, has been amended to exclude *trans* fatty acids containing conjugated double bonds for labelling purposes (Gebauer et al. 2011).

Both human and animal studies have suggested that CLA possesses cardiovascular health benefits (Wannamethee et al. 2018). Recent research on humans suggested that CLA would act to reduce adiposity via modulation of lipid metabolism (Lehnen et al. 2015). The possible mechanisms of action of CLA to alter the body composition involve metabolic changes favoring the reduction of the lipogenesis and the potentiation of lipolysis, together with the increased oxidation of fatty acid in the skeletal muscles because of increased carnitine palmitoyltransferase-1 activity (Churruca et al. 2009). A study on Wistar rats to supplement CLA at a rate of 2.0% and 4.0% with respect to the daily feed intake reported 18% reduction in body fat (Botelho et al. 2005). A recent study has reported an inverse association between serum CLA level and risk of heart failure in older men (Wannamethee et al. 2018). The anticarcinogenic properties of CLA have been studied experimentally. It has been reported that animal studies show a beneficial effect of CLA in mammary cancer (Gebauer et al. 2011).

3.2 Formation of Industrial *Trans* Fats

3.2.1 Partial Hydrogenation of Oils

Hydrogenation is a process of adding hydrogen to the double bonds of unsaturated fatty acids in the presence of catalyst, in general nickel. The intermediate product with only one hydrogen added does not contain double bond, thus rotating freely. Thus, the double bond can re-form as either *cis* or *trans*, of which *trans* is energetically more stable. During complete hydrogenation, all unsaturated fatty acids get saturated leading to hard fat; thus, it does not lead to generation of *trans* fatty acids. As a result, *trans* fat is produced as a by-product of partial hydrogenation only. Partial hydrogenation affects mostly the 18:1 isomers while reducing the amount of polyunsaturated fatty acids (PUFA) in the original oil (Ledoux et al. 2007). Elaidic acid (*trans*-9-octadecenoic acid) and linolelaidic acid (*trans*-9,12-octadecadienoic

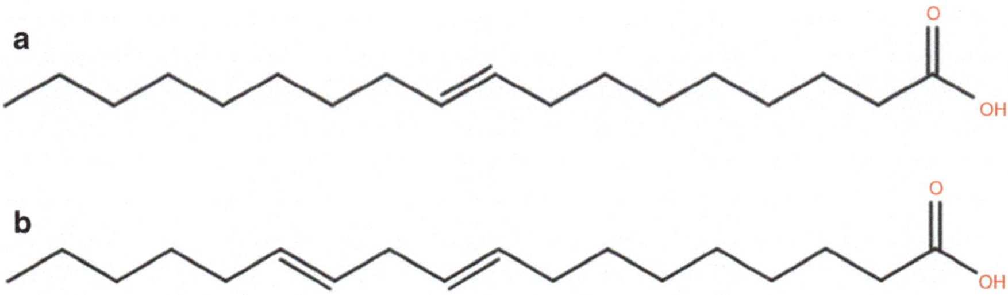

Fig. 4 Structure of elaidic acid (*trans*-9-octadecenoic acid) (**a**) and linolelaidic acid (*trans*- 9,12-octadecadienoic acid) (**b**)

acid) are the major *trans* fatty acids produced during partial hydrogenation, which are estimated to be found in partially hydrogenated vegetable oils at 85–95 and 8–22%, respectively (Żbikowska 2010) (Fig. 4).

During the partial hydrogenation process, oil is hardened, which, when used in food products, enhances its texture and other sensory qualities and increases the shelf life and tolerance to repeated heating. The proportions of industrially produced *trans* fats are usually much higher in processed foods than in natural foods (WHO 2015).

Partially hydrogenated fats may contain up to 50% of *trans* isomers (Guillén and Cabo 1997). Amount of *trans* fat depends on PUFA composition of the native oil, type and amount of the catalyst used in the hydrogenation process, final hardness, and other hydrogenation conditions such as temperature, length of treatment, pressure, and stirring intensity (Ledoux et al. 2007). Thermal treatments of fats and oils carried out during processing of edible oils such as deodorization and cooking and frying operations during preparation of food generate *trans* fats with limited number of double bonds leading to the formation of mainly *trans*-18:2 and *trans*-18:3 (Żbikowska 2010).

3.2.2 Oil Refining

Crude vegetable oils contain non-triacylglycerol compounds which contribute to undesirable physicochemical properties, thus reducing the acceptability and limiting the shelf life. Therefore, oils are refined to remove these unacceptable substances. Refining process includes the steps such as degumming, neutralization, bleaching, and deodorization. During deodorization, oil is heated to 180–270 °C which can result in the formation of *trans* fats. The quantity of *trans* fats formed depends on the temperature and duration of the process (Martin et al. 2007). Thus, refined oils with high level of unsaturation such as sunflower oil may contain *trans* fats.

3.2.3 Frying and Other Culinary Operations

Fried foods are preferred by consumers all over the world due to their unique sensory qualities such as crispy texture, aroma, and golden brown color. However, during the frying process, *trans* fats are generated especially when unsaturated oil is used for frying. Frying foods using oils containing polyunsaturated fatty acids at temperatures above 180 °C for prolonged period of time generates *trans* fats. The extent of formation of *trans* fats during frying depends on the temperature and duration of frying (Martin et al. 2007). Several European countries have determined that the frying temperature should not exceed 180 °C. The ideal frying oil should contain the lowest possible linolenic acid content (less than 2%). In France, it has been established that the commercial frying oil should contain less than 3% α-linolenic acid (Martin et al. 2007). When partially hydrogenated oils are used as frying oils, the formation of *trans* fats is generally low; however, the high initial amount of *trans* fats will result in high amount of *trans* fats in fried food (Martin et al. 2007). A study carried out in the central province of Sri Lanka revealed that over 90% of snack producers use palm oil for frying, and some snack makers use the same oil nearly 40 times during the day (Jayawardena et al. 2014).

3.2.4 Food Irradiation

Irradiation is a promising novel technique used in food preservation. Ionizing radiation causes destruction of pathogenic microorganisms as well as spoilage microorganisms by permanently damaging the DNA (Ganguly et al. 2012). Irradiation of lipids induces the production of free radicals, which react with oxygen, leading to the formation of carbonyls (Brito et al. 2002). Breaking the double bond of unsaturated fatty acids during food irradiation favors the formation of *trans* fatty acids (Martin et al. 2007). Even though, in food preservation, the use of irradiation technology has gained much interest in some countries, it is not so far used in Sri Lanka. Therefore, the *trans* fats produced by the irradiation do not have significant contribution.

4 Sources of Industrially Generated *Trans* Fats

Trans isomers are not found in natural vegetable oils. Partially hydrogenated oils and fats are used for the production of hardened margarine spreads, and these are used for the production of sweet snacks (cookies, cake, doughnuts, pie, crackers and other baked products, and chocolates), salted snacks (French fried potatoes and chips), instant soups, pizzas, and takeaway foods. Thus, these foods are likely to contain significant levels of *trans* fats. In several countries, partial hydrogenation has been replaced by alternative technologies. In Sri Lanka, some food manufacturers, restaurants, and bakers use shortenings made out of partially hydrogenated oils as they are less expensive.

5 Health Concerns of *Trans* Fats

There are evidences to show that consumption of *trans* fat is well associated with the development of cardiovascular diseases (Nestel 2014). Furthermore, *trans* fat intake is also associated with the development of diabetes, dementia, cancer, and infertility and affects fetus. The US-FDA has announced that *trans* fats are not generally recognized as safe (GRAS) for any use in food (Ryan 2014).

Trans fatty acids are present in the diet in esterified form, mainly in triacylglycerols; however, those from ruminant sources may also be present in phospholipids. Before absorption into the body, triacylglycerols must be digested by pancreatic lipase in the small intestine. The hydrolysis and absorption of *trans* fatty acids are similar to that of *cis* fatty acids. Dietary *trans* isomers are incorporated into membranes and other cellular structures. Dietary *trans* fatty acids are incorporated into the brain, liver, adipose tissue, spleen, blood plasma, and milk. *Trans* fatty acids are transported across the placenta and secreted in human milk, and the levels depend on the maternal dietary intake of *trans* fat (Larqué et al. 2001; Sadler 2005; Innis 2006; Żbikowska 2010).

5.1 Cardiovascular Diseases

Cardiovascular diseases remain the principal cause of mortality worldwide (Zapolska et al. 2015). There is strong evidence to support the link between the consumption of *trans* fats and the risk of developing cardiovascular diseases (Chowdhury et al. 2014). The World Health Organization (WHO) points out that *trans* fats increase the risk of cardiovascular diseases more than any other dietary source of energy. *Trans* fats contribute to cardiovascular diseases via multiple risk factors such as increasing plasma concentrations of lipids and lipoproteins and inflammatory markers, impairing endothelial function and possible effects on coagulation, insulin resistance, and displacement of essential fatty acids from membranes, affecting prostanoid-related functions and possibly other key membrane-related functions (Uauy et al. 2009; Gebauer et al. 2011; Takeuchi and Sugano 2017).

Trans fatty acids induce the development of a pro-atherogenic lipid profile leading to increased cardiovascular risk (Revoredo et al. 2017). *Trans* fats mediate the increased risk of cardiovascular diseases by increasing low density cholesterol (LDL-C) levels, total cholesterol, and triacylglycerols and decreasing the levels of high-density lipoprotein cholesterol (HDL-C) in plasma. *Trans* fatty acids create more space for the transport of cholesterol by gathering together inside LDL-C particles. Further, *trans* fats can reduce gene expression of hepatic receptors responsible for the uptake of LDL-C (Hunter et al. 2010; Revoredo et al. 2017). The LDL fraction may be deposited in blood vessels and lead to arteriosclerosis (Remig et al. 2010; Żbikowska 2010; Bendsen et al. 2011a). Thus, *trans* fats increase the LDL/HDL ratio, total cholesterol/HDL apolipoprotein A/apolipoprotein B ratio, and cholesterol content both in LDL and HDL particles in comparison to saturated fatty acids. Moreover, *trans* fatty acids increase hepatic secretion of very low-density lipoprotein (VLDL) and their particle size (Dashti et al. 2002; Chowdhury et al. 2014). Cardiovascular disease is described as an inflammatory disease. C-reactive protein is considered as a better predictor of future cardiovascular incidences than lipid and lipoprotein levels alone. The *trans* fats have been shown to increase inflammatory markers, including C-reactive protein, interleukin-6, and tumor necrosis factor-α, possibly through modulation of monocyte and macrophage activity (Ridker et al. 2002; Remig et al. 2010).

The WHO and the Food and Agriculture Organization (FAO) of the United Nations recommend that daily energy intake from *trans* fats should be less than 1% of total energy intake which is equal to no more than 2 g of *trans* fats per day for a person requiring 2000 kcal (WHO 2003). It is estimated that 2% increase of total energy derived from *trans* fat is associated with an increase in risk of deaths by heart diseases by 23% (Mozaffarian et al. 2003). Moreover, it has been proven that with replacement of partially hydrogenated vegetable oils with alternative fats and oils, the risk of coronary heart diseases may be reduced by as much as 50% (Mozaffarian and Clarke 2009). Removing *trans* fats from the food supply is perhaps one of the most straight forward public health interventions for reducing cardiovascular disease risk (WHO 2015). The US-FDA has estimated that labelling regulations on *trans* fat levels could save up to 500 lives per annum in the United States by reducing the incidence of cardiovascular disease (Resnik 2010). In Denmark, over a period of 20 years, the number of deaths due to coronary heart diseases has been reduced by nearly 50% by the regulations on *trans* fat levels (Iqbal 2014).

5.2 Effects on Early Human Development

Even though earlier studies concluded that human placenta can act as a barrier for *trans* fatty acids, subsequent studies found that *trans*-fatty acid levels in cord blood of term infants are similar to the levels in maternal plasma lipids (Larqué et al. 2001). Dietary *trans* fatty acids are transported to the fetus via placenta and incorporate into fetal tissues, which may lead to infantile birth weight in preterm and healthy term babies as well as reduce the duration of pregnancy (Żbikowska 2010). It has been reported that *trans* fat levels are inversely correlated with infantile birth weight in preterm and healthy term infants (Larqué et al. 2001).

5.3 Diabetes Mellitus

Studies have revealed that *trans* fats may contribute to development of diabetes mellitus. Fourteen-year observation of Nurses' Health Study revealed that the risk of developing type 2 diabetes is associated with *trans* fatty acid intake (Salmeron et al. 2001). Both saturated and *trans* fatty acids increase the risk of developing diabetes; however, the effects of *trans* fatty acids are greater than those of saturated fatty acids (Dhaka et al. 2011). *Trans* fatty acids influence peripheral insulin sensitivity and the risk of developing type 2 diabetes by affecting cell membrane functions (Risérus et al. 2009). *Trans* fat may increase hepatic lipogenesis leading to a greater intolerance to glucose and also induce a reduction in the synthesis of Acylation Stimulating Proteins (ASP) which could contribute to peripheral resistance to the insulin action (Bendsen et al. 2011b; Revoredo et al. 2017).

Studies suggest that there is a differential effect of *trans* fatty acids compared with *cis* fatty acids on the regulation of insulin secretion, with *trans* fatty acids potentiating glucose-stimulated insulin secretion more than *cis* isomers of identical chain length (Thompson et al. 2011). Risérus suggested that replacing *trans* fatty acids with monounsaturated fatty acids and PUFA has beneficial effects on insulin sensitivity (Risérus et al. 2009).

5.4 Cancer

Dietary habits remain one of the leading causes of cancer. Epidemiological studies have indicated that the people who eat saturated and unsaturated *trans* fats, mainly from fried meats, have the highest cancer risk (Bajinka et al. 2017). In the EURAMIC study, an ecological study, commenced in 1997, it has been proven that there is a positive association between *trans* fatty acid intake and the incidence of cancer of the breast and colon (Bakker et al. 1997; Kohlmeier et al. 1997).

5.5 Dementia

Studies have found that intake of *trans* fat is positively associated with the risk of dementia including Alzheimer disease. *Trans* fats tend to elevate plasma total and LDL cholesterol concentrations, which, in turn, may be associated with Alzheimer disease risk (Barnard and Bunner 2014).

5.6 Infertility

Trans fats are believed to cause infertility in men and women. In men, *trans* fatty acids cause adverse effects on reproduction including decrease in fertility, serum testosterone levels and sperm count, motility, and normal morphology of sperms. Extreme levels may lead to the arrest of spermatogenesis and testicular degeneration (Chavarro et al. 2014). Studies on fatty acid composition of semen samples of men undergoing infertility test found that *trans* fatty acids are present in human sperm and are inversely related to sperm concentration (Chavarro et al. 2011; Attaman et al. 2012). It is estimated that when *trans* fatty acid consumption in healthy men aged between 18 and 23 years increased from 0.37% of total calorie intake to 1.03% of total calorie intake, the sperm concentration decreased from 135×10^6 to 94×10^6 (Chavarro et al. 2014; Esmaeili et al. 2015). Thus, *trans* fats affect spermatogenesis. It is also found that *trans* fats may increase the risk of ovulatory infertility (Chavarro et al. 2007; Chavarro et al. 2011).

5.7 Other Negative Effects

There are no evidences to support the relationship between the *trans* fat intake and obesity; however, dietary *trans* fats could be metabolized differently and easily deposited in the adipose tissue, because the melting points of *trans* fats are much higher than those of non-*trans* fats (Ochiai et al. 2013). Further, *trans* fats contain calories in the same quantities as other edible fats. Thus, *trans*-fatty acid may cause excess weight (Stender and Dyerberg 2003). Incidences of allergic reactions (hay fever, atopic disorders, and asthma) are also found to be associated with the intake of *trans* fats (Stender and Dyerberg 2003).

6 Global Scenario of *Trans* Fats

Systemic review of data from 29 countries by Wanders and co-workers (2017) indicated that *trans* fat intakes range from 0.3 to 4.2% of total energy intake of which *trans* fat intake of seven countries is higher than the recommendation by the WHO (less than 1% of the energy intake). In 2010, *trans* fat intakes were within the range of 0.2 to 6.5% of total energy intake worldwide; thus, in the past 17 years, substantial reduction in industrial *trans* fat has been achieved by many countries including the United States, Canada, Netherlands, South Korea, Iran, and Australia (Micha et al. 2014; Downs et al. 2017; Wanders et al. 2017).

6.1 Regulations on *Trans* Fats in Selected Countries

Following the scientific evidence that *trans* fats are strongly associated with the risk of cardiovascular diseases, most countries have already set strict legal limits on *trans* fat content of foods. Currently, more than ten countries have adopted policies including mandatory labelling and bans at both the local and national level targeting the reduction of industrially produced *trans* fats from their food supply (Wu et al. 2017). Denmark was the first country in the world to set regulations related to *trans* fat in 2003 (Pérez-Ferrer et al. 2010). Denmark started to take measures to reduce *trans* fat following the report published in Lancet in 1993 (Willett et al. 1993). In 2003, the Danish Order on *trans* fats in oils and fats became effective (maximum of 2 g *trans* fats/100 g of oils or fats intended for human consumption) after the report recommending legislation on *trans* fats submitted by Danish Nutrition Council in 2001 (Pérez-Ferrer et al. 2010). Since 2006, the US-FDA has required that *trans* fats be declared on the nutrition facts panel of foods. The WHO has called for the elimination of *trans* fats from the global food supply. Recognizing the adverse health effects of *trans* fats, many countries implemented their national rules on *trans* fats, and many food manufacturers and retailers have been systematically removing them from their products in recent years.

Several European countries implemented voluntary strategies aimed at reducing the national consumption of *trans* fat through collaboration between industry and government. Mandatory labelling of *trans* fat content is not currently implemented in any of the WHO European Member States (WHO 2015).

Following the initiatives of Denmark, some other countries such as Switzerland (2008), Austria (2009), Iceland (2011), Hungary (2014), Norway (2014), and Sweden (2011) also implemented legislation on *trans* fats. All these countries have set a legal upper limit of 2 g of industrially produced *trans* fats per 100 g of fat or oil (WHO 2015). Intake of *trans* fat is relatively low in Australia compared to other countries (Wu et al. 2017). The Netherlands achieved a 0.8 g/day reduction in *trans* fat intake from 2.1 to 1.3 g/day by implementing regulation to put a healthier choices logo for food packages since 2006 (Hyseni et al. 2017).

Canada has achieved significant reductions in *trans* fat levels through multifaceted approaches such as mandatory nutrition labelling, the establishment of *Trans* Fat Task Force to develop recommendations and strategies to mitigate *trans* fat from Canadian foods, and monitoring of food products for *trans* fat levels (Pérez-Ferrer et al. 2010; Krenosky et al. 2012). Canada took measures to ban *trans* fat with effect from September 2018 after providing adequate time for the food industries to find suitable alternatives (Downs et al. 2017). *Trans* fat labelling has been mandatory in Hong Kong Special Administrative Region and Taiwan. (Wijesundera et al. 2007; Downs et al. 2013b).

The United States has imposed restrictions on the use of *trans* fats ingredients in food service establishments (US-FDA, 2014). In the United States, the final rule mandating labelling of the *trans* fat content in packaged foods was published on July 11, 2003, with an implementation deadline of January 1, 2006 (Camp et al. 2012). As a consequence, the food industries including large-scale food manufacturing organizations and food service establishments such as McDonald's, KFC, Starbucks, and Burger King have started reducing *trans* fat content in their products. The US-FDA has set a deadline, June 18, 2018, for the food manufacturers to get rid of *trans* fats by removing partially hydrogenated vegetable oils completely from the food supply (Laaninen 2016).

Despite the challenges, there are examples of success in reducing *trans* fat in the food supply by adopting voluntary approaches in Latin American countries such as Costa Rica and Argentina to get an adequate supply of healthy replacement oils for partially hydrogenated vegetable oils (Downs et al. 2015). In Slovenia, voluntary guidelines and regular public communication of the risks related to the *trans* fatty acid consumption alone resulted in significant effect on the *trans* fat levels in the food supply, even though sufficient removal of partially hydrogenated oils has not been achieved from foods (Zupaniˇc et al. 2018).

In India, vanaspati (partially hydrogenated edible oil mixture used as cooking oil and in fried snacks, baked goods, and street-vended foods) remains the major source of *trans* fat. Vanaspati is widely used since it is cheap and oxidatively stable for repeated heating (Downs et al. 2013a). In July 2013, the government of India (Food Safety and Standards Authority) set regulation on *trans* fat levels requiring the manufacturers of vanaspati to limit the *trans* fat content to 10% and also requiring to include *trans* fat content in labelling of packaged food. In December 2014, the regulation was amended to reduce *trans* fat levels to 5% in partially hydrogenated vegetable oils by August 27, 2016, which was then extended up to December 31, 2017 (Downs et al. 2015; Food Safety and Standards Authority of India 2017). Downs et al. (2015) concluded that multi-sectoral food chain approach would be effective in India and other low- and middle-income countries to reduce *trans* fat. This approach requires investment for the development of competitively priced shortenings and incentives for manufacturing foods using healthier oils.

In Malaysia, there are no legislations at the moment to limit the *trans* fat levels present in foods (Khean 2017). However, labelling guidelines are available for *trans* fat content in foods: "Low in *Trans* Fats" (<1.5% of 100 g food or < 0.75 g of 100 mL liquid) and "*Trans* Fat-Free" (<0.1 g of 100 g food or 100 mL liquid) (Azimah et al. 2013).

7 Alternative Methods for Partial Hydrogenation

Finding a replacement for partially hydrogenated vegetable oils is a significant challenge to the food industry to find out healthy replacements that give the same physical and sensory properties of end products as partially hydrogenated vegetable oils. Therefore, food technologists and scientists commenced exploring novel alternative methods for hydrogenation.

7.1 Novel Hydrogenation Techniques

Hydrogenation is a process of adding hydrogen to the double bonds of unsaturated fatty acids in the presence of a catalyst, in general nickel as explained in an earlier section of this chapter. The objective is to increase the oxidative stability of oils by reducing the degree of unsaturation and changing their physical properties, importantly, to change the liquid oils into plastic fats such as shortenings, margarine, and fat spreads (Martin et al. 2007). Modifications in the reaction conditions of hydrogenation process such as pressure, temperature, and catalyst can change the amount of *trans* fatty acids formed and also modify the properties such as melting point and solid fat content of oil (Dhaka et al. 2011).

Three strategies can be used during hydrogenation for reducing *trans* formation. First is increasing the hydrogen pressure so as to increase the amount of hydrogen on the catalyst surface. Second is using a less active catalyst with fewer active sites so as to ensure that the catalyst surface always has enough absorbed hydrogen near the active sites. This can be achieved by producing catalysts with smaller surface area and less activation or by pre-poisoning the catalyst and by blocking most of the active sites with a catalyst poison such as sulfur. Third is slowing the reaction down so as to allow more hydrogen movement to the catalyst surface before the fat molecule is released by the catalyst. This can be done by decreasing the temperature of the reaction (Campbell 2005).

Some novel hydrogenation techniques have also been proposed. Patented procedure is available to prepare low-*trans* shortenings by partial hydrogenation using conditioned catalyst. Edible oils are hydrogenated in a manner to minimize the formation of *trans*-stereoisomers. A catalyst (nickel) conditioned using organic acid phosphates and phosphoric acid discourages *trans*-stereoisomer formation without significantly negatively impacting the length of time required to form solids during hydrogenation (Higgins 2007).

King et al. (2001) have reported a procedure for the hydrogenation of soybean oil in supercritical carbon dioxide, hydrogen, and nickel catalyst with minor formation of *trans* products (King et al. 2001). Wright et al. (2003) have also reported a method for the hydrogenation of canola oil using mixed metal catalysts (nickel and palladium) and lower temperatures that promotes the formation of *cis* isomers and very low production of TFA (11%) (Wright et al. 2003). Lalvani and Mondal (2003) have designed a promising and innovative process for the hydrogenation of edible oils which consists of the electrochemical hydrogenation of oils at low temperature in the presence of formate as the electro-catalyst with nickel and palladium as catalysts. The use of low operating temperatures ($<70\ ^\circ C$) restricts the formation of *trans*-fatty acid isomers to less than 10% of the oil content (Lalvani and Mondal 2003). Novel hydrogenation technique to produce partially hydrogenated vegetable oils rich in conjugated linoleic acids by modifying through pressure, temperature, and catalyst has been proposed (Dijikstra et al. 2008).

Enzymatic hydrogenation techniques are also possible to reduce the *trans* fats in foods using enzymes and pathways similar to that used by rumen microorganisms to produce oils of varying degrees of unsaturation (Loor et al. 2003).

7.2 The Use of Saturates

The natural oils such as coconut, palm, and palm kernel oils are saturated oils. Fractions high in solids can be separated by cooling below its melting point and separating the triglycerides with a higher melting point than the tempering temperature by centrifugation or filtering off from the liquid part. Fractions of palm and palm kernel oils are available commercially (Dhaka et al. 2011). Based on crystallization temperature, cooling rate, and pressure, a number of palm fractions of different composition and functionality can be produced. Saturated stearin palm oil fractions are used to produce cake shortenings, vanaspati (India), pastry margarines, soft and brick margarines, and low-fat spreads (Kaushik and Grewal 2017).

7.3 Structured Lipids

In recent years, edible oil structuring has gained considerable interest. These structuring methods include the creation of structured emulsions by organogelation and interesterification. The interesterified oils are emerging as the most promising alternative for the production of *trans* fat-free oils.

7.3.1 Edible Oleogels (Organogels)

An oleogel or organogel can be defined as a three-dimensional gel network containing an organic liquid entrapped within a thermo-reversible, anhydrous, and structured viscoelastic material, also referred to as oleogels if the organic phase is an edible oil (Kaushik and Grewal 2017). The structural and mechanical properties of structured emulsions and edible oleogels (ethylcellulose oleogels, plant-based wax oleogels, and monoacylglycerol-structured emulsions) can be tailored to mimic that of partially hydrogenated oil-based fat systems. These oil-structuring methods are promising techniques for replacing partially hydrogenated oil (Wang et al. 2016). Oleogelation is the process of converting liquid oil into gel-like material without modifying the chemical characteristics of oil. Oleogels can be applied to various food products to replace partially hydrogenated oils such as baked products, margarines and spreads, and chocolates with the comparable texture, stability, and sensorial properties (Patel and Dewettinck 2016).

Substances that form organogels with edible oils include lecithin, sorbitan tristearate, monoacylglycerides, a mixture of phytosterol and oryzanol, ricinelaidic acid, fatty alcohols, fatty acids, 12-hydroxystearic acid, wax esters, and waxes. Plant waxes are of great interest because of the low cost and availability. The gelation abilities of plant waxes have been studied, and for margarine and spread preparation, wax-based oleogels (sunflower wax, rice bran wax, and candelilla wax) are found to be the most suitable options (Hwang et al. 2013; Hwang et al. 2016; Kaushik and Grewal 2017).

7.3.2 Interesterification

Interesterification is the process of rearranging the distribution of fatty acids either chemically or enzymatically within and between the triacyl glycerol molecules; thus, only the fatty acid distribution is altered while the fatty acid composition remains the same. Interesterification modifies the melting and crystallization behavior of the fat, thus producing fats with the desirable physical properties necessary for the production of margarine or shortening fats without *trans* fats (Dhaka et al. 2011). Thus, interesterification has gained popularity as an alternative for partial hydrogenation to produce *trans*-free margarines, shortenings, and other spreads.

Chemical interesterification process produces full positional randomization of fatty acids on the glycerol backbone. Chemical interesterification is relatively cheap and used in industrial applications, particularly in Europe, to produce low- or zero-*trans* plastic saturated fats. Enzymatic interesterification catalyzed by positional specific lipases offers more control over the reaction products than chemical interesterification. Enzymatic interesterification is carried out under mild processing than chemical interesterification; thus, less by-products are formed (Kaushik and Grewal 2017).

7.4 Other *Trans* Fat Replacement Strategies

Traditional plant breeding or biotechnological methods are used to modify the fatty acid composition of oils, for example, high-oleic acid oils (high-oleic sunflower and canola oils), mid-range-oleic acid oils (mid-oleic sunflower and soybean oils), and low-linolenic acid oils (low-linolenic canola and soybean oils). These modification techniques can be used to formulate *trans*-free hardstocks (Dhaka et al. 2011). It is recommended that salad and frying oils should be developed with moderate levels of oleic acid (< 80%) and low linolenic acid (< 3%). Further, saturated fatty acids should be low (<7–8%) and linoleic acid at least 20–30%. Oils with this fatty acid profile have sufficient oxidative stability, thus eliminating the need for partial hydrogenation; thus, *trans*-fatty acid levels are not increased. Sunflower and canola oils are being modified and commercially available in several countries. Low linolenic canola oil genotypes with less than 3% linolenic acid are available in Canada (Warner et al. 2001).

The National Sunflower Association markets NuSun®, a mid-level (65%) oleic sunflower oil. The NuSun® sunflower germplasm lines were developed at the USDA Agricultural Research Service in Fargo, ND, by traditional plant breeding methods. Various other sunflower seed mutants have been engineered with mid-oleic acid levels (65% to 75% oleic acid) as well as hybrids that produce low levels of palmitic and stearic acids. NuSun® is used in the new zero-*trans* Frito-Lay® snacks and zero-*trans* fat Crisco shortening (Tarrago-Trani et al. 2006).

As a result of the newly introduced regulations, most of the fast-food chains have replaced frying fats by medium- to high-stability vegetable oils (such as high oleic oils), which has resulted in an elimination or reduction of *trans* fats in their products (Dhaka et al. 2011). The use of fat replacers which are ingredients that mimic the functionality and sensory properties of fat is another alternative. Fat replacers can be lipid, protein, or carbohydrate based. Suitable fat replacers can be selected based on the understanding of the food system.

7.4.1 *Trans* Fat Replacement in North America, Europe, China, Malaysia, and India

In North America, *trans* fat replacement includes chemically and enzymatically interesterified margarine/shortening oils, lauric fats for confections, palm-based shortening, baking shortenings made by modified hydrogenation technologies, and high-stability trait-modified oils (List 2014). In Denmark, after the introduction of legal limit of *trans* fat content, saturated fatty acids have been used as the main replacement. Palm oil is typically used in reformulating bakery foods (WHO 2015). In China, palm-based fat has partially replaced hydrogenated fats in most shortening and margarine recipes; however, this replacement has brought about some technical challenges (Zhang et al. 2014).

In India, vanaspati is the major source of *trans* fat (45–50%). Vanaspati is used as an all-purpose fat. Even though the government enforced regulations to reduce the *trans* fat content of partially hydrogenated oils and made labelling mandatory, the replacement solution has not been effective to date. Enzymatic interesterification has been proposed by the Food Safety and Standards Authority of India, and its application is limited and may take time to implement because it is a costly alternative for India (Dhaka et al. 2011).

8 Sri Lankan Perspective

8.1 Possible Sources

As explained earlier in this chapter, some countries such as Denmark, Canada, and the United States have achieved significant reductions in the *trans* fat content of foods.

The use of partially hydrogenated vegetable oils for cooking traditional dishes is common in Sri Lanka. Partially hydrogenated vegetable fats may contain high levels of *trans* fats up to 35–45%. Shortenings used in the production of baked goods in Sri Lanka are supposed to carry high levels of *trans* fats. Food produced using partially hydrogenated oils as a key ingredient such as margarines, shortenings, frying oils, baked products, confectionery products, deep-fried products, frozen pizza, coffee creamers, and many processed snack foods serve as the major sources of *trans* fats.

Foods fried above 180 °C for prolonged time using oils containing polyunsaturated fatty acids in restaurants or at home are a major source of *trans* fats. Furthermore, the reuse of frying oils many a time for frying snacks such as rolls, samosa, *wadei,* and cutlets leads to generation of significant quantities of *trans* fats which are subsequently absorbed by the fried products. Therefore, the snacks and other food items prepared and/or sold by the roadside vendors have become the main source of *trans* fats in Sri Lanka. A recent study carried out in the central province of Sri Lanka revealed that 95% of roadside snack makers use palm oil for frying as palm oil does not become rancid easily upon frying. It was observed that at times, oil is reused nearly 40 times within a single day (Jayawardena et al. 2014). Furthermore, the food items sold by the roadside vendors cost less and are, therefore, most likely to be consumed by people with low income. It is suspected that reused oil from food processing establishments and restaurants are collected by some organized groups for resale after bleaching.

9 Way Forward

9.1 Mitigating the Existence of *Trans* Fats in the Food Supply of Sri Lanka

Different countries have adopted different measures to mitigate the intake of *trans* fats. Data from other countries which have adopted the mitigation strategies show that the content of *trans* fats in their food supply has significantly decreased as a result of these efforts. Information on the extent of *trans* fat consumption in Sri Lanka is very much limited. Most of the Sri Lankan consume *trans* fats from various sources at significantly high levels that increase their risk for heart diseases. Thus, it is necessary to assess the extent of *trans* fat consumption and its relationship to the noncommunicable diseases, importantly heart diseases. According to the Ministry of Health, Nutrition and Indigenous Medicine, Sri Lanka, ischemic heart diseases are the leading cause of death in Sri Lanka since 1995 (Annual Health Bulletin 2015 2016). Thus, there is an urgent need for introducing policy changes. In 2011, the World Bank has urged Sri Lanka to bring laws to encourage good food processing practices and control *trans* fat content in food. Food regulatory authorities in Sri Lanka have not yet formulated any regulations regarding *trans* fat content in processed foods.

Even though the possible level of consumption of *trans* fat by the Sri Lankan population remains high, it is likely that the low-income groups may be exposed to much higher levels of *trans* fat as they consume processed foods available at relatively low cost. Thus, different strategies and policy options targeting different groups should be proposed to reduce the *trans* fat content in the daily diets. As mentioned in the review by Hyseni et al. (2017), these can be described as upstream or downstream interventions. Downstream interventions target individuals and involve behavioral approaches, whereas intermediate interventions target subgroups in workplaces, schools, or communities. Upstream interventions may take place at the population level involving regulatory approaches.

The government of Sri Lanka has prepared a draft labelling regulation which is available for public for comments. According to this draft regulation, the *trans* fat content should be included in the nutritional label (Food Labelling and Advertising Regulations 2015). However, legal limit has not been set for the *trans* fat levels for any food. As evidenced by other countries, labelling regulation can contribute much to achieve reductions in *trans* fat levels. Labelling can help keep the consumers informed of the *trans* fat levels in foods. Thus, health claims such as "*trans* fat-free" may help them select healthy choices. In addition, labelling regulations may accelerate product reformulation by the food industries. *Trans* fats must be declared on the nutrition label of conventional foods and dietary supplements on a separate line immediately underneath the saturated fatty acid content. In order to ensure that the industries are adhering to the labelling regulations, monitoring of *trans* fat content in food products is essential. The Sri Lankan government introduced traffic light labelling system to inform the consumers about the sugar levels in soft drinks. Such system may also be effective in *trans* fat labelling regulation as well.

The most effective technique to mitigate the *trans* fats in foods would be setting a legal limit for the content of *trans* fats in all foods, which can potentially reduce the risks associated with *trans* fats. However, the regulations only can be applied to foods those are labelled. It is quite challenging to regulate the *trans* fat content in food prepared and sold at restaurants, canteens, and eateries and foods prepared by the roadside vendors.

It is important to impose strict regulations on the *trans* fat content of imports. Shortenings are mainly imported from other countries for the bakery and confectionary industries. The shortenings manufactured without using partial hydrogenation are slightly expensive than the hydrogenated counterparts. Therefore, food processors tend to import the low-priced shortenings. The regulatory authorities should develop the capacity to quantify *trans* fat in foods. However, currently, there are only few places which have the capability to quantify *trans* fats. Without developing the capacity to quantify *trans* fats, the enforcement of the regulations would be challenging.

Another issue in enforcing the legal limit is the less availability of alternative fats such as modified fats or interesterified fats. The purpose of partial hydrogenation is to obtain semisolid fats with modified melting and crystallization behavior to be used for different purposes such as production of margarines and shortenings and as ingredients in bakery fats. There are some alternatives used in other countries as explained earlier in this chapter. However, decisions on which alternatives to be used are complicated because they involve availability of resources, facilities for research and development, and the acceptance of sensory quality of the food by consumers. The country has to face several challenges as financial investment and technical facilities; thus, application in Sri Lanka may take time. In order to tackle with these challenges, the country could seek for funds for research and development and reformulation of products. Moreover, most of the processors attempt to keep the cost of production low. The alternative materials are generally expensive; thus, the use of alternatives brings the prices up. Therefore, there will naturally be resistance from the processors against implementation of regulations.

Even though the public is increasingly aware of healthy foods, most of the people in Sri Lanka are not sufficiently aware of the *trans* fats and their health impacts. Thus, there is a need of more public education about saturated fats and *trans* fats. Mandatory food labelling and voluntary labelling and claims also could raise consumers' awareness about the health risks of high intake of *trans* fats and enable consumers to choose products with low-*trans* fat content.

It is high time for the health authorities to carry out nationwide survey on the consumption of *trans* fats. Further, guidelines should be established for local authorities to independently monitor *trans* fat levels at restaurants.

As explained earlier, frying oils also contribute to *trans* fat content in foods. Deep frying using oils containing polyunsaturated fatty acids such as sunflower and soybean oil produces *trans* fats. Therefore, these oils should not be used for deep frying. In Sri Lanka, coconut oil is the most suitable oil based on availability. The risk of formation of *trans* fats in coconut oil is minimum because of its high proportion of saturated fatty acids. Since frying foods at high temperature and oil reuse also produce *trans* fats, as in several European countries, temperature limit can be implemented.

10 Conclusion

Increased prevalence of noncommunicable diseases, especially cardiovascular diseases, among Sri Lankans is at least partially attributed to the high levels of *trans* fats in their diet. Thus, taking industrially produced *trans* fatty acids out of the Sri Lankan food supply will be an effective dietary intervention for reducing the risk of noncommunicable diseases. Many countries have already implemented policies to mitigate the presence of *trans* fats in the food supply and as a result achieved their target significantly to reduce the *trans* fat consumption. It is an urgent need in Sri Lanka to take initiatives to take out the *trans* fat from food supply to ensure that all population is consuming *trans* fat as low as possible which is less than 1% of energy intake. There are no drop-in solutions to be applied easily in order to achieve the targeted reduction in *trans* fat level. The reduction needs a systemic approach involving all sectors. Thus, taking *trans* fat out of the food supply in the country will be time-consuming and a complex issue; however, effective implementation of policies will help achieve the target and assure the welfare of the whole society in the country.

References

Annual Health Bulletin 2015 (2016) Medical statistical unit, Ministry of Health, Nutrition and Indigenous Medicine, Sri Lanka

Attaman JA, Toth TL, Furtado J, Campos H, Hauser R, Chavarro JE (2012) Dietary fat and semen quality among men attending a fertility clinic. Hum Reprod 27(5):1466–1474. https://doi.org/10.1093/humrep/des065

Azimah A, Azrina A, Norhaizan ME, Mohd Sokhini AM, Daud AZ (2013) Industrially produced trans fatty acids: major potential sources in Malaysian diet. Int Food Res J 20(3):1157–1164

Bajinka O, Touray AO, Oyelakin O (2017) Trans-fat food has more risk to cardiovascular disease than having effects as causative factors of Cancer. J Cancer Clin Trials 2(3):134

Bakker N, van't Veer P, Zock PL, Euramic Study Group (1997) Adipose fatty acids and cancers of the breast, prostate and colon: an ecological study. Int J Cancer 72:587–597

Barnard ND, Bunner AE (2014) Saturated and trans fats and dementia: a systematic review. Neurobiol Aging 35:S65–S73

Bendsen NT, Chabanova E, Thomsen HS, Larsen TM, Newman JW, Stender S, Dyerberg J, Haugaard SB, Astrup A (2011a) Effect of *trans* fatty acid intake on abdominal and liver fat deposition and blood lipids: a randomized trial in overweight postmenopausal women. Nutr Diabetes 1:e4. https://doi.org/10.1038/nutd.2010.4

Bendsen NT, Haugaard SB, Larsen TM, Chabanova E, Stender S, Astrup A (2011b) Effect of *trans*-fatty acid intake on insulin sensitivity and intramuscular lipids—a randomized trial in overweight postmenopausal women. Metabolism 60(7):906–913. https://doi.org/10.1016/j.metabol.2011.01.009

Botelho AP, Santos-Zago LF, Reis SMPM, ACd O (2005) Conjugated linoleic acid supplementation decreased the body fat in Wistar rats. Braz J Nutr 18(4):561–565. https://doi.org/10.1590/S1415-52732005000400011

Brito MS, Villavicencio ALCH, Filho J (2002) Effects of irradiation on *trans* fatty acids formation in ground beef. Radiat Phys Chem 63(3–6):337–340. https://doi.org/10.1016/S0969-806X(01)00647-8

Camp DV, Hooker NH, Lin CJ (2012) Changes in fat contents of US snack foods in response to mandatory *trans* fat labelling. Public Health Nutr 15(6):1130–1137. https://doi.org/10.1017/S1368980012000079

Campbell SJ (2005) Methods and opportunities for reducing or eliminating *trans* fats in foods. Canada

Chatgilialoglu C, Ferreri C, Melchiorre M, Sansone A, Torreggiani A (2014) Lipid geometrical isomerism: from chemistry to biology and diagnostics. Chem Rev 114(1):255–284. https://doi.org/10.1021/cr4002287

Chavarro JE, Rich-Edwards JW, Rosner BA, Willett WC (2007) Dietary fatty acid intakes and the risk of ovulatory infertility. Am J Clin Nutr 85(1):231–237

Chavarro JE, Furtado J, Toth TL, Ford J, Keller M, Campos H, Hauser R (2011) *Trans* fatty acid levels in sperm are associated with sperm concentration among men from an infertility clinic. Fertil Steril 95(5):1794–1797

Chavarro JE, Minguez-Alarcon L, Mendiola J, Cutillas-Tolin A, LopezEspin JJ, Torres-Cantero AM (2014) *Trans* fatty acid intake is inversely related to total sperm count in young healthy men. Hum Reprod 29(3):429–440. https://doi.org/10.1093/humrep/det464

Chowdhury R, Johnson L, Steur M (2014) *Trans* fatty acid isomers in mortality and incident coronary heart disease risk. J Am Heart Assoc. https://doi.org/10.1161/JAHA.114.001195

Churruca I, Fernandez-Quintela A, Portillo MP (2009) Conjugated linoleic acid isomers: differences in metabolism and biological effects. BioFactors (Oxford, England) 35 (1):105–111. https://doi.org/10.1002/biof.13

Damodaran S, Parkin KL (2017) Fennema's food chemistry. CRC Press, Boca Raton

Dashti N, Feng Q, Freeman MR, Gandhi M, Franklin FA (2002) *Trans* polyunsaturated fatty acids have more adverse effects than saturated fatty acids on the concentration and composition of lipoproteins secreted by human hepatoma HepG2 cells. J Nutr 132(9):2651–2659

de Souza R, Mente A, Maroleanu A, Cozma A, Ha V, Kishibe T, Uleryk E, Budylowski P, Schunemann H, Beyene J, Anand S (2015) Intake of saturated and *trans* unsaturated fatty acids and risk of all cause mortality, cardiovascular disease, and type 2 diabetes: Systematic review and meta-analysis of observational studies. BMJ 351:h3978. https://doi.org/10.1136/bmj.h3978

Dhaka V, Gulia N, Ahlawat KS, Khatkar BS (2011) Trans fats—sources, health risks and alternative approach - a review. J Food Sci 48(5):534–541. https://doi.org/10.1007/s13197-010-0225-8

Dijikstra AJ, Hamilton RJ, Hamm W (2008) *Trans* fatty acids. Blackwell Publishing, Oxford. https://doi.org/10.1002/9780470697658

Downs SM, Gupta V, Ghosh-Jerath S, Lock K, Thow AM, Singh A (2013a) Reformulating partially hydrogenated vegetable oils to maximise health gains in India: is it feasible and will it meet consumer demand? BMC Public Health 13:1139–1139. https://doi.org/10.1186/1471-2458-13-1139

Downs SM, Thow AM, Leeder SR (2013b) The effectiveness of policies for reducing dietary *trans* fat: a systematic review of the evidence. Bull World Health Organ 91:262–269H. https://doi.org/10.2471/BLT.12.111468

Downs SM, Singh A, Gupta V, Lock K, Ghosh-Jerath S (2015) The need for multisectoral food chain approaches to reduce *trans* fat consumption in India. BMC Public Health 15:693. https://doi.org/10.1186/s12889-015-1988-7

Downs SM, Bloem MZ, Zheng M, Catterall E, Thomas B, Veerman L, Wu JHY (2017) The impact of policies to reduce *trans* fat consumption: a systematic review of the evidence. Current developments in. Nutrition 1(12). https://doi.org/10.3945/cdn.117.000778

Esmaeili V, Shahverdi AH, Moghadasian MH, Alizadeh AR (2015) Dietary fatty acids affect semen quality: a review. Andrology 3:450–461. https://doi.org/10.1111/andr.12024

Fahy E, Cotter D, Sud M, Subramaniam S (2011) Lipid classification, structures and tools. Biochim Biophys Acta 1811(11):637–647. https://doi.org/10.1016/j.bbalip.2011.06.009

Food Labeling and Advertising Regulations (2015) - Draft regulaiton (under review), Food Control Administration Unit, Ministry of Health, Sri Lanka

Food Safety and Standards Authority of India. (2017). http://www.fssai.gov.in/home/WhatsNewAll.html. Accessed 22 Mar 2018

Ganguly S, Mukhopadhayay S, Biswas S (2012) Preservation of food items by irradiation process. Int J Chem Biochem Sci 1:11–13

Gebauer SK, Chardigny J, Jakobsen MU, Lamarche B, Lock AL, Proctor SD, Baer DJ (2011) Effects of ruminant *trans* fatty acids on cardiovascular disease and cancer: a comprehensive review of epidemiological, clinical, and mechanistic studies. Adv Nutr 2:332–354. https://doi.org/10.3945/an.111.000521

Guillén MD, Cabo N (1997) Infrared spectroscopy in the study of edible oils and fats. J Sci Food Agric 75:1–11

Higgins NW (2007) Low *trans*-stereoisomer shortening systems. United States Patent

Hunter JE, Zhang J, Kris-Etherton PM (2010) Cardiovascular disease risk of dietary stearic acid compared with *trans*, other saturated, and unsaturated fatty acids: a systematic review. Am J Clin Nutr 91(1):46–63. https://doi.org/10.3945/ajcn.2009.27661

Hwang H-S, Singh M, Bakota EL, Winkler-Moser JK, Kim S, Liu SX (2013) Margarine from Organogels of plant wax and soybean oil. J Am Oil Chem Soc 90(11):1705–1712. https://doi.org/10.1007/s11746-013-2315-z

Hwang H-S, Singh M, Winkler-Moser JK, Bakota EL, Liu SX (2016) Preparation of margarines from organogels of sunflower wax and vegetable oils. J Food Sci 79(10):C1926–C1932. https://doi.org/10.1111/1750-3841.12596

Hyseni L, Bromley H, Kypridemos C, O'Flaherty M, Lloyd-Williams F, GuzmanCastillo M, Pearson-Stuttard J, Capewell S (2017) Systematic review of dietary *trans*-fat reduction interventions. Bull World Health Organ 95:821–830G

Innis SM (2006) *Trans* fatty intakes during pregnancy, infancy and early childhood. Atheroscler Suppl 7(2):17–20. https://doi.org/10.1016/j.atherosclerosissup.2006.04.005

Iqbal MP (2014) *Trans* fatty acids – A risk factor for cardiovascular disease. Pak J Med Sci 30 (1):194–197. https://doi.org/10.12669/pjms.301.4525

Jayawardena JMSR, Jayasooriya LJPAP, Kulatilake M, Madhujith T (2014) A study on practices used in preparation of selected snacks and their *trans* fatty acid content. In: Proceedings of Peradeniya University International Research Session (iPURSE) Sri Lanka. University of Peradeniya, p 181

Kaushik I, Grewal RB (2017) *Trans* fatty acids: replacement technologies in food. Adv Res 9 (5):1–14. https://doi.org/10.9734/AIR/2017/33297

Khean KY (2017) The evils of *trans* fat: the Malaysian Medical Gazette. http://www.mmgazette.com/the-evils-of-trans-fat-dr-helmy-hazmi/

King JW, Holliday RL, List GR, Snyder JM (2001) Hydrogenation of vegetable oils using mixtures of supercritical carbon dioxide and hydrogen. J Am Oil Chem Soc 78:107–113

Kohlmeier L, Simonsen N, van't Veer P, Strain JJ, Martin-Moreno JM, Margolin B, Huttunen JK, Navajas JF, Martin BC, Thamm M, Kardinaal AFM, Kok FJ (1997) Adipose tissue *trans* fatty acids and breast cancer in the European Community multicenter study on antioxidants, myocardial infarction, and breast cancer. Cancer Epidemiol Biomark Prev 6(9):705–710

Krenosky S, L'Abbé M, Lee N, Underhill L, Vigneault M, Godefroy S, Ratnayake N (2012) Risk assessment of exposure to *trans* fat in Canada. Int Food Risk Anal J 2:1–15

Laaninen T (2016) *Trans* fats – Overview of recent developments. European Parliamentary Research Service, European Union

Lalvani S, Mondal K (2003) Electrochemical hydrogenation of vegetable oils. Unites States Patent,

Lannes SCS, Ignácio RM (2013) Structuring fat foods. In: Muzzalupo I (ed) Food industry. Intech Open Science. doi:https://doi.org/10.5772/53369

Larqué E, Zamoraa S, Gil A (2001) Dietary *trans* fatty acids in early life: a review. Early Hum Dev: S31–S41

Ledoux M, Juanèda P, Sèbedio JL (2007) *Trans* fatty acids: definition and occurrence in foods. Eur J Lipid Sci Technol 109(9):891–900. https://doi.org/10.1002/ejlt.200600276

Lehnen TE, da Silva MR, Camacho A, Marcadenti A, Lehnen AM (2015) A review on effects of conjugated linoleic fatty acid (CLA) upon body composition and energetic metabolism. J Int Soc Sports Nutr 12(1):36

List G (2014) *Trans* fats replacement solutions in North America. https://doi.org/10.1016/B978-0-9830791-5-6.50017-4

Loor JJ, Bandara ABPA, Herbein JH (2003) Characterization of 18:1 and 18:2 isomers produced during microbial biohydrogenation of unsaturated fatty acids from canola or soybean oil in the rumen of lactating cows. J Anim Physiol Anim Nutr 86(11–12):422–432

Martin CA, Milinsk MC, Visentainer JV, Matsushita M, De-Souza ME (2007) *Trans* fatty acid-forming processes in foods: a review. Annal Braz Acad Sci 79(2):343–350

Micha R, Khatibzadeh S, Shi P, Fahimi S, Lim S, Andrews KG, Engell RE, Powles J, Ezzati M, Mozaffarian D (2014) Global, regional, and national consumption levels of dietary fats and oils in 1990 and 2010: a systematic analysis including 266 country-specific nutrition surveys. BMJ 348:g2272. https://doi.org/10.1136/bmj.g2272

Mozaffarian D, Clarke R (2009) Quantitative effects on cardiovascular risk factors and coronary heart disease risk of replacing partially hydrogenated vegetable oils with other fats and oils. Eur J Clin Nutr 63(suppl 2):S22–S33. https://doi.org/10.1038/sj.ejcn.1602976

Mozaffarian D, Aro A, Willett WC (2003) Health effects of trans-fatty acids: experimental and observational evidence. Eur J Clin Nutr 63:S5–S21. https://doi.org/10.1038/sj.ejcn.1602973

Nestel P (2014) *Trans* fatty acids: are its cardiovascular risks fully appreciated? Clin Ther 36 (3):315–321

Ochiai M, Fujii K, Takeuchi H, Matsuo T (2013) Effects of dietary *trans* fatty acids on fat accumulation and metabolic rate in rat. J Oleo Sci 62(2):57–64

Patel AR, Dewettinck K (2016) Edible oil structuring: an overview and recent updates. Food Funct 7:20–29. https://doi.org/10.1039/c5fo01006c

Pérez-Ferrer C, Lock K, Rivera JA (2010) Learning from international policies on *trans* fatty acids to reduce cardiovascular disease in low- and middle-income countries, using Mexico as a case study. Health Policy Plan 25(1):39–49. https://doi.org/10.1093/heapol/czp040

Puligundla P, Variyar PS, Ko S, Obulam VSR (2012) Emerging trends in modification of dietary oils and fats and health implications - a review. Sains Malaysiana 41(7):871–877

Remig V, Franklin B, Margolis S, Kostas G, Nece T, Street JC (2010) *Trans* fats in America: a review of their use, consumption, health implications, and regulation. J Am Diet Assoc 110 (4):585–592

Resnik D (2010) *Trans* fat bans and human freedom. Am J Bioeth 10(3):27–32. https://doi.org/10.1080/15265160903585636

Revoredo CMS, de Araújo CGB, Silva DFS, Rocha JKD, Libânio JA, Santos ACA (2017) Nutritional implications of *trans* fatty acids on people's health: a reflective analysis article. J Nurs 11(2):731–735. https://doi.org/10.5205/reuol.10263-91568-1-RV.1102201729

Ridker PM, Rifai N, Rose L, Buring JE, Cook NR (2002) Comparison of C-reactive protein and low-density lipoprotein cholesterol levels in the prediction of first cardiovascular events. N Engl J Med 347(20):1557–1565. https://doi.org/10.1056/NEJMoa021993

Risérus U, Willett WC, Hu FB (2009) Dietary fats and prevention of type 2 diabetes. Prog Lipid Res 48(1):44–51. https://doi.org/10.1016/j.plipres.2008.10.002

Ryan JG (2014) No longer "GRAS": the *trans* fatty acids debate. Clin Ther 36(3):312–314

Sadler MJ (2005) *Trans* fatty acids. In: Encyclopedia of human nutrition, 2nd edn. Elsevier, Amsterdam

Salmeron J, Hu FB, Manson JE, Stampfer MJ, Colditz GA, Rimm EB, Willett WC (2001) Dietary fat intake and risk of type 2 diabetes in women. Am J Clin Nutr 73:1019–1026

Silva RR, Rodrigues LBO, Lisboa MM, Pereira MMS, de Souza SO (2014) Conjugated linoleic acid (CLA): a review. Int J Appl Sci Technol 4(2):154–170

Stender S, Dyerberg J (2003) The influence of *trans* fatty acids on health, 4th ed. The report of the Danish Nutrition Council, pp 1–84

Takeuchi H, Sugano M (2017) Industrial *trans* fatty acid and serum cholesterol: the allowable dietary level. J Lipids 2017:10. https://doi.org/10.1155/2017/9751756

Tarrago-Trani MT, Phillips KM, Lemar LE, Holden JM (2006) New and existing oils and fats used in products with reduced *trans*-fatty acid content. J Am Diet Assoc 106(6):867–880

Thompson AK, Minihane A-M, Williams CM (2011) *Trans* fatty acids, insulin resistance and diabetes. Eur J Clin Nutr 65:553–564. https://doi.org/10.1038/ejcn.2010.240

Uauy R, Aro A, Clarke R, Ghafoorunissa R, L'Abbé M, Mozaffarian D, Skeaff M, Stender S, Tavella M (2009) WHO scientific update on *trans* fatty acids: summary and conclusions. Eur J Clin Nutr 63:S68–S75. https://doi.org/10.1038/ejcn.2009.15

Valenzuela BR, Valenzuela BV (2013) Overview about lipid structure. In: Valenzuela BR (ed) Lipid metabolism. https://doi.org/10.5772/52306

Wanders AJ, Zock PL, Brouwer IA (2017) *Trans* fat intake and its dietary sources in general populations worldwide: a systematic review. Nutrients 9:840–853. https://doi.org/10.3390/nu9080840

Wang FC, Gravelle AJ, Blake AI, Marangoni AO (2016) Novel *trans* fat replacement strategies. Curr Opin Food Sci 7:27–34. https://doi.org/10.1016/j.cofs.2015.08.006

Wannamethee SG, Jefferis BJ, Lennon L, Papacosta O, Whincup PH, Hingoran AD (2018) Serum conjugated linoleic acid and risk of incident heart failure in older men: the British regional heart study. J Am Heart Assoc. https://doi.org/10.1161/JAHA.117.006653

Warner K, Neff WE, Byrdwell WC, Gardner HW (2001) Effect of oleic and linoleic acids on the production of deep-fried odor in heated triolein and trilinolein. J Agric Food Chem 49:899–905

WHO (2003) Diet, nutrition and the prevention of chronic diseases. Report of a joint WHO/FAO expert consultation, Geneva 2003. WHO Technical Report Series 916

WHO (2015) Eliminating *trans* fats in Europe: a policy brief

Wijesundera C, Richards A, Ceccato C (2007) Industrially produced *trans* fat in foods in Australia. J Am Oil Chem Soc 84:433–442

Willett WC, Stampfer MJ, Manson JE, Colditz GA, Speizer FE, Rosner BA, Sampson LA, Hennekens CH (1993) Intake of *trans* fatty acids and risk of coronary heart diseases among women. Lancet 341:581–585

Wright AJ, Wong A, Diosady LL (2003) Ni catalyst promotion of a Cis-selective Pd catalyst for canola oil hydrogenation. Food Res Int 36:1069–1072

Wu JHY, Downs S, Catterall E, Bloem M, Zheng M, Veerman L, Barendregt J, Thomas B (2017) Levels of *trans* fats in the food supply and consumption in Australia: expert commentary rapid review brokered by the Sax Institute (www.saxinstitute.org.au) for the National Heart Foundation of Australia. Sax Institute

Zapolska DD, Bryk D, Olejarz W (2015) *Trans* fatty acids and atherosclerosis-effects on inflammation and endothelial function. J Nutr Food Sci 5:426. https://doi.org/10.4172/2155-9600.1000426

Żbikowska A (2010) Formation and properties of *trans* fatty acids – A review. Polish J Food Nutr Sci 60(2):107–114

Zhang J, Adhikari P, Yang T, Xia S, Hu P, Jiang Y, Xu X (2014) *Trans* fats replacement solutions in China. https://doi.org/10.1016/B978-0-9830791-5-6.50020-4

Zupanič N, Hribar M, Kupirovič UP, Kušar A, Žmitek K, Pravst I (2018) Limiting *trans* fats in foods: Use of partially hydrogenated vegetable oils in prepacked foods in Slovenia. Nutrients. https://doi.org/10.3390/nu10030355

Functional Foods and Health

Terrence Madhujith and Nishala Wedamulla

1 Introduction

Discovery of the association between food and health opens up a novel research area to find out the significance of functional components naturally present in food to improve health and quality of life. With this knowledge, consumers began to view foods from a new standpoint, in which diets have been identified as the first line of defense in the prevention of various chronic diseases introducing the concept of "functional foods" (Hasler 2000). Functional foods are defined as foods that, by virtue of the presence of physiologically active components, provide a health benefit beyond basic nutrition (Hasler 2000). However, some definitions of functional foods are more complex: "A food can be regarded as 'functional' if it is demonstrated to affect one or more target functions in the body beneficially, beyond adequate nutritional effects, in a way that is relevant to either an improved state of health and well-being and/or reduction of risk of disease. A functional food must remain food and it must demonstrate its effects in amounts that can normally be expected to be consumed in the diet: it is not a pill or a capsule, but part of the normal food pattern" (Diplock et al. 1999).

The concept of functional food originated in Japan in the 1980s (Kaur and Das 2011). However, this concept dates back to thousands of years or more. Many types of foods in ancient China, Japan, and other Asian countries have been associated with specific health benefits. In comparison with those, in western societies, the renowned global brand Coca-Cola® was first introduced as a functional food at the beginning of the twentieth century (Weststrate et al. 2002). Then, the concept of

T. Madhujith (✉)
Department of Food Science and Technology, Faculty of Agriculture, University of Peradeniya, Peradeniya, Sri Lanka

N. Wedamulla
Department of Export Agriculture, Faculty of Animal Science and Export Agriculture, UvaWellassa University, Badulla, Sri Lanka

R. P. De Silva et al. (eds.), *Agricultural Research for Sustainable Food Systems in Sri Lanka*, https://doi.org/10.1007/978-981-15-3673-1_14

foods for specified health uses (FOSHU) was established in 1991. These foods were included as one of the four categories defined in the "Nutrition Improvement Law" as "Foods for special dietary use," i.e., food that are used to improve personal health which were allowed to carry health claims. Moreover, FOSHU were assumed to be consumed as part of a regular diet and also expected to be in the form of ordinary foods, not as pills or capsules (Diplock et al. 1999).

Designer foods, functional foods, fortified foods, and nutraceuticals are some of the well-known terms that appear in the food- and nutrition-related literature. Designer foods refer to foods that are designed to provide some health benefits other than their traditional nutritional value (Rajasekaran and Kalaivani 2013). Designer foods or functional foods are also defined as foods similar in appearance to conventional food consumed as part of the usual diet which contains biologically active components with demonstrated physiological benefits and offers the potential of reducing the risk of chronic diseases beyond basic nutritional functions (FAO 2004). Moreover, fortified foods refer to the food fortified or enriched with nutrient content already present in them or other complementary nutrients (Rajasekaran and Kalaivani 2013). In the literature, nutraceuticals are defined as any substance that may be considered as food or part of food and provides medical or health benefits, including the prevention and treatment of diseases. Further, those products range from isolated nutrients, dietary supplements, and diets to genetically engineered "designer" foods, herbal products, and processed foods such as cereals, soups, and beverages (Keservani et al. 2010). Thus, similarities and dissimilarities in the meanings of aforementioned terms are not clearly defined.

2 Lifestyle Changes and Emergence of Noncommunicable Diseases

Lifestyle and diet patterns were strongly affected by industrialization, urbanization, and globalization. Urbanization resulted in a decline of undernutrition, however, meanwhile giving rise to inappropriate dietary patterns and thereby decreased physical activity which ultimately lead to the spread of diet-related disorders including obesity, diabetes, cardiovascular diseases, hypertension, osteoporosis, and cancer (Lajolo 2002). Accordingly, high salt intake, high saturated and *trans*-fatty acid intake, low fruit and vegetable consumption, and insufficient physical activity are identified as key risk factors for noncommunicable diseases (NCDs) (Lachat et al. 2013). Further, it is reported that consumption of foods containing high saturated fats and industrially produced trans fats, salt, and sugar is the cause of at least 14 million deaths or 40% of all deaths every year from NCDs (Wagner and Brath 2012). In addition, NCDs are identified as the leading cause of global deaths. These diseases were responsible for 38 million (68%) of the world's 56 million deaths in 2012 (WHO 2014). This emergence of NCDs in low- and middle-income countries is projected to be a significant contributor to the overall global burden of disease (Townsend et al. 2017).

Though Sri Lankan public health system is acknowledged internationally for its significant achievements in mitigating communicable diseases (Somanathan et al. 2000), still Sri Lanka has fallen behind in improving its nutritional status (Ministry of Health 2010). The lifestyle changes in Sri Lanka along with the marketing strategies adopted by food manufacturers, convenience factor, and peer pressure lead to unbalanced dietary patterns which ultimately results in nutritional deficiencies and other diet-related NCDs such as diabetes mellitus, cardiovascular diseases, hypertension, and certain types of cancer (Ministry of Health, Sri Lanka, 2010). Furthermore, high selling prices of fruits and vegetables, scarcity of good-quality produce, and safety concerns arising from the misuse of pesticides and the use of improper methods for storage and artificial ripening have led to low intake of fruits and vegetables. This creates an urgent need of producing foods with enhanced functionalities to overcome these challenges. In this backdrop, functional foods offer an excellent opportunity to uplift the current health status as functional foods contain biologically active compounds associated with physiological health benefits for the prevention and management of chronic diseases (Alkhatib et al. 2017).

3 Types of Functional Foods

3.1 Whole Foods as Functional Foods

Whole foods such as fruits and vegetables are considered as the simplest form of functional foods. These are rich in bioactive phytochemicals, mainly polyphenols and carotenoids which act against oxidative damage owing to their antioxidant properties, thereby reducing the risk of developing certain types of cancer (Daya et al. 2009). Furthermore, recent studies reveal that beneficial effects of a diet rich in fruits and vegetables are attributed to the complex mixture of phytochemicals present in the whole foods as additive and the synergistic effect of phytochemicals in fruits and vegetables is responsible for potent antioxidant and anticancer properties (Liu 2003). Besides, whole grains also exhibit health-promoting properties due to the chemical constituents present in them, viz., dietary fiber, inulin, β-glucan, resistant starch, carotenoids, phenolics, tocotrienols, and tocopherols (Borneo and León 2012). Moreover, nuts such as hazelnut, walnut, and almonds also contain chemical components beneficial to health which include phenolic acids, flavonoids, proanthocyanidins, carotenoids, alkaloids, and polyunsaturated fatty acids, among others (Alasalvar and Shahidi 2008). There are more than 60 different fruit species identified in Sri Lanka. Of them, there are many native forest fruits which rapidly disappear due to clearing of forests, the use of trees for timber and firewood, and urbanization. Most of these native fruit species are rich in phytochemicals; however, unfortunately they are not regularly consumed by many Sri Lankans. Pomegranate, one of the fruits rich in antioxidants, was grown in the backyard of many houses in Sri Lanka decades ago; however, its presence is very limited at present. Due to the less availability of these phytochemical-rich fruits, consumers solely depend on 10–15 common fruits. Moreover, the high prices of fruits, less availability of good-quality fruits, and safety concerns, among others, have significantly contributed to the low consumption of fruits in Sri Lanka. Despite the WHO recommendation of 200 g of fruits per day, an average Sri Lankan consumes very little amount of fruits (Bopitiya and Madhujith 2012; Piyathunga et al. 2016).

3.2 Plant-Derived Functional Foods

Bioactive compounds synthesized in plants are classified as primary and secondary metabolites, in which secondary metabolites play a crucial role in plant well-being and exhibit antibiotic, antifungal, and antiviral properties (Lavecchia et al. 2013). Primary metabolites include plant proteins, β-glucans, and omega-3 fatty acids, while secondary metabolites include phytoestrogens, antioxidants, vitamins, tocopherols, steroids, γ-linolenic acid, and phase II enzyme inducers (Vattem and Maitin 2016). Traditionally used food grains, millets, fruits, and vegetables are identified as rich sources of functional ingredients. Food grains and cereals, viz., rice, wheat, maize, sorghum, millets, and oats, are important sources of micronutrients such as folates, phenolic acids, vitamin E, zinc, iron, selenium, copper, manganese, carotenoids, betaine, choline, sulfur amino acids, and dietary fiber (Das et al. 2012). Moreover, (−)-epicatechin, (−)-epicatechin-3-gallate, (−)-epigallocatechin, and (−)-epigallocatechin-3-gallate are bioactive constituents present in green tea which exhibit strong antioxidant properties (Prakash et al. 2017).

3.3 Animal-Derived Functional Foods

Animals serve as rich sources of functional foods as they have bioactive compounds that offer a variety of biological effects on human health. Animal-derived polyunsaturated fatty acids which include omega-3 and omega-6 fatty acids and conjugate linoleic acid, bioactive peptides, glucosamine, and chondroitin sulfate are few well-known examples of animal-derived functional foods (Zhang et al. 2014; Vattem and Maitin 2016). The major polyunsaturated fatty acids are α-linolenic acid (ALA), eicosapentaenoic acid (EPA), and docosahexaenoic acid (DHA). The main sources of EPA and DHA are fatty fish and ruminant milk, while meat serves as the major source of conjugate linoleic acid (Pisulewski and Kostogrys 2003). Many clinical and epidemiological studies have proven the beneficial role of omega-3 fatty acids in the prevention of cancer and cardiovascular diseases, while recent studies have shown the positive role of these fatty acids in mitigating various mental disorders (Riediger et al. 2009). Furthermore, conjugated linoleic acids are also reported to impart anticarcinogenic, antiatherogenic, antidiabetogenic, and immune-modulating properties (Rainer and Heiss 2004). Moreover, milk proteins are identified as one of the most important sources of bioactive peptides, and these milk-derived bioactive peptides play a vital role in health-promoting functional foods. Recent research reveals that bioactive milk peptides are beneficial in decreasing the risk of obesity and the development of type 2 diabetes apart from their beneficial effect on the heart, bones, and the digestive system (Bhat et al. 2015).

3.4 Fortified, Enriched, and Enhanced Foods

Foods are fortified by adding nutrients (fortificants or additives) to the food of interest, which serves as a vehicle for carrying the nutrient. With the tremendous rise of global diseases, there is a growing trend in food fortification with functional ingredients (Ansari and Kumar 2012). Grain products such as flour, breakfast cereals, and fruit juices fortified with folic acid and vitamin C are few examples of fortified foods. Though food fortification is considered as an effective and economical way of improving nutrient quality and provides benefits to consumers, still there are few problems associated with fortified foods in relation to food composition (Spence 2006). Margarine fortified with vitamins A and D has been in the Sri Lankan market for a few decades. Wheat flour fortified with iron and folic acid was introduced into the Sri Lankan market by one of the three wheat processors operating in the country nearly a year ago. The need for introducing iron and folic acid foods into the local market has been a long-felt need in the country. The Ministry of Health prepared guidelines for the fortification of foods with iron and folic acid in 2008; however, the project is yet to be implemented. The Ministry of Health has been working on introducing rice fortified with iron and folic acid into the local market with the assistance of the UN World Food Programme.

Enriched foods are another type of functional foods in which additional components that are not naturally present in required quantities are added to a particular food. Orange juice with added calcium (Spence 2006) and margarine with added plant stanols which exhibit cholesterol-lowering effect are few examples of enriched products (Law 2000). High-lysine corn and fruit and vegetables with enhanced content of vitamins and phytonutrients are few examples of enhanced foods in which nutrient composition is altered by changing the raw commodities. Golden rice and carotenoid-rich potatoes are the best known examples in which functional components are embedded into food plants that do not normally produce those dietary components (Spence 2006).

3.5 Fermented Foods

Fermentation process alters the chemical nature of foods as well as the efficacy of some bioactive constituents. Alcoholic beverages such as beer and wine, vinegar, yoghurt, pickles, soy sauce, and various fermented bean products are the most popular among other fermented products. Tea which is also a fermented product provides a myriad of health benefits. The positive relationship between tea and health is attributed to catechins and their derivatives present in tea (Higdon and Frei 2003). Moreover, yeast fermentation has been introduced recently to enhance the nutritive and therapeutic value of black tea (Pasha and Reddy 2005). Red wine is a rich source of polyphenols such as catechins, stilbenes, anthocyanins, and proanthocyanidins, and these polyphenols are responsible for the health benefits of red wine against many diseases (Shahidi 2009). Beer is the other alcoholic beverage which also exhibits health-promoting properties. The functional components of beer include phenolics, B vitamins, minerals such as selenium, and possibly soluble fiber. Vinegar is another fermented product produced by the fermentation of sugar-rich coconut sap, fruit juices, or similar materials which also claim to possess health-promoting properties. Vinegar that has been used as a preservative and condiment in the past is now considered as a potential functional food due to the presence of phytochemicals in it (Shahidi 2009).

3.6 Functional Beverages

Beverages are popular around the world due to their variety of pleasant taste and sensation. Besides, there is a growing trend of consuming beverages for health reasons. Moreover, constituents present in some beverages exhibit anticarcinogenic, antioxidative, or antimutagenic effects. Polyphenols and flavonoids present in green and black tea exert inhibitory effects on cancer and have antimutagenic and anticlastogenic properties (Zbakha and Abbassib 2012). Different types of commercially available beverages can be grouped into three categories: dairy-based beverages including probiotics and minerals/ω-3-enriched drinks; vegetable and fruit beverages; and sports and energy drinks. Fresh milk, fermented milk, and yoghurt drinks are the most common products of dairy-based beverages which are considered as excellent vehicles for probiotics. Considering the drawbacks associated with milk-based functional beverages, novel products containing probiotics have been launched recently based on fruits, vegetables, cereals, and soybeans. Cranberry, blueberry, pomegranate, apple, blackcurrant, grapes, cherries, kiwifruits, strawberries, feijoa, peach, and plums are the most common fruits used in commercial preparation. Moreover, some nonprobiotic fruit and vegetable beverages with added vitamins and minerals (vitamin D or calcium) are also available in functional food markets (Corbo et al. 2014). Although the consumption of fruit juices such as apple, orange, grape, and berry juices is a common practice in many developed countries, the consumption of fruit juice is quite uncommon in Sri Lanka. Processed fruit juices available in supermarkets are expensive; thus, many do not regularly consume fruit juices in Sri Lanka.

4 Health and Nutritional Benefits of Functional Foods

4.1 Oxidative Stress

Reactive oxygen species (ROS) are believed to be a causative factor for carcinogenesis as ROS are involved in various biological oxidation reactions that lead to DNA and RNA damage (Waris and Ahsan 2006). ROS are radicals, ions, or molecules that

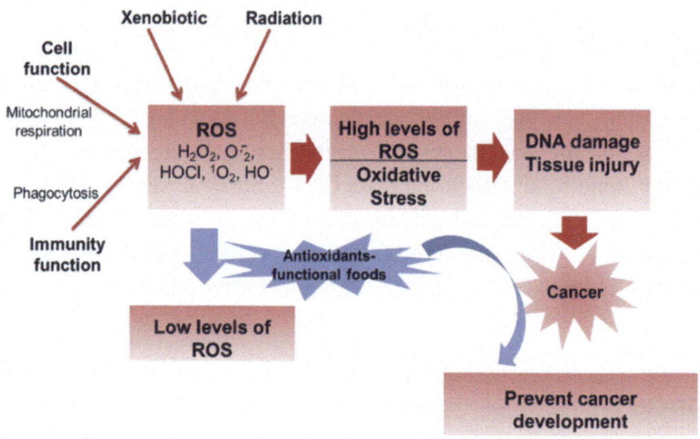

Fig. 1 Illustration depicting the relationship among reactive oxygen species (ROS), antioxidants, and carcinogenesis

are highly reactive; these include superoxide ($O_2^{\bullet-}$), hydroxyl radical ($^{\bullet}OH$), nitric oxide ($NO^{\bullet}$), hydrogen peroxide (H_2O_2), and singlet oxygen ($^1O_2$), among others (Liou and Storz 2010). Figure 1 illustrates the relationship between ROS, cancer, and antioxidants. The mitochondrial respiratory chain (electron transport complexes) is the major source of ROS generation in living cells (Nohl et al. 2003). Phagocytes such as neutrophils and macrophages also produce ROS during phagocytosis (Forman and Torres 2002). When the critical balance between free radical generation and antioxidant defenses is unfavorable, oxidative stress results which is characterized by high levels of ROS (Carocho and Ferreira 2013). Macromolecules such as DNA, proteins, and lipids are subjected to nonspecific damage by elevated levels of ROS. Damage to DNA by ROS has been identified as a major cause of cancer (Waris and Ahsan 2006). Moreover, ROS have been implicated in the induction and complications of diabetes mellitus, age-related eye disease, and neurodegenerative diseases such as Parkinson's disease (Rao et al. 2006).

Antioxidants are defined as any substance that directly scavenges ROS or indirectly acts to upregulate antioxidant defenses or inhibit ROS production (Khlebnikov et al. 2007). These molecules react readily with ROS and inhibit oxidative mechanisms that lead to degenerative diseases (Sre et al. 2012). Thus, dietary antioxidants play a crucial role as functional foods in the management of human diseases (Lobo et al. 2010).

4.2 Cardiovascular Diseases

Cardiovascular diseases (CVD) are one of the leading causes of global deaths, thus becoming a growing health concern. Dietary factors are also equally important in the pathogenesis of CVD. It has been proven that CVD can be prevented by lifestyle changes and improved dietary practices (Stampfer et al. 2000). Functional foods comprise physiologically active components either from plant or animal sources which are capable of reducing heart disease risk. Moreover, functional foods are supposed to exert this cardioprotective effect principally through lipid-lowering effects, antioxidant actions, and/or decreased homocysteine levels. Long-chain omega-3 fatty acids, dietary fibers, and phytochemicals are extensively studied to investigate their beneficial effects against CVD. Several studies have proven that high intake of dietary fish and fish oil supplements is involved in reducing the risk of CVD. Further, influx of literature available to report the beneficial effect of diets rich in fruits and vegetables on CVD risk (Alissa and Ferns 2012).

4.3 Cancer

Following CVD, cancer is the second leading cause of death in the world. As reported by the American Cancer Society, cancer statistics 2015 showed that approximately 1.66 million new cases of cancer are going to be diagnosed and nearly 589,430 mortalities from cancer are estimated to occur in the USA in the year 2015 (American Cancer Society 2015). Vegetables, medicinal herbs, and their extracts or components are used by Americans, Japanese, Chinese, and Europeans, among others, to prevent or treat cancer. Antioxidative compounds naturally present in plant and their extracts exhibit potential chemopreventive properties. These phytochemicals include carotenoids, carotenoid pigments, xanthine, lycopene, and astaxanthin, among others. ROS are believed to be a causative factor in carcinogenesis as ROS are involved in various biological oxidation reactions including DNA damage. Antioxidants are able to reduce free radical damage to DNA which is believed to be the root cause of various cancers (Aghajanpour et al. 2017).

4.4 Obesity

Obesity arises as a result of positive energy balance in the body. Obesity is a multifactorial condition which is influenced by genetic, environmental, behavioral, lifestyle, cultural, and metabolic factors. Moreover, this leads to development of cancer, hypertension, diabetes, dyslipidemia, atherosclerosis, and others. Thus, the alarming rate of increase in obesity needs urgent attention (Sunkara and Verghese 2014).

Since the link between nutrition, biological responses, and diseases is clearly established, functional foods can be effectively utilized to prevent or treat obesity (Riccardi et al. 2005). Phenolic compounds, tannins, and anthocyanins abundantly present in tea, coffee, cocoa, berries, and peas exhibit in vitro lipase inhibition activity. Di-caffeoylquinic acids are coffee polyphenols which exhibit more potent inhibitory activity on digestive enzymes (maltase, sucrase, and lipase), thus playing a vital role in postprandial hyperglycemia and hyperinsulinemia (Sunkara and Verghese 2014).

4.5 Neurodegenerative Diseases

Neurodegenerative diseases are age-related diseases characterized by memory loss, cognitive dysfunction, neuronal damage, and death. They are identified as a major threat to human health especially during the later stages of life. Alzheimer's disease, Parkinson's disease, and Huntington disease are some of the widely present neurodegenerative diseases. Association between age-related diseases and consumption of functional foods is well established. Foods originating from plant sources serve as rich sources of bioactive phytochemicals; thus, their consumption could prevent chronic degenerative diseases. The surge of research carried out in the recent past has confirmed that functional foods such as fruits and vegetables, spices, nuts, and cereals, among others, are potential candidates for preventing neurodegenerative diseases as these foods are rich in a wide array of phytochemicals such as polyphenols, alkaloids, carotenoids, and anthocyanins. These phytochemicals are capable of improving cognitive function, learning, general brain functions, and well-being (Olasehinde et al. 2017).

5 Different Classes of Functional Foods

5.1 Plant Phenolics

Polyphenols are secondary plant metabolites which are involved in the defense mechanism of plants against pests and microorganisms. Polyphenols have drawn much attention from scientific communities owing to their antioxidant properties and potential role in the prevention of various diseases linked to oxidative stress, such as cancer, cardiovascular diseases, and neurodegenerative diseases. These compounds are classified into different groups considering the number of phenol rings that they contain and the structural elements that bind rings to one another. The phenolic groups include phenolic acids, flavonoids, stilbenes, and lignans (Manach et al. 2004). The chemical structures of different phenolic groups are illustrated in Fig. 2.

Derivatives of benzoic and cinnamic acid are the two major classes of phenolic acids. Edible plants generally contain low levels of hydroxybenzoic acid; however, certain red fruits, black radish, and onions contain high concentrations of hydroxybenzoic acid. Tea also is an important source of gallic acid. Hydroxycinnamic acids are more abundant than hydroxybenzoic acids and mainly comprise *p*-coumaric, caffeic, ferulic, and sinapic acids. The bound forms of glycosylated derivatives or esters of quinic acid, shikimic acid, and tartaric acid are more common than free forms. Chlorogenic acid is formed by the combination of caffeic and quinic acids and is found in high concentrations in coffee and many types of fruits (Manach et al. 2004). Hydroxycinnamic acids are found in all parts of fruits, while ferulic acid is present in cereal grains (Gharras 2009).

Flavonoids are classified into six subclasses: flavonols, flavones, isoflavones, flavanones, anthocyanidins, and flavanols (Gharras 2009). The richest sources of flavonols are onions, curly kale, leeks, broccoli, and blueberries. Besides red wine,

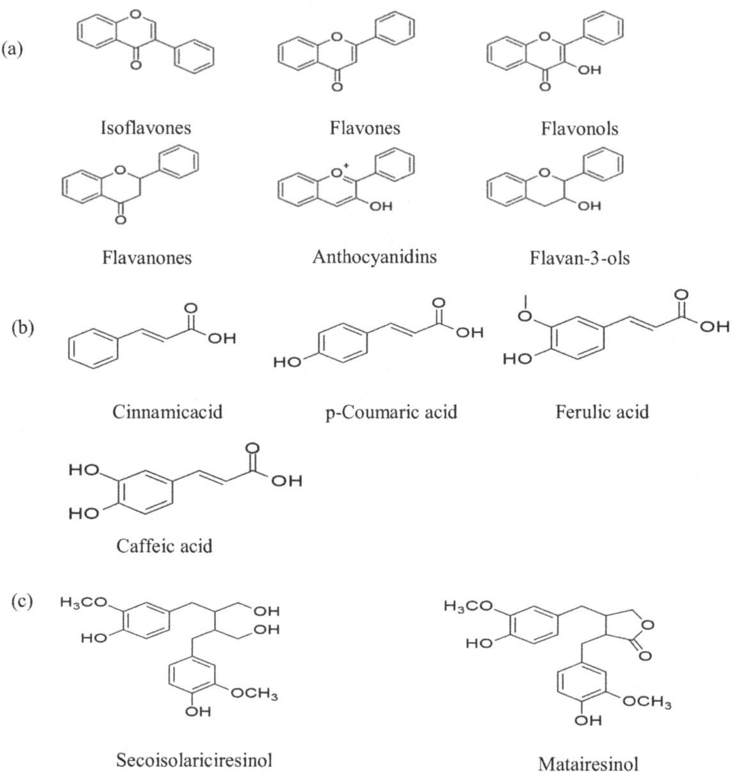

Fig. 2 The chemical structures of (**a**) flavonoid subclasses, (**b**) phenolic acids, (**c**) lignans

tea contains considerable amounts of flavonols. These compounds exist in glycosylated forms, and fruits often contain between 5 and 10 different flavonol glycosides (Manach et al. 2004). Flavones are less common and mainly consist of glycosides of luteolin and apigenin. Cereals such as millet and wheat contain C-glycosides of flavones, while the skin of citrus fruit contains high amounts of polymethoxylated flavones, viz., tangeretin, nobiletin, and sinensetin. Tomatoes and some aromatic plants such as mint contain flavanones. However, only citrus contains high concentrations. Isoflavones are exclusively found in leguminous plants. Thus, soya and its processed products are identified as the major source of isoflavones in the diet. Catechins and proanthocyanidins are monomeric and polymeric forms of flavanols, respectively. Catechins are present in many types of fruits, tea, and red wine. Green tea is considered the richest source of catechins. Tea also contains gallocatechin, epigallocatechin, and epigallocatechin gallate. Anthocyanins are pigments dissolved in the vacuolar sap and found in red wine, certain varieties of cereals, and certain leafy and root vegetables (aubergines, cabbage, beans, onions, radish); however, these pigments are abundantly present in fruits (Manach et al. 2004).

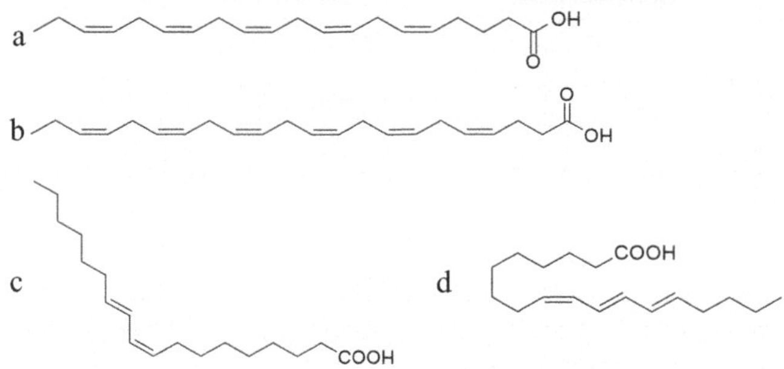

Fig. 3 Chemical structures of (**a**) eicosapentaenoic acid (EPA), (**b**) docosahexaenoic acid (DHA), (**c**) c-9,t-11 conjugated linoleic acid (CLA) isomer, and (**d**) α-eleostearic acid

The major dietary source of lignans is linseed which contains secoisolariciresinol and less quantities of matairesinol. Traces of these lignans are also present in other cereals, grains, fruits, and certain vegetables. However, very low quantities of stilbenes are present in human diets (Manach et al. 2004).

5.2 Omega-3 Fatty Acids

Omega-3 fatty acids are a group of polyunsaturated fatty acids such as eicosapentaenoic acid (EPA) and docosahexaenoic acid (DHA) (Fig. 3) which are the most extensively studied polyunsaturated fatty acids. They are also known as highly unsaturated fatty acids (HUFA) in the new nomenclature. The positive effects of these two fatty acids on cardiovascular diseases are widely acknowledged (Lee et al. 2009). Marine sources such as fish and fish oils and algae serve as sources of EPA and DHA, while dietary oils derived from both plant and animal sources serve as sources of α-linolenic acid (ALA) (Ganesan et al. 2014).

EPA and DHA possess diverse biological effects compared to other fatty acids owing to their longer chain length and the high number and positioning of double bonds. EPA and DHA are capable of altering the structure and function of cell membrane. Furthermore, omega-3 fatty acids and their derivatives play a vital role in immune and inflammatory response. In relation to human health, omega-3 fatty acids aid in improving cognitive development, learning, and visual development in the early stages of life. Moreover, EPA and DHA are involved in decreasing the risk of depression and suicide and delaying the onset of the neurological degeneration of aging. The surge of research in the area of cardiovascular diseases also demonstrates the beneficial effect of omega-3 fatty acids on cardiac health. Consumption of omega-3 fatty acids correlates with reduced cardiac death from arrhythmias, providing antithrombotic effect, and having anti-inflammatory and antihypertensive effects (Deckelbaum and Torrejon 2012). Identifying the significance of DHA in infant nutrition, infant formula manufacturers have started adding DHA into their products since the early 2000s. DHA-enriched infant formulae started to arrive in the

Sri Lankan market after 2005. Babies fed on breast milk get adequate amounts of DHA.

5.3 Conjugated Linoleic Acid (CLA) and Other Conjugated Fatty Acids

Conjugated linoleic acid is a collective name for a mixture of positional and geometric isomers of C 18:2 fatty acids having conjugated double bonds (Eynard and Lopez 2003). Microbial isomerization of dietary linoleic acid results in the synthesis of CLA. These fatty acids are synthesized in the rumen during biohydrogenation of lipids in feed (Blankson et al. 2000). Thus, animal products from ruminants serve as a principal dietary source of CLA. Moreover, ruminant meat contains significantly high amounts of CLA than nonruminant meat: veal (2.7 mg CLA/g fat) has the lowest, while lamb (5.6 mg CLA/g fat) has the highest. Seafood contains less quantities of CLA contrary to dairy products such as milk, butter, and yoghurt which contain considerable amounts of CLA. Natural cheese also serves as a good source of CLA. However, cheese ripened for more than 10 months has shown a low CLA content (Chin et al. 1992). Furthermore, plant oils possess comparatively low levels of CLA which range from 0.1 mg CLA/g fat in coconut oil to 0.7 mg CLA/g fat in safflower oil. Animal (75% of the total CLA) and dairy (90% of the total CLA) products contain considerably high amounts of the c-9,t-11 CLA isomer which is believed to be a biologically active form, while plant oils contain less than 50% of the total CLA (Chin et al. 1992).

The influx of literature has proven the beneficial effects of CLA on various metabolic disorders in humans owing to their anticarcinogenic, anti-obese, antidiabetic, and antihypertensive properties (Koba and Yanagita 2014). In relation to anticarcinogenic effects, it has been reported that CLA inhibits the initiation and incidence of mammary tumors in rodents, and it also interferes with the growth of breast and prostate tumors. Numerous clinical studies have confirmed the anti-obese and hypolipidemic effects of CLA on animals using mice, rats, and pigs (Koba and Yanagita 2014).

5.4 Eleostearic Acid

Besides conjugated linoleic acid, there are some other conjugated beneficial fatty acids. Seed oils of certain plants contain conjugated triene fatty acids such as α-eleostearic acid. The seed oil of bitter gourd (*Momordica charantia*) contains 60% α-eleostearic acid, while its flesh contains a small amount of α-eleostearic acid (Tsuzuki et al. 2004a).

Previous studies on α-eleostearic acid reported it to have anticancer properties. Further studies reported that α-eleostearic acid induced apoptosis via lipid peroxidation (Tsuzuki et al. 2004b). Moreover, DLD-1 colorectal adenocarcinoma cells treated with α-eleostearic acid in vitro were growth inhibited and underwent DNA laddering indicative of apoptosis. Further, Caco-2 and HT-29 colon cancer cells demonstrated decreased viability and increased DNA fragmentation when treated with α-eleostearic acid (Grossmann et al. 2009).

5.5 Dietary Fiber

A variety of plant substances that are resistant to digestion by human gastrointestinal enzymes are collectively known as dietary fiber. Based on water solubility, dietary fiber is classified into two groups: structural or matrix fibers (lignins, celluloses, and some hemicelluloses) which are insoluble and natural gel-forming fibers (pectins, gums, mucilages, and the remainder of the hemicelluloses) which are soluble (Brown et al. 1999). Plants, vegetables, cereal grains, woody plants, fruits, legumes, and leguminous plants are identified as principal sources of dietary fiber (Meyer 2004).

A significant role of dietary fiber in the prevention of several diseases is well acknowledged. Previous studies have proven the effect of fiber polysaccharides on the absorption of lipids; these fibers act as strong inhibitors of the pancreatic lipase that participates in lipid metabolism. Moreover, dietary fibers decrease the levels of total cholesterol and low-density lipoproteins in plasma. A high volume of literature is available to confirm the protective effect of dietary fiber on cardiovascular diseases, diverticulosis, constipation, irritable colon, colon cancer, and diabetes (Rodríguez et al. 2006).

5.6 Phytosterols and Phytostanols

Phytosterols, phytostanols, and their esters are natural constituents of plant cell membrane which belong to a group of steroid alcohols. Phytosterols are structurally similar to cholesterol. The most prominent plant sterols in foods are β-sitosterol, campesterol, and stigmasterol. Plant stanols are the saturated form of plant sterols. Thus, the saturation or absence of double bond in the sterol ring results in sitostanol, campestanol, or stigmastanol (Bacchetti et al. 2011).

Vegetable products such as vegetable oil, fruits, nuts, grains, and pulses are the principal sources of phytosterols. The level of phytosterols in vegetables ranges between 1.1 and 53.7 mg/100 g of edible portion with the highest concentration of phytosterols found in pea, cauliflower, broccoli, and romaine lettuce. On the other hand, fruits contain 1.6–32.6 mg/100 g of phytosterols, and navel orange, tangerine, and mango exhibit the highest concentration. However, the phytosterol content is modified by ripening and postharvest practice in fruits and vegetables (Bacchetti et al. 2011). Moreover, phytosterol content in nuts ranges from 95 to 280 mg/100 g (Chen and Blumberg 2008). Conversely, cereals such as corn, wheat, rice, and triticale serve as the main sources of phytostanols and their ester (Bacchetti et al. 2011).

The association between diet rich in fruits and vegetables and many health implications are well documented. Many epidemiological studies have revealed the inhibitory effect of phytosterols on the uptake of both dietary and endogenously produced (biliary) cholesterol from intestinal cells, leading to lowered serum total and LDL cholesterol levels (AbuMweis et al. 2008). In addition, studies reveal that plant sterols prevent hyperproliferation of vascular smooth muscle cell which plays a vital role in atherosclerosis development. Further, recent studies have proven that plant sterol could exert a modulator effect against oxidative damage (Bacchetti et al. 2011). Moreover, a surge of studies suggest a protective role of phytosterols, especially β-sitosterol, on colon, prostate, and breast cancers (Awad et al. 2000).

5.7 Probiotics and Prebiotics

Probiotics are defined as viable microbial dietary supplements that beneficially affect the host through their effects in the intestinal tract, while prebiotics are defined as "nondigestible food ingredients that beneficially affect the host by selectively stimulating the growth and/or activity of one or a limited number of bacteria in the colon" (Roberfroid 2000). *Lactobacillus* and *Bifidobacterium* are the bacterial genera that are most often used as probiotics. These probiotics are available in fermented dairy products such as yoghurt or freeze-dried cultures; however, fermented vegetables and meats also contain probiotics. The most widely acknowledged prebiotics are galactooligosaccharides, fructooligosaccharides, and inulin in which galactooligosaccharides are nondigestible and are derived from lactose. Inulin and inulin-type fructans are acknowledged as soluble dietary fibers (Roberfroid 2005). In addition, dietary fiber contains several non-starch polysaccharides which include cellulose, dextrins, pectins, β-glucans, waxes, and lignins. Asparagus, chicory, tomatoes, and wheat are well known as natural sources of prebiotics (Al-Sheraji et al. 2013).

The health benefits of probiotics and prebiotics are well established. Probiotics improve lactose digestion by reducing intolerance symptoms as well as by slowing orocecal transit. A strain of *Lactobacillus casei* supplemented in infants is reported to enhance the concentrations of circulating immunoglobulin A which correlates with a shortened duration of rotavirus-induced diarrhea. Further, the consumption of *L. acidophilus* and *Bifidobacterium bifidum* significantly enhances the nonspecific immune phagocytic activity of circulating blood granulocytes. However, rotavirus-induced diarrhea and colon cancer are the only maladies for which there is evidence of disease reduction from probiotic consumption (Roberfroid 2000). Jeygowri and others (2015) studied the isolation of potentially probiotic microorganisms from fermented rice and suggest for its utilization as potential probiotic starter cultures in the food industry. In relation to prebiotics, studies have reported that genotoxic enzyme activity has decreased upon the administration of prebiotics. On the other hand, many studies have demonstrated the effect of fructooligosaccharides on mineral absorption. Moreover, prebiotics may also have an effect on lipid regulation (Al-Sheraji et al. 2013).

5.8 Mushrooms as Functional Foods

Mushrooms have drawn a considerable attention as a functional food owing to their antioxidant, antitumor, and antimicrobial properties. However, mushrooms that are both edible and have functional properties are limited (Chang 1996). Mushrooms serve as a good source of nutrients as well as a rich source of bioactive substances. Mushrooms contain secondary metabolites such as acids, terpenoids, polyphenols, sesquiterpenes, alkaloids, lactones, sterols, metals, chelating agents, nucleotide analogs and vitamins, glycoproteins, and polysaccharides, mainly β-glucans (Kumar 2015). Besides, new proteins with biological activities have also been found in mushrooms: lignocellulose-degrading enzymes, lectins, proteases and protease inhibitors, ribosome-inactivating proteins, and hydrophobins (Erjavec et al. 2012).

Recent work on functional mushrooms has demonstrated their beneficial effects which include modulating the immune system, lowering blood pressure and blood lipid concentration, and inhibiting tumors, inflammation, and microbial action (Chang 1996). Patel and Goyal (2012) reveal that mushrooms that exhibit anticancer properties belong to the genera *Phellinus*, *Pleurotus*, *Agaricus*, *Ganoderma*, *Clitocybe*, *Antrodia*, *Trametes*, *Cordyceps*, *Xerocomus*, *Calvatia*, *Schizophyllum*, *Flammulina*, *Suillus*, *Inonotus*, *Inocybe*, *Funlia*, *Lactarius*, *Albatrellus*, *Russula*, and *Fomes*. Moreover, anticancer compounds play a vital role as reactive oxygen species induce, mitotic kinase inhibitor, antimitotic agent, angiogenesis inhibitor, and topoisomerase inhibitor (Kumar 2015).

5.9 Spices as a Source of Functional Ingredients

Spices and aromatic herbs have long been identified as preservatives, colorants, and flavor enhancers and serve as the basis of traditional medicine in many countries (Viuda-Martos et al. 2010). Many studies have proven the antioxidant, antibacterial, anti-inflammatory, antiviral, and anticancerigenic properties of these substances. These functional properties of spices are attributed to the presence, type, and concentration of phenolic compounds. On the other hand, essential oils present in spices also contribute to functional properties (Viuda-Martos et al. 2010).

Many studies have proposed several mechanisms for the anti-inflammatory activity of spices. The study of Mani and others (2006) reported that the anti-inflammatory capacity of spices is basically due to the presence of flavonoids that inhibit the development of inflammation triggered by a variety of agents. Most interestingly, few studies suggested that the compounds present in spices such as quercetin, curcumin, and silymarin have the same anti-inflammatory effects as indometacin (a nonsteroidal drug) (Volate et al. 2005). Further, spices and their extracts are considered as rich sources of natural antioxidants and thus exert a high degree of antiplatelet activity which leading to lowered cardiovascular disease incidences. Moreover, studies revealed that the main components present in rosemary, carnosol and carnosic acid, show inhibitory activity against human immunodeficiency virus (HIV). Besides, rosemary and its components, ursolic acid, carnosol, etc.; turmeric and its principal component, curcumin; and pepper also exhibit anticarcinogenic properties (Viuda-Martos et al. 2010).

5.10 Algae as a Functional Food

With the growing consumer demand for functional food ingredients, substantial research efforts are being devoted to identifying the functional ingredients which exhibit health-promoting properties. In this backdrop, the marine ecosystem has been identified as a potential source to explore high value-added ingredients with multiple activities. Also, algae are promising organisms which supply essential nutrients and novel bioactive substances with medical and therapeutic values. Pigments, lipids, polyunsaturated fatty acids, proteins, polysaccharides, and phenolics are the algal bioactive compounds of commercial interest (Dominguez 2013). Carrageenan, alginic acid, furcelleran, fucoidan, ascophyllan, and laminarin are the polysaccharides found in seaweeds, while polyuronides are found in green algae. Moreover, brown seaweeds serve as rich sources of carotenoid which is pronounced as a powerful antioxidant. Phlorotannin polyphenols are the most important seaweed-derived phenolic phytochemical which is uniquely found in brown seaweeds. Similar to most tannins, phlorotannins are also reported to have beneficial effects on cardiovascular diseases through an increase of HDL cholesterol and the prevention of atherosclerosis (Vattem and Maitin 2016).

6 Functional Foods of Sri Lankan Origin

The usage of natural health products and health-enhancing foods is becoming popular in Sri Lanka (Chandrasekara and Nelum 2014) introducing the concept of functional foods. As a result, a large volume of research takes place to assess the functional properties of natural products of Sri Lankan origin. These research primarily focus on the properties of functional foods (Bopitiya and Madhujith 2012, 2013, 2014), technologies involved in functional food development (Adikari and Bandara 2015; Herath et al. 2013), isolation of active compounds (Wijayabandara et al. 2015), nutrition and health aspects (Chandrasekara and Peiris 2013), and policies and marketing aspects (De Zoysa et al. 2014) (Fig. 4).

A large number of studies have been directed to assess the variety of properties of functional food, including antioxidant, antimicrobial, anticancer, etc. The study of Bopitiya and Madhujith (2012) has reported the antioxidant potential of pomegranate (*Punica granatum* L.) and identified the particular fruit as an extreme antioxidant source. A surge of studies has been conducted in relation to assessment of the antioxidant potential of many local foods, viz., brinjal (*Solanum melongena*), guava (*Psidium guajava*), watermelon (*Citrullus lanatus*), dragon fruit (*Hylocereus* spp.), tamarind (*Tamarindus indica* L.), himbutu (*Salacia chinensis* L.), palu

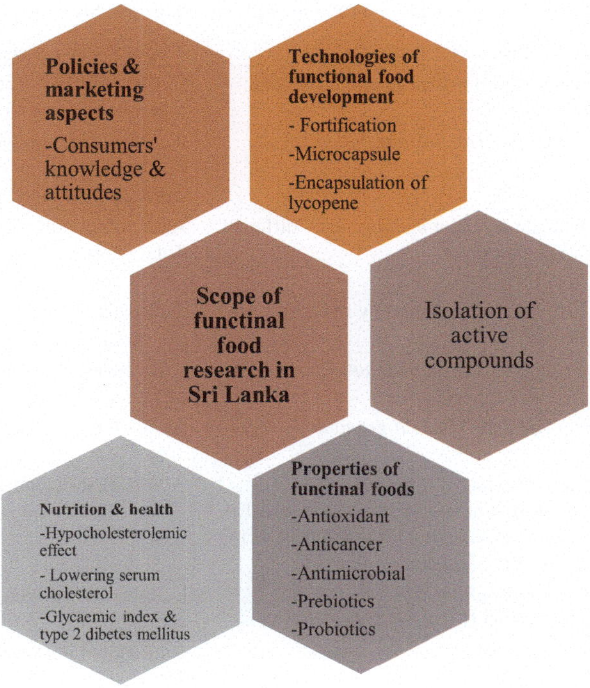

Fig. 4 Scope of functional food research in Sri Lanka

(*Manilkara hexandra* (Roxb.) Dubard), koan (*Schleichera oleosa* (Lour.) Oken), and cherry (*Eugenia uniflora*) (Table 1).

However, the novel functional food research focused more on the assessment of processing effects on functional properties and the development of novel technologies for functional food development. In a recent study, cinnamon oil was encapsulated by the complex coacervation method using chitosan and gum arabic wall material. The study confirmed the controlled-release of encapsulated cinnamon under conditions similar to those found in the gastrointestinal tract thereby introducing a novel method for utilization of cinnamon oil as digestive supplement (Ranasinghe and Perera 2017). The study of Adikari and Bandara (2015) has introduced a novel encapsulation method to improve the releasing of lycopene in its hydrophobic form. Moreover, several studies focus on the assessment of antioxidant activity of underutilized tubers in Sri Lanka under different cooking methods (Herath and Wijerathna 2017).

There is a growing trend toward the identification of the association between diet rich in functional ingredients and chronic diseases. Chandrasekara and Peries (2013) have studied the use of special foods and natural health products among type 2 diabetic patients and reported thebu (*Costus speciosus*), ranawara (*Cassia auriculata*), and bael (*Aegle marmelos*) as the top three most commonly consumed specific foods. Moreover, the hypocholesterolemic activity of *Phyllanthus* species is also extensively studied in Sri Lankan species (Kalpani et al. 2017). Liyanage and

Table 1 Functional foods and their effects studied in Sri Lanka

Functional food	Property(ies) assessed	References
Bitter melon (*Momordica charantia*) seed fat	Cytotoxicity effect of α-eleostearic acid-rich fat	Ranasinghe et al. (2015)
Tamarind (*Tamarindus indica* L.)	Agro-ecological region-based variations in antioxidant activity	Amararathna et al. (2015)
Brinjal (*Solanum melongena*) – different skin colored	Antioxidant potential Total phenolic content (TPC)	Somawathie et al. (2014)
Rice bran oil prepared by red and white rice	Antioxidant activity	Bopitiya and Madhujith (2014)
Sesame (*Sesamum indicum* L.) seed oil	Antioxidant activity and TPC	Bopitiya and Madhujith (2013)
Pomegranate (*Punica granatum* L.)	Antioxidant potential	Bopitiya and Madhujith (2012)
Palu (*Manilkara hexandra* (Roxb.) Dubard), himbutu (*Salacia chinensis* L.), koan (*Schleichera oleosa* (Lour.) Oken), and cherry (*Eugenia uniflora*)	Antioxidant potential and TPC	Piyathunga et al. (2016)
Guava (*Psidium guajava*) and watermelon (*Citrullus lanatus*)	Carotenoid and lycopene content	Chandrika et al. (2009)
Dragon fruit (*Hylocereus* spp.)	Antioxidant potential and TPC	Wedamulla and Madhujith (2013)
Elabatu (*Solanum melongena* var. *insanum*)	Antioxidant activity and TPC	Rizlia et al. (2014)
Virgin coconut oil	Anti-candida effect	Gurusinghe et al. (2016)
Annona (*Annona squamosa*), bale fruit (*Aegle marmelos*), jack fruit (*Artocarpus heterophyllus*), durian (*Durio zibethinus*)	Prebiotic effect	Madhujith and Mahaarachchi (2013)
Fermented rice	Probiotics	Jeygowri et al. (2015)

others (2015) have studied the hypolipidemic and hypoglycemic effects of banana blossom in rat fed with high-cholesterol diet. Further, Lakmini and others (2012) have studied the effect of cowpea (*Vigna unguiculata* L. Walp.)-incorporated diets on serum lipids, cecal bacterial population, and liver antioxidant capacity in Wistar rats (*Rattus norvegicus*).

One of the most demanding research areas is the isolation of active compounds from natural sources. A study carried out suggests to isolate bergenin as the major active constituent from the leaves of *Flueggea leucopyrus* Willd which is an extensively used medicinal agent required for the production of various pharmaceutical formulations (Wijayabandara et al. 2015). However, a few studies have been conducted in the area of consumer knowledge and attitudes related to functional foods. The results of the study revealed that the knowledge on nutrition among people is not up to a considerable level (De Zoysa et al. 2014).

7 Technologies for the Development of Functional Food

7.1 Microencapsulation

Envelopment of small solid particles, liquid droplets, or gases in a coating is known as microencapsulation. This technique provides protection for a wide range of materials of biological interest, from small molecules and protein (enzymes, hormones) to cells of bacterial, yeast, and animal origin (Betoret et al. 2011). Thus, microencapsulation is identified as an effective technique to improve the delivery of bioactive compounds into foods, particularly probiotics, minerals, vitamins, phytosterols, lutein, fatty acids, lycopene, and antioxidants (Champagne and Fustier 2007).

Lipid, proteins, and carbohydrates are commonly known bioactive compounds which are already encapsulated in industrial applications. These lipids include fatty acids, phospholipids, carotenoids, and oil-soluble vitamins which cannot be easily dissolved in food products and thus are highly susceptible to oxidation. However, the encapsulation process can form an effective barrier for oxygen; therefore, these lipids are now widely applied in powdered products (Betoret et al. 2011).

7.2 Edible Films

Any type of material used for enrobing (coating with fat carrying special melting characteristics) various food to extend shelf life of the product that may be eaten together with food or without further removal is known as edible film or coating (Betoret et al. 2011). These films have wide applications as they control moisture migration, gas exchange, or oxidation processes. Edible films are well acknowledged for their ability to carry active ingredients such as anti-browning agents, colorants, flavors, nutrients, spices, and antimicrobial compounds. Much work has been carried out to incorporate minerals, vitamins, and fatty acids, among other bioactives, into edible film and coating formulations with the aim of enhancing the nutritional value of some fruits and vegetables. However, the studies devoted to examine the addition of probiotics to obtain functional edible films and coatings are still scanty (Betoret et al. 2011).

7.3 Interesterification

Interesterification is a novel technique which is gaining considerable attention over the other techniques used to modify physical and chemical properties of natural fats. Interesterification reorganizes fatty acids within and between a triacylglycerol molecule in the presence of a chemical catalyst or an enzyme. Interesterified fats can retain functional properties of food; however, nutrition and health implications of long-term consumption of interesterified fat are not well documented (Mensink et al. 2016). There is a growing trend toward the synthesis of modified starch using the technique of esterification to enhance functional properties. Xin and others (2012) study the esterification of cornstarch with palmitic acid to produce modified starch. Coconut oil is the main oil used in Sri Lanka used for culinary purposes—as cooking oil and frying oil. However, it contains a very high amount of saturated fatty acids which amount to nearly 90%. In accordance with WHO guidelines, the daily intake of saturates should be limited to less than 10% of the total calorie requirement of a person. Moreover, sesame oil which is currently underutilized is rich in monounsaturated fatty acids (MUFA) which bear beneficial health effects. The fatty acid composition of coconut oil was changed by incorporating MUFA from sesame oil through enzymatic interesterification (Kugeswararajah et al. 2015, 2016; Sivakanthan et al. 2016a, b, 2017). Furthermore, α-eleostearic acid possesses a number of health effects including strong anticancer properties as explained in an earlier section of this chapter. Bitter gourd fat which is rich in α-eleostearic acid was successfully incorporated into virgin coconut oil (Demini et al. 2017).

8 Safety of Functional Foods

Safety assessment of functional foods should address the intrinsic hazards associated with the food/food ingredient along with knowledge about the expected exposure that will occur (Bradford 2015). Functional food may be a completely new food or a food component which has already been present in the diet for long. Thus, different strategies are often being adopted to assess the safety of functional foods depending on the type of food/food ingredient. Moreover, a case-by-case assessment should be used to determine the extent of nutritional and toxicological testing required as well as the types of tests to be used (Huggett and Schilter 1996). On the other hand, the application of currently used methodologies and risk assessment strategies to ensure the safety of food components present in certain functional food is problematic. Thus, unique safety assessment strategies are required to address functional foods which are, or which contain dietary macro-components, new protein sources (potential food allergens) and viable microorganisms (Huggett and Schilter 1996).

The growing consumer demand creates a good market opportunity for functional foods while introducing considerably higher prices which lead to adulterations and other deceptions. These include the addition of adulterant substances that are illegal and potentially toxic. Many investigations revealed the presence of pharmaceutical ingredients in functional food products (Jen and Chen 2017). Thus, it is vital to pay attention to the prevention of functional food frauds as well. On account of the high demand and the lucrative nature of business, a large number of novel products continue to mushroom in the international market. They make various attractive health and structure and function claims most of which have not been clinically proven. Unfortunately, many countries have not so far properly regulated functional foods which, thus, can bring about potential safety issues. In many countries, functional foods and ingredients are not categorized under drugs; therefore, they are mostly sold over the counter.

9 Regulations Governing Functional Foods

Functional food labeling is a very crucial aspect from the standpoint of both consumers and manufacturers. Consumers' interest in functional foods has remarkably improved along with the high availability of public information regarding diet-disease concepts and health benefits. Increased consumer awareness in relation to health issues can have a direct influence on consumers' buying decisions. Thus, manufacturers can use labeling or claims to uplift their market share by altering consumers' consciousness (Shimizu 2003).

Depending on regions and countries, there are distinct differences with respect to policies on health claims. In the United States, the Food and Drug Administration (US-FDA) is the governing body that maintains strict policies on food labeling for health and nutrient claims, whereas the European Commission (EC) and the European Food Safety Authority (EFSA) are in authority for countries in the European Union. Both the US and European authorities maintain strict policies on food labeling to ensure consumers' safety and protection from misleading information (Martirosyan and Singharaj 2016).

In terms of food and health claims, the US-FDA categorizes the claims into four comprehensive types: nutrient content claims, health claims, qualified health claims, and structure/function claims. Comparing with FDA, the EC defines two different categories of food claims: nutrition claims and health claims. In accordance with the policies of EC, health claims are divided into three subgroups, namely, functional health claims, risk reduction claims, and health claims referring to children's development (Martirosyan and Singharaj 2016).

The FDA defines health claim as any claim made on the label or in the labeling of food, including a dietary supplement, that expressly or by implication, including third-party references, written statements, symbols, or vignettes, characterizes the relationship of any substance to a disease or health-related condition. Nutrition claim is defined as any claim which states, suggests, or implies that a food has particular beneficial nutritional properties (European Commission 2018).

In relation to the amendments of the Federal Food, Drug, and Cosmetic Act (FFDCA; 1) of the United States, the US-FDA has no authority to establish a formal regulatory category for functional foods due to its incapability of providing a statutory definition of functional food. Functional foods do not exist under the regulation of FDA. Thus, the essence of functional food is stripped away, generalizing it as a food or supplement. Based on the EC regulation, a risk reduction or functional health claim can be made for functional foods as long as the information does not make false claims in order for consumers to purchase a product (Martirosyan and Singharaj 2016). Currently, Sri Lanka does not have clear regulations on functional foods; however, the Ministry of Health, Nutrition and Indigenous Medicine is currently working on developing regulations.

10 Economic and Marketing Aspects of Functional Foods

Since functional foods suffer from lack of generally agreed definition, it is very difficult to obtain a reliable estimate of the market demand for functional foods (Weststrate et al. 2002). However, the degree of familiarity with and the acceptance of functional food by consumers are the major factors which can affect the development of the market in the future (Vergari et al. 2010). Studies reported that the acceptance of specific functional ingredients is linked to the consumers' knowledge of their health effects. Thus, healthy appearance of a functional food product or a specific ingredient is an essential requirement for successful commercialization. However, consumers are not willing to change their daily lifestyle or eating patterns for the consumption of a specific functional food product (Vergari et al. 2010). Moreover, for the market success of functional food, it should be available in sufficient quantities within distribution channels such as supermarkets, general retail stores, or discount retailers as consumers are reluctant to go in specific shops just to buy functional food products (Menrad 2003).

Available future market estimations predict that functional food will increase its market volume in the coming years. However, these foods will not be able to rise up to the mass market in the future. They will instead represent a multi-niche market with a high number of rather limited product segments (Menrad 2003). The functional food market is at its infant stage in Sri Lanka. The price is a sensitive factor that decides the food purchasing behavior of customers in Sri Lanka. Therefore, consumers naturally opt for regular food rather than enhanced food which is generally high priced. As the general prices of food in the country are high compared to the disposable income and nearly 70% of income is spent on food, the general consumer thinks twice before paying extra rupees for a functional food. Furthermore, health consciousness among general consumers remains quite low at the moment in the country, and this may also have influenced their purchasing behavior.

11 Future of Functional Foods

The concept of functional foods moves toward a more individualized profile rather than focused on more generalized profile. The surge of research in the area of nutritional genomics introduced the concept of nutrigenomics where interaction between foods and an individual's genome was taken into consideration. This technological advancement has an enormous impact on healthcare and will create a need for designing a diet specifically targeted to a defined genetic profile (O'Sullivan 1999). However, information flow generated by research is essential to support private investments, consumption decisions, and government regulations in the development of functional foods (Bigliardia and Galati 2013).

12 Conclusions

Functional foods effectively translate the "diet and health relationship" into effective food while maintaining the safety of the modern food supply. Increased consumer interest toward health and nutrition creates a more competitive and diverse market for functional foods. In order to retain the market for functional foods, it is important to prove the effectiveness of functional ingredients. The growing trends toward a functional food market create an urgent need to establish more specific regulations to monitor and control the quality and safety of functional foods, and this will undoubtedly safeguard both the consumer and the manufacturer.

In the Sri Lankan context of functional food research, many researchers focus on the exploration of novel natural sources which aid in the development of functional foods with increased incidence of diet-related NCDs in Sri Lanka. The findings of these research create wealth of knowledge which can effectively be utilized in future applications.

References

Abumweis SS, Barake, R, Jones P (2008) Plant sterols/stanols as cholesterol lowering agents: a meta-analysis of randomized controlled trials. Food Nutr Res:52

Adikari AMNK, Bandara BMR (2015) Cyclodextrin-mediated intercalation of lycopene into cation-exchange montmorillonite clay. In: Proceeding of Postgraduate Institute of Science Research Congress, Peradeniya, Sri Lanka: 9–10 October 2015

Aghajanpour M, Nazer MR, Obeidavi Z, Akbari M, Ezati P, Kor NM (2017) Functional foods and their role in cancer prevention and health promotion: a comprehensive review. Am J Cancer Res 7(4):740–769

Alasalvar C, Shahidi F (eds) (2008) Tree nuts: composition, phytochemicals and health effects. CRC Press, London

Alissa EM, Ferns GA (2012) Functional foods and nutraceuticals in the primary prevention of cardiovascular diseases. J Nutr Metab:569486

Alkhatib A, Tsang C, Tiss A, Bahorun T, Arefanian H, Barake R, Khadir A, Tuomilehto J (2017) Functional foods and lifestyle approaches for diabetes prevention and management. Nutrients 9 (12):1310

Al-Sheraji SH, Ismail A, Manap MY, Mustafa S, Yusof RM, Hassana FA (2013) Prebiotics as functional foods: a review. J Funct Foods 5:1542–1553

Amararathna DIM, Weerakkody WAP, Madhujith T, Eeswara JP (2015) Agro-ecological region based variations in the antioxidant activity of tamarind (*Tamarindus indica* L.) fruit pulp. Acta Hortic 1106:67–72

American Cancer Society (2015) Cancer facts & figures. Atlanta, American Cancer Society

Ansari MM, Kumar DS (2012) Fortification of food and beverages with phytonutrients. Food Public Health 2(6):241–253

Awad AB, Downie A, Fink CS, Kim U (2000) Dietary phytosterol inhibits the growth and metastasis of MDA-MB-231 human breast cancer cells grown in SCID mice. Anticancer Res 20:821–824

Bacchetti T, Masciangelo S, Bicchiega V, Bertoli E, Ferretti G (2011) Phytosterols, phytostanols and their esters: from natural to functional foods. Medi J Nutr Metab 4:165–172

Betoret E, Betoret N, Vidal D, Fito P (2011) Functional foods development: trends and technologies. Trends Food Sci Technol 22:498–508

Bhat ZF, Kumar S, Bhat HF (2015) Bioactive peptides of animal origin: a review. J Food Sci Technol 52(9):5377–5392

Bigliardia B, Galati F (2013) Innovation trends in the food industry: the case of functional foods. Trends Food Sci Technol 31:118–129

Blankson H, Stakkestad JA, Fagertun H, Thom E, Wadstein J, Gudmundsen O (2000) Conjugated linoleic acid reduces body fat mass in overweight and obese humans. J Nutr 130(12):2943–2948

Bopitiya D, Madhujith T (2012) Antioxidant potential of pomegranate (*Punica granatum* L.) cultivars grown in Sri Lanka. Trop Agric Res 24(1):71–81

Bopitiya D, Madhujith T (2013) Antioxidant activity and total phenolic content of sesame (*Sesamum indicum* L.) seed oil. Trop Agric Res 24(3):296–302

Bopitiya D, Madhujith T (2014) Antioxidant potential of rice bran oil prepared from red and white rice. Trop Agric Res 26(1):1–12

Borneo R, León AE (2012) Whole grain cereals: functional component and health benefits. Food Funct 3(2):110–119

Bradford B (2015) Safety assessment of functional foods. In: Hock F (ed) Drug discovery and evaluation: pharmacological assays. Springer, Berlin/Heidelberg

Brown L, Rosner B, Willett WW, Sacks FM (1999) Cholesterol-lowering effects of dietary fiber: a meta-analysis. Am J Clin Nutr 69(1):30–42

Carocho M, Ferreira ICFR (2013) Review on antioxidants, prooxidants and related controversy: natural and synthetic compounds, screening and analysis methodologies and future perspectives. Food Chem Toxicol 51:15–25

Champagne CP, Fustier P (2007) Microencapsulation for the improved delivery of bioactive compounds into foods. Curr Opin Biotechnol 18(2):184–190

Chandrasekara A, Nelum D (2014) Role of functional foods in non communicable disease management, 7th conference on functional foods, Nutraceuticals, Natural Health Products and Dietary Supplements, Isthanbul, Turkey

Chandrasekara GAP, Peiris DLP (2013) Use of specific foods and natural health products among type 2 diabetic patients. In: Silva R (ed) Abstract of annual scientific sessions of Nutrition Society of Sri Lanka, Taj Samudra, Colombo, Sri Lanka, 2–3 February 2013

Chandrika UG, Fernando KSSP, Ranaweera KKDS (2009) Carotenoid content and in vitro bioaccessibility of lycopene from guava (*Psidium guajava*) and watermelon (*Citrullus lanatus*) by high-performance liquid chromatography diode array detection. Int J Food Sci Nutr 60 (7):558–566

Chang R (1996) Functional properties of edible mushrooms. Nutr Res 54(11):91–93

Chen CY, Blumberg JB (2008) Phytochemical composition of nuts. Asia Pacific J Clin Nutr 17 (Suppl 1):329–332

Chin SF, Liu W, Storkson JM, Ha YL, Pariza MW (1992) Dietary sources of conjugated dienoic isomers of linoleic acid, a newly recognized class of anticarcinogens. J Food Compost Anal 5 (3):185–197

Corbo MR, Bevilacqua A, Petruzzi L, Casanova FP, Sinigaglia M (2014) Functional beverages: the emerging side of functional foods, commercial trends. Research, and health implications. Compr Rev Food Sci Food Saf 13:1192–1206

Das A, Raychaudhuri U, Chakraborty R (2012) Cereal based functional food of Indian subcontinent: a review. J Food Sci Technol 49:665–672

Daya L, Seymoura RB, Pittsa KF, Konczaka I, Lundina L (2009) Incorporation of functional ingredients into foods. Trends Food Sci Technol 20:388–395

De Zoysa MPN, Dissanayaka D ME, Gunawardena SDNC, Wijesinghe AB, Shashikala GN, Nipunajith GUD, Dias P (2014) An Assessment of consumers' knowledge, attitudes and habits in relation to functional foods. In: SAITM research symposium on engineering advancements, Colombo, Sri Lanka, pp 192–196

Deckelbaum RJ, Torrejon C (2012) The omega-3 fatty acid nutritional landscape: health benefits and sources. Int J Nutr 142(3):587–591

Demini D, Jayasooriya A, Madhujith T (2017) production of structured lipid by incorporating bitter gourd (*Momordica charantia*) seed oil into virgin coconut oil and evaluation of its physicochemical and antioxidant properties. In: Proceedings of Faculty of Agriculture Undergraduate Research Symposium, University of Peradeniya, Sri Lanka, December 2017
Diplock AT, Aggett PJ, Ashwell M, Bornet F, Fern EB, Roberfroid MB (1999) Scientific concepts of functional foods in Europe: consensus document. Br J Nutr 81:1–27
Dominguez H (ed) (2013) Functional ingredients from algae for foods and nutraceuticals. Woodhead Publishing, pp 2–13
Erjavec J, Kos J, Ravnikar M, Dreo T, Saboti J (2012) Proteins of higher fungi from forest to application. Trends Biotechnol 30(5):259–273
European Commission (2018). http://ec.europa.eu/food/safety/labelling_nutrition/claims/nutrition_claims/index_en.htm. Accessed on 5 May 2018
Eynard AR, Lopez CB (2003) Conjugated linoleic acid (CLA) versus saturated fats/cholesterol: their proportion in fatty and lean meats may affect the risk of developing colon cancer. Lipids Health Dis 2:6
FAO (2004) RAP Publication 2004/33: report of the regional consultation of the Asia-Pacific network for food and nutrition on functional foods and their implications in the daily diet. Bangkok, p 61
Forman HJ, Torres M (2002) Reactive oxygen species and cell signaling-respiratory burst in macrophage signaling. Am J Respir Crit Care Med 66:4–8
Ganesan B, Brothersen C, McMahon DJ (2014) Fortification of foods with omega-3 polyunsaturated fatty acids. Crit Rev Food Sci Nutr 54(1):98–114
Gharras HE (2009) Polyphenols: food sources, properties and applications – a review. J Food Sci Technol 44(12):2512–2518
Grossmann ME, Mizuno NK, Dammen ML, Schuster T, Ray A, Cleary MP (2009) Eleostearic acid inhibits breast cancer proliferation by means of an oxidation-dependent mechanism. Cancer Prev Res 2(10):879–886
Gurusinghe N, Jayatilake JAMS, Madhujith T (2016) *In vitro* anti-candida effect of Sri Lankan virgin coconut oil. In: Proceedings of International Peradeniya University research sessions, November 4–5, Peradeniya, Sri Lanka, p 218
Hasler CM (2000) The changing face of functional foods. J Am Coll Nutr 19(5):499–506
Herath HMAJ, Wijerathna HDJE (2017) Impact of different cooking methods on the antioxidant content of selected underutilized tubers in Sri Lanka. In: Abstract of presentation, 4th Ruhuna international science and technology conference, Matara, Sri Lanka, 26 January, 2017
Herath HMT, Rajapakse D, Aponso DMK (2013) Fortification of iron fortified biscuit using brown rice as a functional food ingredient. In: Silva R (ed) Abstract of annual scientific sessions of Nutrition Society of Sri Lanka, Taj Samudra, Colombo, Sri Lanka, 2–3 February 2013
Higdon JV, Frei B (2003) Tea catechins and polyphenols: health effects, metabolism, and antioxidant functions. Crit Rev Food Sci Nutr 43(1):89–143
Huggett AC, Schilter B (1996) Research needs for establishing the safety of functional foods. Nutr Res 54(11):143–148
Jen JJS, Chen J (eds) (2017) Food safety in China: science, technology, management and regulation. Wiley, Hoboken
Jeygowri N, Parahitiyawa N, Jeyatilake S, Ranadheera S, Madhujith T (2015) Study on isolation of potentially probiotic *Lactobacillus* species from fermented rice. Trop Agric Res 26(3):428–440
Kalpani PGM, Weerakoon SR, Somaratne S, Nilakarawasam N, Ranasinghe CA (2017) preliminary study on the evaluation of hypocholesterolemic activity of some *Phyllanthus* species. In: Abstract of presentation, 4th Ruhuna international science and technology conference, Matara, Sri Lanka, 26 January 2017
Kaur S, Das M (2011) Functional foods: an overview. Food Sci Biotechnol 20(4):861–875
Keservani RK, Kesharwani RK, Vyas N, Jain S, Raghuvanshi R, Sharma AK (2010) Nutraceutical and functional food as future food: a review. Pharm Lett 2(1):106–116

Khlebnikov AI, Schepetkin IA, Domina NG, Kirpotina LN, Quinn MT (2007) Improved quantitative structure-activity relationship models to predict antioxidant activity of flavonoids in chemical, enzymatic and cellular systems. Bioorganic Med Chem 15:1749–1770

Koba K, Yanagita T (2014) Health benefits of conjugated linoleic acid (CLA). Obes Res Clin Pract 8(6):e525-32

Kugeswararajah T, Sivakanthan S, Tachibana H, Jayasooriya LJPAP, Madhujith T (2015) Effect of reaction parameters on enzymatic interesterification of virgin coconut (*Cocos nucifera*) oil and sesame (*Sesamum indicum*) oil blend using lipases from *Thermomyces lanuginosus* and *Rhizomu cormiehei.* (Abs.) In: Third conference on Sri Lanka-Japan collaborative research, University of Peradeniya, Sri Lanka, 18–20 September 2015

Kugeswararajah T, Sivakanthan S, Jayasooriya LJPAP, Madhujith T (2016) Enzymatic interesterification of virgin coconut (*Cocos nucifera*) oil and sesame (*Sesamum indicum*) oil and evaluation of the physico-chemical properties of the interesterified oil. In: Proceedings of Jaffna University International Research Conference (JUICE 2016), Sri Lanka, 12–13 August 2016

Kumar K (2015) Role of edible mushrooms as functional foods – a review. SAJFTE 1:211–218

Lachat C, Otchere S, Roberfroid D, Abdulai A, Seret FMA, Milesevic J, Xuereb G, Candeias V, Kolsteren P (2013) Diet and physical activity for the prevention of non-communicable diseases in low- and middle-income countries: a systematic policy review. PLoS Med 10(6):1001465

Lajolo FM (2002) Functional foods: Latin American perspectives. Br J Nutr 88(2):145–150

Lakmini GWAS, Liyanage R, Vidanarachchi JK, Jayawardana BC, Jayawardana NWIA, Madhujith T, Ranadheera CS (2012) Effect of cowpea (*Vigna unguiculata* L. Walp.) incorporated diets on serum lipids, caecal bacterial population and liver antioxidant capacity in Wistar rats (*Rattus norvegicus*). In: Proceedings of the undergraduate research symposium, Rajarata University of Sri Lanka, p 17

Lavecchia T, Rea G, Antonacci A, Giardi MT (2013) Healthy and adverse effects of plant-derived functional metabolites: the need of revealing their content and bioactivity in a complex food matrix. Crit Rev Food Sci Nutr 53(2):198–213

Law MR (2000) Plant sterol and stanol margarines and health. West J Med 173(1):43–47

Lee JH, O'Keefe JH, Lavie CJ, Harris WS (2009) Omega-3 fatty acids: cardiovascular benefits, sources and sustainability. Nat Rev Cardiol 6:753–758

Liou GY, Storz P (2010) Reactive oxygen species in cancer. Free Radic Res 44(5):1–30

Liu RH (2003) Health benefits of fruit and vegetables are from additive and synergistic combinations of phytochemicals. Am J Clin Nutr 78(3):517–520

Liyanage R, Gunasegaram S, Weththasinghe P, Rizliya V, Jayathilake C, Jayawardana B, Vidanarachchi JK (2015) Effect of banana (*Musa acuminate* Colla) blossom incorporated diets in serum cholesterol and serum glucose levels in rats fed with cholesterol. In: Proceeding of Postgraduate Institute of Science Research Congress, Peradeniya Sri Lanka: 9–10 October 2015

Lobo V, Patil A, Phatak A, Chandra N (2010) Free radicals, antioxidants and functional foods: impact on human health. Pharmacogn Rev 4(8):118–126

Madhujith T, Mahaarachchi TMD (2013) Evaluation of prebiotic effect of selected fruits grown in Sri Lanka. In: Silva R (ed) Abstract of annual scientific sessions of Nutrition Society of Sri Lanka, Taj Samudra, Colombo, Sri Lanka, 2–3 February 2013

Manach C, Scalbert A, Morand C, Rémésy C, Jiménez L (2004) Polyphenols: food sources and bioavailability. Am J Clin Nutr 9(5):727–747

Mani F, Damasceno HCR, Novelli ELB, Martins EAM, Sforcin JM (2006) Propolis: effect of different concentrations, extracts and intake period on seric biochemical variables. J Ethnopharmacol 105(1–2):95–98

Martirosyan DM, Singharaj B (2016) Health claims and functional food: the future of functional foods under FDA and EFSA regulation. Funct Foods Chronic Dis:410–424

Menrad K (2003) Market and marketing of functional food in Europe. J Food Eng 56:181–188

Mensink RP, Sanders TA, Baer DJ, Hayes K, Howles PN, Marangoni A (2016) The increasing use of Interesterified lipids in the food supply and their effects on health parameters. Adv Nutr 7 (4):719–729

Meyer PD (2004) Nondigestible oligosaccharides as dietary fiber. J AOAC Int 87(3):718–726

Ministry of Health (2010) National Nutrition Policyof Sri Lanka, extraordinary gazette No. 1639/5 of Democratic Socialist Republic of Sri Lanka dated 2010.02.02

Nohl H, Kozlov AV, Gille L, Staniek K (2003) Cell respiration and formation of reactive oxygen species: facts and artifacts. Biochem Soc Trans 31:1308–1311

O'Sullivan J (1999) Impact of genetic engineering on the profession of dietetics. Top Clin Nutr 14:1–5

Olasehinde TA, Oyeleye SI, Ogunsuyi OB, Ogunruku O, Oboh G (2017) Functional foods in the management of neurodegenerative diseases, pp 72–81

Pasha C, Reddy G (2005) Nutritional and medicinal improvement of black tea by yeast fermentation. Food Chem 89:449–453

Patel S, Goyal A (2012) Recent developments in mushrooms as anticancer therapeutics: a review. Biotech 2:1–15

Pisulewski PM, Kostogrys RB (2003) Functional properties of foods of animal origin and the methods of their assessment. Pol J Food Nutrsci 12(53):65–73

Piyathunga ALI, Mallawaarachchi MALN, Madhujith WMT (2016) Phenolic content and antioxidant capacity of selected underutilized fruits grown in Sri Lanka. Trop Agric Res 27(3):277–286

Prakash B, Kujur A, Singh P, Kumar A, Yadav A (2017) Plants-derived bioactive compounds as functional food ingredients and food preservative. J Nutr Food Sci 2:005

Rainer L, Heiss CJ (2004) Conjugated linoleic acid: health implications and effects on body composition. J Am Diet Assoc 104(6):963–968

Rajasekaran A, Kalaivani M (2013) Designer foods and their benefits: a review. J Food Sci Technol 50(1):1–16

Ranasinghe JPS, Perera BGK (2017) Preparation of cinnamon oil microcapsules to be used as a digestive supplement for humans. In: Abstract of presentation, 4th Ruhuna international science and technology conference, Matara, Sri Lanka, 26 January 2017

Ranasinghe KNK, Jayasooriya AP, Premachandra TN, Hilmi AJ, Kularatne SAM, Madhujith T, Rajapakse RPVJ (2015) Cytotoxic effects of α-Eleostearic acid-rich fat extracted from seeds of bitter melon (*Momordica charantia*) on Vero 76 cells. In: Proceedings of the international postgraduate research conference 2015, University of Kelaniya, Kelaniya, Sri Lanka, p 177, 339 pp

Rao AL, Bharani M, Pallavi V (2006) Role of antioxidants and free radicals in health and disease. Adv Pharmacol Toxicol 7:29–38

Riccardi G, Capaldo B, Vaccaro O (2005) Functional foods in the management of obesity and type 2 diabetes. Curr Opin Clin Nutr Metab Care 8(6):630–635

Riediger ND, Othman RA, Suh M, Moghadasian MH (2009) A systemic review of the roles of n-3 fatty acids in health and disease. J Am Diet Assoc 109:668–679

Rizlia V, Somawathi KM, Madhujith WMT, WIjesinghe DGNG (2014) Antioxidant activity and total phenolic content of different skin coloured elabatu (*Solanum melongena* var. insanum). In: Proceedings of the Wayamba international conference, 29–30 August 2014

Roberfroid MB (2000) Prebiotics and probiotics: are they functional foods. Am J Clin Nutr 71 (6):1682–1687

Roberfroid MB (2005) Introducing inulin-type fructans. Br J Nutr 93:13–25

Rodríguez R, Jiménez A, Fernández-Bolaños J, Guillén R, Heredia A (2006) Dietary fibre from vegetable products as source of functional ingredients. Trends Food Sci Technol 17:3–15

Shahidi F (2009) Nutraceuticals and functional foods: whole versus processed foods. Trends Food Sci Technol 20:376–387

Shimizu T (2003) Health claims on functional foods: the Japanese regulations and an international comparison. Nutr Res Rev 16(2):241–252

Sivakanthan S, Jayasooriya LJPAP, Madhujith T (2016a) Optimization of the reaction parameters of the production of structured lipid from coconut (*Cocos nucifera*) and sesame (*Sesamumindicum*) oils using response surface methodology (RSM). In: Peradeniya University International Research Sessions (iPURSE), Sri Lanka, 4–5 November 2016

Sivakanthan S, Jayasooriya LJPAP, Madhujith T (2017) Optimization of enzymatic interesterification of coconut (*Cocos nucifera*) and sesame (*Sesamum indicum*) oils using *Thermomyces lanuginosus*lipase by response surface methodology. Trop Agric Res 28 (4):457–471

Sivakanthan S, Jayasooriya LJPAP, Madhujith T (2016b) Optimization of enzymatic interesterification of coconut (*Cocos nucifera*) and sesame (*Sesamum indicum*) oils using *Thermomyces lanuginosus* lipase by response surface methodology. In: 28th Annual Congress, Postgraduate Institute of Sri Lanka, University of Peradeniya, Sri Lanka, 16–17 November 2016

Somanathan A, Hanson K, Dorabawila T, Perera B (2000) Operating efficiency in public sector health facilities in Sri Lanka: measurement and institutional determinants of performance. Small applied research paper no. 12. Bethesda, MD: Partnerships for Health Reform Project, Abt Associates Inc.

Somawathie KM, Rizliya V, Wijesinghe DGNG, Madhujith WMT (2014) Antioxidant activity and total phenolic content of different skin colored brinjal (*Solanum melongena*). Trop Agric Res 26 (1):152–161

Spence JT (2006) Commentary challenges related to the composition of functional foods. J Food Compost Anal 19:4–6

Sre PRR, Sheila T, Murugesan K (2012) Phytochemical screening and "in-vitro" antioxidant activity of methanolic root extract of *Erythrina indica*. Asian Pac J Trop Biomed 2 (3):1696–1700

Stampfer MJ, Hu FB, Manson JE, Rimm EB, Willett WC (2000) Primary prevention of coronary heart disease in women through diet and lifestyle. N Engl J Med 343(1):16–22

Sunkara R, Verghese M (2014) Functional foods for obesity management. Food Nutri Sci 5:1359–1369

Townsend N, Williams J, Wickramasinghe K, Karunarathne W, Olupeliyawa A, Manoharan S, Friel S (2017) Barriers to healthy dietary choice amongst students in Sri Lanka as perceived by school principals and staff. Health Promot Int 32(1):91–101

Tsuzuki T, Tokuyama Y, Igarashi M, Nakagawa K, Ohsaki Y, Komai M, Miyazawa T (2004a) α-Eleostearic acid (9Z11E13E-18, 3) is quickly converted to conjugated linoleic acid (9Z11E-18, 2) in rats. J Nutr 134(10):2634–2639

Tsuzuki T, Tokuyama Y, Igarashi M, Miyazawa T (2004b) Tumor growth suppression by α-eleostearic acid, a linolenic acid isomer with a conjugated triene system, via lipid peroxidation. Carcinogenesis 25(8):1417–1425

Vattem DA, Maitin V (2016) Functional foods, Nutraceuticals and natural products. DEStech Publications Inc., Pennsylvania

Vergari F, Tibuzzi A, Basile G (2010) An overview of the functional food market: from marketing issues and commercial players to future demand from life in space. In: Giardi MT, Rea G, Berra B (eds) Bio-farms for nutraceuticals. Advances in experimental medicine and biology, vol 698. Springer, Boston

Viuda-Martos M, Ruiz-Navajas Y, Fernández-López J, Pérez-Álvarez JA (2010) Spices as functional foods. Crit Rev in Food Sci Nutr 51(1):13–28

Volate SR, Davenport DM, Muga SJ, Wargovich MJ (2005) Modulation of aberrant crypt foci and apoptosis by dietary herbal supplements (quercetin, curcumin, silymarin, ginseng and rutin). Carcinogenesis 26:1450–1456

Wagner K, Brath H (2012) A global view on the development of non-communicable diseases. Prev Med 54:S38–S41

Waris G, Ahsan H (2006) Reactive oxygen species: role in the development of cancer and various chronic conditions. J Carcinogen 5:14

Wedamulla N, Madhujith T (2013) Antioxidant activity of two dragon fruit (*Hylocereus* spp.) species cultivated in Sri Lanka. In: Silva R (ed) Abstract of annual scientific sessions of Nutrition Society of Sri Lanka, Taj Samudra, Colombo, Sri Lanka, 2–3 February 2013

Weststrate JA, van Poppel G, Verschuren PM (2002) Functional foods, trends and future. Br J Nutr 88(Suppl. 2):S233–S235

Wijayabandara MDJ, Choudhary M, Wijayabandara MDLO (2015) Isolation of bergenin from the leaves of *Flueggea leucopyrus* Willd (katupila) – a novel method of obtaining bergenin. Pharm J Sri Lanka 5:10–15

World Health Organization (2014) Global status report on non-communicable diseases. World Health Organization, Geneva

Xin JY, Wang Y, Liu T, Lin K, Chang L, Xia CG (2012) Biosysthesis of corn starch palmitate by lipase Novozym 435. Int J Mol Sci 13(6):7226

Zbakha H, Abbassib AE (2012) Potential use of olive mill wastewater in the preparation of functional beverages: a review. J Funct Foods 4:53–65

Zhang X, Chen F, Wang M (2014) Bioactive substances of animal origin. In: Cheung PCK, Mehta BM (eds) Handbook of food chemistry. Springer, Berlin/Heidelberg, pp 1–21

ICT-Based Information Systems in Agricultural Extension and Their Economic Implications: Sri Lankan Perspectives

Uvasara Dissanayeke, Pahan Prasada, and Heshan Wickramasuriya

1 Background

Agriculture provides employment to more than a quarter of (27%) the population, while the contribution of family members is as high as 60%; thus, agriculture continues to play an important role in Sri Lankan economy despite the decline or stagnation of its contribution to the needed gross domestic production (Department of Census and Statistics 2017). Given the fact that a large number of the country's population is dependent on agriculture as a direct or indirect source of living, it has been always identified as one of the important development priorities in the country.

Agriculture is partly shifting from a labor-intensive to a capital-intensive economic activity. This is fueled, partly, by the phenomenon of labor leaving agriculture. Moreover, the recent developments in terms of information and communication technologies (ICTs) both at the national and international levels have demanded ICTs to be integrated into the agriculture sector. The e-agriculture strategy guide (FAO and ITU 2016) suggested eight important roles of ICTs in agriculture (Fig. 1). This chapter investigates three of the main roles of ICTs, i.e., strengthening agricultural extension and advisory services, enhancing market access, and increasing access to financial services for rural communities, along with ICT adoptions and their economic implications.

U. Dissanayeke (✉) · H. Wickramasuriya
Department of Agricultural Extension, Faculty of Agriculture, University of Peradeniya, Peradeniya, Sri Lanka

P. Prasada
Department of Agricultural Economics and Business Management, Faculty of Agriculture, University of Peradeniya, Peradeniya, Sri Lanka

R. P. De Silva et al. (eds.), *Agricultural Research for Sustainable Food Systems in Sri Lanka*, https://doi.org/10.1007/978-981-15-3673-1_15

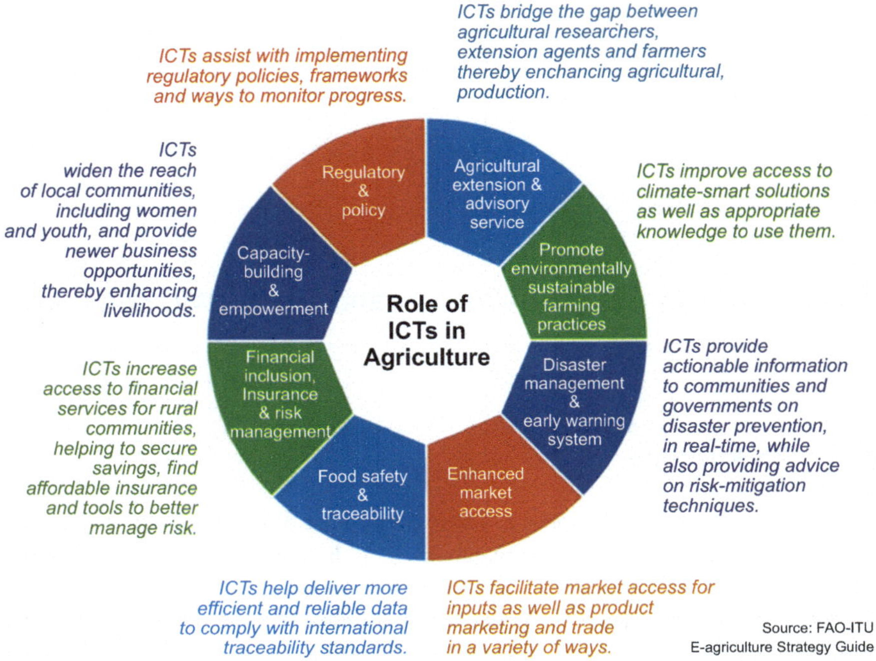

Fig. 1 Role of ICTs in agriculture. (Source: FAO-ITU 2016)

2 ICT Backdrop

Information communication technology has been understood more as an umbrella term to include a broad range of technologies. It further includes media that can be used to store or record information, broadcasting technologies that are used to communicate information on a mass scale, and all other forms of communication platforms that bring across voice, sound, and images (IGI Global 2018). Implementation of ICTs in any sector would be affected by the availability and accessibility of ICTs to a given community. With regard to the network readiness index, Sri Lanka is in a much better position compared to other lower- and middle-income countries. For instance, ICT infrastructure facilities have considerably improved over the recent years (WEF 2016) (Fig. 2), especially in terms of affordability, skills, and government usage rating. Sri Lanka as a country ranks in a favorable position, providing ample opportunities to integrate ICTs in national development.

With this backdrop, agriculture information dissemination systems have become a fertile ground for ICT initiatives. ICT as technology is most useful due to the potential to create information platforms that are not costly. The need for high-frequency information creates a timely role for IT. The initiatives focusing on "ICT in development" has preceded ICT in agriculture as far as low-cost information dissemination and connectivity are concerned. Where ICT in agriculture has made a

	Rank (out of 139)	Value (1–7)
Networked Readiness Index (2016)	63	42
Networked Readiness Index 2015 (out of 143)	65	4.1
Networked Readiness Index 2014 (out of 148)	76	3.9
Networked Readiness Index 2013 (out of 144)	69	3.9
A. Environment subindex	73	3.9
1st pillar: Political and regulatory environment	64	3.8
2nd pillar: Business and innovation environment	81	4.1
B. Readiness subindex	63	4.9
3rd pillar: Infrastructure	103	3.0
4th pillar: Affordability	35	6.0
5th pillar: Skills	32	5.7
C. Usage subindex	**67**	**3.9**
6th pillar: Individual usage	102	2.8
7th pillar: Business usage	49	3.9
8th pillar: Government usage	20	5.0
D. Impact subindex	**49**	**4.0**
9th pillar: Economic impacts	70	3.2
10th pillar: Social impact	42	4.7

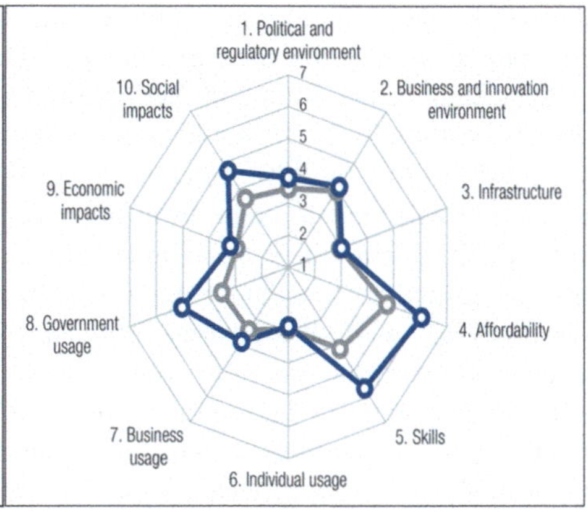

Fig. 2 ICT readiness of Sri Lanka. (Source: Network Readiness Index Country Profile Sri Lanka (WEF 2016))

unique case is the potential for customized and targeted information on prices and market conditions that would technically be chaotic in multiagent spot markets. The dispersion of prices while creating economic opportunities for speculative intermediaries alienates the smallholder from markets, creating the most prevalent limitation to the poor agricultural producer. Solving the market access problem and the removal of frictions in market logistic services is by far the most significant value that ICT has created in agriculture. While digital technology's and ICT's promise is not contested, the potential dividends of ICT in agriculture have not penetrated as fast as the technology itself. The World Development Report of 2016 looks at the slow realization of "digital dividends."

3 Use of ICTs in Agriculture Extension

The primary objective of agriculture extension services is to bridge the gap between agricultural research and the actual practice of agricultural innovations in the field level by various agricultural stakeholders, including farmers. This requires passing down information related to the latest agricultural innovations to the actual users in an effective and efficient manner while also ensuring the identification of field problems and transferring them to researchers. Agricultural extension systems nonetheless go beyond mere technology transfer to include facilitation and suggest several other important aspects such as helping farmers form groups, dealing with marketing issues, and partnering with a broad range of service providers and other agencies (Davis 2008). Thus, in a broad sense, agricultural extension can be defined as the "entire set of organizations that support people engaged in agricultural production and facilitate their efforts to solve problems; link to markets and other players in the agricultural value chain; and obtain information, skills, and technologies to improve their livelihoods" (Davis 2008).

According to the Food and Agriculture Organization, the world population will reach 9.1 billion by 2050, and to feed that number of people, global food production will need to grow by 70%. In countries of the global south, 80% of the necessary production will have to come from improvements in yields and cropping intensity and only 20% from the expansion of arable land. This context makes the case for low-cost initiatives to enhance the efficiency of agriculture as a production imperative process. As in all industries, efficiency gains are predicated upon having an efficient information system to coordinate the process.

ICTs can support agricultural extension in many different ways. The main functions of agriculture extension include identifying farmer problems and opportunities, transfer of technologies to farmers in the form of advisory services, awareness creation, skill development, and education (Farrington 1995; Bell 2015). ICT can play an important role in facilitating most of these functions, enhancing the effectiveness and efficiency of extension services.

ICT tools that are commonly used in extension can be classified into five broad categories based on the *platform* on which they operate: radio, TV and videos, basic cell phones (text, voice) and landlines, smart devices, and computer and the Internet (Bell 2015). There can be various tools that operate on the above platforms, sometimes combining services from two or more platforms. Bell and Payne (2014) suggested a comprehensive analysis of the different forms of ICTs to support agricultural extension. Furthermore, they suggested that it is necessary to combine traditional approaches of agricultural extension together with new ICTs to get the maximum benefits of such integrations. Table 1 summarizes the usability of ICT tools in supporting different functions of agricultural extension (Bell and Payne 2014).

The Bell and Payne (2014) classification discusses three of the main uses of ICTs in agriculture extension, namely, identifying farmers' problems, promoting behavior change, and collecting feedback against the potentials of five ICT tools and technologies, e.g., radio, television, basic mobile phones, smart devices, and computers and the Internet. In the next section, we discuss the applicability of these tools and technologies in agriculture information dissemination in Sri Lanka.

3.1 Radio

Radio is considered one of the most preferred sources of agricultural information for the large majority of smallholder farmers. Radio is both an affordable and an accessible medium for rural communities, making it more popular among them. Other benefits of radio include that can be listened to while working and be carried to the field if desired. Recent developments in the field of agricultural radio include the introduction of a web radio in 2014, which can also be listened to through mobile phones with an Android operating system using an Android app. Presently, the Farm

Table 1 ICT tools and the forms of support for agricultural extension

Extension functions	Information communication technology and tools: Radio	TV and videos	Cell phones (text, voice)	Feature and smart devices	Computers and the Internet
Identifying farmers' problems and opportunities: what do they need and want?					
Diagnose problems	Some potential if dealing with general problems or if capacity for interaction and expertise are available	Visuals are very helpful as "seeing is believing," even better if combined with ways to receive feedback	Some potential if farmers can call or text in and sufficient expertise is available	Additional potential to a simple cell phone as it enables web or app access to special diagnostic tools	Good comprehensive diagnostic tools are available
Collect information	Some potential if capacity for interaction is available		Can be used for data collection	Good for data collection with GPS	Some potential if Internet is available
Promoting behavior change: what is practical and relevant to meet the needs?					
Raise awareness of general opportunities or needs, convince farmers to try something new	Very good especially with persuasive programming	Visuals are usually very helpful as "seeing is believing"	An option if users are registered to receive such messages (SMS)	An option if users are registered to receive such messages (SMS, email)	An option if users are registered to receive such messages (email)
Provide specific information needed for change. What is involved? What are the benefits? Demonstrate or train?	Some potential but limited information delivered, can be enhanced with call in	Good option as "seeing is believing"	Has potential if farmers can call or text in and sufficient expertise is available	Additional potential to a simple cell phone as it enables web access and plays videos	Good option for intermediaries to seek information and videos
Facilitate access to credit and inputs	Can be used to inform about available services, but one-way communication	Can be used to inform about available services, but one-way communication	Allows mobile banking and negotiating directly with suppliers	Allows mobile banking and negotiating directly with suppliers	Allows online banking

(continued)

Table 1 (continued)

Extension functions	Information communication technology and tools				
	Radio	TV and videos	Cell phones (text, voice)	Feature and smart devices	Computers and the Internet
Link farmers to markets	Good for providing general price reports		Provides access to price information (call in, subscription)	Can bring potential buyers and producers together, enables access to price information	Can bring potential buyers and producers together, enables access to price info
Collecting feedback: how can each step be improved?					
Collect and respond to farmer feedback	Good if producers can call or text and sufficient expertise is available	Good if producers can call or text and sufficient expertise is available	Some potential if farmers can call or text in and sufficient expertise is available	Good option for intermediaries to seek information (if optimized for smart devices)	Good option for intermediaries to seek information
Assist with business planning	Some potential	Some potential		Enables access to simple farm management "apps," for record keeping	Enables access to farm management tools, for record keeping

Source: Bell and Payne (2014)

Broadcasting Service broadcasts more than 30 live radio programs per week (DOA 2017).

Radios can be easily coupled with other ICTs, such as telephones, to make the listening experience more interactive for listeners, such as coupling with a call-in or SMS facility for feedback (Kanchana and Dissanayeke 2016; World Bank 2017). Most of the live radio programs open a discussion for questions coming from the listeners, either as short messages or direct calls. These interactions from audiences provide a valuable input for radio programs, making programs more interactive with the audience and fulfilling specific information needs of the listeners.

3.2 Television and Videos

TV is an important medium to convey information to farming communities. More than 80% of the Sri Lankan population has a television set. Television is regarded as the most effective mass media in diffusing scientific information to masses.

Video programs were extensively used in disseminating agricultural information in the Sri Lankan context. The food crop sector has produced more than 100 s of such video documentaries, which were also telecasted on national television channels. Furthermore, videos were used in extension activities conducted by some other sectors, such as the tea cultivation sector, the export agriculture crop sector, and the coconut cultivation sector (Senevirathne and Dissanayake 2014; Rajasinghe and Samansiri 2014; CRI 2017). Agricultural television programs were highly successful and had improved farmers' knowledge on agriculture (Ekanayake 1995; Nazari and Hassan 2011). Television was found to be the most important source of agricultural information for farmers, according to two field studies carried out in Kandy district and Anuradhapura district (Kumari et al. 2009; Madana and Dissanayeke 2009). When compared with the new ICTs such as the Internet and websites, farmers still prefer traditional mass media such as TV and radio (Khan et al. 2010; Hassan et al. 2010).

The most commonly used format for agricultural television programs was the documentary format. The main purposes of such videos were motivational, educational, instructional, and entertainment. For instance, motivational videos such as *Mihikatha Dinuwo* presented stories of successful farmers. Educational videos were mostly for the purpose of technology dissemination; for example, the *Govibimata Arunalu* program presented the latest technologies in the food crop sector, while the *Ketha Batha Kamatha* documentary program was used to disseminate traditional agriculture knowledge (DOA 2016). The *Soora Goviya* television program was both motivational and entertainment oriented, which followed reality television concepts, which was relatively a new approach to agricultural television programs (Wijekoon 2014). This program reported an island-wide competition among the farmers, and one of the main intentions was to popularize agriculture among the young generation. The main problem in terms of using television in agricultural extension was that only a small percentage of farmers watched agricultural TV programs. In the early period, the main problems that prevented farmers from watching agricultural TV programs were low awareness on such programs and the telecasting time (Ekanayake 1995).

3.3 Basic Cell Phones and Telephone

Mobile cellular phones have gained popularity in recent years both in terms of accessibility and affordability, even among the lower income communities. This has led to some of the interesting developments in agricultural extension and advisory services in both public and private sector service providers (World Bank 2017), such as call-line services, interactive voice response services (IVR), and SMS-based information dissemination systems. As a communication device, mobile phones can cater to all levels of communication, including one-to-one and one-to-many communication, thus providing opportunities for enhancing individual, group, and mass extension methods. One of the main advantages of mobile phones is reduction of transaction cost (Box 2).

One of the most successful agricultural advisory call lines in the country is the *Govi Sahana Sarana* (GSS)-1920, which provides cultivation-related advice during office hours. The service handled 40,000–50,000 calls annually during the first stage of implementation (Madana and Dissanayeke 2009). GSS was found to be useful and time saving, and the callers had positive attitudes toward the quality of the service and the advice given (Kumari et al. 2009). The majority of the inquiries were related to plant diseases, pest attacks, and cultivation management practices (Kumari et al. 2009). Furthermore, some farmers called GSS when they needed advice on which crop to cultivate and to solve problems related to harvesting and postharvest management (Wijeratne 2011). Those farmers who called the GSS service were mostly the younger farmers and those less experienced, i.e., with less than 10 years of farming experience (Madana and Dissanayeke 2009). Furthermore, urban dwellers who practiced home gardening as an additional income source found that GSS is useful mainly because they did not have access to other reliable extension services (Kumari et al. 2009). Poor awareness of GSS among farmers was one of the main reasons for not accessing the service. Some farmers felt that they could manage with other information sources, while some were not confident to use ICT-based methods in obtaining agricultural information (Wijeratne 2011).

Short message service (SMS)-based information dissemination systems were another group of initiatives in agricultural extension. Such systems included sending market prices of vegetables to registered farmers (De Silva and Ratnadiwakara 2008), sending information on potential fishing grounds to the fishermen (Wijekoon 2014), and sending auction prices, yield predictions, the schedule for extension training programs, and daily weather forecasts to coconut growers (CRI 2017). SMS systems have the advantage of delivering timely information on the recipient's for a fee or nominal charge. Interested farmers can subscribe to these services by registering with the system. SMS-based systems were also tested for implementation of mobile learning programs using social media platforms (Dissanayeke et al. 2016). The efforts were successful in terms of achieving basic knowledge and cognitive skills such as comprehension; however, there were limited opportunities to design for higher order learning skills.

Interactive voice response systems (IVRS) can be considered as one of the more recent and more successful ICT initiatives developed for a mobile cellular platform. In order to receive the service, the recipient needs to initiate a call to the service, which makes it a demand-driven and an interactive service. IVR systems have been used in disseminating market prices (HARTI 2018), catering to specific information needs (https://www.dialog.lk/govi-mithuru, see Box 1) (Wijeratne and De Silva 2014; Senevirathne and Dissanayake 2014). All these IVR services partnered with mobile service providers to operate the IVR system and consequently were able to cater to a large number of recipients. Poor awareness among the farming community remains to be one of the main barriers to adopting these services (De Silva et al. 2012).

Box 1: *Govi Mithuru* (Dialog Axiata PLC)

Mobile telephony has been long recognized as a cost-effective method for reaching out to remote farmers as state-led visit-based extension systems have failed to keep up with the information needs in the field. The government itself has displayed the need to disseminate information through mobile telephony in previous joint ventures such as the HARTI-Mobitel platform. As of 2016, the global penetration of mobile phones stands at 93%. While South Asia as a region is lagging with 71% penetration, Sri Lanka records mobile penetration values comparable with the global average (in cities, the penetration exceeds 100%).

Govi Mithuru by Dialog Axiata PLC is a bilingual mobile-phone-based information dissemination platform launched in 2015. The service originated through a UK-funded development project coordinated by GSMA international. Content partners include the Ministry of Agriculture of Sri Lanka and Centre for Agriculture and Biosciences International (CABI). The main service contains crop information; however, there is supplementary content offered in partnership with the Ministry of Health, Nutrition and Indigenous Medicine of the Government of Sri Lanka, which provides family nutrition advice.

In the last quarter of 2014, product design and testing started with field surveys of key information sought by farmers, mainly paddy farmers. The focus was to develop a value-added product that was user driven and easily accessible in terms of obtaining timely and verified information as an add-on to Dialog mobile subscription. An initial hurdle was getting farmers on board through registration to the service. The typical farmer is poorly literate and not technology savvy. Thus, the four-step self-registration proved difficult for potential users with low technical literacy. Even at the pilot launch of the service for free, adoption was poor due to registration resistance. This hurdle

(continued)

Box 1 (continued)
had to be met with an operator-facilitated registration and field support through product marketing, where the operator or the marketing personnel guided the farmer through the registration or carried it out themselves on the farmer's behalf.

At first instance, the content carried by the service was on paddy, but it was expanded to six crops in the next season. The service took almost a year to reach the first 10,000 users. Initial radio marketing, coupled with field sales, led to minimal acquisitions at a high cost. The network of clients increased rapidly in the last quarter of 2016 through word-of-mouth promotion and rural marketing efforts, which included a Govi Mithuru branded SIM. In December 2016, 35% of the identified target market for the year had registered for the service.

At the commercial launch in November 2015, the product was priced at 1 LKR per day on top of the usual phone subscription. Service provided offered unlimited access to previous content when users revisited the content on their own.

In 2017, a field survey was conducted to evaluate the progress. By this time, the scale of registered users have risen to 300,000, making *Govi Mithuru* the fifth largest value-added service under Dialog Axiata PLC (Sri Lanka). Four key outcomes were reported, and GSMA attributed the following outcome to the content delivered through the service: 90% of repeat users reported at least one on-farm change of practice, 25% reported changes to planting dates, 12% reported changes to postharvest storage, and 60% reported decreased fertilizer and pesticide use.

3.4 Computers and the Internet

Computer-based information systems rose to prominence in the early years of the twenty-first century when the newly established Information Communication Technology Agency (ICTA) supported government organizations to establish websites under its e-governance initiations. Most of the agricultural institutions developed their websites under this project, and they started sharing various types of extension materials, which they have already developed using the website. Agricultural videos, recordings from radio programs, leaflets, and many other publications were shared on the website, giving easy access to interested stakeholders. Websites of agriculture-related organizations provide good repositories of extension training materials, which ensure wide access and easy reference and also serve as e-learning materials. Latest market prices of agricultural commodities are displayed on websites such as Hector Kobbekaduwa Agrarian Research and Training Institute (http://www.harti.gov.lk/), Coconut Development Authority (http://www.cda.lk), Department of Export Agriculture (http://www.exportagridept.gov.lk), and Sri

Lanka Tea Board (http://www.pureceylontea.com/). Price information given on these websites are frequently updated; thus, users can easily access recent and most up-to-date information.

Only a few websites, however, frequently update their contents, while even fewer would publish sector-specific and timely information on the web, such as agricultural news. The poor layout of websites when viewed through smart devices is another shortcoming, which needs to be considered in future developments.

Another major use of computers in the agriculture sector was the implementation of a computer-based extension service, which seems to be an affordable and convenient mechanism to reach rural farmers (Wijekoon 2014). The cyber-extension mechanism was initially implemented as an offline computer-based learning system, with a series of multimedia CDs carrying cultivation details of various food crops. The system was later improved to a complete online learning system with other interactive facilities like e-discussion forums and agriculture Wikipedia.

Management information systems (MIS) and decision support tools are two important computer-based information systems that have not been fully utilized in the agriculture sector so far. A series of MIS has been identified as potential strategies to implement in the coming years as a component of e-agriculture strategy (Table 2) (DOA 2016).

Table 2 Information systems proposed under e-agriculture strategy

Information system	Potential use
Pesticide registration and pesticide information system	A system to monitor the testing and registration process of pesticides
Plant protection e-service	Mobile and web-based application for correct diagnosing of pest and disease problems
Research information management system	Research information management system to gather, store, and share research work among agriculture scientists
E-agriculture library system	Electronic content management system for digital extension materials, e.g., IMMCDs, videos, etc.
Natural resource management information system	A system to manage soil and weather information for different agro-climatic zones and other legal documents related to natural resource management
Plant quarantine e-service	An information system to monitor plant quarantine procedures at seaports and airports
Weather forecasting and advisory service	An information system to provide daily weather forecasting and advisory services for farmers
Land use and soil conservation mapping and e-information system	An information system to map land use and soil conservation in agricultural areas
Geo-spatial information service	GIS to collect real-time information on agricultural areas to facilitate decision-making
Food crop forecasting and marketing information system	Real-time information system to record cultivation extents and generate yield forecasting

Source: DOA (2016)

3.5 Smart Devices

Smart devices such as smartphones and tablets have been tested for agriculture extension functions such as learning devices (Dissanayeke et al. 2016), networking tools to coordinate agriculture producers to share cultivation details for yield projections (Ginige et al. 2018), live agriculture radio streaming, and information management systems (Atapattu et al. 2017).

One of the important developments related to smart devices is their ability to provide access to real-time content-sharing and social media platforms. High availability of smart devices among the population coupled with lower mobile broadband charges is helpful in this regard. As Sri Lanka has one of the lowest broadband Internet charges and lowest prepaid charges in the world, it provides ample opportunities to integrate ICTs into agriculture information communication when compared to other ICTs. A smart device is used as a personal device, which can be carried out from place to place. Its ability to connect with agriculture stakeholders to access real-time information, its ability to facilitate interactive communications, and its ability to combine with other ICTs are some of its main benefits. Given the high accessibility and affordability of smart devices, there is a high potential to promote social media and open-source software applications on smartphone-based platforms to disseminate agriculture information.

Social media have been extensively used in sharing agricultural information in recent years. Promotion of the appropriate use of social networks and open-source materials to disseminate information among agricultural stakeholders needs to be emphasized (Saravanan 2012). Facebook was found to be the most popular social media platform among agriculture extension and advisory service actors, according to a study that covered 62 countries (Bhattacharjee and Saravanan 2016). Furthermore, some respondents believed that social media have the potential to bridge the gap between stakeholders in agricultural information services.

In Sri Lankan context, Facebook pages have been used by agriculture stakeholders such as government organizations, small and medium agri-business enterprises, extension agents, and farmers. Sharing information related to recent events and developments at individual and institutional levels was the main objective of these pages. Some of these pages showcase a rich source of extension teaching materials such as videos, leaflets, and other publications (https://www.facebook.com/pg/SLKDOA, https://www.facebook.com/departmentofexportagriculture/).

The interactive platform of social media provides ample opportunities for knowledge sharing, giving constructive feedback, and initiating meaningful discussions when compared to traditional ICTs such as radio and television. Another advantage is the ability of social media to couple with traditional ICTs such as video and audio sharing. Contents that are perceived to be interesting and important by the recipients will get shared in their posts, thus having a multiplication effect of extension.

Facebook pages were also used by small and medium agri-business operators to find a market for their products (https://www.facebook.com/haritah.agro/, https://www.facebook.com/sahanakrushipiyasadelgoda/). Furthermore, there were interest groups that used the Facebook page as a community forum to share knowledge and discuss current problems (https://www.facebook.com/bee.srilanka). These forms were similar to the communities of practice approach, where members of the learning community share their experiences in discussion forums. Mobile phones are increasingly being used as the main platform to access social media among agriculture advisory service actors (Bhattacharjee and Saravanan 2016). Microblogging platforms such as Twitter had been tested for initiating mobile learning among young farmer communities and was successful in providing basic cognitive skills (Dissanayeke et al. 2014). Free and open-source software has the potential to be used among agricultural communities on a small scale (Wijekoon 2014; Jayathilake et al. 2018).

4 Economic Returns to Information and ICTs

Information is a factor in the production function of agriculture. With fast-changing production technologies and uncertain market dynamics, real-time access to information may determine the entire viability of a farming venture. Connecting farmers to ICT has a long-lasting impact since the scope of impact on the bottom layer of the pyramid essentially can create changes throughout. ICT has significant sunk costs in reaching out to rural locations, and incremental usage is not costly as in other material inputs.

One of the influential theoretical arguments for ICT in agriculture is related to returns to information. Information/knowledge is an input in the production function of agriculture. Just as labor and capital have respective returns to inputs, information is responsible for its own share of returns or profits. Often on public patronized agricultural systems, the information component is subsidized and not paid for in the revenue. One of the key developments in the recent advent of private investment ICT in agriculture is that information is no longer subsidized and accounted as a legitimate input in the production activity.

Lio and Liu (2006) report that the coefficients of ICT adoption on agricultural output in high-income countries vary between 0.35 and 0.29 (statistically significant). For poorer countries, the corresponding coefficients of the ICT adoption variable are also significant but considerably lower, between 0.092 and 0.181. The richer countries display about two times the estimated elasticity of the ICT variable than poorer countries do.

Muto and Yamano (2009) studied rural Ugandan households when the mobile phone network was increasing rapidly. The proportion of the farmers who sold a perishable commodity (banana) increased in communities more than 20 miles away from district centers, while the impact was not seen in nonperishables (maize).

Box 2: Use of Free and Open-Source Software (FOSS) in Facilitating Agricultural Communications

A 5-year research study undertaken to introduce free and open-source software such as FrontlineSMS for texting, Freedom Fone for voice calls, and Ushahidi for interactive mapping in facilitating agricultural communications through networking farmers with extension agents proved to be effective in supplementing extension communications at field level.

The research team designed eight communication campaigns covering four districts to test the FOSS technologies in evaluating the possibility and effectiveness of knowledge sharing within farming communities. Extension officers were trained on the use of FOSS during the campaigns. Technology stewards (TS) (Wenger et al. 2009) were identified beforehand and trained on using FOSS and designing and running short-term communication campaigns. Then the TS were encouraged to design and implement a campaign targeting a given communication problem of the community.

The use of "FrontlineSMS" for "texting" and Freedom Fone for "voice" has improved the day-to-day agricultural communications by 22% and 8%, respectively, when the process has been assisted by TEs. The study concludes that field-level extension agents can be effectively trained to use FOSS, which can eventually lead to a reduction of the transaction costs associated with the sharing of information with farmers.

(Source: Jayathilake et al. 2017, 2018)

4.1 Reduction of Transaction Costs

The second influential economic argument for ICT in agriculture is the transaction cost perspective. In multistakeholder systems, transaction costs are high due to the many frictions that exist between the interfaces. Many scholars have pointed out that transaction costs in the agricultural value chains are prohibitively high, especially in low-income country settings. These are mainly due to coordination failures and nonsynchronized decision-making in the production activities. As the pressures on markups are passed to the primary producers if the consumers are not absorbing the same, the presence of transaction costs essentially translates to low returns to producers vis-à-vis the other parties in the value chain.

Aker (2008) shows that cellular phones reduce grain price dispersion across markets by a minimum of 6.5% and reduce intra-annual price variation by 10% in Niger. The primary mechanism by which cell phones affect market-level outcomes appears to be a reduction in search costs as grain traders operating in markets with cell phone coverage search over a greater number of markets and sell in more markets (Fig. 3).

Small-scale African farmers have shown significant time and cost savings in using information and communication technology (ICT) for extension services.

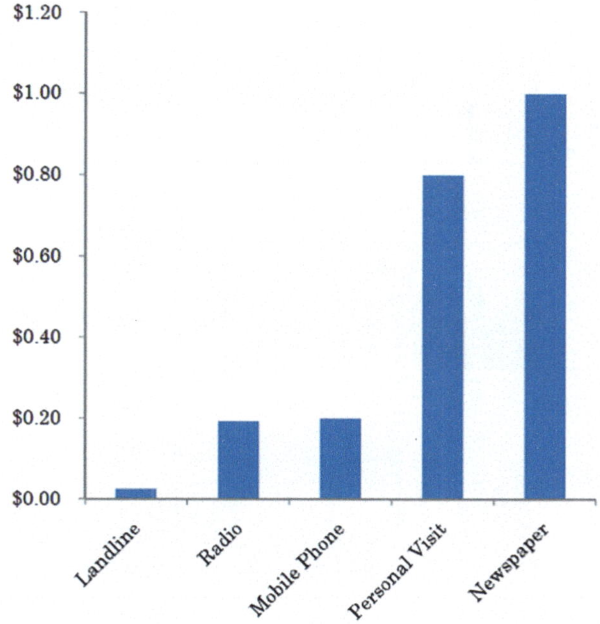

Fig. 3 Relative costs of different dissemination tools

Aker (2011) estimated that instead of traveling to visit a farmer, extension agents may use mobile phones for a comparative cost of one fourth of the price of a visit.

4.2 Improving Market Access and Participation

Access to markets is a long-standing bottleneck in low-income agriculture globally. In smallholder sectors, which do not receive the price and quality signals generated at the markets efficiently and transparently, producers are production oriented. Their failures are due mainly to the lacuna in market information and the modalities and flexibility to investigate markets. Thus, market access still remains the most dominant economic basis for ICT development in agriculture.

In mobile telephony, the delivery of information is mainly through short message service (SMS), although voice messages, interactive voice response (IVR) systems, or mobile applications (apps) are also used.

In the literature, there are examples of the significant impact of mobile phones on earnings and reducing price volatility. Indian fishermen have benefitted from lower price shocks and lower wastage due to better access to information using digital technologies (Jensen 2007) (Fig. 4).

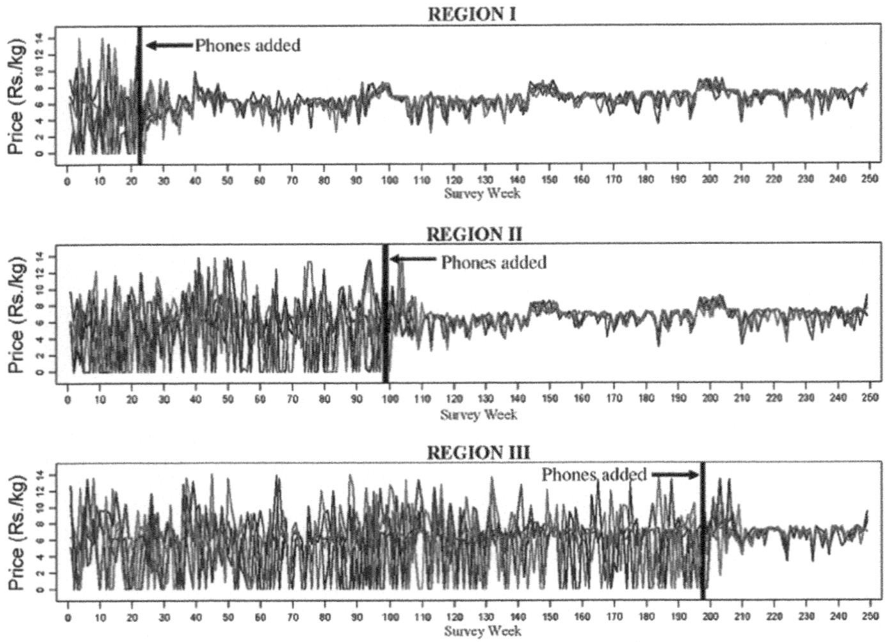

Fig. 4 Price volatility reduction after the introduction of mobile phones in three regions of India

4.3 Market Creation: Certification and Traceability

Verification and documentation of materials and practices have emerged as a key requirement in global value chains. In particular, niche markets demand both certification and traceability as consumer requirements. Information on certification and traceability thus has an economic impact on market creation and sustainability of demand. It also adds to real-time monitoring of food safety.

5 ICT Interventions in Agriculture Extension: A Typology

Looking at the ICT interventions implemented in the field of agriculture information communication, a few major developments can be observed in terms of providing real-time information and advisory services, content creation and sharing, providing interactive platforms to share feedback, combined media models, networking stakeholders, building social capital, and moving for communities of practice.

Recent advances in the telecommunication sector, together with the affordability of mobile communications, enabled real-time communications among agriculture stakeholders. Market information systems such as agri price index (HARTI 2018) offer agriculture commodity prices to interested stakeholders using an IVR system.

Market information exchange platforms for farmers and buyers have been developed as Android-based applications and made available to the community. Daily prices that are collected from several dedicated economic centers are distributed in this intervention. Another important development is the provision of agriculture advisory services over telephone, allowing stakeholders to access extension advice as demand arises.

Sharing agricultural information is regarded as one of the main roles of extension. Websites have been extensively used in sharing important contents. However, the most important development in the field is the availability of user-generated contents such as individual Facebook pages, which were started by independent stakeholders to share information with closed community groups. These independent web pages are found to be sharing specific problems in the field with rich interactions and creating user-generated contents. Social media platforms also provide opportunities for interaction among stakeholders, sharing of feedback, and networking of stakeholders, which eventually lead to communities of practice in agricultural extension.

Thus, three main typologies can be identified in relation to ICT interventions in agriculture extension; the first generation of ICTs, which mainly followed one-way and top-down models; the second generation of ICT tools, which facilitated interactions with key extension and advisory personnel and farmers; and third generation of ICTs, which supports the networking of agricultural stakeholders, leading to communities of practice.

6 Main Challenges and the Way Forward

One challenge in implementing ICTs in the agriculture sector is catering to an extension system that is highly fragmented and complex (World Bank 2007). Extension and advisory services are oriented toward catering to certain commodities, requiring the extension personnel to be highly specific about the technologies they transfer to end users. An integrated extension approach, which combined services from different key sectors, was tested in 1993 but failed mainly due to poor coordination between key departments. Thus, extension systems remain more independent, pursuing traditional, top-down, and supply-driven approaches to the various projects and programs they implement. Furthermore, the extension and advisory work is also subdivided under government, private, and nongovernment sectors. This has given rise to the emergence of a series of sector-specific and highly specialized approaches of ICT interventions. Thus, an average client may have to depend on several services to obtain information.

Use of ICTs at community level is affected by a number of factors such as poor access to ICTs and lack of locally available contents (Kumar and Singh 2012). Further lack of awareness among end users on ICT tools in agricultural extension, poor ICT literacy among extension staff, and administration problems were also identified as main challenges during the implementation of the cyber-extension project (Wijekoon 2014). Web radio implementation faced challenges such as poor awareness; poor access to ICTs, including computers and the Internet; difficulties in listening to programs on computers as they were not very user-friendly; and inconvenience of the broadcasting timing of radio programs (Kanchana and Dissanayeke 2016).

Blanket approaches to introducing ICTs are often excessively top-down and externally driven, and thus the needs of the recipients are not adequately addressed (Unwin 2009). Other countries in the region have faced barriers such as need for energy, poor literacy, and gender-based digital divide; however, none of these were seen as main problems in Sri Lanka (Dissanayeke et al. 2009). Rather, the adoption of ICTs was positively influenced by education level and computer exposure (Dissanayeke et al. 2009).

The lack of independent research to evaluate the real impact of ICT initiatives can be seen as one of the main challenges to be faced in relation to ICTs in agriculture development. Most of the interventions had been evaluated by the institution conducting it as part of the implementation process, while results of such evaluations had been used to improve the process (Wijekoon 2014).

6.1 Future Directions for Investments for ICTs in Subsectors of Agriculture

New ICT products can be used to develop stand-alone tools such as decision support tools, as well as to improve the usability of traditional extension media. There is a potential to combine "traditional extension media such as radio, television with new ICTs (World Bank 2017) and develop mixed media models" (Sharma 2013). For instance, Krushi Radio combines with SMS systems so that information related to the program's schedule can be passed down to farmers as a reminder, while the same SMS system can be used by the farmers to pass their queries to live radio shows. The program schedule and SMS system can be introduced to the farmers during training programs by the extension agents (Kanchana and Dissanayeke 2016). Such models can be used to enhance the extension communication systems by introducing new dimensions such as interactivity to the traditional one-way approach of extension.

The potential of ICT tools in the systematic collection of field information has not been fully utilized yet. Field problems experienced by farmers can be easily reported to extension agencies using ICTs. Similarly, farmer feedback for various extension activities can be gathered using ICT-based systems. Diagnosis of field problems and collecting information are regarded as important functions of extension that can be addressed using ICT tools such as cell phones (Bell 2015).

It is necessary to initiate independent research on the usability of ICT integrations in agriculture and obtain user feedback on such integrations. Most of the recent initiatives have been introduced using project approaches, and the results of the evaluations have been used to further improve the system. These evaluations have been mostly conducted by the same institutions that implemented the system. It would be beneficial to plan, from the inception, for such new initiatives to be implemented as action research projects. Such an approach would provide empirical evidence of greater validity to determine the effectiveness of such initiatives. Further, outside organizations such as independent research groups need to initiate research on the applicability and usefulness of ICTs in agricultural extension which will provide more research-based evidence.

References

Aker JC (2008) Does digital divide or provide? The impact of cell phones on grain Markets in Niger. University of California, Berkeley

Aker J (2011) Dial "a" for agriculture: using ICTs for agricultural extension in developing countries. Agric Econ 42(6):631–647

Atapattu DTM, Silva GLLP, Dissanayeke U, Dunusinghe SG (2017) Development of android based mobile information system for intensive and semi intensive dairy farmers. In: Proceedings of faculty of agriculture undergraduate research symposium –2017. Faculty of Agriculture, Peradeniya, Sri Lanka, p 186

Bell M (2015) ICT–Powering behavior change for a brighter agricultural future. USAID, Washington. [Online]. http://www.fao.org/e-agriculture/sites/default/files/uploads/kb/2016/03/2015-meas-discussion-paper-ict-powering-behavior-change-in-ag-ext.pdf. Accessed 30 May 2018

Bell MA, Payne J (2014) ICT in extension. IPO information for impact series. IPO. UC Davis In Mark Bell (2015) ICT–powering behavior change for a brighter agricultural future. USAID, Washington. [Online]. http://www.fao.org/e-agriculture/sites/default/files/uploads/kb/2016/03/2015-meas-discussion-paper-ict-powering-behavior-change-in-ag-ext.pdf. Accessed 30 May 2018

Bhattacharjee S, Saravanan R (2016) Social media: shaping the future of agricultural extension and advisory services. GFRAS Interest Group on ICT4RAS. [Online].. https://www.g-fras.org. Accessed 28 May 2018

CRI -Coconut Research Institute Sri Lanka (2017) Annual report. Coconut Research Institute, Lunuvila

Davis K (2008) Extension in sub-Saharan Africa: overview and assessment of past and current models and future prospects. J Int Agric Ext Educ 15(3):15–28

De Silva H, Ratnadiwakara D (2008) Using ICT to reduce transaction costs in agriculture through better communication: A case-study from Sri Lanka. [Online]. http://www.lirneasia.net. Accessed 28 May 2018

De Silva LNC, Goonetillake JS, Wikramanayake GN, Ginige A (2012) Towards using ICT to enhance flow of information to aid farmer sustainability in Sri Lanka, in ACIS 2012: Berlin, Heidelberg. In: Proceedings of the 23rd Australasian conference on information systems 2012, ACIS, Geelong, VIC, pp 1–10. http://dro.deakin.edu.au/view/DU:30049096

Department of Census and Statistics (2017) Sri Lanka Labour Force statistics: Quarterly Bulletin. Ministry of National Policies and Economic Affairs, Colombo. Sri Lanka. Issue no. 76. [Online]. http://www.statistics.gov.lk/samplesurvey/LFS_Q1_Bulletin_WEB_2017_final.pdf. Accessed 28 May 2018

Dissanayeke UI, Wickramasuriya HVA, Wijekoon R (2009) Evaluation of computer based learning materials in agricultural information dissemination in Sri Lanka. Trop Agric Res 21(1)*

Dissanayeke U, Hewagamage KP, Ramberg R, Wikramanayake GN (2014) Creating m-learning opportunities to facilitate collaborative learning: a mobile SMS based twitter implementation. In: International conference on advances in ICT for emerging regions: Colombo, Sri Lanka, vol 14. https://doi.org/10.1109/ICTER.2014.7083878

Dissanayeke U, Hewagamage KP, Ramberg R, Wikramanayake G (2016) Developing and testing an m-Learning tool to facilitate guided-informal learning in agriculture. Int J Adv ICT Emerg Reg 8(3)

DOA – Department of Agriculture Sri Lanka (2016) Sri Lanka E-agriculture strategy. [Online]. https://www.doa.gov.lk/ICC/images/publication/Sri_Lanka_e_agri_strategy_-June2016.pdf. Accessed 20 May 2017

DOA – Department of Agriculture Sri Lanka (2017) Farm Broadcasting service. [Online]. Available at: https://doa.gov.lk/ICC/index.php/en/2016-05-09-08-39-8/29-farm-broadcasting-ser vice. Accessed 2017.05.20

Ekanayake EMHB (1995) Study of agricultural television programme of the Department of Agriculture (Unpublished undergraduate dissertation). University of Peradeniya, Peradeniya, Sri Lanka

FAO – Food and Agriculture Organization of the United Nations, and ITU – International Telecommunication Union (2016) E-agriculture strategy guide: piloted in Asia-Pacific countries. [Online]. http://www.fao.org/3/a-i5564e.pdf. Accessed 30 May 2018

Farrington J (1995) The changing public role in agricultural extension. Food Policy 20(6):537–544

Ginige T, De Silva L, Indika A, Ginige A (2018) Extending DSR with sub cycles to develop a digital knowledge ecosystem for coordinating agriculture domain in developing countries, pp 268–282. https://doi.org/10.1007/978-3-319-91800-6_18

HARTI – Hector Kobbekaduwa Agrarian Research and Training Institute (2018) Mobitel "6666" Agri Price Information Index. [Online]. http://www.harti.gov.lk/index.php/en/news-events/mobitel-6666-agri-price-information-index. Accessed 30 May 2018

Hassan SM, Shaffril HAM, Ali MSS, Ramli NS (2010) Agriculture agency, mass media and farmers: a combination for creating knowledgeable agriculture community. Afr J Agric Res 5 (24):3500–3513

IGI Global (2018) What is Information and Communication Technology (ICT). [Online]. https://www.igi-global.com/dictionary/information-and-communication-technology-ict/14316. Accessed 28 May 2018.

Jayathilake HACK, Jayasinghe-Mudalige UK, Perera LDRD, Gow GA, Waidyanatha N (2017) Fostering technology stewardship approach to promote knowledge sharing among farming communities in Sri Lanka. Trop Agric Res 28(3):238–246

Jayathilake HACK, Jayasinghe-Mudalige UK, Perera LDRD, Gow GA, Waidyanatha N (2018) Use of free open source software technologies to enhance knowledge mobilization in smallholder agricultural communities in Sri Lanka. Trop Agric Res 29(2):147–156

Jensen R (2007) The digital provide: information (technology), market performance, and welfare in the South Indian fisheries sector. Q J Econ 122(3):879–924

Kanchana KGI, Dissanayeke UI (2016) A mechanism to promote Krushi FM – web radio among farming communities. In: Proceedings of the International Peradeniya University Research Sessions (iPURSE). University of Peradeniya

Khan GA, Muhammad S, Chaudhry KM, Khan MA (2010) Present status and future preferences of electronic media as agricultural information sources by the farmers. Pak J Agric Sci 47 (2):166–172

Kumar D, Singh K (2012) Information communication technologies. In: Agricultural extension: innovations and dimensions, 1st edn. Satish Serial Publishing House, Delhi, pp 247–288

Kumari BSA, Sivayoganathan C, Sisira Kumara WAG (2009) Effectiveness of toll free agricultural advisory service to vegetable growers in Kandy district. In: Proceedings of agricultural extension conference, 27–28th August 2009, Kandy, Sri Lanka, pp 355–371

Lio M, Liu MC (2006) ICT and agricultural productivity: evidence from cross-country data. Agric Econ 34(3):221–228

Madana MHBPH, Dissanayeke DMLB (2009) Field assessment of toll free information delivery service: case study in Anuradhapura District of Sri Lanka. In: Proceedings of agricultural extension conference, 27–28th August 2009, Kandy, Sri Lanka, pp 341–354

Muto M, Yamano T (2009) The impact of mobile phone coverage expansion on market participation: panel data evidence from Ugandaî. World Dev 37(12):1887–1896

Nazari MR, Hassan SBH (2011) The role of television in the enhancement of farmers ' agricultural knowledge. J Agric Res 6(4):931–936

Rajasinghe JCK, Samansiri BAD (2014) Evolvement of agricultural in tea research institute and present applications. In: Wijeratne M, Mahaliyanarachchi RP (eds) Agricultural extension: recent interventions for development in Sri Lanka, 1st edn. Sri Lanka Agricultural Extension Association (SLAEA), Kandy, pp 111–128
Saravanan R (2012) ICTs for agricultural extension in India: Policy implications for developing countries. In: Proceedings of the 8th Asian conference for Information Technology in Agriculture, AFITA, pp 1–11
Senevirathne MAPK, Dissanayake DMDSK (2014) Change of extension interventions to suit the present needs of the export agriculture sector. In: Wijeratne M, Mahaliyanarachchi RP (eds) Agricultural extension: recent interventions for development in Sri Lanka, 1st edn. Sri Lanka Agricultural Extension Association (SLAEA), Kandy, pp 99–110
Sharma A (2013) Community radio as an alternative tier of broadcasting: scripting a new role for development. In: Singh V, Kahsyap SK, Papnai G (eds) Agricultural communication: opportunities for sustainable agriculture and rural development. Biotech Books, New Delhi, pp 8–25
Unwin T (2009) ICT4D: information communication technology for development, 1st edn. Cambridge University Press, Cambridge, pp 360–376
WEF- World Economic Forum (2016) The global information technology report 2016. [Online].: http://reports.weforum.org/global-information-technology-report-2016/networked-readiness-index/. Accessed 30 May 2018
Wenger E, White N, Smith JD(2009) Digital habitats: stewarding technology for communities. CPsquare, Portland OR
Wijekoon R (2014) Use of information communication technology for rural development in Sri Lanka. In: Wijeratne M, Mahaliyanarachchi RP (eds) Agricultural extension: recent interventions for development in Sri Lanka, 1st edn. Sri Lanka Agricultural Extension Association (SLAEA), Kandy, pp 29–54
Wijeratne S (2011) Mobile telephony for agricultural development of Sri Lanka. [Online]. https://ssrn.com/abstract=1976180. Accessed 22 May 2018
Wijeratne M, De Silva N (2014) Agricultural extension in practice: a contribution to enhance small scale mushroom production. In: Wijeratne M, Mahaliyanarachchi RP (eds) Agricultural extension: recent interventions for development in Sri Lanka, 1st edn. Sri Lanka Agricultural Extension Association (SLAEA), Kandy, pp 159–172
World Bank (2007) Reviving Sri Lanka's agricultural research and extension system: towards more innovation and market orientation. [Online]. https://openknowledge.worldbank.org/bitstream/handle/10986/13044/694660ESW0P0870and0Extension0System.pdf?sequence=1&isAllowed=y. Accessed 5/1/2018
World Bank (2017) ICT in agriculture: connecting smallholders to knowledge, networks, and institutions, Updated edn. World Bank, Washington, DC. https://doi.org/10.1596/978-1-4648-1002-2

Public–Private–Producer (PPP) Partnerships in Sri Lankan Agriculture

Pahan Prasada

Market access and value chain innovation are key components of a modernized agricultural sector. Recent developments in private sector initiatives and public–private partnership initiatives have contributed substantially to the modernization of agriculture. This chapter considers the performance of partnership ventures and discusses the role of such partnerships in local agriculture. Sri Lankan experience with agribusiness PPP model is studied with reference to the National Agribusiness Development Program (NADeP), a public–private–producer (PPP) framework operational under the government.

1 Partnerships as Organizational Innovation

In the era of globalization, multinational development institutions championed privatization of state and para-state institutions as part of the overall agenda of economic liberalization. The large state institutions that carried a mandate in production and service provision were identified as potential candidates for privatization. The stated motives were gains in efficiency and the hope of capital injection from local and foreign investors. While some of the privatization initiatives were established voluntarily, others were imposed as conditions accompanying international lending. Privatization was seen as an effective instrument in the restructuring of economies to reduce the fiscal burden that large state institutions entailed. On the other hand, it eliminated bottlenecks in labor and capital markets, helping these institutions integrate with global economy without friction.

The wave of privatization lost its novelty, in certain instances, after the realization that expected social objectives were not adequately met and, in other instances, after

P. Prasada (✉)
Department of Agricultural Economics and Business Management, Faculty of Agriculture, University of Peradeniya, Peradeniya, Sri Lanka

R. P. De Silva et al. (eds.), *Agricultural Research for Sustainable Food Systems in Sri Lanka*, https://doi.org/10.1007/978-981-15-3673-1_16

the realization that private management was not substantively more efficient than state mechanism (Prager 1994). There are issues of capacity that beset private actors from carrying out large projects efficiently. While there always was the option to commit an outright transfer of management responsibility to international corporations, which certainly had the capacity, sovereign states were hesitant to go as far in fear of public disapproval.

A public–private partnership is a framework that, while engaging the private sector, acknowledges and structures the role of the government in ensuring that social obligations are met and successful sector reforms and public sector investments are achieved (Asian Development Bank 2008). While partnerships are considered a positive development, in most instances, what passes as a partnership may not be a true partnership that shares costs, risks, and returns. Certain so-called partnerships are just cases where a private actor enters into joint operations where traditionally only public actors operated, such as energy, transportation, and hard infrastructure.

PPPs overwhelmingly specialize in the provision of public goods and services in both high-income and low-income settings. This is partially due to the fact that PPPs are not recognized as replacements of private sector investment but are promoted as an efficient alternative to the supply of public goods and services. This, however, does not limit the possibility of a PPP platform specializing in providing private goods in conjunction with public goods. The literature has almost no examples on the supply of exclusively private goods by PPPs. Agricultural PPPs have an impact on the supply of both public and private goods. Customized extension services, vocational training, ICT-based services, and machinery inputs qualify as private goods in many agricultural PPPs.

Partnerships are essentially an organizational solution to the resource coordination problem. As such, these ventures are investment vehicles and sector modernization tools. The appeal of public–private partnerships in low-income countries and emerging economies is borne out of the operating scales that only partnerships can access without being burdened by the administrating costs that large government entities carry. In Sri Lanka, public–private partnerships are still nascent institutions, especially so in the agriculture sector. In rural development models advocated by the International Fund for Agricultural Development (IFAD) and the Food and Agricultural Organization (FAO) of the UN, PPPs feature as structures capable of delivering commercial and development objectives alike. Agri-PPPs are broadly promoted as having the potential to help modernize the agriculture sector and contribute toward the pursuit of sustainable and inclusive agricultural development (WEF 2011). World Economic Forum's New Vision for Agriculture (NVA) primarily aims to modernize agriculture inclusively through initiatives of partnership by 28 global leading private firms.

While public–private joint ventures have a long history, the addition of "producer" to the loop is evident only in recent years. It is mainly practiced as a means of inclusion of the poor and the smallholders directly into the institutional framework. In the context of agricultural projects, the conversion of established outgrower systems into partnership ventures has provided a natural pathway for including the producer to the PPP apparatus.

2 Challenges in Agribusiness Sector and Relevance of PPP Objectives

2.1 Challenges in Agricultural Value Chains

Agriculture traditionally provided an institutional basis for the rural sector, in addition to the provision of subsistence and livelihood. Key functions of agriculture were interlinked with land ownership, use of water, access to community resources, excess food supplies for social safety, and cultural activities. With the transformation of agriculture into a private enterprise (rather than a social activity), the institutional structures built around agriculture rapidly underwent change. In particular, agriculture was organized around value chains directed toward different types of product markets, often beyond national borders.

Still, the predominance of smallholder activity as the bottom rung of the value chain persisted due to land user rights held by farming households across the rural sector. The production function was carried through the collective efforts of the large rural community in most low-income countries. Agriculture provided subsistence to these farmers, and surplus output was collected to channel to markets in the urban sector. Technology and know-how were disseminated by government extension and research services. Labor was locally sourced, and other external inputs were purchased from suppliers or the public sector. As such, at the stage of production, smallholder agriculture remained a collective activity with large public sector involvement.

The delivery of surplus to the market and processing activities tend to be a difficult proposition to the public sector since numerous producers were acting independently in small units. Therefore, the collection of surplus, processing, and channeling to end users attracted a number of private actors. However, because of poor transport infrastructure in many low-income countries and high entry costs (and economies of scale of distribution), collecting and marketing activities have been vulnerable to monopolistic or oligopolistic practices, which in turn reduces competitiveness in the markets. Ultimately, these irregularities lead to price fluctuations. To address these anticompetitive and collusive possibilities, many countries utilized public sector marketing boards.

The performance of marketing boards was a mixed success and varied depending on other operational and environmental factors. The combination of corruption and bureaucratic inefficiency of many of the boards failed to extend support to value chains as readily as expected. Both governments and development agencies have therefore looked for alternative mechanisms that can increase the share of value obtained by a primary producer while integrating the primary producer to the value chain. Simultaneously, attempts were made to make agriculture market friendly with less state control in line with liberalization initiatives.

Sri Lankan experience in supporting smallholder participation in value chains has seen most of the classic interventions. Land redistributions, high-quality planting material development, free extension and training, farm subsidies, and marketing boards have all been attempted. Support has also been given in developing wholesale and special purpose market and distribution facilities. However, credit subsidization and business development activities are the frontiers that have not received adequate attention.

2.2 Key Functions of PPPs

Farming occurs in seasonal cycles. Farmers purchase (or borrow) inputs and farm for a season and settle the dues at harvest. If integrated into a business contract, the cyclic operation can function without being affected by financial bottlenecks. In other words, farmers need not be facing liquidity issues if inputs are available through contracts. This is a key function for a PPP. In addition, where financing becomes critical is value addition to harvest. Storage and processing technology and market development are also challenges that are interlinked to financing at this stage. As will be discussed in the next section, many PPPs have been driven to cater to this stage specifically. The largest limitation on smallholder agriculture is capital and credit shortage. While this is not the exclusive focus of any PPP, it remains a priority challenge for all PPPs. As a result, almost all PPPs have some form of credit subsidization built into their operating framework. Usually, absence of collateral in the rural sector and high costs of loan management have made banks reluctant to lend to agricultural enterprises. The usual policy initiative to encourage more lending by commercial banks is the implementation of refinancing programs with the central banks, which result in subsidized credit to farmers.

2.2.1 Financing Gaps

The magnitude of financing needed in the agricultural sector in low-income countries is often beyond the capacity of one investor. The traditional investor in (agricultural) infrastructure is the government. But most low-income country treasuries are beset with budgetary bottlenecks that prevent investing in agriculture (FAO 2012). A relative rise in public sector finance for agriculture was observed globally during 2000–2008 (Beintema et al. 2012) but has not reached the sufficient levels to reach the needed scales. Therefore, a primary goal of agricultural PPPs is the procurement of needed capital.

2.2.2 Risk Sharing

Apart from climate and environmental risks in crop production, structural risks such as low returns on investment; limited access to factors of production, including land; high transaction costs and production risk associated with heterogeneity among small-scale producers; and government interference in agricultural markets compound the uncertainty in agriculture. PPPs provide sufficiently diversified platforms that can absorb risk impacts and respond constructively. Therefore, PPPs act as risk-smoothing devices. Potential for agricultural insurance in smallholder settings (Prasada 2020a, b; Yallarawa and Prasada 2020) can easily be linked to the risk smoothing role of PPPs, especially when agribusinesses or the public sector act as insurers. In fact, PPPs are seen as a facilitators for popularising agricultural insurance among smallholders.

2.2.3 Innovation and Market Access

PPP projects create the possibility of benefiting from the innovation capacity and efficiency criteria of the private sector while promoting the pursuit of sustainable agricultural policy objectives. The private sector is specialized in logistics and marketing, which public apparatus lacks. The market infrastructure available to the private sector in terms of cost-effective logistics makes PPPs attractive to governments for reaching their development objectives.

2.2.4 Food Security and Inclusion

To date, almost all low-income country PPPs in agriculture have been motivated at least partly by both food security at national level and pro-poor growth. Inclusion of the marginalized and the poor in mainstream economic activity is one of the key justifications of international grant funding for PPPs. FAO and IFAD are among the institutions that have initialized food security and inclusion objectives in PPPs through outright grants, matching grants, or specialized credit lines. In all low-income country PPPs initiated through global grant funding, inclusion of youth and women has been stated as a priority.

3 A Typology of PPPs in Agriculture

All partnerships share the abovementioned broad objectives, but their operation priorities may differ. Based on the current information in Asia, Africa, and South America (FAO 2013), PPPs can be classified into four distinct but related categories:

(a) Partnerships for promoting agricultural value chains toward achieving higher local value addition
(b) Partnerships for joint agricultural research and technology transfer
(c) Partnerships for building and upgrading market infrastructure
(d) Partnerships for the delivery of business development services to farmers and small enterprises (BD)

It is important to comparatively understand the Asian context with reference to Africa and Latin (South) America. There is a clear variation in the priorities of agricultural PPPs in Asia compared to those in the other two settings (Fig. 1). Value chain development and technology transfer goals dominate the PPPs in Asia. African experience in agricultural PPPs displays a predominant focus on value chains only. Compared to Asia, there is no significant attention to market infrastructure and business development services. This may be due to the more localized value chains present in Africa (Fig. 2).

Apart from the key orientation, the second characterizing aspect of PPPs is financing arrangement. In this instance also, Asian experience stands out from

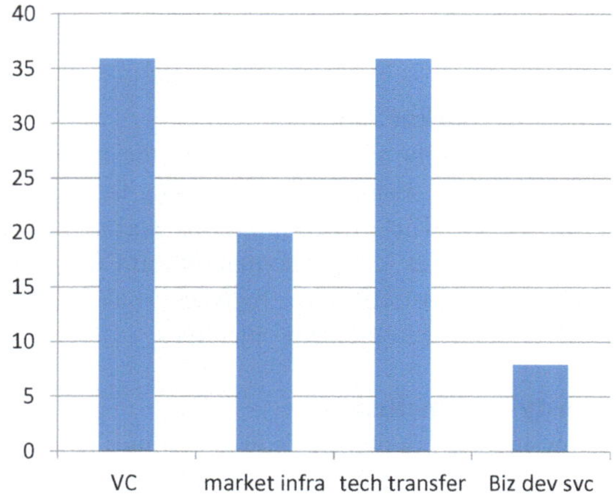

Fig. 1 Different orientations of Asian agriculture sector PPPs as percentages. (Source: FAO 2013)

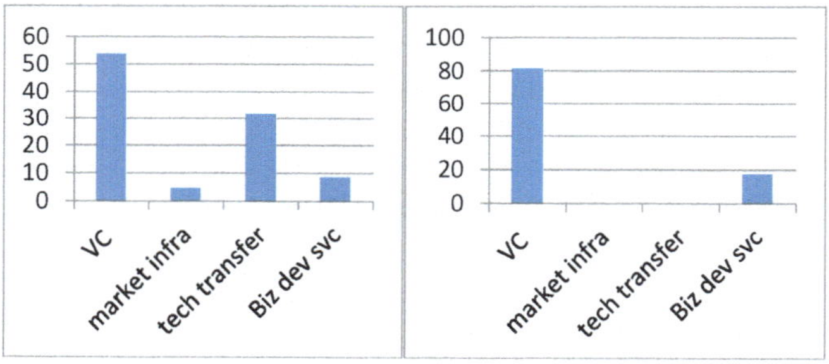

Fig. 2 Different orientations of African (left panel) and South American (right) agricultural PPPs as percentages. (Source FAO 2013)

African and South American circumstances (Fig. 3). Approximately, 65% of all agricultural PPP finance is co-equity investments. Nearly 15% each of the funds are from development projects and grants. In contrast, African PPPs source funds mainly from development projects. South American countries have a unique reliance on sovereign/state funds for the purpose of financing PPPs and a weaker reliance on co-equity investments (Fig. 4). Development project funds and grants are virtually absent in South American PPPs (FAO 2013).

4 Modalities of Partnerships

The differential roles and responsibilities of public partners and private partners are key to the sustainability of PPPs. Producers remain the grassroots implementers and constitute the observable aspect of PPPs. In terms of governance, the producers will be contractually bound to the objectives and evaluation criteria of the partnership

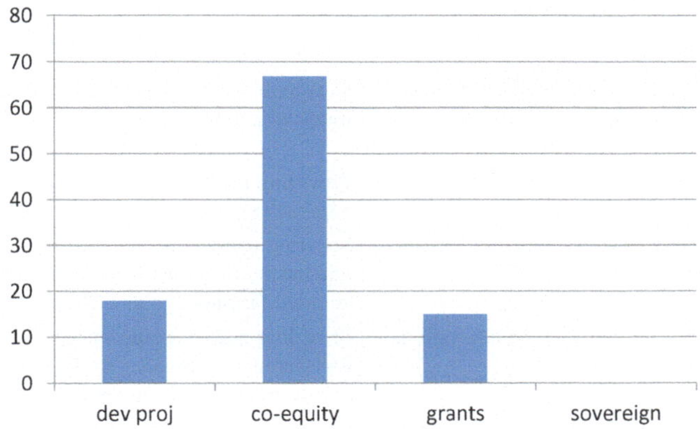

Fig. 3 Percentage decomposition of financing arrangement in Asia. (Source: FAO 2013)

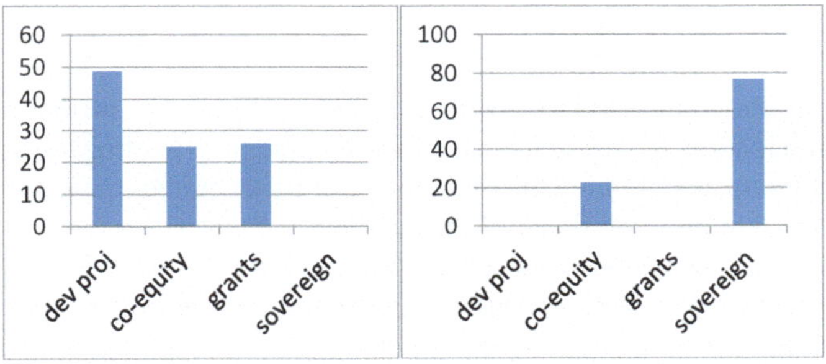

Fig. 4 Percentage decomposition of financing arrangements in African (left) and South American PPPs. (Source: FAO 2013)

agreement. However, the scope and function of public and private partners need to be assessed at the initial stages before the creation of a partnership since there is no functional space for correcting mismatches in objectives while the partnership is in operation. This is mainly due to the lack of process and capacity for renegotiation once the agricultural partnership is initialized (Table 1).

5 Cases and Concerns in PPP Adoption in Sri Lankan Agriculture

5.1 National Agribusiness Development Program (NADeP) and Smallholder Agribusiness Partnership Program (SAPP)

In 2012, the government of Sri Lanka approached the International Fund for Agricultural Development (IFAD) for financing an agriculture sector development proposal. This proposal led to the creation of the National Agribusiness

Table 1 Alternative roles for the institutional partners

Public sector responsibilities	Private partner contributions
Creating a supportive regulatory environment with appropriate incentives for private sector investment	Developing business plans with thorough financial and market analysis
Ensuring the inclusion of smallholders	Providing incentives and outreach devices to approach smallholders
Alignment with national socioeconomic and sector development priorities	Securing markets for end products and purchasing materials from farmers through contract farming agreements
Determining private partner selection criteria	Providing documentation necessary for the transparency of projects to the public; maintaining fairness in producer selection
Training and technical assistance for program management units	Providing technical assistance and business management training to producers
Coordinating negotiation and contract signing	Generating links with financial sector partners for capital and other business development services

Development Program (NADeP) with three key objectives: increasing income through participation in marketing chain development, providing microfinancing to the target group, and training the rural youth for better employment opportunities. NADeP is the first program financed through public funds (loan funds) to promote public–private partnerships in the development of rural value chains. By 2016, 17 PPP business plans were operational under the program, at various stages of implementation, reaching out to 14,910 beneficiary households. The total investment amounted to USD 16.35 million (USD 1125 per producer). The investment per beneficiary approximates to one third cofinancing by NADeP, the private sector (i.e., promoter company) partner, and the beneficiary (including in-kind contribution and contribution through NADeP-facilitated credit). Under the microfinance and youth training component, the number of households mobilized and receiving support is approx. 18,000, of which approx. 7000 have received credit. The Central Bank of Sri Lanka (CBSL) and eight commercial banks implemented the lending facilities to meet producer needs at a negotiated interest rate of 6.5%. The loan product includes a maximum loan size of LKR 150,000 with a maximum 6-month grace period and repayment within a 6-month to 3-year loan term. The overall project evaluation by IFAD in late 2016 reports an average incremental income arising from PPP investments of LKR 92,000 per annum or LKR 7700 per month.

NADeP had a slow start due to the institutional constraints built around its initial program management unit under the Central Bank. The procurement of business proposal did not perform with adequate speed due to the bureaucratic restrictions of being placed under CBSL. In 2015, NADeP was repositioned under the special program unit of the Presidential Secretariat. The institutional change led to a speedy canvassing of proposals and timely disbursements. IFAD reports that 17 PPP agreements were reached within 12 months from this transition.

SAPP originated as a second phase of NADeP, and objectives were very much in line with the original expectations. Scaling up NADeP and partnering with third-party financiers such as banks were an added objective in the transition to SAPP. More consequentially, SAPP considered the proponent of the partnership to be either a private sector company or a farmer organization that can show evidence of a confirmed commercial arrangement with a private sector company. This eventually opened up space for farmer companies to benefit from partnerships.

The first-order challenge in a partnership is the screening and selection of partners. NADeP's process for seeking partners operated as a call for business proposals from private promoter companies. The proposals were evaluated by the program office under the supervision of a national steering committee for advisory and guidance. The approval of the proposals led to a financing agreement of the project. The main investors were a public party, a promoter company, and a donor (IFAD) as a third party. The public funds came through the IFAD-mediated loan facility. Figure 5 considers 12 PPP projects that were operational during 2015–2017 in terms of how the investment commitments are shared between mainly public and private parties to the agreement. Government contribution (state inv), private sector investment (pvtinv), and foreign investment (extinv) are stacked for each partnership venture in the diagram.

Each of the 12 projects had a different coverage of client smallholder producers. Based on the commitments of funds, Fig. 6 displays the relative investments by public vs. private parties in million LKR calculated on a per-producer basis. Except in two projects, all other projects feature higher per producer investment by the public party. The two partnerships where private investment exceeded public investment per producer were two small-scale ventures on beekeeping and dairy.

Figure 7 displays the variation in the focus area of PPPs under NADeP, weighted by the number of producers targeted as part of the partnership. In terms of numbers of producers, the dairy, organic corn, and gherkin projects stand out.

It is important to understand what industry has received attention in terms of business promise based on the realized funding commitments. Contrasting per producer investment by the different agricultural activities reveals that government and private priorities were more or less aligned in terms of the relative size of per

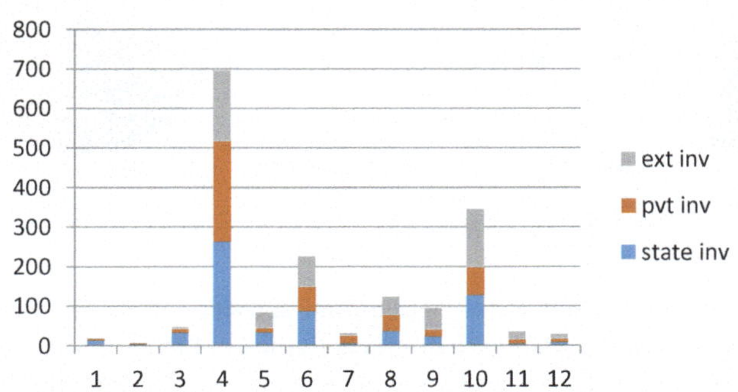

Fig. 5 Sharing of investment between partners (in million LKR)

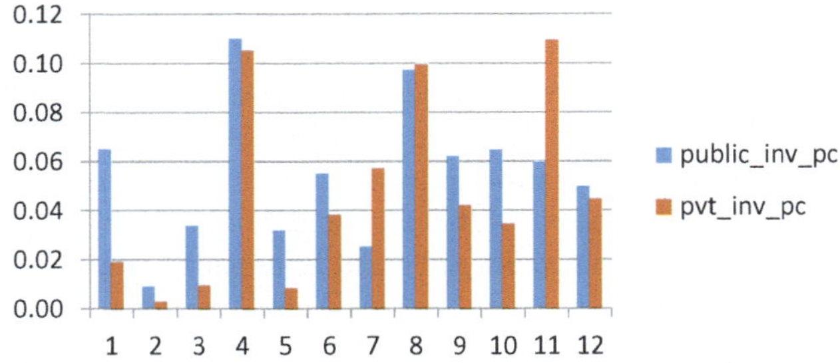

Fig. 6 Per producer investment sharing by public vs. private actors (in LKR million). (Note: public investment per producer is shown in blue, while private sector investment per producer is shown in orange color)

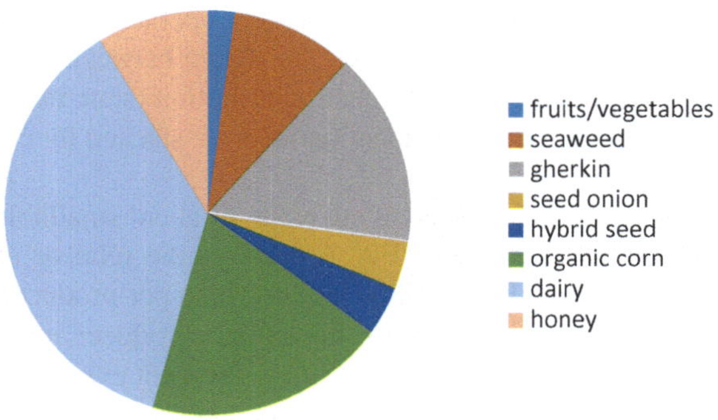

Fig. 7 Producer coverage weights of PPP activities

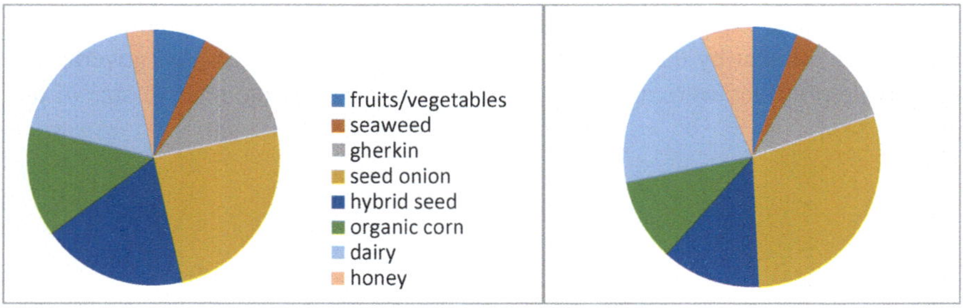

Fig. 8 Per producer investment by the state (left) vs. private promoters (right)

producer investment in each of the agricultural activities (Fig. 8). On the left-side panel of Fig. 8, the color-coded agricultural activities in Fig. 7 (above) are presented indicating the relative size of government investment. On the right, the same statistic corresponding to private investment is displayed. If a colored area for a particular activity was significantly different from left to right, it implies that public and private parties observed or perceived the economic promise of the given agricultural activity differently.

According to Fig. 8, differences in commitments by the state and private entities are observable mainly in seed onion, hybrid seed, organic corn, and dairy. In seed onion and dairy, it can be argued that the private party sees more promise, while in hybrid seed and organic corn, the state perceives more potential. An alternative argument may be that the private parties chose the profitable sectors, while the state picked the residual activities. But there is no evidence to substantiate the motives of selection. All other activities receive approximately the same commitment by the state and private parties.

5.2 Potential Negative Impacts of PPP Projects in Sri Lankan Agriculture

The drive toward PPPs in agriculture is not without its share of issues. Partnership ventures operate at a scale that is larger compared to traditional agriculture. Large scales imply pressure on resource use, especially open-access resources. PPPs often prioritize crops and livestock that may not necessarily be among the traditionally adopted. Even if they are the same crops, the scales favored by PPPs will differ widely with the scales that are practiced otherwise. To date, there had been no Sri Lankan study to substantiate any selective preference of certain scale of production by PPPs. In addition, they make use of new production linkages and networks that deviate from the norms of the locality, such as village collection by small-scale collectors and farmer organization level decision-making. All these possibilities imply that there needs to be sufficient sensitization of the community and the establishment of safeguards that could buffer negative implications and resistance.

The potential threat to traditional social structures and community interests is often highlighted as a widespread implication of agricultural PPPs. Farmer organizations are usually formed on the lines of power hierarchy in the community. Social organization under a value chain often follows an economic logic, favoring the more entrepreneurial members for leadership. There are cases that community divisions could surface under different social groupings under PPPs.[1]

Another motive under partnership ventures is the possibility of farmer groups under the partnership to be used to test different technologies based on an external interest. In particular, the farmer groups under a given partnership venture have loyalty toward the value-chain-promoting company. They are bound to the interests of the company, indirectly limiting individual or community discretion in substantive decisions relating to resource use (and even managing environmental consequences).

[1]In one of the rural settings in North Central Province of Sri Lanka, such a division of community surfaced in association with a PPP when part of the community was interested in following the 'organic' practices prescribed by an external initiative while others were interested in status quo.

Table 2 List of threats from PPPs

Activity	Potential negative impacts
Private-company-centered farmer collectives	Social cohesion within the village may be weakened due to business-motivated collective formation in contrast to the more traditional farmer collectives based on irrigation infrastructure and cultural norms
	The role of women (and the youth) in the created business collectives may deteriorate under a possible male-led structure
Expansion of agricultural activity for commercial purposes using subsidized/open-access resources	Irrigation infrastructure and water resources are largely publicly owned and, in most cases, they are open-access resources. Under business-led models, the exploitation of such resources may exacerbate
Expansion of dairy collection	Exploitation of open–access pastures
Intensive and high-input agricultural models	Potential increase in soil erosion, surface, and subsurface water pollution
Promoter-company-driven farming technologies	Potential negative impacts from the introduction of inappropriate technologies and gradual disappearance of traditional knowledge
Incorporation of high-value alien crop varieties and livestock breeds	Potential negative impacts on biodiversity
	Vulnerability of introduced species to climate-related pest and disease outbreaks
High agrochemical use in cash crops	Potential negative impacts from the overuse of agrochemicals on human health, surface and groundwater contamination, and increased resistance to pests/disease
Vertical integration of agricultural value chains	Elimination of small-scale traditional farmers and vendors due to organized collection and retailing

Disappearance of agro biodiversity is also likely under PPP ventures in agriculture. Though it is not a certainty (since certain initiatives promote diversified farming), this is a likely outcome. Most PPPs aggressively promote export-driven crops and intensify farming exclusively for foreign market interests. In doing so, partnership ventures may disregard the needs of local community and the environment and prioritze foreign market interests instead. Table 2 summarizes several consequences.

5.3 Parallel Institutional Developments

Public–private–producer partnerships are not isolated initiatives in the wider realm of agriculture sector reform. There are other concurrent developments that have been initiated in Sri Lanka under the aegis of development loans and grants.

There is a wave of agriculture sector modernization projects under the World Bank lending that has reached Sri Lanka as well. The modality of the modernization project setting is to seek private sector investment in large-scale agricultural activity in identified crops and livestock with high foreign market potential. These models also target the long-standing land market rigidities in low-income countries and credit shortages. An explicit focus on value chains is also present. Modernization projects operate through the Ministry of Agriculture and create state-led organizational structures for policy formulation toward increased market access. Investment vehicles such as the facilitation of direct investment by foreign or transnational firms are considered in those areas of agriculture where technology and distributional channels can benefit from global standards. One example is dairy farming.

Another development that completes and complements with PPPs are the institutional infrastructure promoting farmer companies created by transforming farmer organizations or farmer cooperatives. Farmer companies were a slightly dated initiative conceived in 1995 but legislated under the Companies Act of 2007. By 2003, there were 92 farmer companies registered with the Registrar of Companies in Sri Lanka (Esham and Usami 2007). Despite the potential advantages of these institutions over their traditional counterparts (i.e., cooperatives), Senanayake (2004) argues that many farmer companies had failed due to (among other reasons) inadequate recruitment of managerial personnel, weak planning and governance, inappropriate mechanisms for monitoring and evaluation, and lack of trust between farmer members and their company. Judging by the record of growth since its 1995 initiation, the farmer company initiative has failed to take off as of 2018.

The last competing development is the wide spread of microfinance services. Leasing and finance companies have developed a wide array of loan and finance products especially directed toward farmers and rural enterprises. These operate through aggressive marketing efforts of the outreach staff of these firms and capitalize on the demand for capital and durables.

6 Governance and Policy Issues in Partnerships

The advent of PPPs brings about novel challenges to administration and policy. The relationship of state actors with PPPs is one of enthusiasm, followed by some disenchantment. There is a tendency for such partnerships to reach subsequent consolidation, depending on the capacity of the farmers. Different countries were caught up in the waves at different times. What accounts for this lack of consistent opinion on the merits of partnerships? Klein (2015) claims that PPPs can outperform the public sector, even though evidence suggests that well-run public firms tend to match the performance of private firms in regulated sectors.

The accountability and transparency requirements that form the basis of the public sector may not hold as tightly in PPPs, which are fundamentally investment vehicles that attract capital (often not the cheapest capital available). In contrast, in most agricultural PPPs, at least some of the core objectives are welfare oriented, requiring a public or civil sector approach. Therefore, in practice, promoting institutional frameworks applicable to agricultural PPPs is a continuing priority. In the African and Asian contexts, new PPP laws and policies that envisage the application of the PPP model to the agribusiness sector have been developed. To date, there is no "best practice" model for regulating PPPs, but the tendency is to administer PPPs through the formation of a PPP management unit within the Ministries of Agriculture or as special task units under the treasury or planning departments. Often the involvement of public sector officials as members in PPP committees occurs when PPP units are established independently.

The Sri Lankan experience in this regard is particularly informative. The PPP unit was initially set up by a Central Bank framework but was revealed to be ineffective in carrying out dynamic partner selection and disbursement under regulatory delays. In response, the unit was transferred to operate under the special project division of the Presidential Secretariat. While the external stakeholders and donors may want to see rapid outcomes, short circuiting of procedures of procurement, commissioning, and lending may lead to a lack of transparency and unequal treatment. This observation is evident from the fact that only large agribusiness companies dominated the first stage of the program.

While it is not possible to discuss such implications in reference to a specific PPP or a class of PPPs, reviews of the comparative performance of PPPs across countries shed light on inconsistencies at the country level. The documentation on Procuring Infrastructure PPPs (2018) reports that the lower the income level of the country, the lesser the information on PPP procurement process disclosed (ranging from 100% in OECD high-income economies to 50% in low-income economies). For instance, Sri Lanka is among the countries that provide bidders only with the result of the procurement process, without disclosing the grounds for the selection of the winning bid.

The governance implications of PPPs can be analyzed at stages of its performance cycle. Comparative evaluations of PPPs by international lending institutions highlight three stages.

(a) Preparation of the PPP

The key governance elements of this stage are quite similar to those relevant to large projects. Starting from the approval by the central budgetary authority (i.e., parliament and treasury, etc.), environmental, social, and risk screening practices are considered prerequisite. Economic and financial analysis and fiscal benchmarking also provide for a sound footing of the venture. Operational guidelines and accountability criteria are stipulated to streamline implementation. According to the 2017 cross-country evaluation of infrastructure PPPs, Sri Lanka was assigned 52% success at this stage. This is below par compared to the European average of 78% success rating. However, among South Asian countries, Sri Lanka is marginally above average.

(b) Procurement

At the outset, procurement includes recruiting qualified private partners if the venture is initiated by a state agency. The more routine elements of procurement involve laying down the guidelines and enforcement mechanism for bidding, tender practices, and contract documentation. If the PPP entails specialized inputs such as research and development, establishment of the process of research and development is also part of the procurement. Sri Lanka's standing (as per cross-country assessment) at this stage is 76%. It is widely believed that Sri Lanka has the political and legislative will to encourage private sector participation, and higher performance at this stage can be attributed to this tendency.

(c) Contract Management

In large PPP ventures with high external investment, this stage entails legal inputs toward recording contractual obligations of various stakeholders. In most agricultural PPPs, the core contract is an agreement on various resource shared, evaluating progress in terms of key performance criteria and how the returns are divided. If companies are formed between stakeholders as part of PPP or if equity/asset sharing is incorporated, these aspects also fall under this stage of the PPP. The space for various parties to deviate from contractual agreements and the eventual termination of contracts is also an important part of PPPs that involve large operational and sharing risks. The other key role of governance at this point is a solid dispute resolution mechanism and a grievance redress framework. Sri Lanka's standing on this aspect is noted as 51% by the global comparative survey.

(d) Independent Regulation and Dedicated Authority for PPPs

The consideration of PPPs as a different category to be regulated and codified by law (business/civil) is often discussed as an issue. While a separate legal status for PPP governance will provide clarity and transparency to all parties concerned, most countries have opted to consider PPPs as special projects. The World Bank records that the percentage of countries governing PPPs under a dedicated management unit amounts to 81% as of 2017. Among these, the dedicated unit serves only an advisory role in 77% while the unit is mandated with complete authority only in 4% of the countries.

Among the South Asian countries, Pakistan recently enacted Public Private Partnership Authority Bill of 2017 intending to "provide a regulatory and enabling environment for private participation in provision of public infrastructure and related services." In addition to establishing the independent Public Private Partnership Authority (PPPA) with advising and gatekeeping roles for the development of PPPs, it also provides a stronger legal basis for many of the areas only contemplated in the 2010 Pakistan Policy on PPPs. For example, the requirement to ensure fiscal affordability is now legally codified, and several steps in the procurement process (including the publication of the procurement notice), as well as dispute resolution mechanisms, are now specifically regulated for PPPs.

7 Conclusion

The limited experience that Sri Lankan agriculture has in PPPs and the wider experience from foreign examples show certain common features of PPPs that are operational especially in low- to middle-income countries. Agricultural PPPs are unique in that they contain explicit involvement of a smallholder community. In the Sri Lankan model, smallholders are linked indirectly to the PPP via a promoting private company (as opposed to direct linkages observable in farmer company models). A limitation in the existing model under NADeP and currently under SAPP is that producers that are currently or historically in partnership with the private company are selected when constituting the PPP. Such a practice essentially crowds out the partnerships that farmers can have with the private sector through barriers to entry and may also increase dependence on state or donor funding in activities that private actors may willingly participate in. A potential solution may be to screen producers and select producers through objective and observable criteria.

PPPs bring about its own share of challenges in all spheres: social, economic, environmental, and political. Managing these concerns and professionally managing a partnership venture will be a test to Sri Lanka and other low-income country governments, given the lapses in governance such as bureaucratic delays, high transaction costs, and lack of business development services and market support systems, which are embedded in our institutional structure. The capacity of the local agriculture sector to benefit from PPPs hinges on the space for correcting mismatches in objectives while the partnership is in operation. Such mismatches emerge mainly due to the lack of process and capacity for renegotiation once the agricultural partnership is initialized.

References

Asian Development Bank (2008) Public-private partnership handbook. ADB, Manila

Beintema N, Stads G-J, Fugile K, Heisey P (2012) ASTI global assessment of agricultural R&D spending: developing countries accelerate investment. International Food Policy Research Institute, Washington, DC

Esham M, Usami K (2007) Evaluating the performance of farmer companies in Sri Lanka: a case study of RidiBendi Ela Farmer Company. J Agric Sci 3(2):86–100

FAO (2013) Agribusiness public–private partnerships: country reports. Rome. Africa: Ghana, Kenya, Nigeria, Uganda, United Republic of the Tanzania; Latin America (in Spanish): Chile, Colombia, Ecuador, Guatemala, Peru; Asia: Indonesia, Pakistan, Thailand. Available at: http://www.fao.org/ag/ags/ags-division/publications/country-case-studies/en/

Food and Agriculture Organization (FAO) (2012) State of food and agriculture: investing in agriculture for a better future. FAO, Rome

Klein M (2015) Public-private partnerships: promise and hype (no. 7340). World Bank, Washington, DC

Prager J (1994) Contracting out government services: lessons from the private sector. Public Adm Rev 54:176–184

Prasada DVP (2020a) Climate-indexed insurance as a climate service to drought-prone farmers: evidence from a discrete choice experiment in Sri Lanka. In: Handbook of climate services. Springer, Cham, pp 423–445

Prasada DVP (2020b) Performance and potential of agricultural insurance: global and Sri Lankan perspectives. In: Agricultural research for sustainable food systems in Sri Lanka. Springer, Singapore, pp 369–387

Procuring Infrastructure PPPs (2018) Public-Private-Partnership In Infrastructure Resource Center (PPPIRC). World Bank, Washington, DC

Senanayake MS (2004) What is ailing farmer companies in their transformation into successful business entities? Paper presented at the third international conference of the Japan Economic Policy Association, November 13–14, 2004. Meiji University, Tokyo, Japan

World Economic Forum (WEF) (2011) Realizing a new vision for agriculture: a roadmap for stakeholders. Available at: http://www.weforum.org/reports/realizingnew-vision-agriculture-roadmap-stakeholders

Yallarawa YSM, Prasada DVP (2020) Demand for crop insurance by tea smallholders in Badulla district: an analysis of willingness-to-pay. Trop Agric Res 31(3):01–10

Impact of Emotional Intelligence in Managing the Outcomes of Agricultural Extension and Advisory Services

L. N. A. C. Jayawardena

1 Development of Agricultural Extension and Advisory Services in Sri Lanka

Sri Lanka has been based on agriculture as a country. The historical legends of Sri Lanka reveal that the agriculture sector was established approximately 2000 years back. Irrigation system has developed gradually adopting agriculture as a means of livelihood. The successive rulers of the country have given a prominent place to the agriculture sector in their political agendas. With the invasion of Europeans, a commercial-level agriculture economy based on plantation crops was established in the country, and rice and cereal had to be imported. Plantation sector was developed as the major source of income earning, and an institutional setup was established. The British colonial rulers established a separate institutional setup in Sri Lanka to ensure food security. Agricultural extension officers were appointed to popularize agriculture crops in the country. Agricultural extension and advisory services (AEAS) consist of a number of different activities providing the information and services needed and demanded primarily by farmers and other stakeholders to help them develop their competencies and practices to improve livelihood and well-being (GFRAS 2012). The agriculture crop production programme was further strengthened by appointing food production overseers to assist agricultural extension officers. The Department of Agriculture, established in 1912, was mandated to develop food crops in the country. A separate Agriculture Research Institute was established subsequently. The Research Institutes for Tea and Rubber were established in 1925, and the Rice Research Institute in 1930. An agriculture officer was appointed for every province of the country, and 74 agriculture diploma holders were appointed to the post of agriculture instructor in 1939 to strengthen the agriculture extension system in the country. Methods such as mobile cinema units

L. N. A. C. Jayawardena (✉)
Department of Agricultural Extension, Faculty of Agriculture, University of Peradeniya, Peradeniya, Sri Lanka

R. P. De Silva et al. (eds.), *Agricultural Research for Sustainable Food Systems in Sri Lanka*, https://doi.org/10.1007/978-981-15-3673-1_17

and field demonstrations were gradually introduced into the agricultural extension system. The appointment of district agricultural extension officers in 1957 further established the agricultural extension system in Sri Lanka. A dedicated division for agricultural extension was established in 1964 with the new position of deputy director of agriculture (extension). Establishment of Agrarian Development Centres, to provide services for the agrarian community under one roof, in 1970 was another important step in the evolution of agricultural extension in Sri Lanka. Creation of the Department of Minor Export Crops in 1972 saw a specific focus on the agricultural extension activities of minor export crops in addition to the food crops in the country.

The 'Training and Visit' extension system, introduced in 1979, improved the coordination among extension officers, research institutes, and training institutes substantially. The 'Second Agricultural Extension Project' was implemented in 1993 to resurrect the agricultural extension system in Sri Lanka. It was terminated in 1998 and the 'Yaya' demonstration programme was implemented, focusing on the paddy growing sector. Government appointed 25 district directors in 2012. These district directors, stationed in District Secretariats, were assigned to focus on agricultural extension. The linkage between central government and provincial system strengthened the agricultural extension system through this approach. The appointment of technical assistants under the supervision of agriculture instructors in 2015 was a notable step in the empowerment of agricultural extension system in the country. The emergence of private sector agricultural extension practices indicates the need to develop the AEAS in Sri Lanka. For example, corporate sector tea estates use different parties such as advisers from head office, private consultants, agricultural-input-based private-extension service providers, tea-related organizations, and non-governmental organizations (NGOs) (Rajasinghe 2009). Similar practices can be observed in rubber cultivation and other sectors. Farmers frequently interact with agro-input suppliers and private sector organizations for information and know-how (Dissanayake et al. 2005; Wanigasundera 2015).

2 Role of Agricultural Extension and Advisory Services

Despite the many stages of agricultural extension systems, namely, training and visit system, Second Agriculture Extension Project, new agriculture extension approach, etc., Sri Lanka does not have a proper agricultural extension system for existing global agricultural trends (Arunapriya et al. 2018; Wanigasundera 2015). Sri Lanka does not have an established policy on agriculture, and that has affected especially the fate of the public sector extension and advisory services. The merits of many of the above agricultural extension approaches could not be realized mainly due to the issues in implementing them. The departmentalization of AEAS rather than offering them in packages, and lack in understanding of priorities by leadership groups have not helped the situation either (Arunapriya et al. 2018; Wanigasundera 2015). There seem to be many inefficiencies in the transfer of information at the field level. There have been issues regarding the responsibility of planning and implementing AEAS interventions due to lack of competencies of those involved (Wanigasundera 2015). Agricultural extension is a complex process, consisting of diverse structures and interdependent relationships (Arunapriya et al. 2018; Wanigasundera 2015). The nature of AEAS demands much more than having technical competence of agricultural practices. It encompasses the challenge of handling a plethora of responsibilities on and off the field. AEAS functions not only include policy engagement and advocacy at senior levels but also leadership, organizational development, networking, partnership development, negotiation, and mentoring (AESA 2016). Gunawardena (2005) indicates effective communication with stakeholder groups (especially the farming community) as the essence of AEAS, providing knowledge and information to rural people to improve their behaviors for sustainable benefits. The approach and the nature of interaction with farmers greatly contribute to the effectiveness of the overall outcome in farming. What has not been properly documented is the expertise categories in field-level interactions with farmers. The capacity of agricultural extension professionals to deliver through appropriate human interactions plays a significant role in the effectiveness of message delivery. Field-level AEAS staff needs to have not only sound technical knowledge and information communication technology skills. Functional skills related to the need assessment of farmers, communication (oral and written), and the facilitation of community mobilization are also vital in AEAS (AESA 2016). Understanding and valuing the nature of interactions among stakeholders in agricultural extension and advisory services play an important role. There is an increasing need to enhance the capacity of agricultural extension personnel to facilitate the application of new knowledge (Wanigasundera 2015). Professional competence enriches the relationship among the extension professionals and farmers. Capacity building through the development of human resources of key stakeholders in extension and advisory services and conducting related research is a priority (AESA 2016). Understanding appropriate methodologies to empower field-level employees to enable timely solutions for the problems encountered at the grassroot levels is essential. Capacity development within AEAS is often misunderstood as "enhancing the technical capacity" of the agricultural extension staff on current and existing technologies (Prasad et al. 2015). Recognition of agricultural extension as a distinct discipline and a profession is long overdue in Sri Lanka. The mindset that anybody is capable of conducting AEAS needs to be changed and challenged (Wanigasundera 2015). Conceptual clarity on AEAS and the appreciation that they need a specialized skill set in their own right have to be institutionalized at a national level. This recognition is imperative to enhance the self-efficacy of professionals involved in AEAS. It will facilitate the capacity development interventions of agricultural extension staff. Developing policies and systems on human resource development of extension staff and promoting cross-learning opportunities within the AEAS setup for other service providers would benefit both the process and system (AESA 2016).

3 Human Capital in the Performance of Organizations

Agricultural extension services are being carried out largely by public sector organizations and, to some extent, by private sector organizations in Sri Lanka. Intellectual capital plays a key role in diverse organizational structures and systems (Bullen et al. 2006) and is applicable to all those public and private sector organizations involved in AEAS in Sri Lanka as well. Intellectual capital consists of human, structural, and relational capital (Ferrier 2001). It consists of all of the knowledge resources contributing to the competitiveness of a working setup or an organizational structure (Guthrie 2001; Vallejo-Alonso et al. 2013). It encompasses the intangible value in the know-how of employees and their internal and external relationships, management staff, and other stakeholders. Intellectual capital represents not only the capacities of employees' (or members') minds but also the complex intangible structure that surrounds them. It plays a significant role toward the functionality of organizations and systems (Fazlagic 2007). Accordingly, intellectual capital contributes significantly to the performance of AEAS in Sri Lanka. The synergistic role effects of human, structural, and relational capitals have been proven at work in varying contexts (Carson et al. 2004). Human capital has been identified as "an inventory of skills and knowledge" of individual employees/members in an organization (Lynn 2000). Organizations benefit through their human capital in a combination of members'/employees' capabilities, commitment, and ideas (Snell and Bohlander 2007). Structural capital (at times referred to as organizational capital) consists of explicit knowledge within the organization, its culture, and process. It includes intangible assets, viz., information systems, distribution networks, work team strategies, market intelligence, and knowledge of structures, systems, and the market (Mouritsen et al. 2001). Relational capital consists of organizational networks and the goodwill of the organization. It refers to the value generated through external relationships with whom that it transacts (Carson et al. 2004). Human capital, structural capital, and relational capital are inherently formulated and driven in organizations by employees/members. In Sri Lankan agricultural extension systems, many resources are owned by the public sector. However, their conversion into intellectual capital (especially into structural and relational capital) has been poor. On the other hand, private sector does a better job in converting limited resources and generating higher portions of intellectual capital. This can be understood through analyzing the deployment of human capital in AEAS. Organizations offering AEAS should not only have a developed system to manage their staff; they also need an organizational culture that supports "collaboration, learning, and networking" (AESA 2016). In addition to that, promotion of cross-learning within the agricultural extension systems is needed to improve the effectiveness of extension systems and services (AESA 2016).

Literature suggests that human capital can be categorized according to personal attributes. These personal attributes consist of personality traits and psychological attributes of intelligence and skills. Structural capital (processes and procedures) forms in to the organization in context. Structural capital in essence is "captured human capital, which depends on the knowledge and skills of organization's employees/members (Carson et al. 2004). This relates to the development of systems and procedures to facilitate the delivery of agricultural extension systems. This indirectly reflects the expertise and the commitment of employees. It is noteworthy that the Department of Agriculture has developed innovations integrated with information communication technology (Wijekoon and Sisira Kumara 2018). Relational capital depends on the knowledge and skills of employees in developing and sustaining the relationship networks with key stakeholders. Public sector organizations are poorly structured with regard to the generation of relational capital in AEAS. The absence of an established agricultural extension policy has also contributed to this. The conflicting (and sometimes overlapping, due to central and provincial government setups) structures in public sector organizations have negatively impacted the generation of relational capital. The loss of phase (and promise) of relational capital has significantly reduced the potential of public sector AEAS in the country.

It indicates that human capital is inherently involved in intellectual capital and in each and every component of it, namely, human, relational, and structural capital. Human capital plays a significant role in the effectiveness of organizations and membership groups. The knowledge and skills of agricultural extension officers and their relationships with key stakeholders make a big impact on the quality of agricultural extension services provided. An important component of human capital is the "tacit knowledge" of employees/members. The nature of AEAS is such that it is not pragmatic to document all the practices. An agricultural extension officer enriches the transmission of know-how based on his or her tacit knowledge. Not only does that supplement the agricultural practices recommended, it also increases the output of farmers. The competencies of agricultural extension staff contribute to the credibility of the extension and advisory systems and their acceptance by the farming community and other stakeholders.

4 Human Capital and Emotional Intelligence

Organizations have been keen to measure human capital (or human intelligence) to identify and recognize the relative value of employees/members to reap the benefits to the organization. Intelligence is a type of mental ability pertaining to the handling and reasoning of different information (Carroll 1993). There are differences in the interpretation of intelligence. Broader, cohesive, groups of abilities are at the middle level, and the highest hierarchy level (of intelligence) involves abstract reasoning across domains (Mayer et al. 2007).

The term "Intelligence Quotient" (IQ), by Alfred Binet, has captivated the world's attention from the beginning of the twentieth century. Many IQ factor models represent cognitive abilities as a three-level hierarchy. Scholars and practitioners started to use IQ as the ultimate yardstick with growing acceptance and reputation over the years. Many Sri Lankan organizations (public, private, and nongovernmental sectors) have been using IQ tests for the selection of executive cadre employees. The pluralistic views of intelligences gathered momentum and contributed to the increased attention among psychologists for delimitations of human intelligence. Amid this interest and enthusiasm, there were many psychologists experimenting on different components of intelligence. Mayer and Salovey are credited with the pioneering work on emotional intelligence (EI). Daniel Goleman made it popular among a wider audience through his book *Emotional Intelligence* (1995). "Emotion" has been viewed as a state of integrated feeling involving physiological changes, motor preparedness, cognition about action, and inner experiences (Mayer et al. 2007). There are alternative views of emotion (Averill and Nunley 1993). Overall, EI approaches can be broadly categorized into two models: (i) the ability model and (ii) the mixed model. Mayer et al. (2007) opined that EI represents abilities combining "intelligence and emotion" to enhance thought. Thereby, the primary focus of EI was identified as to reason about emotions and to use emotions to enhance thought. Mayer and Salovey (1997) have defined EI as *the ability to perceive emotions, to access and generate emotions so as to assist thought, to understand emotions and emotional knowledge, and to regulate emotions so as to promote emotional and intellectual growth.* Further, EI ability model consists of two components based on the focus on specific abilities. Those abilities either focus on a particular skill that is fundamental to EI or are based on the integrative approach (consisting of all the specific EI skills) identifying EI as a cohesive, global ability (Mayer et al. 2007). The four-branch model of EI (Mayer and Salovey 1997; Salovey and Mayer 1990) views overall EI as joining abilities from four areas: accurately (a) perceiving emotion (b) using emotions to facilitate thought, (c) understanding emotion, and (d) managing emotion (Mayer and Salovey 1997; Mayer et al. 2003). In the interaction of an agricultural extension officer with farmers, the ability of the extension officer to recognize the basic emotions in the faces of the farmers is likely to precede the ability to detect the faking of emotional expressions by farmers. As skills of the extension officer grows in one area (such as perceiving of farmers' emotions), his/her skills in other areas (i.e., understanding of farmers' emotions) continue to develop. In other words, having a higher level of EI provides an agricultural extension officer the opportunity to identify the farmers' emotions, to use identified emotions to think of the context of interaction, to understand the emotions encountered by farmers, and to manage the emotions attached to the interaction with farmers at an improved level. It facilitates an effective approach in farmer interactions.

The EI mixed model (Goleman 1995) framework focuses on a combination of cognitive, personality, and affective attributes (Papadogiannis et al. 2009). These models encompass broader definitions of EI that include "non-cognitive capability, competency, or skill" (Bar-On 1997). They also accommodate behaviors that are seen "emotionally and socially intelligent" as components of the model (Bar-On 2003). Mixed models include "dispositions from the personality domain" (Petrides and Furnham 2003). Most measures in this category assess primary EI attributes such as *accurate emotional perception*, in combination with broader scales that are related to EI. For example, the behavioral characters of an agricultural extension officer, such as the ability to empathize with farmers in a given situation, relate to his/her EI. Having socially acceptable mannerisms and behavioral patterns in the interaction with farmers naturally contribute to improve AEAS and to obtain better outcomes.

There are many scientific evidence of the human anatomy relating to the existence of specific and specialized brain areas to regulate EI. Studies conducted on mapping the human brain areas have also confirmed that EI is associated with distinct brain areas (Takeuchi et al. 2011; Takeuchi et al. 2013). Reuven Bar-on and a brain research group used neuropsychology to identify the brain areas for specific behaviors and mental functions. They have identified brain areas specific for emotional abilities (Bar-On 2003). Handling of emotions can lead to positive or negative outcomes. They are inherently related to the anatomy and functioning of the brain. Goleman (2011) stated that the *prefrontal cortex* is the key neural area for self-regulation that guides people when they perform at their best. The *dorsolateral zone* of the *prefrontal area* regulates attention, cognitive control, reasoning, decision-making, voluntary action, and response flexibility. Conversely, the *amygdala* is a trigger point for emotional distress, fear, and anger and can lead to regretful actions. There are empirical findings to support the impact of brain injury related to the deprivation of emotional behaviors of people from many walks of life (BIRC 1998; Goleman 2011). Human brain consists of circuitry involving emotional centers, which contribute to superior performance and excellence at work, such as self-mastery. Research on the anatomy of the human brain has strongly established the existence of EI with numerous supportive evidence. There is sufficient evidence to say that EI has been established as an important component of human intelligences/human capital, along with IQ and other intelligences. Thus, EI has a definite contribution to the intellectual capital of organizations, be it in agriculture or another sector. EI serves as an important indicator of the quality of human capital and is a potential tool in the assessment of human capital (intelligences) in organizations (Goleman 2006). There are many progressive organizations across the world that are deploying EI-related measurement instruments in the selection of executive grade employees and service sector employees.

5 Emotional Intelligence in Organizational Performance

Many researchers have credited EI with occupational success (Kleinberg 2000; Rozell et al. 2002). EI has been interlinked to leadership effectiveness. Jensen et al. (2007) have indicated that EI distinguishes leaders among those with a similar level of IQ. Fielder's "contingency model of leadership effectiveness" has identified "the leader-member relationship" as the most critical variable in determining the favorableness of a situation among its three empirically derived dimensions (Luthans 2011). The conceptual works of Goleman (2001) and Boyatzis et al. (2006) have identified the "emotional competencies" that may be associated with effective leadership. Goleman (2001) has listed four characteristics of emotionally intelligent leaders: (a) they are articulate and arouse enthusiasm for a shared vision and mission; (b) they step forward to lead as needed, regardless of position; (c) they guide the performance of others while holding them accountable; and (d) they lead by example. The four characteristics indicate that emotional intelligence is associated with some skills or competencies of leadership. A positive relationship has been found between EI and transformational leadership style (Jayawardena 2012; Jayawardena 2014; Leban and Zulauf 2004). In extension systems, the extension officers have to play leadership roles, especially in the field. These leadership roles vary from major to minor decisions, and most of them are contextual. The ability of extension officers to lead in given contexts plays a decisive role in the acceptance of them by farmers and their effectiveness in eliciting the desired behavioral patterns from farmers. Decision-making has a significant effect on the achievement of goals and on the success of organizations. Decision-making patterns/styles on the field has a bearing on the outcome of extension and related efforts. It has been found that leaders make decisions largely through unconscious processes called pattern recognition and emotional tagging (Campbell et al. 2009). This is an inherent feature associated with people. Organizations/work settings could avoid flawed decisions, understanding how emotional associations are attached to decision patterns. Having managers high on EI facilitates this process. In providing AEAS, an extension officer interacts with many farmers and other stakeholders. Each interaction could trigger many other decisions, contributing to the overall effectiveness of the process. A manager's emotional style has an impact on the moods and behaviors of colleagues and subordinates. Emotions play their role in mood contagion, driving everyone else's moods and behaviors (Goleman et al. 2001). In agricultural extension systems, the emotional style of extension or advisory service officers impacts the behavior of farmers, invariably. It has an impact on the decision-making patterns of subordinates and fellow colleagues. There are many uncertainties with agricultural produce (mainly due to environmental factors), and the nature of the processes calls for high EI behavior to manage those situations. In developing countries like Sri Lanka, the majority of the farmers spend average to marginalized livelihood that is surrounded by failed expectations and frustrations. It demands higher levels of maturity and understanding from extension and advisory staff to be effective in that type of environment and work setting. The success of AEAS depends on the formation of effective work groups among farmers. EI has a big impact on developing groups, teams, and systems, be it in agriculture or otherwise. Druskat and Wolff (2001) have opined that a group's EI comes from norms that support awareness and regulation of emotions within and outside the team. Group EI norms build the foundation for collaboration and cooperation, helping teams to realize higher potential. Interaction with farmer groups and teams is the key for a successful planning and implementation of appropriate practices. For example, that may involve designing the planting and livestock practices during seasons. EI is linked to citizenship behavior (Jayawardena 2014) and social interactions. Goleman (2011) has identified EI as a differentiator of professionals. He opined that a person's intellect loses its power to determine the desired behavioral patterns that differentiates him (from colleagues) as a productive employee or a manager at a given IQ position. The way an employee handles himself and his relationships (EI skill set) with others determines the outcome. The bearing of EI is high, especially in the social context of providing AEAS. The success of agricultural extension exercises depends largely on the way of approaching or interacting with farmers than the technical know-how of innovative practices (Wanigasundera 2015). Thus the level of EI of agricultural extension professionals has been a decisive factor in the overall effectiveness of AEAS.

Majority of government AEAS providers have seemingly failed to develop that close bonding needed to succeed in farming communities in Sri Lanka. Agro-chemical vendors have established close relationships, so that the farmers tend to believe their recommendations (that often include the usage of commercial agro-chemicals marketed by the companies they are employed) in farming. The relationship between local agro-chemical vendors and farmers has a substantial impact in this context. Research has suggested the dark side of emotional intelligence, in using emotional manipulation skills for selfish gains (Kilduff et al. 2010; Bariso 2018) at work. Agro-chemical companies thrive on the above relationships to achieve their commercial objectives. This affects the income of farmers and the sustainability of agricultural practices in the long run. It is commonsensical that agriculture cannot thrive without an effective AEAS system established through proper recognition of specialized skills and approaches.

6 Social Interactions in Extension and Advisory Services

Agriculture is a social enterprise and AEAS demand (and involve) continuous social interactions. EI has been credited with desirable leadership patterns (specially transformational style) and social behaviors. There are indirect evidence related to the impact of having staff high in EI for AEAS through their behaviour on leadership and social interactions. Leadership in general can be understood as a process wherein "an individual influences a group of individuals to achieve a common goal" (Northouse 2013). Transformational leadership has been found to have improved followers' commitment by influencing their needs, values, and self-esteem (Bass and Avolio 2004). Its' approach was found to be morally uplifting (Avolio 1999). Higher levels of performance, extra effort, and higher satisfaction can be expected from subordinates when managers display transformational leadership. Jung, Chow, and Wu found a direct relationship between transformational leadership and organizational innovation in 32 Taiwanese companies (2003). For agriculture to flourish, it needs to try innovative approaches and practices. Possessing a transformational leadership style by extension officers elicits that kind of innovative behavior among the farmers. Transformational leadership contributes to subordinate satisfaction, motivation, and job performance (Yukl 2010). Northouse (2013) found its effectiveness in a variety of situations. Transformational leaders have aroused and inspired colleagues at work, sharing with them a vision of what can be accomplished through extra effort (Bass and Avolio 2004). Employees get a large amount of information at work from a variety of sources. Managers win the confidence and trust of subordinates when they effectively communicate their vision (Pavitt 1999). This type of communication with field-level extension officers and farmers has the potential to significantly improve professions and services such as AEAS. The absence of an established vision (and a policy) on agricultural extension in Sri Lanka has not helped the process. Sharbrough et al. (2006) found a positive relationship between a manager's use of motivational language and perceived effectiveness in the eyes of his or her subordinates. This is applicable to the relationship between field-level agricultural extension officers and farmers. The EI displayed by the extension and advisory staff plays a significant role to enhance the performances of farmers and the effectiveness of AEAS outcomes.

Self-efficacy has an impact on the empowerment of employees (i.e., extension and advisory officers) and farmers. Self-efficacy is defined as the "people's judgments of their capabilities to organize and execute courses of action required to attain designated types of performances" (Bandura 1986). Self-efficacy has been directly related to job performance (Judge and Bono 2001; Stajkovic and Luthans 1998). Two kinds of relationships between transformational leadership and self-efficacy are found in leadership research. Schyns has defined self-efficacy as a precondition for employees to be able to embrace transformational leadership (2001). Kirkpatrick and Locke (1996) have asserted that transformational leaders influence their subordinates' self-efficacy. Schyns (2001) found a positive relationship between transformational leadership and occupational self-efficacy. Schyns has dual interpretations: "Self-efficacy could be a precondition for the perception of transformational leadership, and transformational leaders might enhance their followers' self-efficacy" (2001). Transformational leadership is found to have different effects on self-efficacy, depending on the organizational climate. Thus, leadership has enhanced the self-efficacy of employees in different organizational climates. Bandura has identified mastery (experience) as a factor influencing self-efficacy (1986). Self-awareness and self-management (dimensions of EI) have been identified as the basis of self-mastery. The positive correlation between task demands and occupational self-efficacy confirmed the above theory (Schyns 2001). Task demands have played a dominant role in the occupational self-efficacy of employees. The moderating effect of task and climate on the relationship between leadership style and occupational self-efficacy revealed that the effect of transformational leadership on self-efficacy is negative in low task demands. Findings implied the benefits of having transformational leadership in building effective agricultural extension systems. The leadership styles of senior management and officials contribute a lot toward empowerment at field-level decision-making practices. The recognition received by the extension and advisory profession also contributes to the effectiveness in adoption of practices. Individuals hold self-beliefs that allow them to apply self-control over who they are and what they want to be. Awareness and management of an individual's internal states lead to outstanding performances. Emotional self-management leads to competences with a focused drive to achieve goals, taking initiatives, and developing adaptability (Goleman. 2011).

Relationship between EI and self-efficacy has been documented through the effects of transformational leadership style. Transformational leaders increase the self-efficacy of followers by expressing their confidence on them. The resulting conducive organizational climates enhance the self-efficacy of employees. Occupational self-efficacy (OSE) reflects the conviction or the confidence of a person to fulfill his or her job-related behavior at a perfectly acceptable level to the employer. Schyns and von Collani (2002) observed a positive correlation between OSE and job satisfaction, and organizational commitment. Rigotti et al. (2008) found positive correlations among OSE, job satisfaction, and job performance. The level of OSE before entering the labour market impact the career development of professionals (Abele and Spurk 2009). The OSE of farmers and their preparedness for change facilitate the adoption of novel agricultural practices promoted through AEAS. Self-efficacy had influenced adaptation to change as a positive predictor of self-initiated change in organizations (Armenakis et al. 2000). Farmers need self-initiated change to thrive in modern climate smart agricultural practices. It has been found that people with a higher sense of self-efficacy persist longer in the face of obstacles and set challenging goals for themselves. The transformational leadership style displayed in AEAS initiatives could promote the self-efficacy of farmers.

It is pertinent to examine the impact of leadership roles of AEAS staff on the performance of farmers. A taxonomic approach helps to reflect the multifaceted nature of leadership situations. It provides a balanced understanding of the leadership process at work. Leadership can be assessed in multiple domains: the leader (charisma), the follower (follower innovative role expectations, follower's attitude toward innovation), and the dyadic leadership relationship (LMX). The combination of these three variables has generated a significant variation of innovative behavior (leadership outcome) beyond any of the three taken alone (Graen and Uhl-Bien 1995). The follower is an integral component in the leadership process (Kelley 1988). LMX operationalizes a relationship-based approach to leadership. Effective leadership processes occur when leaders and followers are able to develop mature leadership relationships that bring many benefits (Graen and Uhl-Bien 1991). The LMX theory "conceptualizes leadership as a process that is centered on the interactions between leaders and followers" (Northouse 2013). Leadership models (Graen and Uhl-Bien 1991) have recognized the utility of high-quality relationships in organizations and described a process to accomplish this through dyadic partnership building. The LMX theory has focused on the differences that might exist in the relationship between the leader and each of his followers. Each linkage or relationship between the superior and his subordinate differs in quality. Thus, the same superior "may have poor interpersonal relations with some subordinates and open and trusting relations with others." "The relationships within these pairings, or dyads, may be of a predominantly in-group or out-group nature" (Lunenburg 2010). The superior initiates either an in-group or an out-group exchange with his/her subordinate during the initial phase of the dyadic relationship. Sometimes this can evolve after a while in their relationship. Subordinates having secured a place in the in-group are more likely to be invited to participate in the decision-making process and are given more flexibility in their roles with added responsibility. "In-group members enjoy the benefits of job latitude (influence in decision making, open communications, and confidence in and consideration for the member), and their subordinates typically reciprocate with greater than required expenditures of time and effort, the assumption of greater responsibility, and commitment to the success of the organization" (Lunenburg 2010). Findings suggest that the perception of similarity felt by employees (subordinates) is a more important factor than actual demographic similarities (age, gender, and ethnicity) (Murphy and Ensher 1999). A sharp distinction between the employees belonging to the in-group and the out-group may not be desirable as the out-group subordinates might resent their relatively inferior status and differential treatment (Yukl 2010). It highlights the significance of having a favorable perception of the follower in a subordinate's relationship with the immediate superior. Thus, the relationships among agricultural extension staff, other field officers, and farmers contribute to the outcome of extension efforts. Improved social relations (through self awareness and social awareness) among the staff deployed in AEAS with proper understanding of field-level realities is vital in this context.

Supervisor-subordinate relationship motivates employees (Kazoleas and Wright 2001) and influences their job satisfaction (Lamb and Mckee 2005; Madlock 2008). It has enhanced their perceived external prestige (Smidts et al. 2001) and trust (Thomas et al. 2009). Establishing a mutual communication channel with the immediate superior has enhanced job satisfaction (Sias 2005). Madlock explored the influence of a manager's communication competence and relational leadership on the job satisfaction of his/her subordinates (2008). A manager's willingness to communicate effectively with his/her subordinates contributes to organizational success (Bartoo and Sias 2004). Failures in the communication system hamper motivation at work (Kazoleas and Wright 2001). Subordinates' satisfaction with communication is an important ingredient of the psychological contract and is closely linked to whether employees feel valued by their managers and organizations (De Ridder 2006). Satisfaction in the relationship between supervisors and subordinates has been a "major indicator of efficiency and organisational effectiveness" (Brunetto and Farr-Wharton 2004). Subordinates view their immediate supervisor as one of the most important of all informational sources (Bartoo and Sias 2004). Subordinates who have effective interactions with their immediate superiors have shown greater motivation, satisfaction, productivity, and commitment to the organization (Gupta and Sharma 2008). Farmers are used to rely on and view government AEAS staff similarly to an immediate supervisory role, and the nature of this relationship impacts the perception (and satisfaction) of farmers. The relationship with field-level extension officers greatly contributes to the performance of farmers and to the outcome of AEAS.

In establishing an effective AEAS, it is necessary to promote citizenship performance behavior (CPB) in extension and advisory setup/organizations. CPB explores performance-related behaviors, going beyond the assigned tasks and responsibilities for which employees are usually held responsible. CPB has been conceptualized by Smith and Organ (1983) as the "discretionary behavior that goes beyond one's official role" (as cited in Borman 2004). Further, it is intended to help other people in the organization to show conscientiousness and support toward the organization. Organ et al. (2006) defined CPB as the "individual behavior that is discretionary, not directly or explicitly recognized by the formal reward system, and in the aggregate promotes the efficient and effective functioning of the organization." For an effective AEAS, it is necessary to have staff members who are willing and capable of going beyond the assigned roles based on the contextual factors triggered by farmers and field-level challenges. In other words, staff should be passionate of their involvements (and impact) in AEAS. CPB goes beyond task performance and technical proficiency. It supports the organizational, social, and psychological contexts that serve as the critical catalyst for the accomplishment of tasks. Research has also linked job task characteristics with the CPB of an employee (Maarleveld 2009). Borman (2004) deliberated on four major themes of CPB, viz., supervisor's use of CPB in making judments about subordinates, personality as a predictor of CPB, link betweeen CPB and organizational effectiveness, and influence of organizational characteristics on citizenship performance. Podsakoff et al. (2000) noted four major categories of antecedents of CPB, viz., task characteristics, organizational characteristics, leadership behaviors, and individual characteristics. The above facts further establish the importance of maintaining sound interpersonal relationships with empathetic communication in organizational setups for effective AEAS. These competencies have to be given detailed attention in the selection of agricultural extension staff and needs to be followed up with training and development efforts. The nature of AEAS is such that it needs greater commitment going beyond their official boundaries from employees for an effective functioning of the practices of the agricultural sector. It is imperative to create a background to recognize the performances and careers of agricultural extension officers. The more they get identified with their careers, the more they are likely to think and act on behalf of their careers. Thus, it creates a catalytic environment for dedicated employees to strive in respective careers. Agricultural extension needs conceptual clarity, and professionals deserve due recognition for the specialized expertise demanded by the nature of tasks. Appreciation of the impact of EI and social interactions in developing the above expertise is fundamental in the success of AEAS. Training programmes; protocols to conduct collective actions; and individual, group, and mass methods of AEAS can be enhanced by mainstreaming EI into the processes. EI offers a significant promise to manage the outcomes of AEAS in local and global contexts.

References

Abele AE, Spurk D (2009) The longitudinal impact of self-efficacy and career goals on objective and subjective career success. J Vocat Behav 74(1):53–62

Agricultural Extension in South Asia (2016) Assessing capacity needs of extension and advisory services in South Asia: synthesis report. CRISP, India

Armenakis AA, Harris SG, Field HS (2000) Paradigms in organizational change: change agent and change target perspectives. In: Golembiewski RT (ed) Handbook of organizational behavior. Marcel Dekker, New York, pp 631–658

Arunapriya SA, Sudasinghe JR, Weerakoon WRWMSNP, Wijayathunga RSCWMABM (2018) Historical evolution of agricultural extension system in Sri Lanka. In: Jayawardena LNAC, Suryamani M, Sivayoganathan S (eds) Transforming agricultural extension systems: towards achieving the relevant Sustainable Development Goals for global impact. Kandy, Sri Lanka, pp 1–6

Averill JR, Nunley EP (1993) Grief as an emotion and as a disease: a social-constructionist perspective. In: Stroebe MS, Stroebe W, Hansson RO (eds) Grief as an emotion and as a disease: a social-constructionist perspective. Cambridge University Press, New York, pp 77–90
Avolio BJ (1999) Full leadership development: building the vital forces in organizations. Sage, Thousand Oaks
Bandura A (1986) Social foundations of thought and action: a social cognitive theory. Prentice-Hall, Englewood Cliffs
Bariso J (2018) EQ applied. Borough Hall, Munich
Bar-On R (1997) Bar on emotional quotient inventory: technical manual. Multi-Health Systems, Toronto
Bar-On R (2003) Exploring the neurological substrate of emotional and social intelligence. Brain 126:1790–1800
Bartoo H, Sias PM (2004) When enough is too much: communication apprehension and employee information experiences. Commun Q 52(1):15–26
Bass BM, Avolio BJ (2004) MLQ multifactor leadership questionnaire, (form 5X- short). Technical report. Mind Garden, Redwood City, pp 1–6
Borman WC (2004) The concept of organizational citizenship. Curr Dir Psychol Sci 14(6):238–241
Boyatzis RE, Smith ML, Blaize N (2006) Developing sustainable leaders through coaching and compassion. Acad Manag Learn Educ 5(1):8–24
Brain Injury Resource Center, United States (1998) Brain map. Retrieved from the website http://www.headinjury.com/brainmap.htm
Brunetto Y, Farr-Wharton R (2004) Does the talk affect your decision to walk? A comparative pilot study examining the effect of communication practices on employee commitment post-managerialism. Manag Decis 42(3):579–600
Bullen E, Fahey J, Kenway J (2006) The knowledge economy and innovation: certain uncertainty and the risk economy. Discourse 27(1):53–68
Campbell A, Whitehead J, Finklestein S (2009) Why good leaders make bad decisions. In: HBR's 10 must reads on emotional intelligence. Harvard Business Review Press, Boston, pp 59–70
Carroll JB (1993) Human cognitive abilities: a survey of factor analytic studies. Cambridge University Press, New York
Carson E, Ranzijn R, Winefield A, Marsden H (2004) Intellectual capital: mapping employee and work group attributes. J Intellect Cap 5(3):443–463. https://doi.org/10.1108/14691930410550390
De Ridder JA (2006) Organisational communication and supportive employees. Hum Resour Manag J 14(3):20–30
Dissanayake DMAP, Wijesuriya W, Edirisinghe JC (2005) Smallholder rubber sector in Moneragala district: potentials and constraints. Bull Rubber Res Inst Sri Lanka 46:25–31
Druskat VU, Wolff SB (2001) Building the emotional intelligence of groups. In: HBR's 10 must reads on emotional intelligence. Harvard Business Review Press, Boston, pp 71–91
Fazlagic A (2007) Measuring the intellectual capital of a university. Paper presented at the conference on trends in the Management of Human Resources in Higher Education. Retrieved from http://www.oecd.org/edu/imhe/35322785.pdf
Ferrier F (2001) Managing, measuring and maximising intangible assets: an eclectic view of some of the latest international developments. Paper presented at the Intellectual Property: Managing, Measuring and Maximising Intangible Assets, Sydney
GFRAS (2012) 'The new extentionist': roles, strategies, and capacities to strengthen extension and advisory services. CRISP, India
Goleman D (1995) Emotional intelligence. Bantam, New York
Goleman D (2006) Working with emotional intelligence. Bantam, New York. (Original work published 1998)
Goleman D (2011) The brain and emotional intelligence: new insights, 1st edn. More than Sound LLC, Northampton

Goleman D, Boyatzis R, McKee A (2001) Primal leadership. In: HBR's 10 must reads on emotional intelligence. Harvard Business Review Press, Boston, pp 23–42
Graen GB, Uhl-Bien M (1991) The transformation of professionals into self-managing and partially self-designing contributions: toward a theory of leader-making. J Manag Syst 3(3):33–48
Graen GB, Uhl-Bien M (1995) Relationship-based approach to leadership: development of leader-member exchange (LMX) theory of leadership over 25 years: applying a multi-level multi-domain perspective. Leadersh Q 6(2):219–242
Gunawardena AMAPG (2005) Communication behaviour of farmers on improved farm practices in Udaipur District of Rajasthan. Unpublished MSc thesis. MPUAT, Udaipur
Gupta B, Sharma NK (2008) Compliance with based of power and subordinates' perception of superiors: moderating effect of quality of interaction. Singap Manag Rev 30(1):1–24
Guthrie J (2001) The management, measurement and the reporting of intellectual capital. J Intellect Cap 2(1):27–41
Jayawardena LNAC (2012) Transformational leadership and emotional intelligence of graduate managers. Manag Prod Eng Rev 13(3):28–33. https://doi.org/10.2478/v10270-012-0021-1
Jayawardena LNAC (2014) Impact of strategic emotional intelligence to managerial self efficacy. Unpublished PhD dissertation. Tomas Bata University in Zlin, Zlin
Jensen S, Kohn C, Rilea S, Hannon R, Howells GN (2007) Emotional intelligence: a literature review. Publication of Department of Psychology, University of the Pacific, pp 10–85
Judge TA, Bono JE (2001) Relationship of core self-evaluations trait self-esteem, generalized self-efficacy, locus of control, and emotional stability with job satisfaction and job performance: a meta-analysis. J Appl Psychol 86:80–92
Jung DI, Chow C, Wu A (2003) The role of transformational leadership in enhancing organizational innovation: hypotheses and some preliminary findings. Leadersh Q 14(4):525–544
Kazoleas D, Wright A (2001) Improving corporate and organisational communication: a new look at developing and implementing the communication audit. In: Heath RL, Vasquez GM (eds) Handbook of public relations. Sage, Thousand Oaks, pp 471–478
Kelley RE (1988) In praise of followers. Harvard Business Review, Boston, pp 142–149
Kilduff M, Chiaburu DS, Menges JI (2010) Strategic use of emotional intelligence in organizational settings: exploring the dark side. Res Organ Behav 30:129–152
Kirkpatrick SS, Locke EA (1996) Direct and indirect effects of three core charismatic leadership components on performance and attitudes. J Appl Psychol 81:36–51
Kleinberg JL (2000) Beyond emotional intelligence at work: adding insight to injury through group psychotherapy. Group 24(4):261–278
Lamb LF, Mckee KB (2005) Applied public relations: cases in stakeholder management. Lawrence Erlbaum Associates Inc., Mahwah
Leban W, Zulauf C (2004) Linking emotional abilities and transformational leadership styles. Leadersh Org Dev J 25(7):554–564
Lunenburg FC (2010) Leader-member exchange theory: another perspective on the leadership process. Int J Manag Bus Admin 13:1–9
Luthans F (2011) Organizational behaviour: an evidence based approach, 12th edn. McGraw-Hill, Boston
Lynn BE (2000) Intellectual capital: unearthing hidden value by managing intellectual assets. Ivey Bus J 64(3):48–52
Maarleveld M (2009) A study on identity orientation and citizenship performance behaviour. Unpublished Master thesis in Human Resource Development. University of Twente, Enschede, Netherlands, pp 2–34
Madlock PE (2008) The link between leadership style, communicator competence, and employee satisfaction. J Bus Commun 45(1):61–78
Mayer JD, Salovey P (1997) What is emotional intelligence? In: Salovey P, Sluyter D (eds) Emotional development and emotional intelligence: educational implications. Basic Books, New York, pp 3–31

Mayer JD, Salovey P, Caruso DR, Sitarenios G (2003) Measuring emotional intelligence with the MSCEIT V2.0. Emotion 3(1):97–105
Mayer JD, Roberts RD, Barsade SG (2007) Emerging research in emotional intelligence. Annu Rev Psychol 59
Mouritsen J, Larsden HT, Bukh PND (2001) Intellectual capital and the capable firm: narrating, visualising and numbering for managing knowledge. Acc Organ Soc 26:735
Murphy SE, Ensher EA (1999) The effects of leader and subordinate characteristics in the development of leader-member exchange quality. J Appl Psychol 29(7):1371–1394
Northouse PG (2013) Leadership: theory and practice, 6th edn. Sage, New York
Organ DO, Podsakoff PM, Mackenzie SB (2006) Organizational behavior: its nature, antecedents, and consequences. Sage, Thousand Oaks
Papadogiannis PK, Logan D, Sitarenios G (2009) An ability model of emotional intelligence: a rationale, description, and application of the Mayer Salovey Caruso Emotional Intelligence Test (MSCEIT). In: Stough C, Saklofske DH, Parker JDA (eds) Assessing emotional intelligence. Springer, London. https://doi.org/10.1007/978-0-387-88370-0
Pavitt C (1999) Theorizing about the group communication-leadership: input-process-output and functional models. In: Frey LR, Gouran DS, Poole MS (eds) The handbook of group communication theory and research. Sage, Thousand Oaks, pp 313–334
Petrides KV, Furnham A (2003) Trait emotional intelligence: behavioural validation in two studies of emotion recognition and reactivity to mood induction. Eur J Personal 17:39–75
Podsakoff PM, MacKenzie SB, Paine JB, Bachrach DG (2000) Organizational citizenship behaviours: a critical review of the theoretical and empirical literature and suggestions for future research. J Manag 26:513–563
Prasad RM, Sulaiman VR, Mittal N (2015) Assessing capacity development needs of extension and advisory services (EAS): a review CRISP working paper 2015
Rajasinghe BA (2009) Technology dissemination in the corporate tea sector in Sri Lanka
Rigotti T, Schyns B, Mohr G (2008) A short version of the occupational self-efficacy scale: structural and construct validity across five countries. J Career Assess 16:238–251. https://doi.org/10.1177/1069072707305763
Rozell EJ, Pettijohn CE, Parker RS (2002) An empirical evaluation of emotional intelligence: the impact on management development. J Manag Dev 21(4):272–289
Salovey P, Mayer JD (1990) Emotional intelligence. Imagin Cogn Pers 9:185–211
Schyns B (2001) Self-monitoring and occupational self-efficacy of employees and their relation to perceived transformational leadership. Curr Res Soc Psychol 7:30–42
Schyns B, von Collani G (2002) A new occupational self-efficacy scale and its relation to personality constructs and organisational variables. Eur J Work Organ Psy 11:219–241
Sharbrough WC, Simmons SA, Cantrill DA (2006) Motivating language in industry: impact on job satisfaction and perceived supervisor effectiveness. J Bus Commun 43(4):322–343
Sias PM (2005) Workplace relationship quality and employee information experiences. Commun Stud 56(4):375–395
Smidts A, Pruyn ATH, Van Riel CBM (2001) The impact of employee communication and perceived external prestige on organisational identification. Acad Manag J 49(5):1051–1062
Snell S, Bohlander G (2007) Human resource management. Thomson South-Western, Mason
Stajkovic AD, Luthans F (1998) Self-efficacy and work-related performance: a meta-analysis. Psychol Bull 124:240–261
Takeuchi H, Taki Y, Sassa Y, Hashizume H, Sekiguchi A, Fukushima A, Kawashima R (2011) Regional gray matter density associated with emotional intelligence: evidence from voxel-based morphometry. Hum Brain Mapp 32(9):1497–1510. https://doi.org/10.1002/hbm.21122
Takeuchi H, Taki Y, Sassa Y, Hashizume H, Sekiguchi A, Nagase T, Nouchi R, Fukushima A, Kawashima R (2013) White matter structures associated with emotional intelligence: evidence from diffusion tensor imaging.Human. Brain Mapp 34(5):1025–1034. https://doi.org/10.1002/hbm.21492

Thomas GF, Zolin R, Hartman JL (2009) The central role of communication in developing trust and its effect on employee involvement. J Bus Commun 46(3):287–310

Vallejo-Alonso B, Gerardo-Arregui-Ayastuy G, Rodriguez-Castellanos A, Garcia-Merino D (2013) Real options in the valuation of intangibles: managers perception. Electron J Knowl Manag 11 (2):168–182

Wanigasundera WADP (2015) Status of extension and advisory services in Sri Lanka, AESA working paper 2015 – 001. CRISP, India

Wijekoon R, Kumara GS (2018) An overview of education and learning to meet the SDGs with special reference to recent innovations of the agriculture sector in Sri Lanka. In: Jayawardena LNAC, Suryamani M, Sivayoganathan S (eds) Transforming agricultural extension systems: towards achieving the relevant Sustainable Development Goals for global impact. Kandy, Sri Lanka, pp 20–26

Yukl GA (2010) Leadership in organizations, 7th edn. Prentice Hall, Upper Saddle River

Goleman, D. (2006). Working with Emotional Intelligence. New York: Bantam. (Original work published 1998)

Podsakoff, P.M., MacKenzie, S.B., Paine, J.B., & Bachrach, D.G. (2000). Organizational citizenship behaviours: A critical review of the theoretical and empirical literature and suggestions for future research. Journal of Management, 26, 513–563.

"Communication" in the Context of Agricultural Extension: Past, Present and Way Forward in Achieving Sustainable Food Systems in Sri Lanka

Madhavi Wijerathna and W. A. D. P. Wanigasundera

1 Introduction

Agriculture is an important sector of the Sri Lankan economy, contributing 7.42% to gross domestic product (GDP) (Central Bank 2019). A considerable percentage (28.7%) of the labor force of the country is employed in the agricultural sector, and a majority of them are living in rural areas. Even though the contribution of agriculture to GDP has been reducing over time, the sector is vital for the national economy and enhances the household income of the people who engage directly and indirectly in agriculture while ensuring the food security of the country. Although the production of rice in Sri Lanka has increased during the last two decades, the overall food production in the country has not increased substantially. A large proportion of foodstuff, which can be grown in the country, are still being imported. Sri Lanka is endowed with a wide variation of climatic and agro-ecological zones, making it possible for the cultivation of a number of different types of crops and rearing livestock. The research and development organizations have been developing a good deal of promising technologies capable of achieving high productivity. Also, the new technologies have more concern on sustainable production than earlier. However, the productivity and profitability of many subsectors remain low, mainly due to declining soil fertility, lack of good planting materials, high cost of inputs, and insufficient postharvest enterprises and markets.

The earnings from subsectors of agriculture such as plantation crops and spices have recently been increasing. Therefore, it is necessary to improve agriculture-based small-, medium-, and large-scale industries as a component of industrial development of the country in a wider perspective. The population of the world as well as in Sri Lanka is increasing. Therefore, it is essential to intensify food production to meet the increasing food demand. However, the extent of arable

M. Wijerathna (✉) · W. A. D. P. Wanigasundera
Department of Agricultural Extension, Faculty of Agriculture, University of Peradeniya, Peradeniya, Sri Lanka

R. P. De Silva et al. (eds.), *Agricultural Research for Sustainable Food Systems in Sri Lanka*, https://doi.org/10.1007/978-981-15-3673-1_18

lands in the country, which can be used to increase agricultural production, is decreasing. According to a strategic review (World Food Programme (WFP) 2017), it has clearly been identified that food and nutrition security in Sri Lanka remains an unachieved social and economic goal despite the different government and other interventions. Therefore, it is a serious challenge to the country at present to achieve sustainable food production while meeting the demand.

Earlier attempts at intensification of agricultural production have contributed largely to meet the global food demand, but these have also created problems of sustainability. Decrease in soil health, increase in soil erosion, excessive water use, biodiversity degradation, and overuse of pesticides and fertilizer are only some of them (Sontakki et al. 2018). Chronic kidney disease of unknown (CKDu) etiology has become a serious environmental and health issue prevailing in the farming community of the North Central Province of Sri Lanka as a consequence of unhealthy production practices as identified by some researchers (Gunatilake et al. 2014). Therefore, increasing the production sustainably to meet the demand is a critical challenge to be addressed. In other words, a sustainable transition is needed. Sustainability transition can be defined as "long- term, multi-dimensional and fundamental transformation processes through which established socio-technical systems shift to more sustainable modes of production and consumption" (Markarda et al. 2012). It is clear that the agricultural extension is important in this transition to introduce sustainable production and consumption practices where communication plays a major role in creating awareness and improving knowledge and skills leading to a behavioral change in farmers and consumers.

Even though agriculture still remains as a major employment provider in Sri Lanka, its income-earning potential is not increasing when compared to other sectors. Consequently, even rural agricultural communities tend to move away from agriculture (Wanigasundera 2015). Going in accordance with sustainable development goals, the Sri Lankan government has planned by 2030 to end hunger and ensure food security (Department of Census and statistics 2017). Therefore, dissemination of sustainable agricultural knowledge and technology using different communication approaches within the agricultural extension system plays an important role in achieving sustainable food systems while meeting the food demand of the county.

Agricultural extension is the bridge between research and farmers while playing a role in human resource development of the stakeholders involved in the value chain. Communication plays a critical role in disseminating agricultural extension messages including new technology and market information and also motivating the farming communities to adopt appropriate farming systems with sustainability. The objective of this chapter is to review the "communication" in the context of agricultural extension with a theoretical insight and highlight the Sri Lankan experience while linking to the global perspectives. Furthermore, the chapter discusses the problems and challenges of communication in agricultural extension in developing sustainable food systems in Sri Lanka and finally proposes means of facing the challenges and the way forward. The literature on the Sri Lankan context was mostly drawn from peer-reviewed journals, conference proceedings, and symposiums held in Sri Lanka and also from undergraduate and graduate thesis to supplement the local facts. In addition, the standard books on the domain were also reviewed.

1.1 Agricultural Extension System in Sri Lanka

According to the literature, it can be identified that communication is not a part of but embedded in agricultural extension. Therefore, understanding the agricultural extension system of the country helps to understand how the communication process is taking place within the agricultural extension system.

According to Arunapriya et al. (2018), recruiting agriculture extension officers to the government of Sri Lanka was started in 1820. The Department of Agriculture was established in Sri Lanka in 1912 to cater to the food crop sector.

Agricultural extension, which is basically a public service, has become a fully devolved subject under the 13th Amendment to the Constitution of Sri Lanka, making it a responsibility of the provincial councils (PCs). Thus, in the current context, the Central and Provincial Departments of Agriculture and the Department of Animal Production and Health provide the extension services for the crop and livestock sectors, respectively. The food crop sector plays an important role in agriculture and the economy. The Department of Agriculture is the apex body which provides extension services to the food crop sector. The agriculture instructor is the ground-level officer of the Department of Agriculture who works with the rural community. This officer is responsible for communicating extension messages to the clients who are mainly farmers. The Department has followed a series of approaches for agricultural extension including conventional training and visit system and farmer participatory and integrated approaches. The present approach is a farmer participatory and commodity development mixed approach (Emitiyagoda 2009). The recently established National Agriculture Information and Communication Centre (NAICC) of the Department of Agriculture is important to facilitate the extension communication of the food crop sector in Sri Lanka. The objectives of NAICC are to collect, compile, and disseminate agricultural information through electronic and print media, develop e-solutions for information management, and also conduct exhibition and maintain the agro technological park. To adopt information and communications technology (ICT) for making food crop sector more efficient and effective is part of the mission of NAICC. There are five subunits that come under this center, namely, (i) Agriculture Publication Unit (APU), (ii) Chamal Rajaphaksha Agro Technology and Tourism Park, (iii) Farm Broadcasting Service (FBS), (iv) Agro Technological Park (GATP) Gannoruwa, and (v) Central Library (https://doa.gov.lk/ICC/en/), which plays a major role in the agricultural communication of the country. These roles are important to have an effective extension service.

Livestock is an important sector in Sri Lankan agriculture. The Department of Animal Production and Health provides the extension services for the livestock sector. The service units at the veterinary range level, the Veterinary Surgeon's

Offices (287 island wide), are staffed by veterinary surgeons and livestock development officers. According to Wickramasooriya et al. (2009), as a result of the devolution of powers to the provincial system, the grassroots-level extension officers have no direct relationship with the central department which implements the training-of-trainers programs.

The Department of Export Agriculture provides the extension services for the export agricultural crop sector. Export agricultural crops represent export-oriented perennials including spices, beverage crops, essential oil-producing crops, and stimulants where half of the produce is exported. The present extension approach is mostly the general extension approach, while participatory, cost-sharing, and educational-institution approaches also operate simultaneously in a few situations (Senevirathna and Gunasinghe 2009). The educational institution approach is important since it links with higher education institutes.

Extension services for the plantation sector are provided through separate institutional arrangements via the commodity research institutes and services, such as the Tea Research Institute, Tea Small Holdings Development Authority, Coconut Research Institute, Coconut Cultivation Board, Rubber Research Institute, Rubber Development Department, and Sugarcane Research Institute. The Cashew Development Board provides the extension services for the cashew sector.

The Mahaweli Authority of Sri Lanka engages in the management of the Mahaweli irrigation schemes. The goal of agricultural extension of the Mahaweli Authority is to improve the income level and living standards of the Mahaweli settler families. A holistic approach to agricultural extension is adopted by the Mahaweli Authority, and extension methods such as "yaya demonstrations" and farmer field schools have contributed to the adoption of new technologies by farmers in the Mahaweli areas (Dayarathna 2009). Field assistants and agriculture officers are employed as the extension officers.

The Department of Agrarian Development is also an important state institution, which administers nearly 500 Agrarian Service Centers (ASCs) island wide, providing services to the farming communities at the grassroots level. All the government agricultural extension services, together with the provision of certain input supplies, are concentrated at the ASCs, which have become the main development center, as field-level extension officers from different departments are stationed there (Waduge 2006). Therefore, the different extension needs of the grassroots level farmers can be provided at one-stop.

Nongovernmental organizations (NGO) and the private sector are also involved in agricultural extension to a limited extent. According to Wanigasundera (2015), there are three types of nonstate sector entities that are involved in agricultural extension in the country, namely, (i) those that deal in the supply of agricultural inputs and/or engage in producing agricultural commodities, (ii) development agencies, and (iii) farmer associations, cooperatives, and companies. The private sector companies provide most of the input supplies. Sales agents and input dealers are the prominent extension informants who represent the private sector (Weerakkody et al. 2004).

1.1.1 Problems and Limitations of the Present Agricultural Extension System

As identified by Wanigasundera (2015), inadequate capacities of extension personnel, limited recognition of extension personnel, inability to build on pluralism in extension delivery, and problems related to downsizing and decentralizing public extension are some of the generic issues of the present agricultural extension service in Sri Lanka. According to him, the present system is ineffective in providing extension services to the entrepreneurial farmers and modern agribusiness firms. According to Weerakkody et al. (2004), the process of technology transfer in traditional extension models is basically one way, following the top-down approaches. Further, many essential technical information and messages are not efficiently transferred to the end users due to various bottlenecks in the system. A significant change in the agricultural extension system of the country occurred through the decentralization of the extension services to provincial councils and project-based extension programs, but these efforts fell below expectations, resulting in a poorer quality of extension services provided. As different government departments and institutions and the private sector are involved in providing agricultural extension services in the country, the linkages and coordination between these institutions and sectors are important to provide an efficient extension service while maximizing the use of available resources (Waduge 2006). According to Wijekoon and Rizwan (2009), the weak intersectoral and intrasectoral linkages among extension, research, training, and service sectors have led to a decline in the effectiveness of these functions, which has impacted much the agricultural output of the country. According to Emitiyagoda (2009), also the extension system of the country is weak and characterized by weak research-extension linkages. There is no formal mechanism for extension coordination among the different entities at the national level (Wickramasooriya et al. 2009). However, there are some coordination mechanisms available at district, divisional, and agrarian service center levels. Such coordination is done through the district agriculture committee, divisional agriculture committee, and agrarian service committees. At present, there is no national policy for agricultural extension service in the country. According to Sivayoganathan (1999), the Council for Agricultural Research (CARP) had attempted to develop a national policy for agricultural extension, but it had not been successful. CARP was established to coordinate the fragmented agricultural research and extension system in Sri Lanka, but CARP is principally engaged with research and lacks a mandate for extension (World Bank 2007).

According to the literature, the agricultural extension system in the country is basically led by the government, and it is supply driven. Private and NGO sectors are involved in agricultural extension to a limited extent. All the government extension departments have field-level extension officers who communicate directly with the farmers, but their capacities are inadequate, they lack motivation, and they are small in number to provide an effective extension service. Coordination among the organizations and the link between research and extension are poor. Under these circumstances, communication has to play a vital role to fulfill the task of extension service. Therefore, organizational, policy, and communication strategies should be developed to link the agricultural organizations as well as to increase the effectiveness of the agricultural extension system of the country.

1.2 "Communication" in the Context of Agricultural Extension

Understanding the term extension is important to realize the relationship of communication with agricultural extension which is essential in agricultural development. There is no universally accepted single definition for agricultural extension, and the term extension is widely open for different other contexts and disciplines such as health and rural development. However, the definition given by Maunder (1972) for agricultural extension is widely accepted. According to him, "Agricultural extension can be defined as a service or a system which assist farmers through educational procedures in improving farming methods and techniques, increasing production efficiency and income, bettering their standard of living and lifting the social and educational standard of the rural people." Accordingly, it can be identified that communication is directly related to the educational process of agricultural extension. Extension becomes most effective when there is a three-way interactive communication between research organizations, frontline fieldworkers, and farmers (Fox 1990). According to the definition of State Extension Leaders Network (SELN, 2006) agricultural extension is the process of enabling change in individuals, communities and industries involve in the primary industry sector and in natural resource management.This definition is important for the presentday context since it covers the role and target group of extension in a broader perspective. The definition of Van den Ban and Hawkins (1996) for agricultural extension explicate the relationship of extension and communication. Their definition for agricultural extension is that "extension involves the conscious use of communication of information to help people form sound openions and make good decisions.

In order to understand the role of communication in agricultural extension and thereby the agricultural development, it is important to understand the meaning of the term communication. There is no universally accepted single definition for communication, but there are widely accepted several definitions. According to the Oxford dictionary, the general meaning of communication is imparting or exchanging of information by speaking, writing, or using some other media. An early definition for communication has been given by Stevens in 1950 from a biological perspective. Accordingly, communication is the discriminatory response of an organism to a stimulus. Mahaliyanaarachchi (2003) has defined communication as a process which occurs by sharing ideas, information, and feelings among two or more people, and it provides an interaction between two or more people.

In terms of a more philosophical perspective, Carey (1989) defines communication as a symbolic process whereby reality is produced, maintained, and transformed. It can be agreed that communication is a symbolic process as Carey (1989) has given in this definition, but at the same time it can be argued that the reality is not always produced and transmitted. For instance, some sensitive cases are exaggerated and communicated by journalists. Griffin et al. (2015) explicated that communication is a relational process of creating and interpreting messages that elicit a response.

Based on these definitions, communication can be understood in a more general perspective as a process which takes place within or between two or more people, sharing information, ideas, attitudes, and feelings and providing the opportunity to have interactions among each other.

According to Mosher (1966), "agriculture" is a special kind of production based on the growth processes of plants and animals. "Farmers" are the persons who manage and stimulate plant and animal growth on farms. It is important to understand the term agricultural development in order to comprehend the role of communication in agricultural extension toward agricultural development. Agricultural development implies the shift from traditional methods of production to new science-based methods of production (Swanson and Claar 1984). The process of development should contain three elements: economic (producing goods and materials), social (provision of social amenities and services), and human (development of people to realize their full potential) (Oakley and Garforth 1985). According to Mosher (1966), the essentials for agricultural development are market for farm products, constantly changing technology, local availability of supplies and equipment, production intensives for farmers, and transportation. Communication plays a vital role in all these elements of agricultural development. Especially, the technologies are generated at research stations and transmitted to the farming community through the extension system. Communication is vital in this transition.

Based on the above discussion, it can be considered that communication and extension are two related disciplines, and communication is a way to achieve the objectives of extension. Moreover, it can also be stated that communication is embedded in agricultural extension.

2 Types of Communications Important in Agricultural Extension

Over the last few decades, there has been an increasing trend to study and practice communication, resulting in new knowledge and varying perspectives of communication. The diversified and multifaceted nature of communication includes mass communication, interpersonal communication, health communication, international communication, speech communication, intercultural communication, communication education, applied communication, organizational communication, and political communication and also includes journalism, media production, information and communication technologies, public relations, corporate communication, and development communication (Mefalopulos 2008). Although the basic process and principles of communication are common for many types of communication, the specific principles, guidelines, and procedures can be identified in different types of communication in different contexts. Moreover, these different types of communication efforts are focused on achieving more specific objectives in different contexts. Out of this multifaceted concept of communication, it can be identified that agricultural communication, development communication, climate change communication, and participatory communication are directly related to agricultural development. Environmental communication and science communication also have a high relative importance.

2.1 Agricultural Communication

Communication means differently for different people in different disciplines. Agricultural communication is offered in many agricultural schools, departments, and universities as an academic discipline for undergraduate and postgraduate degrees. A question can be raised as to what communication is agricultural. Communication in an agricultural context can be identified as agricultural communication in general. Kurtzo et al. (2016) emphasise the difficulty of providing a comprehensive definition for agricultural communication. However, they have identified agricultural communication as a field of emerging interest. According to them, skills and characteristics needed by agricultural communicators ranged from general agriculture and policy understanding to technical communication and science knowledge. Zumalt (2007) has described agricultural communication using a Venn diagram (Fig. 1). According to him, agricultural communication seeks to connect the two well-established streams (agriculture and communication) effectively.

Further, Zumalt (2007) suggests that communication touches and serves all aspects of agriculture and every sector of the food and agricultural enterprise has a dimension of human interaction. Accordingly, a combination of communication tools, techniques, and methods used in the agriculture development process is known as agricultural communication. It can be differentiated from other forms of general communications since communication messages are carefully designed to expect a behavioral change of the stakeholders toward agricultural development.

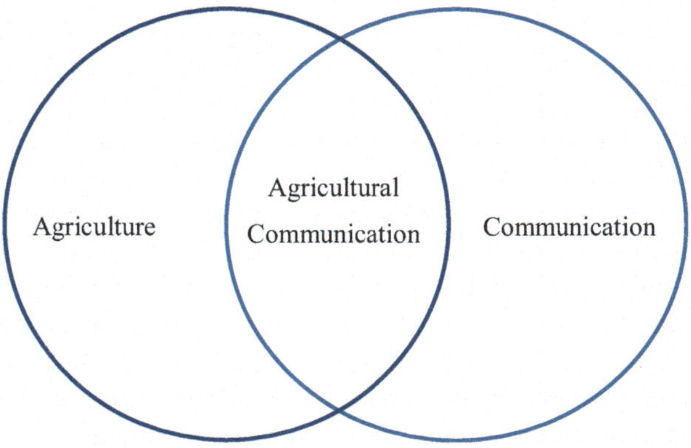

Fig. 1 Venn diagram to illustrate the intersection of two sets (agriculture and communication). (Source: Zumalt 2007)

2.2 Development Communication

Development communication especially focuses on communication to achieve development objectives. It is a relatively newly developed branch of communication. Mefalopulos (2008) gives the definition of development communication as interpreted by the Communication Department of the World Bank as follows: an interdisciplinary field based on empirical research that helps to build consensus while facilitating the sharing of knowledge to achieve a positive change in development initiatives. More importantly, from a development perspective, it is not only dissemination of information effectively but also using two-way communication among stakeholders. According to this definition, development communication efforts would not be finished once the dissemination is completed. It engages with the stakeholders until the desired development objective is achieved.

Development communication often engages with the mainstream media and focuses on issues that are not given space in mainstream media channels. Anand (2014) indicates that development agricultural journalism has several characteristics in general, such as the following:

> (i) It is a key to good governance—power can be misused if the media is not critical about government laws and regulations. (ii) Imposing pressure on decision-making processes and rectifying check and balances of centrally funded schemes. (iii) Information power and right to know—development communicators have the right to information and to know how public money is spent on public development and interest. (iv) Editors and producers give more attention to the corruption aspects of development projects than being appreciative of progress. (v) The media is expected to be critical and objective. (vi) Developmental media is not expected to work along with but expected to serve for common goals of the government. (vii) Participatory citizenship is a mandate of development communication.

Most of the agricultural and rural communication methods are expected to be intensive in community participation as communities carry a significant portion of local knowledge.

Depending on the characteristics of development communication, it can be clearly stated that development communication and agricultural communication go hand in hand. Therefore, incorporating the development communication principles and guidelines in agricultural extension communication is important to achieve the development objectives of the field of agriculture including the sustainable development of the food sector.

2.3 Participatory Communication

Participatory communication is another branch of communication. According to Servaes and Malikhao (2005), the word participatory implies a higher level of public involvement in communication systems. It includes the involvement of the public in the production process and also in the management and planning of communication systems. Participatory communication is an approach based on dialogue which allows the sharing of information, perception, and opinions among the various stakeholders and thereby facilitates their empowerment especially for those who are vulnerable and marginalized. Therefore, this type of communication is especially relevant when agricultural development projects are linked with rural development projects.

The traditional communication of technology transfer in agricultural extension is criticized as top down/one way, and receivers may feel the messages as strange and culturally excluded. Therefore, participatory communication strategies are useful in farming system approach in agricultural extension. In this approach, the assumption is technology which fits the needs of farmers, particularly small farmers, is not available and needs to be generated locally (Axinn 1988). Participatory communication strategies are well fitted to achieve the objectives of farming system approach. Moreover, depending on the characteristics of participatory communication, it can be stated that this type of communication is important to create a sense of belongingness in the target audience.

2.4 Climate Change Communication

World climate is changing and will continue to change. Although climate change is happening, it is difficult to communicate climate change to the public due to lack of visibility and immediacy of this phenomenon (Moser 2010).

Agriculture is one of the most vulnerable sectors for climate change effects. Therefore, climate-smart agriculture (CSA) concept was introduced by the Food and Agriculture Organization. CSA is an approach to developing the technical, policy, and investment conditions to achieve sustainable agricultural development for food security under climate change (FAO 2013). Adaptation and mitigation are the two possible strategies for facing climate change challenges. Being a small developing country, the most appropriate response to climate change would be general adaptation (World Food Programme (WFP) 2017). The priority adaptation measures identified by the Climate Change Secretariat in Sri Lanka are as follows: (i) increase climate change awareness at all levels, (ii) build capacity for climate change adaptation in key economic sectors, and (iii) engage the wider stakeholders in a dialogue on climate adaptation (Climate Change Secretariat of Sri Lanka 2010). It can be clearly identified that communication has to play an important role in implementing these identified strategies for climate change adaptation. Therefore, communicating CSA in agricultural extension is timely needed. An important way to address the combined issues of food security and climate change is to ensure investment on disaster mitigation and climate adaptation that includes bridging the information gap between farmers and policy makers which is lacking at present (World Food Programme (WFP) 2017).

In conclusion, addressing climate change issues is timely needed, and therefore climate change communication strategies should be incorporated and integrated into agricultural extension. All relevant stakeholders at different levels should be included in this strategies.

3 Extension Communication Methods: Experience of Sri Lanka

Basically, there are three main extension methods, viz., individual, group, and mass methods. The individual type of extension can be effective but not cost-effective. As discussed in Sect. 1.1, the agriculture-related government departments in Sri Lanka have a well-structured bureaucratic mechanism to reach the grassroots-level farmers. They often conduct programs to communicate with groups of farmers using group extension methods such as farmer trainings, demonstrations, and field days. However, mass media plays an important role in mass agricultural extension methods specially to reach a large number of people in the country at once. This is important in transferring quick messages on time as well as in terms of coverage.

Table 1 shows the agricultural information sources and the preferences of farmers according to a study conducted in four provinces in Sri Lanka (Sabaragamuwa, North Central, North Western, and Central) out of nine by Adikari (2014).

According to Table 1, a considerable number of farmers use mass media to obtain agriculture information. However, the majority (80%) of the farmers rely on field-level extension officers.

Communication is important to provide information needed by farmers. Farmers need information on different things during different stages of the production process. A case study has been conducted in rural Sri Lanka by Silva and Ratnadiwakara (2008) to assess the information search cost of vegetable farmers along the value chain (Fig. 2). Accordingly, the farmers spend more cost on searching information at the growing stage. They may need more information on

Table 1 Farmers' agricultural information sources and their preferences to use mass media

Variables	Percentage %
1. Agricultural information sources	
Television	52
Radio	31
Newspapers	22
Journals	13
Magazines	
Others	
Extension officer	80
Parents	71
Fellow farmers	41
Friends and relatives	2
Marketing agencies	37
2. Farmers' preference to use mass media (as the sum of 1st, 2nd, and 3rd preferences)	
Television	64
Radio	44
Newspapers, journals, and magazines	36

Source: Adikari (2014)

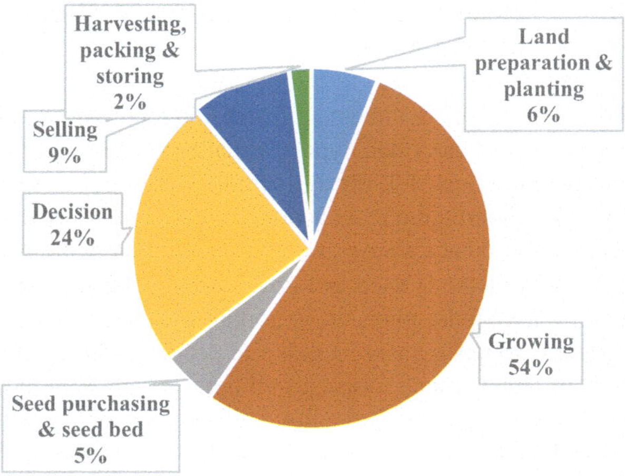

Fig. 2 Information search cost of farmers by growth stage. (Source: Silva and Ratnadiwakara 2008)

pest and disease control and fertilizer application. Farmers should receive information on sustainable agricultural practices at correct time to make sound decisions.

3.1 Use of Print Media

Print media is the oldest form of mass media compared to radio and television. It has gained a significant growth since the late nineteenth and early twentieth centuries as it addresses news and mass popular entertainment items. Print media comes in the form of newspapers, journals, magazines, tabloids, bulletins, broaches, etc. All forms of print media have increasingly commercialized and are facing tremendous competition in the twenty-first century. A specific feature of newspapers and magazines is that they provide words and images that ignite and sustain sociocultural human dynamics. Print media items are created by a variety of producers ranging from most professional publications that adhere to high journalistic standards to fan-produced flyers printed on the fly with everything in between (Guschwan 2016). In addition to that in search of wide circulations, print media grows into scandal spheres while setting the tone of issues and the directionality of local and national issues (Popović and Popović 2014).

The use of print media in Sri Lanka to disseminate agricultural extension dates back to 1820. Leaflets containing agricultural information had been printed and distributed as a means of communication at that time, and therefore this can be considered as the first kind of formal agricultural communication effort in this nature (Arunapriya et al. 2018). Compared to some other developing countries, print media can be effectively used in the Sri Lankan context to disseminate agricultural information since the literacy rate of the Sri Lankans is as high as 93.1% (Central Bank 2017). A regional study in Sri Lanka conducted by Wijerathna et al. (2013) found that 79% of farmers read newspapers and pay special attention to agricultural information. Therefore, newspapers can be used to reach a large audience in agricultural communication. However, 21% do not read newspapers due to various circumstances. One of the reasons may be the cost of the newspapers.

The print media has been forced to pay more attention to standardization, sensationalism, superficial consideration of relatively important issues, and high space for entertainments, sports, and crimes (Guschwan 2016).

When considering the readers' perspective, according to a study conducted in Sri Lanka by Nanayakkara, Wanigasundera, and Dissanayake (2015), respondents prefer to have photographs, success stories of farmers, and subject matter specialists' contribution in newspaper feature articles. Therefore, this suggestion would be important in increasing the readership of farmers on agricultural articles.

3.2 Use of Radio

Radio is a powerful means of communication. Although it is only an audio media compared to the television, some of its unique characteristics are an advantage for agricultural communication. People can listen to the radio while they are working. Especially the rural farmers can listen to the radio even when they are at the field. According to Nazari and Hasbullah (2010), radio has proved to be the most effective media in promoting agriculture and development in rural areas, particularly as a tool for the delivery of quick information. The terms rural radio and community radio have come to be used interchangeably to describe FM stations established to broadcast to a local and predominantly rural audience (Chapman et al. 2003). According to Pringle and David (2002), Sri Lanka Broadcasting Corporation's community radio programming and local FM stations, from their origins in the Mahaweli community radio initiatives in the 1980s to the Kothmale community radio project in the 1990s, are an unusual example of community media.

Radio broadcasting was introduced to Sri Lanka as Radio Ceylon in the early 1920s during the colonial period (Jayarathna et al. 2007). Presently, radio in Sri Lanka is both a public and private service. In Sri Lanka, there is no fully devoted radio channel for agriculture, but agricultural programs are broadcasted via national and few regional radio stations. However, there is a web radio called *Krushi* FM which is devoted to agriculture and is run by the Department of Agriculture. A recent study (Silva et al. 2019) indicates that the Farm Broadcasting Service of the Department of Agriculture plays a key role in producing agricultural radio programs and 41% of farmers listen to these programs regularly. Further, the busy time schedule of the famers and less awareness about the programs are some of the reasons for less listening behavior. Although radio is an old and an audio media compared to the television, still it can be used in agricultural communication to reach a considerably larger audience (Fig. 3).

Although the Kothmale community radio has proven a considerable success, it is not functioning at present. The Mahaweli community radio is also not functioning at present due to various political, administrative, and financial reasons.

Fig. 3 The e-tuktuk, an auto rickshaw of the Kothmale community radio station that served as a mobile radio station and multimedia center. (Source: google/www.flickr.com/photos/10159126@N08/1330828946/ww.flickr.com/photos/10159126@N08/1330828946/)

3.3 Use of Television and Agricultural Video

Video is important as an audiovisual media. Carpenter (1983) argues that television is one of the most effective media in agricultural technology transfer among farmers. According to him, the unique combination of sight, sound, and motion is the reason for its effectiveness, and thereby it can change the behavior of people. Video is an effective tool for awareness raising. Video can be screened in a variety of ways (e.g., on television, at mass events, in local venues, via the Internet) (Lie and Mandler 2009), and it is a relative advantage of this media in agriculture communication. *Govibimata Arunalu* and *Mihikatha Dinuwo* are two popular agriculture video programs produced by the Department of Agriculture and telecasted via the National Television Corporation in Sri Lanka to disseminate agriculture information and technology. *Rividina Arunella* is another popular program on home gardening. This program is produced and telecasted by the National Television Corporation.

Kamatha (traditional threshing floor) was an agricultural television program produced and telecasted by the Department of Agriculture, Ministry of Agriculture, in Sri Lanka and telecasted through a private media channel, namely, "Sirasa." It was popular and was awarded the best development-oriented variety program in Sri

Lanka in 2003. According to a study conducted by Silva (2007) in 8 districts using 1800 households in Sri Lanka, viewership was 50%. The program format included both instructional and entertainment components; an embedded entertainment was an important component to increase the viewership and effectiveness of the program. This video format is known as "edutainment TV" refers to entertaining TV programmes intended priliminary for educational purposes which has a proven success as a global success in rural advisory service initiatives (Kiptot et al. 2016). However, the edutainment format is not widely used at present in Sri Lanka, therefore need to pay attention as it is an effective way of communicating extension messages.

Video programs produced by video technology can be telecasted through media channels or else can be used for projection using multimedia and other electronic devices. Telecasting through media channels is important to reach a large audience at once, and projection using multimedia or other devices is effective, especially in group extension. It has been proven by several studies conducted in Sri Lanka that the use of different video formats is effective in communicating several kinds of agricultural messages that are expecting to change knowledge and especially attitudes.

Wickramaarachchi and Wanigasundera (2011) developed a motivational video for changing the negative disposition of rural youth toward agriculture. Authors found that the video story with motivational, dramatic, and entertaining treatment dimensions can be effectively used to transmit positive messages and thereby change the negative disposition of rural youth in Sri Lanka toward agriculture. A study conducted by Aberathne, Wijerathna, and Wanigasundera (2014) has proven that a short film format can be used effectively to sensitize the farming community and the general public to chronic kidney disease of unknown etiology, a severe environmental and health issues in Sri Lanka at present. Out of all the respondents, 97% had accepted that the short film was successful specially in changing their attitudes.

Basically, agricultural extension is a nonformal adult education process. It is also related to social interventions in agricultural development. Therefore, there is a potential to use participatory video in agricultural extension. However, in a study conducted by Samarakoon and Manthilake (2009), in Sri Lanka, it was found that it is difficult to train farmers in participatory video making. Further efforts should be taken to implement this relatively novel approach in communication in agricultural extension.

3.4 Use of Drama and Theater

Use of drama and theater and role play in agricultural extension is a traditional form of agriculture communication. It is different from mainstream media that is print and electronic. According to Jinadasa (2016), traditional folk media can be utilized effectively in the process of social mobilization for development. However, the authors claimed that the minimal studies in Sri Lanka have been conducted to investigate the sociological and communicative abilities of folk media for development communication. Sri Lanka has a long history in drama and theater. *Sokari* is a good example of traditional drama in Sri Lanka (Swarnasinghe 2000). Based on *Sokari* and Café Theater, a French dramatic style, a new drama format was developed by Samudra and Wanigasundera in 2002. According to an evaluation of its effectiveness, the mean knowledge of 5.8 marks had increased up to 8.3 after watching the drama. However, the use of drama and theater for agricultural communication in the country is limited in history as well as at present. Wijerathna et al. (2004) had developed an interactive drama format to communicate the messages of a community development project. The media treatment incorporating dramatic components was scientifically valid and socially acceptable.

There are different types of drama. With the participation of the audience as characters in the drama, the effectiveness of the communication can be improved. The participatory theater is an example of this type of drama. The Brazilian dramatist Augusto Boal invented this participatory drama format called the "Theatre of the Oppressed," and it shows very successful results in the fields of social conflict resolution, gender issues in society, and health and sexual educational programs (Boal 2008). This type of drama is also known as alternative drama. In India, the forum theater experience shows an increase of awareness of the development issues among civilians, and it implies that the theater cannot change the whole system but can increase the awareness and the critical evaluation of issues (Mills 2009). However, research evidences for the effectiveness of forum theaters in Sri Lanka as a method for disseminating agriculture-related knowledge and technology transfer and attitudinal change is meager.

3.5 Information and Communication Technology (ICT) for Agriculture Communication

The use of ICT in agriculture is the newest form of technology-enabled agriculture communication. It is also known as e-agriculture. It is timely needed to improve ICT for agriculture in Sri Lanka while using efficient and effective traditional communication strategies in agricultural extension. One chapter in this volume, by Dissanayake et al., discusses the use and development of ICT in agricultural communication in Sri Lanka.

4 Key Issues, Challenges, and the Way Forward

Communication in the process of agricultural extension plays an important role in achieving sustainable food systems. However, there are many issues and challenges regarding the effective use of communication in agricultural extension toward the sustainable agricultural development of the country. About 80 percent of the world's extension is publicly funded and delivered by civil servants, providing a diverse range of services to the general population, commercial producers, and disadvantaged target groups through a variety of approaches (Singh et al. 2015). In Sri Lanka also, the agricultural extension is basically a public service. Therefore, as a developing nation, it is challenging to get the required amount of government budgetary allocations to finance agricultural extension. Mostly, face-to-face contact with the farmers are not frequent and adequate due to the small number of extension officers. On the other hand, it is not financially viable to increase the field-level extension officers to the required level. Therefore, effective and efficient communication strategies should be followed to support field-level extension. Traditional communication strategies are not sufficient to reach all clients of agricultural extension fulfilling their extension needs while achieving sustainable development of food systems. In this context, it is important to develop ICT-enabled communication in agricultural extension to face the challenges. Technology stewardship of the extension officers should be developed.

Most of the communication strategies used in agricultural extension in the world as well as in Sri Lanka are mostly linear. The technology transfer model of communication is used as the general agricultural extension approach. This approach is not adequate to address the present challenges of agricultural communication like climate change communication. Therefore, participatory communication approaches are mostly effective in climate change adaptation in agriculture. They are also useful in agricultural development projects, such as the Mahaweli development project (the largest irrigation infrastructure development project) in Sri Lanka which link agricultural development with rural/community development. Furthermore, it would be able to suggest using this branch of communication in farming system approach in agricultural extension where agricultural technologies are supposed to develop locally.

Less priority given in mass media for agriculture and related concerns and issues is one of the problems in mass media communication. Therefore, a reasonable space should be allocated at least by national government newspapers to discuss agricultural issues of the country and provide the necessary knowledge and information. On the other hand, all newspapers circulating within the country are national but not regional. Bandara and Sivayoganathan (2005) had reported that nonavailability of publication and improper distribution are a problem for farmers in Mathale district. Regional newspapers and magazines are useful to discuss regional issues. It is important to make available the different print media, including newspapers, magazines, books, and leaflets to the farmers for their use free of charge or at an affordable price.

Although some national agricultural television programs are available, most of the farmers do not watch the program due to inconvenient time, lack of awareness, and no access to the channel (Udayakanthi et al. 2015). Farmers prefer to watch in the late evening, but it is a challenge to have a time slot during the commercially important peak hours. According to a study conducted by Ekanayake and Wanigasundera (2005), farmers prefer to watch programs between 7 p.m. and 8 p. m. Considering its national importance, it is recommended to change the time of telecasting to late evening, improve the coverage, and raise the awareness of farmers in order to increase the viewership of agricultural programs. At the same time, based on the success stories like the *Kamatha* program (Silva 2007), program formats should be developed in order to capture the attention of both farmers and the general audience. The potential use of community/participatory videos for agricultural communication in Sri Lanka should be further explored.

Community radio model has been identified as an effective participatory means for rural development including agriculture, but sustainability is a key issue. Political, administrative, and financial issues have been identified as the reasons for unsustainability. Carter (2009) has stated the *Kothmale* community radio project as a successful example of a community radio, when its continued existence is in the hands of a succession of political administrations and international funding agencies with their own economic agendas. According to the author, the biggest lesson that the *Kothmale* model teaches is that control of community radio must be in the hands of the community exclusively if it is to succeed in an ongoing, educational, and culturally sensitive manner. Furthermore, according to Pringle and David (2002), Kothmale's potential growth as a rural communication vehicle, for radio or for the Internet, is limited by the lack of independence. In this context, the respective governments should be focused on community radio, especially one that fuels development rather than seeks to fulfill profit making and political objectives. The potential of sustainable use of radio and especially community radio in communicating extension messages to achieve sustainable food systems in Sri Lanka should be further explored.

Drama and theater can be used effectively to communicate agricultural messages, especially in group extension. The recorded version of the drama can be used in television communication and social media like modern communication platforms. Forum theater as a new form of participatory drama to engage with the audience has shown success in other countries. However, this effective method of communication has not yet been used in the country for agriculture communication. Samudra and Wanigasundera (2002) reported that it is hard to find any evidence that stage drama, tele drama, or dialogued cartoons have been used to transmit agricultural information at a large scale. Provisions for these types of communication in a bureaucratic public extension system in Sri Lanka are very limited. However, it is important to incorporate and facilitate this effective method of communication in the agricultural extension system of the country where possible. On the otherhand, although the edutainment media format has a proven success locally, and globally as a format of communicating extension messages, the competent professionals rich in both technical knowledge and creativity in entertainment are lacking. In this context, joint efforts of subject matter specialists with creative art professionals would be reccommended as a way forward.

4.1 Climate Change Communication: A New Paradigm

Traditionally, agricultural extension and communication had worked to promote new technologies and management techniques to increase the production. However, presently, the need of a more sustainable production along with adaptation to climate change challenges has been identified as key issues in agricultural development. The concept and strategies for climate-smart agriculture (CSA) should be incorporated in communication in agricultural extension. Therefore, the suggestions of Sontakki et al. (2018) are important to transform agricultural extension and communication toward more sustainable food ecosystems. Accordingly, the traditional role of extension and communication should be extended toward promoting ecosystem sustainability, as well as the integration of agricultural extension with emerging concepts like social marketing and mindful consumption. Wanigasundera and Carvelho (2014) highlighted the fact that countries like Sri Lanka have a long way to go in bettering media coverage of the topics on climate change. Supporting this recommendation, Kalahinayake et al. (2015) had reported that the awareness of paddy farmers in Hambanthota district of Sri Lanka on climate change and subsequent adaptation strategies are at a low level, and therefore there is a need for conducting awareness programs and media campaigns.

Climate change challenges in Sri Lanka should be addressed. It is timely needed to integrate media strategies into agricultural extension programs in the country to mitigate climate change issues on sustainable food systems.

4.2 Policy Concerns

"A national communication policy should be drawn up to articulate the role of communication in development" (Fraser and Villet 1994). According to Padhy (2010), the government should consider the establishment of a communication policy to support agricultural extension for rural development. According to Bandara and Sivayoganathan (2005), the need for an effective communication system to transfer agricultural information to farmers has been felt since the independence of Sri Lanka. Agricultural communication is embedded in agricultural extension, but there is no agricultural extension policy at present for Sri Lanka. In this context, it is rather new to propose a communication policy for developing sustainable food systems in Sri Lanka. Arunapriya et al. (2018) highlighted the fact that after passing through many different extension systems, still the country has not adapted a mechanism suitable for existing global agricultural trends, agricultural entrepreneurship, and ICT development. In this context, although it is too premature or hard to think of a communication policy for agricultural extension, effective and efficient communication strategies for agricultural development will be established and improved if the recent attempts to develop an agricultural extension policy become a reality.

4.3 Research and Development

The scientific literature specially published on peer-reviewed local and international journals on communication toward agricultural development in Sri Lanka is limited. Many of the research studies have been limited to undergraduate/postgraduate research in universities, but publishing on peer-reviewed journals is not adequate.

In general, the scope of the studies is narrow. A majority of the research studies are regional, and island-wide studies on the discipline are very limited. On the other hand, it is important to publicize the research findings for the relevant authorities to practice/implement. Although research and interventions on ICT for agricultural communication are emerging, the contribution of traditional media and emerging methods like forum theater for effective communication has been neglected. Therefore, research and development on theater and other folk/traditional media for agricultural communication should be a concern. In conclusion, the scientific studies on communication toward developing sustainable food systems should be facilitated within the relevant scientific, academic community in Sri Lanka, and findings should be communicated for practice. This would be more important for the transition of conventional food production systems into becoming more sustainable.

References

Abeyratne KMVV, Wijerathna M, Wanigasundera WADP (2014) Development of a short film on chronic kidney disease to sensitize farming community and general public. In: Proceedings of the Peradeniya University International Research Sessions, Sri Lanka, July 2014, vol 18, p 79

Adikari P (2014) Usage of mass media by farmers in Sri Lanka. International knowledge sharing platform. Dev Country Stud 4(4):1–4. www.iiste.org. ISSN:2224-607X (Paper) ISSN:2225-0565

Anand VE (2014) Development journalism: a catalyst for positive change. Proc Soc Behav Scie 157 (2014):210–225

Arunapriya SA, Sudasinghe JR, Weerakoon WRWMSNP, Wijayathunga RSCWMABM (2018) Historical evolution of agriculture extension system in Sri Lanka. In: International conference on transforming agricultural extension systems: towards achieving the relevant sustainable development goals for global impact, 10–12, May 2018. University of Peradeniya, Sri Lanka

Axinn GH (1988) Guide on alternative extension approaches. Food and Agriculture Organization of the United Nations, Rome

Bandara MRH, Sivayoganathan C (2005) Effectiveness of the extension booklets and leaflets produced by the Department of Agriculture on Big Onion. Agricultural Extension Student projects abstract. Department of Agricultural Extension, Faculty of Agriculture, University of Peradeniya, Sri Lanka

Boal A (2008) Theater of the oppressed. Pluto Press, London

Carey JW (1989) Communication as culture: essay on media and society. Routledge, New York/London

Carpenter WL (1983) Communication handbook. The Interstate Printers and Publishers, Inc, Danville

Carter LFRH (2009) Kothmale community radio interorg project: true community radio or feel-good propaganda? International review of research in open and distance learning. 10(1). https://doi.org/10.19173/irrodl.v10i1.555. Accessed on 18 Dec 2018

Central Bank Report (2017) Department of senses and statistics, Colombo, Sri Lanka

Central Bank Report (2019) Department of senses and statistics, Colombo, Sri Lanka

Chapman R, Blench R, Kranjac-Berisavljevic G, Zakariah ABT (2003) Rural radio in agricultural extension: the example of vernacular radio programmes on soil and Water conservation in Ghana. Agricultural Research and Extension Network. https://www.odi.org/sites/odi.org.uk/files/odi-assets/publications-opinion-files/5200.pdf. Accessed on 25 Feb 2019

Climate Change Secretariat Sri Lanka (2010) Information, education and communication strategy for climate change adaptation in Sri Lanka. Ministry of Environment, Sri Lanka. http://www.climatechange.lk/adaptation/Files/IEC_Strategy.pdf. Accessed on 09 Nov 2018

Dayarathna MS (2009) Mahaweli experiences in agricultural extension. In: Agriculture extension conference. Plant Genetic Resource Centre, Peradeniya, Sri Lanka, pp 110–122

Department of Census and Statistics (2017) Status of sustainable development goals indicators in Sri Lanka. Ministry of National Policies and Economic Affairs, Sri Lanka. http://www.statistics.gov.lk/sdg/application/publications/book.pdf. Accessed on 3 June 2018

Ekanayake EMHB, Wanigasundera WADP (2005) Study of agriculture television programmes of the Department of Agriculture. Agricultural Extension Student projects abstract. Sri Lanka, Department of Agricultural Extension, Faculty of Agriculture, University of Peradeniya

Emitiyagoda SS (2009) Extension services with special reference to the functions of the Department of Agriculture, Sri Lanka: historical development, current position and future directions. In: Agriculture extension conference. Plant Genetic Resource Centre, Peradeniya, Sri Lanka, pp 77–97

FAO (2013) Climate-smart agriculture source book. www.fao.org/docrep/018/i3325e/i3325e.pdf. Accessed on 10 Oct 2018

Fox J. Extension Communication (1990) Manual for frontline agricultural extension staff. Manhattan, Kansas, USA

Fraser C, Villet J (1994) A new agenda for communication in development. FAO, Rome. http://www.fao.org/docrep/t1815e/t1815e03.htm Accessed on 07 Jul 2018

Griffin E, Ledbetter A, Sparks G (2015) A first look at communication theory, 9th edn. McGraw-Hill Education, New York

Gunatilake SK, Samaratunga SS, Rubasinghe RT (2014) Chronic kidney disease (CKD) in Sri Lanka - current research evidence justification: a review. Sabaragamuwa Univ J 13(2):31–58

Guschwan M (2016) Print media: circulating fandom. Soccer Soc 17(3):317–331. https://doi.org/10.1080/14660970.2015.1082759

Jayarathna T, Jayawardena KP, Gunarathne JDA, Silva S (2007) Legal challenges and practical constraints: comparative study in community radio in Sri Lanka. Law Soc Trust Fund 18:28–35

Jinadasa M (2016) *Bali* rituals and therapeutic communication in the traditional rural society in Sri Lanka. Journal Mass Commun 6(11):679–699. https://doi.org/10.17265/2160-6579/2016.11.003

Kalahinayake KAD, Wanigasundera WADP, Punyawardena BVR (2015) Proceedings of the 2nd undergraduate symposium held in Faculty of Agriculture, University of Peradeniya, Sri Lanka, 1 December 2015

Kiptot E, Franzel S, Nora C, Steyn A-M (2016) Edutainment TV for disseminating information about agriculture. Note 22. GFRAS Good Practice Notes for Extension and Advisory Services. GFRAS, Lausanne. https://www.g-fras.org/en/good-practice-notes/22-edutainment-tv-for-disseminating-information-about-agriculture.html?start=1

Kurtzo F, Hansen MJ, Rucker KJ, Edgar LD (2016) Agricultural communications: perspectives from the experts. J Appl Commun 100(1):33–45

Lie R, Mandler A (2009) Video in development. Filming for rural change. http://www.fao.org/uploads/media/Video%20in%20Development_1.pdf. Accessed 2 June 2018

Mahaliyanaarachchi RP (2003) Basics of agricultural extension. Godage International Publishers, Wellampitiya

Markarda J, Ravenb R, Truffer B (2012) Sustainability transitions: an emerging field of research and its prospects. Res Policy 41:955–967

Maunder (1972) Agricultural extension: a reference manual. https://files.eric.ed.gov/fulltext/ED075628.pdf. Accessed 28 May 2018

Mefalopulos P (2008) Development communication source book: broadening the boundaries of communication. The World Bank, Washington, DC. http://siteresources.worldbank.org/EXTDEVCOMMENG/Resources/DevelopmentCommSourcebook.pdf. Accessed 5 July 2018

Mills S (2009) Theatre for transformation and empowerment: a case study of Jana Sanskriti theatre of the oppressed. Dev Pract 19(4/5):550–559

Moser SC (2010) Communicating climate change: history, challenges, process and future directions. WIREs Clim Change 1(1):31–53. https://doi.org/10.1002/wcc.11. Accessed 8 July 2018

Mosher AT (1966) Getting agriculture moving. Fredrick A. Praeger Pulishers, New York

Nanayakkara TPKM, Wanigasundera WADP, Dissanayake UI (2015) Development of a model format for newspaper feature articles for agricultural information dissemination. In: Proceedings of the 2nd Faculty of Agriculture Undergraduate Symposium, Peradeniya, Sri Lanka, p 210. 1 December 2015

Nazari MR, Hasbullah AH (2010) Radio as an educational media: impact on agricultural development for communication and humanities. J South East Asia Res Centre 2:13–20

Oakley P, Garforth C (1985) Guide to extension training. FAO, Rome

Padhy MK (2010) Communication policy for sustainable development: an Indian experiment. In: International conference on future imperatives of communication and information for development and social change, Bangkok, 20–22 December 2010

Popović V, Popović P (2014) The twenty-first century, the religion of tabloid journalism. Proc Soc Behav Sci 163(2014):12–18

Pringle I, David MJR (2002) Rural community ICT applications: the Kothmale model. Electron J Inf Syst Dev Countries 8(4):1–14. https://doi.org/10.1002/j.1681-4835.2002.tb00048.x. Accessed on 01.06.2018

Samarakoon A, Manthilake S (2009) Problems and solutions in training farmers in participatory video making: a case study. Experience and challenges in agricultural extension. In: Agriculture extension conference, Plant Genetic Resource Centre, Peradeniya, Sri Lanka, p 330

Samudra B, Wanigasundera WADP (2002) Development of a dramatized format for agricultural communication. Undergraduate research thesis, Department of Agricultural Extension, Faculty of Agriculture, University of Peradeniya, Sri Lanka

Senevirathna MAPK, Gunasinghe UM (2009) Extension approaches adopted in the export agricultural sector in Sri Lanka. In: Agriculture extension conference, Plant Genetic Resource Centre, Peradeniya, Sri Lanka, pp 122–138

Servaes J, Malikhao P (2005) Development communication approaches in an international perspective. Communication for Development and Social Change. https://doi.org/10.4135/9788132108474.n8

Silva ND (2007) Kamatha as a strategic media approach for agricultural communication. Annals of Sri Lanka Department of Agriculture. https://doa.gov.lk/~agridept/images/ASDA/Asdabook/2007_Asda.pdf. Accessed 10 Feb 2018

Silva DH, Ratnadiwakara D (2008) Using ICT to reduce transaction costs in agriculture through better communication: a case-study from Sri Lanka. LIRNEasia, Colombo, Sri Lanka

Silva DSDN, Dissanayake UI, Peries WLH (2019) Effectiveness of agricultural radio programmes broadcasted by the Department of Agriculture: a case study. In: Proceedings of the 5th Faculty of Agriculture Undergraduate Research Symposium held in Faculty of Agriculture, University of Peradeniya, Peradeniya, Sri Lanka, 21 February, p 188

Singh KM, Shekhar D, Meena MS (2015) Modern extension approaches for livelihood improvement for resource poor farmers. https://mpra.ub.uni-muenchen.de/68414/1/MPRA_paper_68414.pdf. Accessed 4 June 2018

Sivayoganathan C (1999) Agricultural research and extension interface in Asia. Country paper of Sri Lanka. Asian Productivity Organization, Tokyo

Sontakki BS, Rao IS, Rao B (2018) Extension approaches for sustainability of agriculture and food eco systems. In: Compendium of the international conference on transforming agricultural extension systems: towards achieving the relevant sustainable development goals for global impact. Peradeniya, Sri Lanka

State Extension Leaders Network (2006) Enabling change in rural and regional Australia

Stevens SS (1950) Introduction: a definition of communication. J Acoust Soc Am 22(6):689

Swanson BE, Claar JB (1984) The history and development of agricultural extension. Agricultural extension: a reference manual. FAO, Rome

Swarnasinghe KMI (2000) Sanvardhana sannivedanaya saha sokari. Godage International Publishers, Wellampitiya

Udayakanthi DN, Wanigasundera WADP, Kumara WGS (2015) Evaluation of "Govibimata Arunalu" agricultural television programme. In: Proceedings of the 2nd undergraduate symposium Faculty of Agriculture, University of Peradeniya, Sri Lanka

Van den Ban AW, Hawkins HS (1996) Agricultural extension, 2nd edn. Blackwell Publishing, Oxford

Waduge P (2006) Enhancement of extension systems in agriculture. Country paper, Sri Lanka. Asian Productivity Organization, Tokyo

Wanigasundera WADP (2015) Status of extension & advisory services in Sri Lanka. AESA working paper no 1, Agricultural Extension in South Asia

Wanigasundera WADP, Carvelho A (2014) Role of extension and communication to promote climate-smart agriculture in Sri Lanka: agricultural extension: recent interventions for development in Sri Lanka. Sri Lanka Agricultural Extension Association, Department of Agricultural Extension, Faculty of Agriculture, University of Peradeniya, Sri Lanka, pp 1–29

Weerakkody PR, Kumara SK, Rathnayake RMGKB (2004) The privatization of extension services in the non-plantation agricultural sector in Sri Lanka: scope and limitations. Hector Kobbekaduwa Agrarian Research and Training Institute, No. 114, Wijerama Mawatha, Colombo 07

Wickramaarachchi WASV, Wanigasundera WADP (2011) Development of a motivational video for changing negative disposition of rural youth towards agriculture. In: Proceedings of the national research symposium on Agricultural Extension and Organizational Management, pp 12–14

Wickramasooriya BC, Kendaragama KMT, Wijayathilaka TP (2009) Approaches adopted in rural transformation related to livestock sector. In: Agriculture extension conference, Plant Genetic Resource Centre, Peradeniya, Sri Lanka, pp 77–95

Wijekoon R, Rizwan MFM (2009) ICT initiative for agriculture extension of the Department of Agriculture, Agriculture Extension Conference, Plant Genetic Resource Centre, Peradeniya, Sri Lanka, pp 297–329

Wijerathna RMS, Wanigasundera WADP, Chandrasiri J (2004) Development of a drama format for interactive communication with special reference to community development and livelihood improvement project. In: Proceedings of the 60th annual sessions, Sri Lanka Association for the Advancement of Science, Colombo, Sri Lanka, p 255

Wijerathna RMS, Perera K, Suriyagoda LDB (2013) Agricultural information need and the most popular sources of information of the farmers. In: Proceedings of the international symposium on agriculture and environment, University of Ruhuna, Sri Lanka

World Bank (2007) Reviving Sri Lanka's agricultural research and Extension system: towards more innovation and market orientation. Washington, DC. © World Bank https://openknowledgeworldbankorg/handle/10986/13044. License: CC BY 30 IGO. Accessed 12 Dec 2018

World Food Programme (WFP) (2017) National strategic review of food security and nutrition towards zero hunger. https://docs.wfp.org/api/documents/WFP-0000039591/download/?_ga=2.212255377.40445852.1550941043-792398579.1550941043. Accessed 23 Feb 2019

Zumalt JR (2007) Identifying the core periodical literature of the agricultural communications. Documentation Center. University of Illinois at Urbana-Champaign, 200, Library Information and Alumni Center, MC-633, 1101 South Goodwin Avenue, Urbana, IL. https://core.ac.uk/download/pdf/4813649.pdf. Accessed 5 May 2018

A Psychological Lens to Conceptualize Sri Lankan Farmers' Adaptation Behavior in the Face of Significant Environmental Stressors

J. M. P. N. Anuradha

1 Introduction

The recent drought that hit Sri Lanka in early 2016 and lasted through mid-2017 provides alarming evidence of the extent of the vulnerability of Sri Lankan farm households, even to environmental adversities that have been frequent phenomena for ages. The recent drought resulted in a record low rice cultivation during the last main paddy growing season, whereby only 35% of the paddy lands could be brought under cultivation (Ministry of Disaster Management [MDM] and World Food Program [WFP] 2017). The most disturbing development surrounding the recent drought is not merely the dramatic fall in rice production; rather, it concerns the figures revealed with respect to the economic resilience of the rural farm households in drought-prone areas. According to a joint report by the Food and Agriculture Organization [FAO] and WFP (2017), the household income of one-third of the drought-affected population was reduced by more than half. The same report shows how the drastic decline in the production of food crops coupled with deteriorating household economies led to over 200,000 households reaching borderline food insecurity toward the end of the drought spell. Although Sri Lanka has been exposed to such droughts continuously over its history and most distinctively over the last two decades (FAO and WFP 2017), the ability of the drought-prone population to cope with and adapt to such recurring stressful events still seems underdeveloped. Adding more fear to this situation, several other environmental events (i.e., frequent floods (MDM 2014), elephant crop raiding (Fernando 2015; Perera 2009), and diminished access to potable water) have also emerged as significant stressors[1]

[1]A stressor is an "event or circumstance that taxes normal environmental transactions and relationships and initiates and motivates adaptation responses and stress and coping processes" (Reser and Swim 2011).

J. M. P. N. Anuradha (✉)
Department of Agricultural Extension, Faculty of Agriculture, University of Peradeniya, Peradeniya, Sri Lanka

R. P. De Silva et al. (eds.), *Agricultural Research for Sustainable Food Systems in Sri Lanka*, https://doi.org/10.1007/978-981-15-3673-1_19

that critically hamper the livelihoods and well-being of farm households in drought-prone areas and to which many of the sufferers have not been able to adapt[2] well.

Many of these wicked environmental stressors, whether they are natural or human induced, cannot practically be eradicated or even kept under control in some cases. The human-elephant conflict, for which humans are more responsible (Bandara and Gunaratne 2006), is one such stressor on smallholder farmers in the dry zone that is quite difficult to address, unless either the elephant population is brought down to make room for the growing human population and their demands or human activities are kept to a minimum in elephant habitats. Both of these options, given the sociocultural, political, and current economic context of the country, are seemingly inviable, at least in the near future. Thus, from a national-level perspective, reducing the exposure of such a large population to these stressors is out of reach for the government.

However, effective and accurately targeted policy interventions can enhance the adaptation capacity of the suffering to make their own households and their economies more resilient, even when the stressor is present or likely to recur. Well-formulated and well-equipped policies can enhance the level of tolerance of farm households to adversities, facilitate adaptation of countermeasures by the suffering within their own household or at the community level, and even motivate them to transform their livelihoods to reduce their exposure to environmental stressors. The recent drought, as well as other environmental stressors that make frequent news regarding the magnitude of the havoc that they can create for Sri Lankan farm households, however, cast into doubt the ability of the existing policy interventions to motivate farm households to adapt to recurring stressors. World Bank (2015), while commending the comprehensive and well-structured Sri Lankan government policy interventions to introduce "climate-smart agricultural (CSA) practices" to farmers, found that only a minority (<30%) of rice and vegetable farmers have adopted the listed CSA practices. Williams and Carrico (2017) found that the cultivation of developed short-duration paddies was an effective drought adaptation strategy adopted by farmers in the dry zone of Sri Lanka. However, only 27% of their sample respondents had adopted this strategy.

Although Sri Lanka has a policy environment that fosters climate change adaptation, it has been challenging to implement policy directives and guide farmers toward adaptation of climate-resilient strategies (Marambe et al. 2015). The situation remains more or less the same, even with regard to the adaptation support policies related to other critical environmental stressors. Since success of most adaptation support policies affecting agriculture sector depends on voluntary efforts by individual farmers, it is important for such policies to address the factors that either motivate or constrain farmers to adopt recommended adaptation measures. Thus, any

[2] "Adaptation refers to a wide range of responses that an individual can make in difficult circumstances, including initial understanding, affective responses to situations, behavioral responses to situations, the process of selecting responses, and the reciprocating impact of responses to individuals, communities, and the physical environment" (Swim et al. 2009).

adaptation support policy initiative to be effective, it must have a multifaceted understanding of the interplay between contextual and intrapersonal drivers of adaptation to the respective stressor by those suffering from it (Swim et al. 2009). This understanding can only be acquired through a multidisciplinary insight into human adaptation to environmental stressors related to a given context (Reser and Swim 2011; Simonet 2010), and this insight can only be achieved through well-structured research built upon ecological, economic, sociological, anthropological, and psychological principles of adaptation.

In Sri Lankan research, which has a strong adherence to provide directives to policy makers, a large gap exists with regard to our understanding of the role of psychological drivers deemed salient in the process of adapting to environmental stressors. Among the local research studies, which have been predominantly focused on climate change stressors and of which only a few have been published in peer-reviewed and indexed journals, the primary concern has been to capture the role of factors, such as income level, demographics, access to technology, and institutional support, on adaptation decisions and behavior. Only a handful of researchers have ventured into revealing how psychological factors, such as perceptions, attitudes, beliefs, subjective norms, and mental biases, affect the adaptation of the suffering to environmental challenges in Sri Lanka. This is not surprising, as this neglect of psychological factors in favor of more objective financial, technical, and institutional factors has been a common issue, even in the global adaptation literature (Le Dang et al. 2014; Reser and Swim 2011). One reason psychological factors are ignored is that, in comparison to other nonpsychological factors that often involve objective measurements, such factors are difficult to capture and measure in empirical studies (Fresque-Baxter and Armitage 2012; Grothmann and Patt 2005). Another plausible reason, particularly with relevance to local Sri Lankan research, is that there is less understanding of the role of psychological factors in human decision-making processes related to adaptation. The knowledge developed within psychological disciplines, particularly in social psychology, environmental psychology, and cognitive psychology, can contribute immensely to fill the above research gap and subsequently provide policy directives to make effective policies.

This chapter attempts to provide insight into the psychological drivers of human adaptation and, with reference to local and global research, provides a fundamental understanding as how to incorporate those factors in research to study Sri Lankan farmers' adaptation process in the face of various environmental challenges within a much broader spectrum.

2 Why Adopt a Psychological Lens to Understand Human Adaptation?

"Give a man a fish, and you feed him for a day; teach a man to fish, and you feed him for a lifetime." I refer to the above proverb known to most of us and often quoted in the poverty literature to underline the rationale of this study. To adapt this proverb to signify a context of environmental perturbation instead of poverty or hunger, however, I have made slight changes, with certain additions to elaborate on the motive of this chapter. Under the context of environmental hardship, giving people welfare and post-recovery support, as if giving a fish to a man to quench his hunger, would likely serve short-term benefits. As the proverb says, policy initiatives must aim at guiding and facilitating people toward sustainable adaptation strategies to overcome environmental hardships, even without providing continuous external support, as if teaching a man to catch fish by himself to alleviate his own hunger.

However, the proverb is still incomplete, with some important information missing. Even if the man is taught all of the fishing methods, given all of the fishing gear that is needed, and the fish is found to be a very effective and abundant food, the assurance of him continuing to fish is still doubtful. Some simple questions can bring the validity of the popular proverb into question. If the man is too concerned about his neighbors, whom he perceives to be conservationists that generally demean those who are not, what is the chance that the man will still continue to fish, in conflict with the social norms to which he is subject? If the man lives in a conservative neighborhood that rejects innovative ideas and has not practiced or seen fishing before, would the man dare to withstand social repulsion and continue fishing? What if the man perceives that periodic hunger is not a large problem and that fishing is an extra burden not worth investing his time in? If the man fears that the government will cease welfare food rations to which he is already entitled if he is found to be self-sustainable with fish, would he opt to risk fishing at the likely cost of losing what he already has on hand? What if the man has been brought up in a culture that does not embrace fish as a common staple? Would it be easy to motivate him to adopt a new food habit? Finally, if the man refrains from killing animals due to his own firmly held personal attitudes or religious beliefs, would it be easy for him to find his own fish? As long as alternatives and tradeoffs are available that would prevent him from experiencing near death without "fish" as the last resort to survive, the above questions raised by intrapersonal psychological barriers may carry validity. If any policy-led training, motivation, or facilitation program aimed at helping the man continue fishing does not consider these socially constructed and transmitted intrapersonal factors, there is a strong likelihood that these efforts would be wasted. Thus, it can be misleading to try to understand human behavior (or more specifically adaptation behavior) exclusively as a function of utilities and objective external factors (e.g., physical, economic, social, and technological). Nevertheless, more integrative theories of human behavior that consider the interplay between both intrapersonal factors and external factors have proven to be more useful than models that only consider either intrapersonal or external factors for understanding drivers and conditions that can influence behavioral change (Prager 2012). Hence, consideration of the psychological aspects of adaptation by policy makers has an utmost importance (Grothmann and Patt 2005), and the responsibility of local researchers is to provide policy makers with insight into the psychological understanding of adaptation processes.

3 Adaptation to Environmental Perturbation Through a Psychological Lens

A psychological perspective on adaptation to environmental stress is built upon the premise that human (adaptation) behavior results from a mix of conscious and unconscious mental processes (Baumeister et al. 2011). As psychologists acknowledge that behavioral dispositions belong to the individual and human mind alone, adaptation in psychology is solely conceptualized at the individual level (van der Linden 2014). However, psychology does not ignore the impact of contextual influences, as psychological theories often relate the roots of psychological constructs that drive human behavior to sociocultural origins (Maio et al. 2007) and structural factors (e.g., access to resources) that are often highlighted as the drivers of human behavior in many social science disciplines. Thus, psychological constructs can also be considered as actors that mediate the relationship between demographic, sociocultural, economic, and other contextual factors and adaptation responses. In other words, particularly with respect to applied research, psychological constructs such as perceptions can be used as symptomatic manifestations of intrapersonal as well as objective and structural barriers to adaptation.

Backed by empirical evidence from a large body of research dating back several decades, psychological disciplines have identified a number of salient conscious and unconscious psychological constructs that drive human responses to stimuli from environmental perturbation. The following section briefly reviews some of the highly emphasized psychological constructs of human adaptation behavior and a theoretical model that combines those constructs to conceptualize the complex human adaptation responses to environmental stressors.

3.1 Conscious Psychological Determinants of Adaptation Intentions and Behavior

Among the conscious psychological drivers of human adaptation to stressors, two cognitive processes have gained the most notable attention: risk perception and coping appraisal. Risk perception, elsewhere termed risk appraisal, has been considered a major cognitive process that drives human behavior (Arbuckle et al. 2015; Kroemker and Mosler 2002; Nigg and Mileti 2002). Particularly in the climate change literature, risk perception has been consistently found to be one of the strongest predictors of adaptation intentions and subsequent adaptation actions by individuals prone to climate change-induced stressors (Bradford et al. 2012; Grothmann and Patt 2005; Wachinger et al. 2013). Risk perception is the subjective assessment by an individual regarding the probability and severity of a particular threat on him/her and the things he/she values if no adaptation action is taken to alter the vulnerability to the threat concerned (Sjoberg et al. 2004). Here, the threat is measured in terms of the potential health, economic, social, and general environmental consequences of a particular stressor on an individual (Leiserowitz 2006; Zahran et al. 2006).

Risk perception has been found to even outweigh objective risk factors as determinants of adaptation policy support, particularly in climate change sector (Arbuckle et al. 2015). Some studies have found that reaction to a risk can be based on subjective risk but does not necessarily relate to the physical proximity to the objective risk (Silver et al. 2002). Thus, even if the objective risk remains the same, subjective risk perceptions can result in variability in human adaptation behavior in response to a specific stressor.

This discrepancy between the objective risk and subjective risk perception can result from uncertainty[3] surrounding the outcomes and predictability of an environmental stressor that can make people interpret and experience the risk posed by the stressor differently. On the other hand, apart from the objective risk factors such as proximity to the hazard, risk perception can depend on a number of personal and external factors: personal factors (e.g., age and gender), information factors (e.g., awareness through mass media), and contextual factors (e.g., economic and community characteristics) (Wachinger et al. 2013; Kellens et al. 2011; Keller et al. 2006). Nauges and Van Den Berg (2009) found that external factors, such as education and access to information on hygiene practices, drive Sri Lankan urban dwellers' risk perception regarding unsafe drinking water. The authors also found that a higher perceived risk regarding unsafe water increases the probability that individuals will boil and filter water before drinking.

However, certain studies have failed to establish a relationship between risk perception and behavioral responses (Miceli et al. 2008). Wachinger et al. (2013) claims that risk perception may not directly predict adaptation intentions or actions when the relationship between risk perception and adaptation intentions/actions is moderated by intrapersonal (e.g., trust in public adaptation, risk experience, and perceived risk of adaptation actions) and external factors (e.g., economic and physical resources required to adopt adaptation strategies). This implies that effect of risk perception on adaptation behavior can be masked by other relatively stronger psychological constructs as well as objective contextual factors.

Coping appraisal, the other cognitive factor, elsewhere referred to as adaptation appraisal, is the perceptual evaluation of an individual on three aspects of intended adaptation action against a particular threat: his/her ability to avert the threat through the intended adaptation action (self-efficacy), the ability of the adaptation action to be effective in protecting oneself and the people and things he/she values (response efficacy), and finally the cost entailed to realize the adaptation action. As with risk appraisal, studies have found that objective and subjective ability to adapt can differ. For instance, religious people in Bangladesh believe that floods, tornadoes, and cyclones are God's will, and hence, they appraise that they are incapable (or not allowed by God) of adopting countermeasures, and in some cases, they even have to be forced to move to safe places (Schmuck 2000). This example implies that even

[3]Uncertainty, known to be an important mediator of human responses in situations with unknown outcomes (Sjoberg et al. 2004), is a psychological construct (Windschitl and Wells 1996) that limits human cognition.

when people have access to adequate resources to adopt adaptation strategies in response to a stressor, they may perceive their objective capabilities as inadequate to respond to the stressor due to cognitive limitations.

Apart from risk and coping appraisal, there are several other psychological drivers of human adaptation behavior that are frequently highlighted in the literature. Studies have shown that people's willingness to respond to a stressor is driven by the perceived cause of the stressor (Weiner 1995). When people attribute an anthropogenic stressor to distant others, it is likely that they will also attribute the responsibility to mitigate the stressor to agencies outside of their control (Reser and Swim 2011). Subjective norms, a concept central to theories of human behavior in social psychology (Ajzen 1991; Cialdini et al. 1990), are another cognitive factor found to predict people's adaptation responses in the face of environmental stressors (Le Dang et al. 2014; Esham and Garforth 2013). Subjective norms or normative beliefs refer to one's beliefs about the expectations of others who are significant to him/her and the motivation to comply with those expectations. The relevance of subjective norms in human behavior arises out of the general tendency of humans to observe and copy others' behaviors and to engage only in those behavior that others accept and approve (Dawnay and Shah 2005). Social norms are more likely to guide behavioral responses in a social group for which social identity is more central and salient (Fielding et al. 2008). Thus, this psychological construct can play a salient role in traditional rural societies where social ties are much stronger. The literature also cites risk attitude as a significant determinant of how people respond to risks (Keil et al. 2000; van Winsen et al. 2014). Risk attitude, which refers to the propensity or tendency of people to avoid or take risk, can predict farmers' behavior under conditions such as climate uncertainty (Lucas and Pabuayon 2011).

Apart from the cognitive factors, affective responses, such as fear, worry, anxiety, and hope, are also found to be significant determinants of adaptation behavior (Reser and Swim 2011). For instance, emotions such as fear and anxiety can elicit immediate behavioral response to an acute environmental stressor. Nevertheless, it is believed that emotions such as fear can amplify the effect of risk perception on behavioral responses (van der Linden 2014).

3.2 Unconscious Psychological Determinants of Adaptation Behavior

A large body of research acknowledges that humans often rely on unconscious and intuitive thinking, rather than deliberate and rational thinking, to arrive at decisions when making responses to environmental stimuli (Gifford 2011; Kahneman 2011; Thaler 2015). People use heuristics, or mental shortcuts, to simplify the decision-making process when information are not adequately available to make an informed decision. Although heuristics can be effective, particularly when making simple and everyday decisions, they can lead to serious errors when making decisions with relatively large and significant consequences (Sklad and Diekstra 2014). Thus, under contexts of uncertainty and limited information, these unconscious psychological constructs can make people susceptible to cognitive biases and result in erroneous decisions. For instance, in the face of onset of a likely drought, without even bothering to refer to expert opinion or seek for information on climate forecast, a farmer may decide to abandon the cultivation season based on some of the bitter memories on past droughts that rush into his mind.

Apart from heuristics, imparting a direct effect on adaptation behavior or maladaptation, they can also implicitly affect adaptation/maladaptation through their effect on risk perceptions (Wachinger et al. 2013) and coping appraisals (Cervone and Peake 1986). For instance, according to Tversky and Kahneman (1982), the availability heuristic can act as a key determinant of risk appraisal. The availability heuristic is a mental shortcut that can help an individual to judge a situation with ease based on an immediate example (or experience) that comes to mind. A recent incident or an experience that works as an availability heuristic for judging a risk can be powerful, not only due to its high recallability with ease but also because those mental images are often connected with emotions (Tversky and Kahneman 1982). Applying this notion to the long-standing human-elephant conflict in Sri Lanka, it can be assumed that people in a community in which a human has been killed by a wild elephant would perceive a high-risk context compared to people in other communities with the same objective risk. Thus, the propensity for adopting strategies to cope up with the human-elephant conflict would be higher among the people of the community with a strong availability heuristic when they judge the risk of the problem.

Habits[4] have also proven to act as an unconscious psychological barrier on human adaptation to changing environmental conditions. Studies have shown that intentions, which are sometimes even used as a proxy for actual human behavior (Ajzen et al. 2004), may not always translate into actions, as habits can prevent people from carrying out their intentions (van der Linden 2014; Sheeran 2002). For instance, a dry zone farmer, who may report that he intends to drink an adequate amount of water to prevent himself from dehydrating while working in the field, would not necessarily put his intention into action. As stated earlier, his habit of "working continuously for long hours in the field" may prevent him from taking breaks to drink water even without his conscious awareness.

3.3 Conceptualizing Adaptation Behavior: With Reference to Protection Motivation Theory

Psychologists, integrating the conscious and unconscious psychological processes related to human behavior in response to stressful conditions, have developed a number of conceptual models to understand complex human responses to

[4]Habit is defined as frequency of past behavior in psychology (Maio et al. 2007). However, not all frequent behavior is habitual. For a frequent behavior to qualify as a habit, it should be conducted with minimal conscious awareness and intention (automaticity) (Verplanken and Wood 2006).

environmental stressors. These models also provide an interdisciplinary bridge to integrate psychological constructs of behavior with nonpsychological factors to understand the human adaptation process with a wider perspective.

Protection motivation theory (PMT) (Rogers 1983; Rogers and Prentice-Dunn 1997), which was originally proposed as a model to understand the conscious cognitive processes related to health behavior, provides one of the highly cross-culturally validated behavioral models for understanding human responses in the face of environmental stressors (Grothmann and Patt 2005; Mulilis and Lippa 1990; Osberghaus et al. 2010; Poussin et al. 2014; Vaughan 1993). PMT models behavioral intentions (protection motivation) and subsequent behavioral responses as a function of the two major perceptual processes mentioned earlier: risk appraisal and adaptation appraisal. First, if an individual perceives a higher degree of exposure to a threat in terms of probability and frequency of its occurrence and the damage that it is likely to cause, he/she may appraise the risk to be high. Second, if he/she believes that he/she has the resources and capability to adopt a measure against a particular threat, that the intended measure can effectively help him/her to overcome the adversity of the threat, and that the cost of the intended measure is affordable, he/she may perceive a high adaptation capacity. According to PMT, these two appraisals of threat and possible adaptation responses against the stressor can generate one of two opposing responses: adaptation and maladaptation. If the risk perception and the perceived adaptation capacity are high, it is more likely that an individual will engage in an adaptation behavior to prevent damage from the threat. However, if an individual perceives a low risk and adaptation capacity against a threat, maladaptation is more likely. Maladaptation includes avoidant reactions, such as denial of the threat, wishful thinking, and fatalism (Grothmann and Patt 2005).

Although Rogers and Prentice-Dunn (1997) use the term "maladaptation" more as an antonym to adaptive coping strategies, Grothmann and Reusswig (2006) have a different conceptual view on the term. They consider maladaptation as another adaptive coping strategy that may best fit to certain objective situations. For instance, maladaptation can be the most feasible coping strategy for an ill and poor person living in a flood-prone area. Although maladaptation cannot physically protect the victim from physical damages of the threat, it can help the victim to overcome psychological consequences of the threat, such as fear or depression. Thus, maladaptation itself can be a way of adapting to environmental stresses for the people who are socioeconomically incapable of adopting measures to protect themselves against environmental stresses. Maladaptation beliefs, such as fatalism or reliance on "karma," on the other hand, can even distance people from protective measures that they are capable of adopting. Thus, in certain extended PMT models, maladaptation beliefs are considered as predictors of adaptation intentions (Le Dang et al. 2014).

The PMT model has been found to predict human behavior under contexts of perturbation (Milne et al. 2002). However, the behavioral models developed as extensions to PMT, particularly in the climate change literature, have proven to explain more of the variance in adaptation behavior, providing broader insight into the adaptation process. Grothmann and Patt (2005) modified the PMT model by adding new cognitive and noncognitive factors into the model. They added an objective adaptation capacity factor (comprising resources such as time, money, power, knowledge, entitlements, and institutional support) as a predictor of adaptation appraisal and a moderator of adaptation intentions and adaptation actions relationship. Cognitive biases, heuristics, risk experience appraisal, and reliance on public adaptation were also added to their extended PMT model as predictors of both risk and adaptation appraisals, whereas perception of adaptation incentives was added to the model as an explicit predictor of adaptation intentions. The explanatory power of the extended model has been validated by field studies (see Grothmann and Reusswig 2006; Patt and Gwata 2002). Le Dang et al. (2014) added subjective norms, habits, and beliefs in climate change to the PMT model. Their extended model, which was empirically validated through a study in Vietnam, predicted farmers' intention to adopt climate-resilient agricultural practices, and subjective norms were statistically significant predictors along with risk and adaptation appraisals. Swim et al. (2009) extended the PMT model by adding attributions, affective responses (such as fear and hope), and motivational processes as predictors of coping responses. Like Grothmann and Patt (2005), Swim et al. (2009) used the PMT model to integrate psychological behavioral drivers with nonpsychological drivers (e.g., social capital and characteristics of the physical environment) by adding nonpsychological factors into the model as moderators of each step relating psychological constructs to behavioral responses.

4 Psychological Orientation in Adaptation Studies of Sri Lanka: A Review

Using three research search engines, namely, *Scopus*, *PubMed*, and *Google Scholar*, this chapter searched for empirical studies that have been conducted in Sri Lanka with regard to drivers of farmer adaptation to environmental hardships. In the initial stage, the search was not limited to studies focusing on psychological drivers but indiscriminately extended to all studies that have investigated both psychological and nonpsychological drivers of Sri Lankan farmers' adaptation to environmental stressors. To broaden the search, seven key words and their modified versions with suffixes were used to represent three important search filters: adaptation as behavior, agriculture as sector, and Sri Lanka as location. The key words were searched within the title, abstract, and key words to filter out the most relevant studies. To specify the location, the term "Sri Lanka" was included in the search criterion. To specify adaptation, a group of words ("adapt"/"decision"/"adopt"/"behavior" or "intention") and their modified versions with suffixes were included. To specify the sector, the terms "agriculture" and "farm," along with their modified terms, were included. The period searched was 1998–2018. Of the papers filtered by the research search engines, the most relevant ones were manually selected, and the reference section of each selected paper was scanned to look for studies that otherwise would have been left undetected by the search engines. Table 1 specifies the list of empirical studies that were retrieved through the online survey. The survey indicates that only

Table 1 Local studies on determinants of Sri Lankan farmers' adaptation behavior in the face of environmental stressors

Author(S)	Context/ perturbation	Sample size/area	Method	Dependent variable(s)	Key explanatory variables
Truelove et al. (2015)	Climate change-induced drought	192 paddy farmers in five villages of the dry zone	Hierarchical logistic regression	Intention to adopt drought-resilient farming behaviors	Drought risk perception (S)∗; efficacy beliefs (S)∗; village identification (S)∗; FO membership (R/DC); perceived descriptive norms (S)∗; demographic factors (income∗/village∗); norm~∗identification; efficacy~ risk perception
Esham and Garforth (2013)	Climate change-induced drought	126 farmers in four local divisions	Multiple linear regression	Level of adaptation of given adaptation practices	Demographics (age/farming status); farm extent; climate change perception (S)∗; risk perception (S)∗; social networking (S)∗; effectiveness of adaptation measures (S)∗
Illukpitiya and Gopalakrishnan (2004)	Soil erosion	204 potato farmers in the upper Uma Oya watershed	Contingent Valuation method with generalized least square model	Farmers' willingness to invest in soil conservation	Personal factors (education∗/experience∗/age); institutional factors (land tenure∗); economic factors (off-farm income∗/credit∗/debt∗)
Menike and Arachchi (2015)	Climate change-induced water stress	125 farmers in four agro-ecological zones	Logistic regression	Probability of undertaking adaptation to climate change	Social economic factors (household size∗/education∗/group member∗); environmental factors (crop∗); institutional factors (media∗/distance to point of acqusition of inputs∗/location∗/loan access∗); economic structure (income∗)
Jayasooriya and Aheeyar (2015)	Insect pests	290 vegetable farmers in four districts	Stepwise regression	Adoption score	Demographics (age/years of schooling/number of household labor∗/experience in farming); economic (farm income∗); institutional (extension visits/information channels on IPM); social

(continued)

Table 1 (continued)

Author(S)	Context/ perturbation	Sample size/area	Method	Dependent variable(s)	Key explanatory variables
					(involvement in farmer group); knowledge score
Burchfield and Gilligan (2016)	Drought	38 interviews (farmers and other stakeholders) and four farmer focus groups	Qualitative analysis	A remotely sensed metric; enhanced vegetation index (EVI) as a proxy for rice growth during the drought	Structural factors (infrastructural capacity/institutions/physical environment); dynamic factors (local control of water/ perceived risk/ community cohesion/farmer experience)
Williams and Carrico (2017)	Water stress	190 household heads/spouses in 12 water-stressed dry zone communities	Binomial logistic model	Adaptation of five different drought resilient strategies	Demographic (age, gender, wealth, household labor, education level∗); institutional (participation in the farmer organization∗, attending drought meetings∗, received drought information∗, contacted ag. advisory services, proportion of paddy land in major irrigation system∗); agro-ecological (total paddy holdings, presence of agro-well, proportion of paddy land at tail-end of canal∗, proportion of rain-fed paddy land)
Niranjan et al. (2013)	Climate change-induced stressors	50 farmers in Mahagalwewa village in the dry zone	Descriptive statistics	Factors determining the capacity to adapt to climate change	Financial wealth; ownership of sources of water; good health; occupation; water storage capability; attitudes; age

Note: ∗ Variables that were significant at $P < 0.05$; (S) Variables with subjective assessment; ~ Interactions a few empirical studies have been conducted and published in the public domain on farmer adaptation under contexts of environmental adversity in Sri Lanka. Even among the list of eight studies, six were focused on farmer adaptation under water-stress conditions. Demographic factors, economic factors, institutional factors, and ecological factors have been commonly identified as the drivers of farmer adaptation to environmental adversities such as drought, insect pests, and soil erosion.

Among the handful of local empirical research on farmer adaptation to environmental stressors, only two studies have attempted to validate the influence of psychological factors on farmer adaptation in Sri Lanka, whereas only one study is built upon a psychology-based theoretical approach to human behavior. Truelove et al. (2015), drawing on PMT, empirically verified that a socio-psychological model can more accurately predict Sri Lankan farmers' intention to adopt adaptation strategies in response to drought than a strictly demographic model can. Using hierarchical logistic regression, the authors attempted to predict the impact of socio-psychological factors, in the presence of demographic covariates, on farmers' intention to adopt four different adaptation strategies in response to drought. Truelove et al. (2015) concluded that efficacy beliefs were the strongest predictor of behavioral intentions. The authors also established a significant role of risk perception and beliefs in community integration on Sri Lankan farmers' agricultural adaptation intentions. The other study (Esham and Garforth 2013), with a sample of 126 farmers prone to drought in four localities in the wet zone of Sri Lanka, attempted to analyze the effect of cognitive factors (climate change perception, risk perception, and adaptation efficacy) on farmers' level of adaptation to drought. Their dependent variable represented the number of adaptation practices adopted by each farmer from a predetermined list of adaptation strategies. Using multiple linear regression, Esham and Garforth (2013) found that all of the aforementioned cognitive factors significantly predicted the dependent variable. However, their study had some methodical flaws. One major issue was that all of the cognitive variables were single item subjective measures (measured on a Likert scale). Hence, it is doubtful whether each item could capture all of the aspects of each cognitive factor.

5 Recommendations for Future Empirical Research

A lack of studies, particularly ones that are vigorous and have a sound theoretical foundation and methodology, highlights the need to conduct studies to unveil how Sri Lankan farmers make sense of environmental stressors and how this sense-making relates to farmers' adaptation processes. The aims of such research, with respect to various significant environmental stressors in Sri Lanka, may focus on the following areas:

1. To understand how Sri Lankan farmers experience and interpret environmental stressors, their adaptation responses, and their own capacities to adapt
2. To understand the conscious and unconscious psychological factors and their interactive effect on Sri Lankan farmers' adaptation responses

3. To understand the conscious and unconscious psychological factors moderating the relationship between farmers' behavioral intentions and actual adaptation behavior
4. To capture spatial and temporal patterns of psychological processes and their effects on adaptation intentions and behavior
5. To understand how psychological factors can best fit in multidisciplinary approaches to broaden our understanding and capacity to conceptualize Sri Lankan farmers' adaptation behavior
6. To assess the effectiveness of policy interventions and the effect of media communications to address the conscious and unconscious psychological barriers to Sri Lankan farmers' adaptation responses

5.1 Some Methodological Concerns

To empirically study mental processes and human behavior, psychology applies experimental or psychometric approaches or a mix of both paradigms. Usually, experimental studies are conducted in controlled environments in which human subjects are exposed to different treatment conditions to obtain variation. By contrast, psychometric research, without altering the internal or external conditions of the human subjects, examines differences in their psychological constructs to obtain variation (Bindra and Scheier 1954). The latter has been the dominant approach in adaptation research involving environmental stressors and moderate/large samples of human subjects.

Psychometric paradigms often involve factor analytical research conducted with psychometric scales to measure cognitive, affective, and behavioral constructs (Kamphaus 1987). In adaptation studies, psychometric scales are used to quantify constructs of psychological processes, such as risk perception and adaptation appraisal. However, prior to applying established psychometric scales to a particular population, concurrent validity studies are required to assess the psychometric properties of those scales within the sociocultural context of the population concerned. On the other hand, local research may be required to develop new/modified psychometric scales that are more reliable and consistent.

In psychometric research, proper construction of data collection instruments is indispensable at any cost. One such concern is that careful attention regarding choice of terminology when translating or phrasing a new psychometric scale must be used in a local survey. For instance, certain studies report that the two terms global warming and climate change, which are often interchangeably used in climate change communication, can produce different meanings to different people (Schuldt et al. 2011). Thus, selecting the appropriate terms that are more familiar and sensible to a given population is important (van der Linden 2014). Conducting a pre-test and a subjective assessment of the survey instrument by a panel of experts (assuring face validity) can help to select and accurately word the questions/statements in a psychometric tool intended to measure a particular psychological construct (Hardesty and Bearden 2004; van der Linden 2014).

Sample selection and sampling techniques are also important methodological concerns. In particular, samples should be representative of the study population and consist of a sufficient number of respondents to run multivariate statistical tests. For instance, structural equation modeling, a popular multivariate analytical method used by psychologists to test for complex multivariate relationships between psychological constructs (and nonpsychological constructs) in behavioral models, desires at least a sample of 200 survey respondents or five to ten respondents per estimated parameter (Kline 2011).

Another important methodological concern is to reduce measurement errors that can occur when collecting data with a psychometric instrument. One such serious and systematic error in behavioral research is the common method bias (Podsakoff et al. 2003). Common method bias is the variance caused by the instrument or the measurement method rather than the constructs the measures represent. Common method bias can be overcome by various ways. One of the most common strategies is to temporally separate the assessment of dependent variable from the explanatory variables by maintaining a time lag between the measurement of the predictor and explanatory variables (Podsakoff et al. 2003; van der Linden 2014). Other measurement errors that may arise due to the mode of data collection (i.e., respondent satisficing, social desirability bias, and opt-in selection bias) may also lead to false conclusion if remedies are not taken (*see* Roberts 2007).

6 Conclusions

In the face of the most critical and wicked environmental stressors, rural Sri Lankans, who are mostly dependent upon farming, must cope with and adapt to these stressors both individually and/or the community level to sustain and develop their living standards. To create strategic adaptation support policies to provide effective adaptation guidance and assistance to farmers at the household and community levels, a broader understanding of the drivers of farmer adaptation behavior, beyond a limited focus on the objective demographic and contextual drivers, is required. Most importantly, an insight into the conscious and unconscious mental processes that are the most proximal determinants of human behavior is essential to conceptualize how farmers sense and respond to environmental stressors. A large body of literature in the domain of psychology has empirically demonstrated the salient role of psychological processes, such as risk perception, adaptation appraisal, subjective norms, emotions, and mental biases, in human decision-making with respect to environmental adaptation. On the other hand, integrative models of adaptation behavior that capture both psychological and nonpsychological factors have proven to explain much of the variability in farmers' adaptation responses to environmental adversities. However, a psychological understanding of farmers' adaptation to environmental stressors in Sri Lanka remains a barely addressed gap in the research. A lack of local empirical studies on the psychological drivers of farmers' adaptation behavior emphasizes the necessity for future studies. While encouraging more local research to uncover the role of psychological drivers on Sri Lankan farmers' adaptation behavior in the face of significant environmental stressors, this chapter also emphasizes on the importance of designing and executing such research with careful attention on methodological concerns.

References

Ajzen I (1991) The theory of planned behavior. Organ Behav Hum Decis Process 50(2):179–211
Ajzen I, Brown TC, Carvajal F (2004) Explaining the discrepancy between intentions and actions: the case of hypothetical bias in contingent valuation. Personal Soc Psychol Bull 30 (9):1108–1121
Arbuckle JG, Morton LW, Hobbs J (2015) Understanding farmer perspectives on climate change adaptation and mitigation: the roles of trust in sources of climate information, climate change beliefs, and perceived risk. Environ Behav 47(2):205–234
Bandara KADKSD, Gunaratne LHP (2006) Economic potential of the tourism industry to conserve the endangered elephants in Sri Lanka. Wildlanka 1(1):17–32
Baumeister RF, Masicampo EJ, Vohs KD (2011) Do conscious thoughts cause behavior? Annu Rev Psychol 62(1):331–361
Bindra D, Scheier IH (1954) The relation between psychometric and experimental research in psychology. Am Psychol 9(2):69
Bradford RA, O'Sullivan JJ, Van Der Craats IM, Krywkow J, Rotko P, Aaltonen J, Schelfaut K (2012) Risk perception – issues for flood management in Europe. Nat Hazards Earth Syst Sci 12 (7):2299–2309
Burchfield EK, Gilligan J (2016) Agricultural adaptation to drought in the Sri Lankan dry zone. Appl Geogr 77:92–100
Cervone D, Peake PK (1986) Anchoring, efficacy, and action: the influence of judgmental heuristics on self-efficacy judgments and behavior. J Pers Soc Psychol 50(3):492
Cialdini RB, Reno RR, Kallgren CA (1990) A focus theory of normative conduct: recycling the concept of norms to reduce littering in public places. J Pers Soc Psychol 58(6):1015
Dawnay E, Shah H (2005) Behavioral economics: seven principles for policy-makers. New Economics Foundation, London. http://www.i-r-e.org/bdf/docs/a005_behavioural-economics-7-principles-for-policy-makers.pdf. Accessed 2 Mar 2018
Esham M, Garforth C (2013) Agricultural adaptation to climate change: insights from a farming community in Sri Lanka. Mitig Adapt Strateg Glob Chang 18(5):535–549
FAO/WFP Crop and Food Security Assessment Mission to Sri Lanka (2017) Special report. FAO/WFP, Rome
Fernando P (2015) Managing elephants in Sri Lanka: where we are and where we need to be. Ceylon J Sci Bio Sci 44(1):1–11
Fielding KS, Terry DJ, Masser BM, Hogg MA (2008) Integrating social identity theory and the theory of planned behaviour to explain decisions to engage in sustainable agricultural practices. Br J Soc Psychol 47(1):23–48. https://doi.org/10.1348/014466607X206792
Fresque-Baxter JA, Armitage D (2012) Place identity and climate change adaptation: a synthesis and framework for understanding. Wiley Interdiscip Rev Clim Chang 3(3):251–266
Gifford R (2011) The dragons of inaction: psychological barriers that limit climate change mitigation and adaptation. Am Psychol 66(4):290
Grothmann T, Patt A (2005) Adaptive capacity and human cognition: the process of individual adaptation to climate change. Glob Environ Chang 15(3):199–213
Grothmann T, Reusswig F (2006) People at risk of flooding: why some residents take precautionary action while others do not. Nat Hazards 38(1–2):101–120
Hardesty DM, Bearden WO (2004) The use of expert judges in scale development: implications for improving face validity of measures of unobservable constructs. J Bus Res 57(2):98–107

Illukpitiya P, Gopalakrishnan C (2004) Decision-making in soil conservation: application of a behavioral model to potato farmers in Sri Lanka. Land Use Policy 21(4):321–331

Jayasooriya HJC, Aheeyar MM (2015) Adoption and factors affecting on adoption of integrated pest management among vegetable farmers in Sri Lanka. Procedia Food Sci 6:208–212

Kahneman D (2011) Thinking, fast and slow. Farrar, Straus and Giroux, New York

Kamphaus RW (1987) Conceptual and psychometric issues in the assessment of adaptive behavior. J Spec Educ 21(1):27–35

Keil M, Wallace L, Turk D, Dixon-Randall G, Nulden U (2000) An investigation of risk perception and risk propensity on the decision to continue a software development project. J Syst Softw 53 (2):145–157. https://doi.org/10.1016/S0164-1212(00)00010-8

Kellens W, Zaalberg R, Neutens T, Vanneuville W, De Maeyer P (2011) An analysis of the public perception of flood risk on the Belgian coast. Risk Anal 31(7):1055–1068

Keller C, Siegrist M, Gutscher H (2006) The role of the affect and availability heuristics in risk communication. Risk Anal 26(3):631–639

Kline RB (2011) Principles and practice of structural equation modeling, 3rd edn. Guilford Press, New York

Kroemker D, Mosler HJ (2002) Human vulnerability – factors influencing the implementation of prevention and protection measures: an agent based approach. In: Steininger K, Weck-Hannemann H (eds) Global environmental change in alpine regions. Impact, recognition, adaptation, and mitigation. Edward Elgar, Cheltenham, pp 95–114

Le Dang H, Li E, Nuberg I, Bruwer J (2014) Understanding farmers' adaptation intention to climate change: a structural equation modelling study in the Mekong Delta, Vietnam. Environ Sci Pol 41:11–22

Leiserowitz A (2006) Climate change risk perception and policy preferences: the role of affect, imagery, and values. Clim Chang 77(1–2):45–72

Lucas MP, Pabuayon IM (2011) Risk perceptions, attitudes, and influential factors of rainfed lowland rice farmers in Ilocos Norte, Philippines. Asian J Agric Dev 8(2):61–77

Maio GR, Verplanken B, Manstead ASR, Stroebe W, Abraham C, Sheeran P, Conner M (2007) Social psychological factors in lifestyle change and their relevance to policy. Soc Issues Policy Rev 1(1):99–137

Marambe B, Silva P, Weerahewa J, Pushpakumara G, Punyawardena R, Pallawala R (2015) Enabling policies for agricultural adaptations to climate change in Sri Lanka. In: Leal Filho W (ed) Handbook of climate change adaptation. Springer, Berlin/Heidelberg, pp 901–927

Menike LMCS, Arachchi KK (2015) Adaptation to climate change by smallholder farmers in rural communities: evidence from Sri Lanka. Procedia Food Sci 6:288–292

Miceli R, Sotgiu I, Settanni M (2008) Disaster preparedness and perception of flood risk: a study in an alpine valley in Italy. J Environ Psychol 28(2):164–173

Milne S, Orbell S, Sheeran P (2002) Combining motivational and volitional interventions to promote exercise participation: protection motivation theory and implementation intentions. Br J Health Psychol 7(2):163–184

Ministry of Disaster Management (2014) Sri Lanka comprehensive disaster management program 2014–2018. http://www.disastermin.gov.lk/web/images/pdf/slcdmp%20english.pdf. Accessed 22 Feb 2018

Ministry of Disaster Management & World Food Programme (2017) Sri Lanka initial rapid assessment on drought, January. https://www.wfp.org/sites/default/files/SLA_Drought_20170119_updated.pdf. Accessed 22 Feb 2018

Mulilis JP, Lippa R (1990) Behavioral change in earthquake preparedness due to negative threat appeals: a test of protection motivation theory. J Appl Soc Psychol 20(8):619–638

Nauges C, Van Den Berg C (2009) Perception of health risk and averting behavior : an analysis of household water consumption in Southwest Sri Lanka. Tsewp 33

Nigg JM, Mileti D (2002) Natural hazards and disasters. In: Dunlap RE, Michelson W (eds) Handbook of environmental sociology. Greenwood Press, Westport, pp 272–294

Niranjan F, Jayathilake MWAP, Uddika NPC, Dhananjani T, Bantilan C, Singh NP (2013) Vulnerability to climate change: adaptation strategies and layers of resilience farmers' perceptions of climate change in Sri Lanka: qualitative analysis. Research report no. 17. ICRISAT, Patancheru, Telangana, India.http://oar.icrisat.org/8737/1/2012-304%20Res%20Rep%2017%20Srilanka.pdf. Accessed 20 Feb 2018

Osberghaus D, Finkel E,Phl M (2010) Individual adaptation to climate change: the role of information and perceived risk, Discussion paper no. 10-061. Center for Eurpian Economic Research

Patt A, Gwata C (2002) Effective seasonal climate forecast applications: examining constraints for subsistence farmers in Zimbabwe. Glob Environ Chang 12(3):185–195

Perera BMAO (2009) The human-elephant conflict. Gajaha 30:41–52

Podsakoff PM, MacKenzie SB, Lee JY, Podsakoff NP (2003) Common method biases in behavioral research: a critical review of the literature and recommended remedies. J Appl Psychol 88 (5):879

Poussin JK, Botzen WW, Aerts JC (2014) Factors of influence on flood damage mitigation behaviour by households. Environ Sci Pol 40:69–77. https://doi.org/10.1016/j.envsci.2014.01.013

Prager K (2012) Understanding behaviour change: how to apply theories of behaviour change to SEWeb and related public engagement activities. James Hutton Institute. http://www.environment.scotland.gov.uk//media/16539/Understanding-Behaviour-Change.pdf. Accessed 17 Feb 2018

Reser J, Swim J (2011) Adapting to and coping with the treat and impacts of climate change. Am Psychol 66(4):277–289

Roberts C (2007) Mixing modes of data collection in surveys: a methodological review. http://eprints.ncrm.ac.uk/418/1/MethodsReviewPaperNCRM-008.pdf. Accessed 12 Feb 2018

Rogers RW (1983) Cognitive and physiological processes in fear appeals and attitude change: a revised theory of protection motivation. In: Cacioppo BL, Petty LL (eds) Social psychophysiology: a sourcebook. Guilford Press, London, pp 153–176

Rogers RW, Prentice-Dunn S (1997) Protection motivation theory. In: Gochman DS (ed) Handbook of health behaviour research. I: personal and social determinants. Plenum Press, New York, pp 113–132

Schmuck H (2000) An act of Allah: religious explanations for floods in Bangladesh as survival strategy. Int J Mass Emerg Disasters 18(1):85–96

Schuldt JP, Konrath SH, Schwarz N (2011) Global warming or climate change? Whether the planet is warming depends on question wording. Public Opin Q 75(1):115–124

Sheeran P (2002) Intention-behavior relations: a conceptual and empirical review. Eur Rev Soc Psychol 12(1):1–36

Silver RC, Holman EA, McIntosh DN, Poulin M, Gil-Rivas V (2002) Nationwide longitudinal study of psychological responses to September 11. JAMA 288(10):1235–1244

Simonet G (2010) The concept of adaptation: interdisciplinary scope and involvement in climate change. Sapiens 3(1):1–14

Sjoberg L, Moen BE, Rundmo T (2004) Explaining risk perception. An evaluation of the psychometric paradigm in risk perception research. Rotunde Publication, Trondheim

Sklad M, Diekstra R (2014) The development of the heuristics and biases scale (HBS). Procedia Soc Behav Sci 112:710–718

Swim J, Clayton S, Doherty T, Gifford R, Howard G, Reser J, Weber E (2009) Psychology and global climate change: addressing a multi-faceted phenomenon and set of challenges. A report by the American Psychological Association's Task Force on the interface between psychology and Global Climate Change. http://www.apa.org/science/about/publications/climate- change.aspx. Accessed 12 April 2018

Thaler RH (2015) Misbehaving: the making of behavioral economics, 1st edn. W.W. Norton, New York

Truelove HB, Carrico AR, Thabrew L (2015) A socio-psychological model for analyzing climate change adaptation: a case study of Sri Lankan paddy farmers. Glob Environ Chang 31:85–97
Tversky A, Kahneman D (1982) Availability: a heuristic for judging frequency and probability. In: Kahneman D, Slovic P, Tversky A (eds) Judgment under uncertainty: heuristics and biases. Cambridge University Press, Cambridge, pp 163–189
van der Linden S (2014) The social-psychological determinants of climate change risk perceptions, intentions and behaviours: a national study. Doctoral dissertation, The London School of Economics and Political Science (LSE)
van Winsen F, de Mey Y, Lauwers L, Van Passel S, Vancauteren M, Wauters E (2014) Determinants of risk behaviour: effects of perceived risks and risk attitude on farmer's adoption of risk management strategies. J Risk Res 19(1):56–78. https://doi.org/10.1080/13669877.2014.940597
Vaughan E (1993) Chronic exposure to an environmental hazard: risk perceptions and self-protective behavior. Health Psychol 12(1):74
Verplanken B, Wood W (2006) Interventions to break and create consumer habits. J Public Policy Mark 25(1):90–103
Wachinger G, Renn O, Begg C, Kuhlicke C (2013) The risk perception paradox –implications for governance and communication of natural hazards. Risk Anal 33(6):1049–1065. https://doi.org/10.1111/j.1539-6924.2012.01942.x
Weiner B (1995) Judgements of responsibility: a foundation for a theory of social conduct. Guilford Press, New York
Williams NE, Carrico A (2017) Examining adaptations to water stress among farming households in Sri Lanka's dry zone. Ambio 46(5):532–542. https://doi.org/10.1007/s13280-017-0904-z
Windschitl PD, Wells GL (1996) Measuring psychological uncertainty: verbal versus numeric methods. J Exp Psychol Appl 2(4):343–364
World Bank (2015) Climate-smart agriculture in Sri Lanka, CSA country profiles for Africa, Asia, and Latin America and the Caribbean series. The World Bank Group, Washington DC
Zahran S, Brody SD, Grover H, Vedlitz A (2006) Climate change vulnerability and policy support. Soc Nat Resour 19(9):771–789

Empirical Applications of Theory of Firm in Agriculture Research in Sri Lanka: A Review of the Literature

Senal A. Weerasooriya and Dilini Hemachandra

1 Introduction

Economic theory involves choices of the two decision makers in an economy: the consumer and the producer. The decisions made by producers are explained in the economic theory of firm where the producers are seen as making decisions in order to maximize profits given the constraints (Varian 1984; Beattie and Taylor 1993). Producers of agricultural commodities seek to maximize farm profits or in the least maximize revenue subject to resource constraints such as land, labor, and farm machinery. For many of the agricultural commodities, the markets could be assumed to be perfectly competitive or nearly perfectly competitive as there are a large number of small-scale farmers (Debertin 2004). The decisions of a farmer involve the choice of output to be produced and allocation of resources among outputs so that the profits would be maximized.

In any production process, the output depends on the factors used and the production technology in which they are combined. These input–output relationships expressed in a production function provide the foundation for economic theory from a production perspective. A production function relates physical output of a production process to physical inputs or factors of production, i.e., it is a mathematical function that relates the maximum amount of output that can be obtained from a given number of inputs (Shepherd 2015). There are many objectives of using production functions in empirical research: to derive the physical relationships of inputs and outputs, to derive marginal productivities of inputs and resource use efficiency of decision-making units (e.g., technical and economic efficiency), to estimate technical/technological change of a production process over time, and to test economic theories (e.g., diminishing marginal returns). Due

S. A. Weerasooriya · D. Hemachandra (✉)
Department of Agricultural Economics and Business Management, Faculty of Agriculture, University of Peradeniya, Peradeniya, Sri Lanka

R. P. De Silva et al. (eds.), *Agricultural Research for Sustainable Food Systems in Sri Lanka*, https://doi.org/10.1007/978-981-15-3673-1_20

to their appealing properties, production functions are used widely in a wide array of areas (Beattie and Taylor 1993).

In addition to production functions, the concept of duality in production theory introduced by McFadden (1978) brought about a breakthrough in empirical application of production theories. The term dual used in this context means that all of the information needed to obtain the corresponding cost function is contained in the production function, and, conversely, the cost function contains all of the information needed to derive the underlying production function. Hence, according to the dual concept, any constrained maximization problem can be converted into a corresponding constrained minimization problem, and vice versa. The use of the inputs becomes the function to be minimized, and the revenue function becomes the constraint. Since cost/price data are easily available than the quantity data, estimation of parameters of the production function indirectly from the profit/cost function data became increasingly popular (Beattie and Taylor 1993).

The application of the theory of firm in the agricultural commodities involves production functions, profit functions or cost functions, derived input demand functions, and output supply functions. Within the context of Sri Lanka, use of production, profit, supply, and input demand functions in agriculture has received some level of interest. However, no attempt has been made to collectively review the literature concerning the use of above mentioned approaches for Sri Lankan agriculture. In this study, a systematic review of literature is conducted on studies dealing with the derivation of any of the aforementioned functional forms from primary and secondary data. It goes from early use of it to the evolution in its use in agricultural research in Sri Lanka. This review focuses on estimating the output elasticities of major agricultural inputs: fertilizer, labor, and land which constitute a large portion of the cost of cultivation in Sri Lanka and supply and input demand elasticities for own price and fertilizer price. Inclusion of other inputs such as capital in empirical studies is sparse. A reason could be that use of capital in annuals such as paddy is not common and hence difficult to find data. For perennial crops, even though capital investment is high, annual service flows have to be separated out. Studies were not found which incorporate service flow of capital in the context of plantation crops in Sri Lanka. Specifically, this study will focus on two types of crops: paddy and plantation crops, which include tea, rubber, and coconut. Both paddy and planation crops are socioeconomically and culturally integral components of Sri Lankan agriculture. Rice being the staple food of the people of the island, no other crops have received research and policy interest as much as paddy. Plantation crops remained economically significant in the past as the largest foreign exchange earning agricultural commodities for the country. Tea still remains as the top most export income earning agricultural commodity. Therefore, these crops have received and will continue to receive priority in agricultural research in the country. Within this milieu, the main objectives of this study are to compile research with empirical production function estimations in paddy and plantation crops in Sri Lanka with emphasis on fertilizer, labor, and land inputs and to compile research with empirical profit, supply, and input demand function estimations in paddy and plantation crops in Sri Lanka with emphasis on own price elasticities and input price elasticities.

2 Theoretical Background

The producer faces an allocation problem similar to that faced by the consumer. The consumer frequently is interested in allocating income such that utility or satisfaction is maximized. The producer frequently is interested in allocating resources such that profits are maximized and eventually his utility is maximized. Profit is the difference between the revenues obtained from what is sold and the costs incurred in producing the goods. However, producers face constraints, too. The set of choices faced by the producer of goods and services is constrained by resource availability. Economic theory of firm consists of factor–product relationships, factor–factor relationships, and product–product relationships that explain the decisions made by a rational producer. In factor–product relationships, the relationship of input to that of a particular output is explained. The technical relationship could be expressed mathematically as follows.

$$y = f(x_1, \ldots, x_n | x_{n+1}, \ldots, x_m)$$

where y is output, $x_1, \ldots, x_n$ are variable inputs, and $x_{n+1}, \ldots, x_m$ are fixed inputs. The objective of a rational producer is maximizing profits (π). The producer could decide on the level of inputs that maximize profits or level of output that gives maximum profit. Both these approaches yield the same profits. In input perspective, total profit is explained as the difference between total value product (TVP) and total factor cost (TFC). For a perfectly competitive producer who receives constant output price (p^*), TVP $= p^*y$. Hence, profit becomes

$$\pi = p^*y - \text{TFC}$$

where y is a function of inputs. In output perspective, in a perfectly competitive market, producer can sell all the output he/she produces at the current market price. The total income is the revenue of the producer (TR). According to the output perspective, the maximum profit is given by the highest difference between TR and total cost (TC), where TC comprises variable cost (VC) and fixed costs (FC).

$$\pi = \text{TR} - \text{TC}$$

The VC is closely linked to the production function that underlies it. If input prices are constant, all the information about the VC function is contained in the equation underlying production function. Furthermore, the underlying production function could be derived from the VC function if the prices for the inputs are known (Debertin 2004). Note that the maximum profit is achieved at the point where the slope of the profit function is equal to zero. The corresponding level of output is the amount at which an additional unit of output adds zero value to the total profit. Hence,

$$\frac{d\pi}{dy} = \frac{dTR}{dy} - \frac{dTC}{dy} = 0$$

where $\frac{dTR}{dy}$ yields the slope of TR (otherwise known as marginal revenue or MR) and $\frac{dTC}{dy}$ yields the slope of TC (otherwise known as marginal cost or MC). Thus, we have $MR - MC = 0$ from which the below result is obtained:

$$MR = MC$$

Inverse functions of production functions for some simple functional forms can be readily derived. All that is required is to solve the function in terms of the x instead of y. The inverse function contains all the coefficients contained in the original production function and can be converted into true variable cost functions by multiplying by the constant price of the input x. The advantage is, it is therefore not necessary to know the physical quantities of the inputs that are used in the production process in order to determine the coefficients of the production function. If the cost function is known, it is frequently possible to determine the underlying production function. In a nutshell, any production function has an underlying dual cost function or correspondence. A general rule is that if the production function is

$$y = f(x)$$

Then the corresponding inverse production function is

$$x = f^{-1}(y)$$

As shown earlier, the profit-maximizing firm will equate MC with MR. If the firm operates under conditions of pure competition, MR will be the same as the constant price of the output. If the firm produces only one output, the MC curve that lies above average variable cost will be the supply curve for the firm. Each point on the MC curve above average variable cost is a point of profit maximization if the output sells for the price corresponding to the point. Hence, under alternative assumptions with respect to MR or the price of the product, the supply curve for the firm will consist of the series of profit-maximizing points. Similar to the relationship between the production function and the cost function, the product supply and factor demand equations consistent with a firm's optimizing behavior can be obtained by two different but equivalent approaches: the primal and dual approaches. In primal approach, an optimization problem (e.g., output/profit/revenue maximization or cost minimization) is explicitly solved. In dual approach, product supply and factor demand equations are obtained by partial differentiation of an indirect objective function.

Empirical estimation of the production relationships involves selection of a functional form that best describes the relationship between the variables. The most common functional specification is Cobb–Douglas (C–D) type.

$$y = x_1^{\beta 1} x_2^{\beta 2}$$

The common use of the C–D type specification is attributed to several characteristics of the C–D function. Above all is the computational ease. The input elasticities could easily be obtained by the log–log specification of C–D production function where the input coefficients (βs) are the input elasticities. However, C–D type functional specification has a few weaknesses that limit its ability to closely relate to the true functional form of the technical relationship of interest. Its inherent assumption of constant elasticity of substitution is one limitation of its empirical application. Elasticity of substitution is defined as the percentage change in the input ratio divided by the percentage change in the marginal rate of substitution. Due to these limitations, much more flexible functional forms such as constant elasticity of substitution (CES) production functions and transcendental logarithmic (translog)-type specifications were used. Specifically, the translog production function[1] introduced by Christensen et al. (1973) became popular due to its many advantages over C–D or CES specifications. First, it does not assume rigid premises such as smooth substitution between factors or perfect competition of the factor markets (Klacek et al. 2007). Second, it permits nonlinear relationship between the output and the input. In addition, due to its properties, the translog production function can be used for the second-order approximation of a linear-homogenous production, the estimation of Allen elasticities, the estimation of production frontier, or the estimation of the total factor productivity dynamics (Pavelescu 2011). The general form of the translog production function for *i* inputs is given below.

$$\ln y = \ln \alpha + \sum_{i=1}^{n} \beta_i \ln x_i + \left(\frac{1}{2}\right) \sum_{i=1}^{n} \sum_{j=1}^{n} \beta_{ij} \ln x_i \ln x_j$$

Estimation of a production function for an annual crop is different from a perennial crop due to several reasons. One is how capital expenditure on perennial crops is incorporated in the estimation. A grown perennial crop accumulates capital over a period of time. Production functions are defined for a short period of time usually for a year. Siphoning out the annual service flow from capital is required when estimating production functions for perennial crops. Some studies deal with this by using lag variables.

[1]First form of a translog production function involved the approximation of a CES production function with a second-order Taylor series, when the elasticity of substitution is very close to one.

3 Method

This review was conducted covering production functions related studies carried out on paddy and plantation crops in Sri Lanka. More specifically, the review covered studies related to production function estimations and/or supply, demand, and profit function estimations. Note that only the primary production is considered in this review and studies on the processed agricultural commodities were not considered. Systematic literature review is a method/process where a body of literature is aggregated, reviewed, and assessed while making use of predetermined techniques (Sampaio and Mancini 2007). The process of conducting a systematic literature review entails candid questions, search strategy definition, establishment of inclusion or exclusion criteria, and careful analysis of the selected literature (Beatty et al. 2011; Sampaio and Mancini 2007). As for the search strategy, searching for published economic studies on paddy and plantation crop production in Sri Lanka was carried out.

The search was done in the following databases without restriction to the publication year: Google Scholar, ResearchGate, and EconLit and reverse search methods. The search criteria also combined the terms "production function," "profit function," "supply function," "input/derived demand function," "output elasticities," "factor demand elasticities," "own price elasticities," "Cobb-Douglas," "translog," "OLS/Ordinary Least Square Estimator," "MLE/Maximum Likelihood Estimator," "technical efficiency," "paddy," "rice," "tea," "rubber," "coconut," "Sri Lanka," "fertilizer," "land use/extent," "labor use," "cost of cultivation," "fertilizer price," and "paddy price." Unpublished work were also used if they were identified in a database or referenced in a publication. Specifically, this included PhD and master's theses and full papers submitted for conference proceedings. However, abstracts of conference proceedings and unpublished undergraduate theses were excluded from the search. Further, reviews, editorials, and studies involving previously published work were also excluded.

The search resulted in a large number of studies, yet after applying exclusion and inclusion criteria, only 38 studies were selected. From this, 26 are studies concerning estimation of production functions. The remaining 12 are studies concerning estimation of profit, supply, and factor demand functions. From the 38 studies, Data on output elasticities concerning inputs such as land/land extent, labor usage, and fertilizer usage were ollected from the 38 studies. In addition, elasticities for own price and fertilizer price were also obtained from studies which were conducted using profit, supply, and factor demand functions. Data sources included both primary and secondary sources and were time series or cross sectional or panel data by structure.

4 Results and Discussion

The results and discussion is organized into two main sections: production function estimation and supply, demand, and profit function estimation. Within these two main sections, paddy/rice and plantation crops are discussed separately.

4.1 Production Function Estimation

Altogether, 26 studies which have estimated produciton fucntions were found by employing the exclusion and inclusion criteria which have estimated production functions. From this, 14 are on paddy, whereas the rest are for plantation crops. Below sections separately describe the results obtained for paddy and plantation crops.

I. Paddy/Rice

Out of the 14 studies on paddy, ten studies are articles in peer-reviewed journals, two are full papers from conference proceedings, and another two are unpublished Master's theses. Results of the systematic literature review are given in Table 1. Time periods in which the studies have been conducted ranges from 1972 to 2015. Most studies have used data from Yala and Maha, the two cultivating seasons in Sri Lanka. Some studies investigated both seasons together, whereas some other studies investigated a single season. In terms of study location, there is some level of heterogeneity although most studies were concentrated in major paddy growing areas. There are some studies on in Low-Country Wet Zone. In terms of functional specification, most studies have used the Cobb–Douglas specification. It should be noted that in many of the studies, estimating a production function was not the main objective. For instance, production function estimations were used as a first step in estimating technical and allocative efficiencies (Gedara et al. 2012). Most of the studies have used either OLS (Ordinary Least Squares) or MLE (Maximum Likelihood Estimate) as the estimation approach.

Elasticities were collected for three main inputs in paddy, namely, fertilizer use, land use/extent, and labor use. For each input, 26 elasticity measures were collected. Elasticity estimates showed a wide range of heterogeneity. Elasticity estimates for fertilizer ranged from 0.003 to about 0.598. Few values were negative although most were statistically insignificant. Tiruchelvam (2005) found a couple of elasticity estimates which were negative and significant at 10%. Elasticities for labor use ranged from 0.005 to 0.652, and elasticity for land use ranged from 0.257 to 0.874. Output elasticities for fertilizer were found to be more inelastic in comparison to elasticities for labor and land. Interestingly, in all instances, the elasticities were found to be inelastic, i.e., absolute value less than one.

All in all, several observations can be made from the findings. First, there is a wide range of heterogeneity in terms of the elasticity values for the three inputs. This may have been attributed to the study location, data type and structure, method of

Table 1 Output elasticities for paddy production

Author/s	Study location	Time period	Functional specification	Estimation method	Dependent variable	Elasticities		
						Fertilizer	Labor	Land
Abeysekara (1980)	Polonnaruwa, Hambantota, Kurunegala, Kandy, and Colombo	1972/73 Maha	Cobb–Douglas	OLS	Paddy yield	0.598***	0.203***	0.257***
Hafi (1985)[a]	Kurunegala	1983/84 Maha	Cobb–Douglas stochastic frontier	OLS	Paddy production	0.003***	0.098***	0.329**
			Cobb–Douglas stochastic frontier	MLE		0.003***	0.116***	0.333***
Kotagama (1986)[a]	Rajangana irrigation scheme	1981 Yala and 1981/82 Maha	Cobb–Douglas	OLS	Paddy output			
					High access to water	0.002**	0.005	0.661*
					Low access to water	0.272**	0.593**	0.801
					Pooled	0.174**	0.420*	0.463
Ekayanake (1987)	Block 313 of the Mahaweli system H	1984/85 Maha	Cobb–Douglas stochastic frontier	MLE	Paddy production			
					Head reach	0.120***	0.408***	0.624***
					Tail reach	0.262***	0.145*	0.760***
Karunaratne and Herath (1989)	Mahaweli system H	1986 Yala and 1986/87 Maha	Cobb–Douglas stochastic frontier	OLS	Paddy production			
					1986/87 Maha	0.028**	0.018	0.874***
					1986 Yala	−0.065	0.652***	0.512***

Gunaratne and Thiruchelvam (2002)	Anuradhapura	2000/01 Maha	Cobb–Douglas stochastic frontier	MLE	Paddy production			
					Rajangana	−0.145	0.239	0.337**
					Elayapattuwa	0.198	−0.119	0.265*
Illukpitiya and Yanagida (2004)	Badulla	1998/99 Maha	Cobb–Douglas	OLS	Rice output	0.347*	0.062	0.471*
				MLE		0.320*	−0.024	0.522*
Thiruchelvam (2005)	Polonnaruwa and Anuradhapura	2000/01 Maha	Cobb–Douglas stochastic frontier	MLE	Paddy production			
					Mahaweli H and Rajangana	−0.191*	0.099*	0.418***
					Elaypattuwa minor tanks	0.198	−0.119	0.265*
					Mahaweli B and C and Parakrama S	−0.117*	0.054	0.651***
Udayanganie et al. (2006)[b]	Mahaweli system H	2003/04 Maha	Cobb–Douglas stochastic frontier	MLE	Paddy yield	0.010*	−0.001	0.680*
Gedara et al. (2012)	Galgamuwa Division in Kurunegala	2009/10 Maha	Translog	MLE	Paddy production	–	0.160	–
Shantha et al. (2012)	Trincomalee	2009/10 Maha	Cobb–Douglas stochastic frontier	MLE	Paddy output	0.231***	0.224***	0.281***
Shantha et al. (2013)	Mahiyanganaya	2010 Yala	Cobb–Douglas stochastic frontier	MLE	Paddy output	0.396***	0.463***	0.403***
			Translog	MLE		0.282	0.350***	0.347***

(continued)

Table 1 (continued)

Author/s	Study location	Time period	Functional specification	Estimation method	Dependent variable	Elasticities		
						Fertilizer	Labor	Land
Kanthilanka and Weerahewa (2016)[b]	Aggregate data	2005–2015	Cobb–Douglas	OLS	Average paddy yield	Urea 0.229**	−0.006	–
						MOP 0.030	–	–
						TSP −0.075	–	–
Warnakulasooriya and Athukorale (2016)	Gampaha and Kalutara	2009 Yala and 2009/10 Maha	Cobb–Douglas stochastic frontier	MLE	Paddy production			
					2009 Yala	0.145***	0.122	0.681***
					2009/10 Maha	0.219***	0.089	0.441***

[a]Unpublished Master's theses
[b]Full papers from conference proceedings
***Significant at 1% or less
**Significant at 1–5%
*Significant at 5–10%

analysis, time period, and underlying policies governing inputs such as fertilizer and land. Specifically, the fertilizer subsidy program and land policies have been found to be influencing the use of respective inputs in a production process (Zhu and Lansink 2010). In addition, in Sri Lankan context, these policies particularly involving fertilizer and land have changed over time, which might explain some of the heterogeneity in the elasticity estimates. Second, among the three inputs, fertilizer was found to be more inelastic followed by labor and land. Third, the elasticity values were found to be positive albeit few studies in which negative values were reported. This may have been due to the stage in the production process (stage III). Fourth, all the elasticity estimates were found to be inelastic. Fifth, the elasticities for all three inputs seem to depict a decreasing trend (i.e., become more inelastic) with time. Sixth, among the studies which included both seasons versus studies which used Yala and Maha separately, elasticity numbers were different albeit showing no identifiable pattern. Finally, no consistent patterns emerged across data structure (time series, cross-sectional, and panel) and data source (primary and secondary).

II. Plantation Crops: Tea, Rubber, and Coconut

Studies that involved production function estimations with respect to plantation crops, tea, rubber, and coconut, were also reviewed. Altogether there were 12 research publications used for this review comprising nine peer-reviewed journal articles, one Master's thesis, and two conference proceedings. Of them, five, four, and three studies are related to the tea, rubber, and coconut sectors respectively. Even among research conducted on a particular crop, there was variation in the type of data used ranging from experimental data to aggregate national level data. Production function estimation to understand the output response was found more common in the older years. They commonly used the OLS estimator and used C–D functional form for computational and interpretational ease. Later studies have used translog models together with C–D form since real data is unlikely to behave as stipulated by the C–D functional form. In the papers reviewed, Cobb–Douglas and translog are the only functional forms used in production function estimations in. This is true for studies conducted in other countries as well. More recent research has estimated production functions as a first step in measuring technical efficiencies of farms. Majority of these research estimated stochastic production frontiers using Maximum Likelihood Estimator. The findings are given in Table 2.

The research on tea that involved production function estimations dates back to 1984, and the most recent paper is published in 2018. The production functions are estimated for several reasons: to understand the physical relationship between inputs and tea yield (output) and to estimate the returns to scale in production and, as a first step in measuring technical efficiencies. The studies have used aggregate production data from secondary sources or primary data. The majority of papers involved primary data collected from tea smallholders in Sri Lanka. Most of the papers employed C–D production function in log–log functional specification for its easiness in interpretation. The second most popular functional form is the translog function. Majority of the papers used OLS estimator, and those whose objective

Table 2 Output elasticities for plantation crops

Author/s	Study location	Time period	Functional specification	Estimation method	Dependent variable	Elasticities: Fertilizer	Labor	Land
Tea								
Bogahawatte and Herath (1984)	Kotmale, Nuwara Eliya	1983	Cobb–Douglas	OLS	Tea production	−0.088*	0.314	0.570*
Mendis (1990)	Nuwara Eliya Ratnapura	1989	Cobb–Douglas	OLS	Tea production			
					Nuwara Eliya	0.062	0.027	0.801***
					Ratnapura	−0.092	0.352	0.809***
Basnayake and Gunaratne (2002)	Yatinuwara DS, Kandy district	2001	Cobb–Douglas, translog stochastic frontier	OLS, MLE	Tea production	0.004**	0.088*	1.11**
Dharmadasa (2014)[b]	Kalawana and Ratnapura DS	2013	Cobb–Douglas stochastic frontier	OLS, MLE	Tea yield	0.414***	0.038***	0.402***
Premaratne et al. (2018)	Liyangahawela and Alakolawewa	2016	Translog stochastic frontier	MLE	Tea production	0.454*	0.202*	−0.042*
Rubber								
Choo-Kian (1976)[a]	Mathugama, Agalawatte	1975	Cobb–Douglas	OLS	Latex yield	–	–	0.86* (clone1)
						–	–	0.908* (clone 2)
Chandrasiri et al. (1977)	Mathugama Agalawatte	1976	Cobb–Douglas	OLS	Latex production	–	–	0.91***
Wijesuriya et al. (2011)	Moneragala	2008	Cobb–Douglas stochastic frontier	MLE	Latex yield/ha	0.347*	0.062	0.471*
Waduge et al. (2013)	Kalutara	2005–2015	Cobb–Douglas	OLS	Rubber production	0.051	42.75***	56.76***

Coconut								
Loganathan and Balakrishnamurti (1975)	Madampe	1961–1975	Quadratic	OLS	Number of nuts or kg copra/palm/year	–	–	–
De Silva and Tisdell (1981)	Aggregate data	1956–1981	Log–log inverse function	OLS	Coconut production/year	–	–	–
				MLE		–	–	–
Mangika (2009)[b]	North-Western Province	2009	Cobb–Douglas stochastic frontier	MLE	Coconut productivity	1.640**	5.750**	−0.924**

[a]Unpublished master's theses
[b]Full papers from conference proceedings
***Significant at 1% or less
**Significant at 1–5%
*Significant at 5–10%

was to measure technical efficiencies of tea growers using stochastic frontier approach employed Maximum Likelihood Estimator. The average output elasticities estimated for land, labor, and fertilizer are depicted in Table 2. The output elasticities of the land, labor, and fertilizer were positive and significant for tea in all papers reviewed except in Mendis (1990). The estimations in Mendis (1990) show signs for the presence of multicollinearity with high R^2 values and yet with insignificant elasticities.

The research on rubber that involved production function estimations dates back to 1976, and the most recent paper is published in 2013. The production functions are estimated for several reasons: to understand the physical relationship between inputs and rubber yield (output) and to compare yield of different rubber clones and, as a first step in measuring technical efficiencies. Papers involved primary data collected from rubber growers, mainly small growers in Sri Lanka. Most of the papers employed C–D production function in log–log functional specification for its easiness in interpretation. One paper used Just and Pope stochastic production function (Just and Pope 1978) and estimated the stochastic production function using Feasible Generalized Least Squares (FGLS) estimator. FGLS estimator was preferred as it incorporated the labor and weather-related risk in rubber production. However, majority of the papers used OLS estimator, and those whose objective was to measure technical efficiencies of rubber growers using stochastic frontier approach employed Maximum Likelihood Estimator. The average output elasticities estimated for land, labor, and fertilizer are depicted in Table 2. It is difficult to conclude on the output elasticities due to the less number of studies found for the review. Even among the few available, there are mixed results for output elasticities except for land, which shows a significant positive elasticity of production.

The research studies on coconut that involved production function estimations dates back to 1975, and the most recent paper is published in 2009. In early years, the production functions are estimated under experimental condition (field experiments). Later research used non-experimental data and aggregate data to estimate production functions. Production functions are estimated either to understand the physical relationship between inputs and coconut yield (output) or as a first step in measuring technical efficiencies. Early papers used quadratic functional form in their efforts to understand the response of coconut plant to NPK (urea, triple superphosphate, and muriate of potash) fertilizers, and others used C–D production function in log–log functional specification for its easiness in interpretation. The majority of the papers used the OLS estimator. However, if the objective was to measure technical efficiencies of coconut growers using a stochastic frontier approach, many employed a Maximum Likelihood Estimator. The average output elasticities estimated for land, labor, and fertilizer are depicted in Table 2. It is difficult to conclude on the output elasticities due to the less number of studies used in the review. Even among the few available, there are mixed results for output elasticities.

All in all, a few observations can be made from the findings. First, a wide range of heterogeneity is observed in terms of the elasticity estimates for all three inputs (land, labor and fertilizer). This is observed across the three plantation crops: tea, rubber, and coconut. Second, most elasticity estimates were found to be inelasatic except some studies which found large numbers such as Waduge et al. (2013). Third, the elasticity estimates were found to be theoretically consistent for most part, i.e., a positive number. Fourth, among inputs, fertilizer was found to be the most inelastic followed by labor and then land. There are differences in the way fertilizer is used as an explanatory variable. The perennial nature of the plantation crops poses challenges in estimating the output elasticities of variable inputs such as fertilizers. One study has addressed this issue by incorporating lagged fertilizer variable, but others have simply ignored it. Further, most of the studies did not consider the heterogeneity of the requirement of different types of fertilizers (e.g., N, P, K). Therefore, the findings of output elasticities of fertilizer should be used carefully. The high elasticity for land input is notable for plantation crops. In addition, it was difficult to make any patterns across crop type, data source, or data structure due to lack of enough research.

4.2 Supply, Demand, and Profit Function Estimation

Altogether, only 12 studies which involved estimation of profit, supply, and input demand functions were selected after employing the exclusion criteria. From these studies, six were carried out on paddy, whereas the rest were for plantation crops. Results of these studies are discussed below separately for paddy and plantation crops.

I. Paddy/Rice

All six studies are journal articles from peer-reviewed journals. Results of the systematic review are given in Table 3. Time periods in which the studies were conducted range from 1952 to 2012. All studies have used annual data in their respective studies. Heterogeneity was observed in terms of location. Three studies used aggregated secondary data, whereas two studies used primary data collected from major paddy growing areas which included Anuradhapura, Polonnaruwa, Hambantota, and Kurunegala. In addition, Wet-Zone areas such as Kalutara and Northern areas such as Vavuniya have been studied. As for the functional specification, translog, C–D, quadratic, log–log, and log-linear specifications were observed. Most studies used OLS as their estimation technique, and others used Seemingly Unrelated Regression Estimator (SURE).

Own price and factor demand elasticities were collected for variables, namely, paddy price and fertilizer price. For fertilizer and paddy price, 25 and 21 elasticity measures were collected respectively from the aforementioned six studies. Elasticity estimates showed a wide range of heterogeneity. Similar to the findings in Tables 1 and 2, these may have been attributed to the study location, data structure, data source, method of analysis, and time period among other reasons. Elasticity estimates for fertilizer price ranged from relatively inelastic −0.009 to elastic −2.760 values. None of the studies reported positive elasticity estimates, and most

Table 3 Output supply and factor demand elasticities for paddy

Author(s)	Study location	Time period	Functional specification	Estimation method	Dependent variable	Elasticities	
						Fertilizer price	Paddy price
Shumway et al. (1987)	Vavuniya	1969–82	Quadratic (fertilizer demand function)	Nonlinear LS	Quantity of fertilizer	−2.760	−0.250
Gunawardana and Oczkowski (1992)[a]	Aggregate data	1952–87	Log-linear	OLS	Paddy output		
					Short run	0.090**	–
					Long run	0.110**	–
Ekanayake (2006)	Aggregate data	1981–2004	Log–log (fertilizer demand function)	OLS on first difference	Quantity of fertilizer used (NPK)		
					N (urea)	−0.154*	0.331**
					P (TSP)	−0.075	0.415
					K (MOP)	−0.213*	0.792***
Rajapaksha and Karunagoda (2009)	Polonnaruwa, Hambantota, Kurunegala, and Kalutara	1990–2006	Translog (fertilizer demand function)	SURE	Fertilizer quantity		
					Hambantota	−1.027*	2.846**
					Polonnaruwa	−1.333	2.945
					Kurunegala	−1.562	4.869**
					Kalutara	−1.092**	5.049
			Translog (profit function)	SURE	Paddy supply		
					Hambantota	−0.227*	0.954**
					Polonnaruwa	−0.262	0.850**
					Kurunegala	−0.466	2.265**
					Kalutara	−0.383**	2.368**

Rodrigo and Abeysekera (2015)	Anuradhapura, Polonnaruwa, Hambantota, Kurunegala, Kalutara	1990–2011	Cobb–Douglas (fertilizer demand function)	OLS	Fertilizer usage		
					FE	−0.232***	–
					RE	−0.224***	–
Wijetunga (2016)	Anuradhapura, Polonnaruwa, Hambantota, Kurunegala	1990–2012	Translog (fertilizer demand function)	SURE	Fertilizer quantity		
					Anuradhapura	−0.578**	0.596*
					Hambantota	−0.374**	0.100
					Kurunegala	−0.618**	0.607**
					Polonnaruwa	−0.679**	1.143**
			Translog (profit function)		Paddy supply		
					Anuradhapura	−0.054*	0.478**
					Hambantota	−0.009	0.422*
					Kurunegala	−0.060**	0.462**
					Polonnaruwa	−0.113**	0.653**

[a]Authors calculated the elasticity for the ratio between paddy and fertilizer prices
***Significant at 1% or less
**Significant at 1–5%
*Significant at 5–10%

were found to be relatively inelastic, i.e., absolute value to be less than one. Elasticities for paddy price ranged from 0.100 to as large as 5.049.

Several observations can be made from the findings. First, we observed a wide range of heterogeneity in terms of the elasticity values (as depicted in Table 3). This may have been caused due to the differences in the estimation technique, study location, data type and structure, and time of analysis. Second, the elasticities were found to be theoretically consistent in terms of the sign (positive for paddy price and negative for fertilizer price). Third, with the exception of few instances, most studies found that paddy prices had more influence on paddy supply and demand for fertilizer than fertilizer prices. In other words, elasticity for paddy price was larger in comparison to the elasticity for fertilizer price. Fourth, in many cases, the elasticities were found to be inelastic albeit some results with elastic estimates despite them being statistically insignificant.

II. Plantation Crops: Tea, Rubber, and Coconut

Application of duality theory for plantation crops is sparse. The first study we came across dates back to 1984. It involved estimating supply functions for tea (Ramanujam 1984). Further, Ramanujam (1984) estimated supply functions for several countries and reports positive own price elasticity of supply. However, the coefficients were insignificant. Weerahewa et al. (1997) estimated a linear supply function using aggregate data and find that the own price elasticity for raw tea is 0.097. Even though they found a negative relationship with fertilizer price, it is insignificant. Two studies were found with application of duality theory for rubber production. Of them, one study involved estimating supply functions for rubber and the other involved estimating a profit function for rubber. Hartley et al. (1984) found that long-run response of fertilizer price to re-planning of rubber is 1.7. Edirisignhe et al. (2010) estimated a profit function using C–D functional form and found that fertilizer price elasticity is −0.099 for rubber. They found that the C–D form better applies to data than the translog form. Smit (1984) used the vintage approach in estimating rubber supply function. There was only one study which involved the use of duality theory in estimating production function relations with respect to coconut production. Tisdell and De Silva (1985) estimated a coconut supply response function with government expenditure and found that there is a positive effect of government expenditure in 3 year lags.

Several observations could be made from the findings. First, there are very few studies concerning plantation crops in Sri Lanka. Second, even among the few studies, the elasticity estimates showed a high level of heterogeneity mainly due to changes in the type of function (profit, supply, or factor demand), estimation technique, data, and time of analysis. Third, most of the estimates were found to be theoretically consistent in terms of the sign.

5 Conclusions

In this study, an attempt was made to compile research with empirical production function and profit, supply, and input demand function estimations in paddy and plantation crops in Sri Lanka. In production functions, special emphasis was given to output elasticities for fertilizer, labor, and land given their central role as inputs in agricultural context. In profit, supply, and input demand functions, special emphasis was given to own price and fertilizer input price elasticities. Based on the findings, some conclusions are drawn.

In economic theory of firm, the study of production technologies is fundamental in decision-making. Empirical estimation of production functions yields the nature of input–output relationships. In agricultural setting, empirical estimation of production functions is much more challenging than in industry setting as it is very difficult to control for all the exogenous factors affecting the output. However, with the advancement of data collection methods and econometric techniques, the estimations have become more accurate. In Sri Lankan agricultural context, there are not many research studies involving production function estimations. Most studies used for this review were on paddy. In some studies, estimation of production function and obtaining output elasticities were performed as a first step to estimate the inefficiencies in terms of both technical and allocative efficiencies. These estimates were even sparser with respect to perennial crops than annuals. Response to fertilizer is a major concern with respect to production technologies of annuals. It is less so in the case of perennials, and hence, even in the research involving production function estimations for perennials, fertilizer was not used as an input. All in all, for both paddy and perennial crops, a wide range of heterogeneity among output elasticities was observed as a result of the differences in study location, data type and structure, method of analysis, time period, and underlying policies governing inputs.

Even though there is a wide range in the elasticity values for paddy, the own price elasticity of supply yielded the expected positive sign, whereas the fertilizer price elasticity yielded the expected negative sign. It is difficult to conclude on the nature of the input demand elasticity or output supply elasticity as the values vary over a large range. In some studies, the own price elasticity of paddy is highly elastic, whereas in some others, it is inelastic. Similarly, own price elasticity of demand for fertilizer is inelastic in some studies, whereas it is elastic in certain others. Input price elasticity of paddy supply is also inelastic in some studies whereas elastic in others. The inelastic nature of fertilizer price in paddy supply could be attributed to the fertilizer subsidy policies where a fixed quantity of fertilizer was given to paddy farmers at a subsidized price. In terms of plantation crops, it was difficult to draw any conclusions on own price and input elasticities of fertilizer due to less amount of literature available.

Translog functional form is more popular in supply function and input demand function estimations. Parameters of the production function are estimated indirectly from the cost function data. Thus, a more common research approach is to rely on duality to estimate important parameters of the underlying production function by working with a cost function having a translog form.

Overall, there is only a limited application of dual cost or supply functions in paddy and plantation crops in Sri Lanka even though duality has been a very useful concept in understanding profit-maximizing choices made by firms. The application of duality in plantation crops is extremely rare. It could be due to the difficulty in using cross-sectional data for plantation crops. However, with the improvement in record keeping and availability of secondary data, it should be less difficult than before to study the responsiveness of inputs and price elasticities in a dynamic setting.

Even though there are quite a number of research studies looking at the input–output relationships and estimating own and input price elasticities of supply with respect to paddy, most of the studies were limited to specific localities, and hence, results seem scattered. Therefore, it is not meaningful to derive average values for elasticities. It is important that research studies are designed to cover larger geographical areas representing agro-ecological zones or cropping systems: rain-fed and irrigated. There is a lot of scope for supply response studies with respect to plantation crops with time series data now made available. Very little is known about the price elasticities of plantation crops, and this would be a potential area which could be investigated. Further, none of the research aimed at providing a supply response model for any of the plantation crops. They rather focused on the relationship/elasticity of a particular variable of interest such as research and climate parameters. There is a scope for future research to estimate a full supply response model for the plantation crops in Sri Lanka. In doing so, it is better to focus on broader cropping systems (e.g., up-country, mid-country, and low country) rather than on small geographical areas.

References

Abeysekara WAT (1980) Production efficiency in paddy farming. Sri Lanka J Agrarian Stud 1 (1):12–19

Basnayake BMJK, Gunaratne LHP (2002) Estimation of technical efficiency and it's determinants in the tea small holding sector in the mid country wet zone of Sri Lanka. Sri Lankan J Agric Econ 4(1):137–150

Beattie BR, Taylor CR (1993) The economics of production. Krieger Publishing, Malabar

Beatty ME, Beutels P, Meltzer MI, Shepard DS, Hombach J, Hutubessy R, Dessis D, Coudeville L, Dervaux B, Wichmann O, Margolis HS, Kuritsky JN (2011) Health economics of dengue: a systematic literature review and expert panel's assessment. Am J Trop Med Hyg 84(3):473–488

Bogahawatte C, Herath MALB (1984) Pricing and technical efficiencies of smallholder green tea production: evidence from Kotmale area in the Nuwara Eliya district, Sri Lanka. J Agrarian Stud 5(2):37–51

Chandrasiri CR, Carrad B, Teo CK, Weerasinghe S (1977) The specification and estimation of a production function for smallholding rubber in Sri Lanka. J Rubber Res Inst Sri Lanka 54:398–416

Choo-Kian T (1976) Production function analysis of small rubber farms in Sri Lanka. MS thesis. Australian National University, Australia

Christensen LR, Jorgenson DW, Lau LJ (1973) Transcendental logarithmic production frontiers. Rev Econ Stat 55(1):28–45
De Silva NTMH, Tisdell CA (1981) Response of coconuts to fertilizer and advice to Sri Lankan growers: an aggregative approach. Ceylon Coconut Q 32:72–79
Debertin DL (2004) Agricultural production economics, 3rd edn. Macmillan, New York
Dharmadasa RAPIS (2014) Impact of labor out-migration on technical efficiency of tea smallholders in low country wet zone. In: 11th International conference on business management, University of Sri Jayewardenepura, Sri Lanka
Edirisignhe J, Wijesuriya W, Bogahawatte C (2010) Profit efficiency of small holder rubber farms in Kegalle, Kalutara and Ratnapura districts. J Rubber Res Inst Sri Lanka 90:64–77
Ekanayake HKJ (2006) Impact of fertilizer subsidy on paddy cultivation in Sri Lanka. Staff Stud Cent Bank Sri Lanka 36(1&2):73–92
Ekayanake SAB (1987) Location specificity, settler type and productive efficiency: a study of the Mahaweli project in Sri Lanka. J Dev Stud 23(4):509–521
Gedara KM, Wilson C, Pascoe S, Robinson T (2012) Factors affecting technical efficiency of rice farmers in village reservoir irrigation systems of Sri Lanka. J Agric Econ 63(3):627–638
Gunaratne RUMS, Thiruchelvam S (2002) Comparative study on technical efficiency of paddy production under major and minor irrigation schemes in Anuradhapura district. Trop Agric Res 14:341–350
Gunawardana PJ, Oczkowski EA (1992) Government policies and agricultural supply response: paddy in Sri Lanka. J Agric Econ 43(2):231–242
Hafi AAB (1985) Technical efficiency in rice cultivation in Kurunegala district of Sri Lanka. MS thesis, Australian National University, Australia
Hartley M, Nerlov M, Peters Jr RK (1984) The supply response for rubber in Sri Lanka. World Bank staff working papers, 657
Illukpitiya P, Yanagida JF (2004) Improving agricultural production through technical efficiency: a case study of small-holder paddy farming in Sri Lanka. Trop Agric (Trinidad) 81(2):1–7
Just RE, Pope RD (1978) Stochastic specification of production functions and economic implications. J Econom 7(1):67–86
Kanthilanka H, Weerahewa J (2016) Resource-use pattern in paddy cultivation in Sri Lanka: a production function approach. Sri Lanka Forum of University Economists (SLFUE), Department of Economics, Faculty of Social Sciences, University of Kelaniya, Sri Lanka
Karunaratne MAKHSS, Herath HMG (1989) Efficiency of rice production under major irrigation conditions: a frontier production function approach. Trop Agric Res 1:143–158
Klacek J, Vošvrda M, Schlosser Š (2007) KLE translog production function and total factor productivity. Statistika 87(4):261–274
Kotagama HB (1986) An economic analysis on the effect of differential access to water on paddy cultivation in a gravity irrigation system in Sri Lanka. MS thesis. University of the Philippines at Los Banos, Philippines
Loganathan P, Balakrishnamurti TS (1975) Response of coconut *(Cocos nucifera)* to N,P and K fertilizer application. Ceylon Coconut Q 26:89–98
Mangika TPS, Jayasinghe-Mudalige UK, Jayalath KVNN, Pathiraja PMEK (2009) Exploring the efficiency of coconut production in Sri Lanka: where the lazy man's crop lies? In: Agricultural research symposium, pp 43–48
McFadden D (1978) Cost, revenue and profit functions. In: Fuss M, McFadden D (eds) Production economics: a dual approach to theory and application, vol 1. North-Holland Publishing, Amsterdam
Mendis P (1990) Production function analysis of tea estates in Sri Lanka. Econ Rev 16(2):14–29
Pavelescu FM (2011) Some aspects of the translog production function estimation. Rom J Econ 32 (1):131–150
Premaratne SP, Priyanath HMS, Yoosuf A, Maurice D (2018) Technical efficiency for tea smallholder farmers under UTZ certification system in Sri Lanka: a stochastic frontier approach. J Manag 1(2):1–19

Rajapaksha RDDP, Karunagoda KS (2009) Fertilizer demand for paddy cultivation in Sri Lanka with special reference to fertilizer subsidy program. Sri Lanka J Agrarian Stud 13(2):25–38

Ramanujam P (1984) The world tea economy: Supply, demand and market structure. PhD thesis. Australian National University, Australia

Rodrigo C, Abeysekera L (2015) Why the fertilizer subsidy should be removed: key factors that actually derive the fertilizer demand in paddy sector of Sri Lanka. Sri Lanka J Econ Res 3 (2):71–98

Sampaio RF, Mancini MC (2007) Systematic review studies: a guide for careful synthesis of scientific evidence. Braz J Phys Ther 11(1):77–82

Shantha A, Ali ABGH, Bandara RAG (2012) Efficiency and managerial ability of paddy farming under minor irrigation conditions: a frontier production function approach. J Agric Sci 7 (3):145–158

Shantha A, Ali ABGH, Bandara RAG (2013) Technical efficiency of Paddy farming under major irrigation conditions in the dry-zone of Sri Lanka: a parametric approach. Aust J Basic Appl Sci 7(6):104–112

Shepherd RW (2015) Theory of cost and production functions. Princeton University Press, Princeton

Shumway CR, Jegasothy K, Alexander WP (1987) Production interrelationships in Sri Lankan peasant agriculture. Aust J Agric Resour Econ 31(1):16–28

Smit HP (1984) Natural rubber supply in Sri Lanka: analysis and projections using vintage approach. In: Proceedings of the International rubber conference, vol. 1, pp 597–623

Tiruchelvam S (2005) Efficiency of rice production and issues relating to cost of production in the district of Anuradhapura and Polonnaruwa. J Natl Sci Found 33(4):247–256

Tisdell C, De Silva NTMH (1985) Increasing coconut production in Sri Lanka: the potential strategies and the government's role. Q J Int Agric 24:355–368

Udayanganie ADD, Prasada DVP, Kodithuwakku KASS, Weerahewa J, Little DC (2006) Efficiency of the agrochemical input usage in the paddy farming systems in the dry zone of Sri Lanka. In: Prepared for the annual meeting of The Canadian Agricultural Economics Society, Quebec, Canada

Varian RH (1984) Microeconomics analysis, 2nd edn. W.W. Norton, New York

Waduge TD, Edirisnghe JC, Fernando APS, Herath HMLK, Jayasinghe-Mudalige UK (2013) Labor and weather related risks in smallholder rubber production: evidence from Kalutara district. Trop Agric Res Ext 16(3):87–91

Warnakulasooriya HU, Athukorale W (2016) Productive efficiency of rice farming under rainfed conditions in the Gampaha and Kalutara districts of Sri Lanka. Sri Lanka J Food Agric 2 (1):51–64

Weerahewa J, Goddard E, Perera GMS (1997) Impact of research on tea production in Sri Lanka. Trop Agric Res 9:96–107

Wijesuriya W, Disssanayake DMAP, Herath HMLK, Gunaratne PKKS (2011) Constraints in sustainable smallholder rubber farming in the Moneragala District. J Rubber Res Inst Sri Lanka 91:61–73

Wijetunga CS (2016) Rice production structures in Sri Lanka: the normalized translog profit function approach. Asian J Agric Rural Dev 6(2):21–35

Zhu X, Lansink AO (2010) Impact of CAP subsidies on technical efficiency of crop farms in Germany, the Netherlands and Sweden. J Agric Econ 61(3):545–564

Embracing Entrepreneurship in Sri Lankan Agribusiness Research: A Review and a Research Agenda

S. S. Kodithuwakku and C. Weerakoon

1 Chapter Objectives

The objective of this chapter is to map and discuss the intellectual roots and trends of agribusiness entrepreneurship research. This chapter firstly explores the knowledge domain of agribusiness management research to uncover the intellectual roots given intermittent evolution of agribusiness research. Secondly, this chapter assesses the emphasis on entrepreneurship in agribusiness research linking both world and Sri Lankan contexts. Finally, this chapter proposes and discusses possible future research avenues in this stream with a special focus on Sri Lankan agribusiness research.

2 Background

Agribusiness management research has been evolving into a multifaceted body of knowledge over the last 50 years. However, in their review on the evolution of agricultural economics and agribusiness management, Cook and Chaddad (2000) identify a sporadic evolution in agribusiness research. This irregularity in agribusiness research seems to be stemmed from the basic philosophical challenges faced by agribusiness researchers given the strong influence of agricultural economic research (Harling 1995). Further, what constitutes in agribusiness research is predominantly dictated by its definition as there has been multiple definitions proliferated into the field such as agro-industrialization (Boehlje 1999; Cook and Chaddad 2000), value

S. S. Kodithuwakku (✉)
Department of Agricultural Economics and Business Management, Faculty of Agriculture, University of Peradeniya, Peradeniya, Sri Lanka

C. Weerakoon
Department of Business Technology and Entrepreneurship, Swinburne University of Technology, Melbourne, Australia

R. P. De Silva et al. (eds.), *Agricultural Research for Sustainable Food Systems in Sri Lanka*, https://doi.org/10.1007/978-981-15-3673-1_21

or net chains (Lazzarini et al. 2001), and agriceuticals (Goldberg 1999). While these works make important contributions to the field, the reviews seem to be very subjective in their assessments. Additionally, many agribusiness research works are concerned with managerial decision-making of firms (Sterns et al. 1998).

Yet, as the field matures, it is important to divert the interests and focus of agribusiness research to address a wide range of new and more complex entrepreneurship issues. Especially, in the Sri Lankan agribusiness research, there is a paucity of studies focusing on entrepreneurship, and the studies tend to follow the trends found above in relation to international level.

Therefore, in this study, we aim to uncover, map, and discuss the intellectual foundations of agribusiness management research (with entrepreneurship emphasis) as a whole and also with a special attention to Sri Lankan context. By doing so, we offer insights into future research focuses with entrepreneurship emphasis. To our knowledge, this is the first attempt to examine the intellectual structure of the agribusiness research in the field. Moreover, this systematic review is an objective assessment of examining the knowledge base that serves as the knowledge roots in agribusiness.

Accordingly, in this chapter, we first review the past empirical work on agribusiness management research and analyze its theoretical foundations building on the entrepreneurial typologies proposed by Wortman Jr (1987). A special focus will be paid to the contributions of the Sri Lankan agribusiness management research. By doing so, we arrive at an assessment of the current status of agribusiness knowledge base with entrepreneurship emphasis as the field in general and also specifically in the Sri Lankan context. This has important implications to the new researchers to understand the foundations of agribusiness research. Further, we also contribute to the methodological advancement in the field by conducting a review with entrepreneurial typologies outlined by Wortman Jr (1987). Third, we outline the potential entrepreneurship research focuses in agribusiness research aiming to improve the rigor and relevance of such research. Hence, we advance the agribusiness management research by providing future areas of research.

3 Review Methodology

3.1 Literature Search Method

A systematic literature review (Pittaway et al. 2014) was conducted to obtain the insights into the evolution and structure of Sri Lankan agribusiness research. This systematic literature search was performed executing Boolean operations of the terms "entrepreneur∗" and "agribusiness" in the Scopus database and also separately on the key agribusiness-related journals: *International Food and Agribusiness Management Review*, *Agribusiness*, *Journal of Food Distribution Research*, *Journal of International Food and Agribusiness Marketing*, and *Journal of Agribusiness in Developing and Emerging Economies*. This search generated 171 scholarly work on agribusiness entrepreneurship. In addition, the search was also performed in the Sri

Lankan Agriculture sector-based academic journals: *Journal of Agricultural Sciences* and *Sri Lanka Journal of Agrarian Studies* to include all the possible research outputs related to Sri Lankan Agribusiness sector. This search added another 21 research outputs to the sample. A pyramid search/reverse search of the references in the initially identified sample was carried out to draw the literature to review sample and found eight more scholarly publications on agribusiness entrepreneurship. Further, the reverse search was also performed in Google Scholar with the above search terms to draw as many publications as possible to the sample and could expand the sample by another 22 publications. Following van der Have and Rubalcaba (2016), this study was restricted to peer-reviewed scholarly work of journal papers and conference papers. In addition, the sample was restricted to the publications in English language. All the references were added on to an Endnote referencing file. Eighteen duplicated references were deleted, and the abstract of every reference was read to verify the relevance for the review. Fourteen non-related references were deleted, and the remaining 180 publications were utilized for the review.

3.2 Literature Analysis Method

The literature analysis was mainly based on categorizing every reference in terms of entrepreneurship typologies proposed by Wortman Jr (1987). This framework of entrepreneurship typologies is presented as follows. The respective categories were decided by reading the introduction and obtaining an understanding about the focus of each of the scholarly work included in the sample. The basic bibliometric information and thematic areas of each reference were tabulated accordingly. Summaries of the information were investigated, and patterns were identified and analyzed. Future areas for investigation were suggested by identifying the missing research focuses (Fig. 1).

4 How Agribusiness Management Research Embraces Entrepreneurship?

4.1 Profiling the Growth of Agribusiness Entrepreneurship Literature

The basic level analysis of this study aims to provide an overview of agribusiness scholarly work with entrepreneurship emphasis published during 1979–2018. The total aggregated number of publications during this period seems to follow an upward trend especially after 2002, giving rise to three distinctive periods (Fig. 2).

While the initial 30-year slow growth period from 1979 to 2009 produced only 24% of total 180 agribusiness research works with entrepreneurship emphasis, the next 5-year period from 2010 to 2014 accounts for around 28% of the total publications. It appears that nearly 37% of the agribusiness research with

Fig. 1 Entrepreneurship research typologies framework by Wortman Jr. (1987)

1. **Theoretical**
   - A. Theories of entrepreneurship
     - i) Corporate
     - ii) Individual
   - B. Frameworks
     - i) Comprehensive
     - ii) Psychological
     - iii) Sociological
     - iv) Economic
     - v) Political
     - vi) Social
   - C. Definitions of entrepreneurship
   - D. Models of Entrepreneurship
2. **Historical**
   - A. corporate entrepreneurship
   - B. Individual entrepreneurship
3. **Environmental**
   - A. Macro environmental
     - i) International
     - ii) National
   - B. Micro environmental
4. **Organizational**
   - A. Business
   - B. Government
   - C. Non-profit
   - D. International
5. **Functional**
   - A. Corporate entrepreneurship
     - i) Structure
       - a. New venture formation
       - b. Venture management
       - c. Innovation
       - d. Technology
       - e. Public policy
     - ii) Behavior (total organization)
   - B. Individual entrepreneurship
     - i) Structure
       - a. New venture formation
       - b. Operations
       - c. Associations and external groups
     - ii) Behavior
       - a. Careers
       - b. Personal characteristics
       - c. Job characteristics
6. **Future entrepreneurship**

entrepreneurship have been generated during 2015–2018 period, suggesting that agribusiness research with entrepreneurship focus gained its prominence after 2009. The Sri Lankan literature included 20 scholarly works out of the total 180 publications. Yet, it is hard to observe a substantial growth but instead follow a more constant pattern (Fig. 1).

There have been a large number of journals carrying these 180 publications. *International Food and Agribusiness Management Review* (30) was the major publication vehicle identified in this sample. In addition, other journals such as *Scientific Papers Series Management, Economic Engineering in Agriculture and*

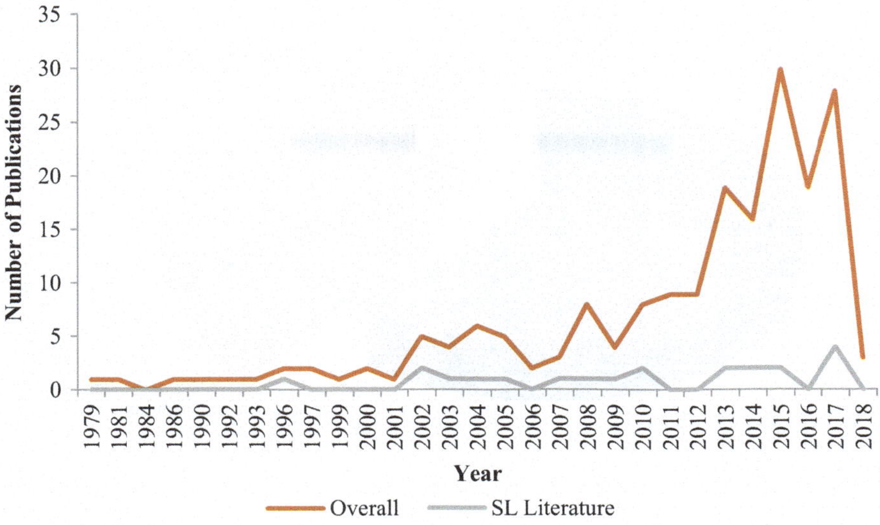

Fig. 2 Growth of entrepreneurship-related agribusiness research

Rural Development (7), *Journal of International Food and Agribusiness Marketing* (2), *Journal of Agriculture and Rural Development in the Tropics and Subtropics, Supplement* (2), and *Journal of Agribusiness in Developing and Emerging Economies* (2) were among the top five main journals sheltering the scholarly work of agribusiness with entrepreneurship emphasis. In relation to the Sri Lankan studies, *Tropical Agricultural Research* is one of the main journals in the field.

These 180 publications included 163 journal articles and 17 conference proceedings. The total of 20 Sri Lankan scientific work included 15 journal articles and five conference proceedings.

4.2 Thematic Areas of Agribusiness Entrepreneurship Literature

The following section will examine the scholarly work in the field of agribusiness with an entrepreneurship focus. During the period from 1979 to 2018, there is an uneven distribution in the issues examined by agribusiness-based entrepreneurship studies. However, there is a substantial diversity in the topics studied by these scholarly works. Figure 3 depicts the distribution of scholarly work among the major typologies.

A majority of studies (60%) have focused on functional typology of entrepreneurship with 59% examining the corporate entrepreneurship-based functional matters while 41% studied individual entrepreneurship. A significantly higher percentage, 26%, of the scholarly work has focused on environmental aspects of agribusiness management. Another 4%, 3%, and 0% have investigated about organizational, historical, and future aspects of agribusiness entrepreneurship, respectively.

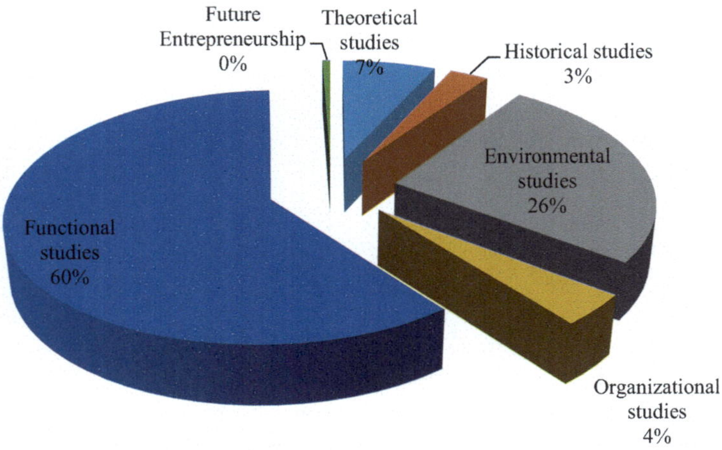

Fig. 3 Distribution of scholarly work among major typologies

4.2.1 Theoretical Studies

There is a paucity of theoretical models examining agribusiness entrepreneurship. Among the handful of theoretical entrepreneurship model-related studies, Poláková, Koláčková, and Tichá (2015) define a business model for agribusiness sector based on Osterwalder and Pigneur's (2010) business model canvas and Holloway and Sebastiao's (2010) and Shafer et al.'s (2005) business model approaches. Poláková et al. (2015) adjust the existing business model approaches by incorporating the role of the owners, their expectations, and influence of the farm environment on the final offer to the customer given the less attention paid to the peculiarities of the sector of small and individual farms by the existing business models.

In another study, Banson et al. (2013) propose a system thinking-based business model for agribusiness sustainability. They have used causal loop modelling to determine the components and interactions between the policy, social, environmental, and economic dimensions to provide insights into potential system behavior and to facilitate leverage points and systemic intervention strategies that are required for sustainable development of the agricultural industry. Moreover, in a very recent study, Santini (2017) provides a detailed literature review uncovering the research outputs and trends in ecopreneurship linking to agribusiness entrepreneurship. Further, this study distinguishes the traits of ecopreneurship from those of traditional entrepreneurship.

4.2.2 Historical Studies

Only four studies out of 180 scholarly works had reported a historical account of agribusiness management. A recent study by Widmar et al. (2015) illustrates the challenges faced by food and agribusiness firms that are transitioning from small, entrepreneurial businesses to larger, more professionally managed enterprises following a fictional case study approach. Another historical narration was found in the work of Geraci (2000) where the author provides the evidence of business evolution from grape grower, to family wine farm, to wine business illuminated through a case study of the Santa Barbara, California, wine industry. Furthermore, with a more macroeconomic perspective, Vázquez (2016) looked into the historical evolution of the strategic activities of commercializing fine grapes in the economy by specifically analyzing the state agents' involvement in advocacy and promotion and the product positioning strategies by the entrepreneurs. However, historical accounts related to individual entrepreneurship are non-existing in this sample.

4.2.3 Environmental Studies

Most of the agribusiness entrepreneurship studies can be categorized as environmental and functional studies. Environmental studies are the second largest composition out of the major six classification areas of the typology. A substantial number of studies (31) had examined the micro-environment-related research issues. In the meantime, while only one study addressed international environmental aspects of agribusiness, 13 studies had looked into the national environmental matters. In national macro-environmental studies, agribusiness entrepreneurship studies are usually oriented toward agricultural education (Wallace 1997), credit/financial service provision for agribusinesses (Hussain et al. 1996; Kuhn et al. 2000), competition and innovation in agribusiness sector (Becvarova 2008; Kilian et al. 2004; Kristiansen 2002), and political economy and interventions (Iglécias 2007; Ioris 2016). Although this subtheme is a popular area in the sample, there was only one study after 2008.

In micro-environmental terms, investigations related to supply chains and collaboration arrangements and value chains and resource provision under market conditions (Aman 2015; Farafonova 2013; Kassa et al. 2011; Zaridis et al. 2015) take a prominent interest (V. R. Amanor-Boadu et al. 2009; Anastasiadis and Poole 2015; Andri et al. 2011). A majority of the studies examining the micro-environmental issues related to a particular sector in the economy or an industry of a specific country. The other major themes investigated in this classification include market orientation and innovation of a particular industry (Micheels and Gow 2008; Pitt and Nelle 2008), business incubators and business facilitation (Sardana and Dasanayaka 2008), human capital education (Katon-Kovácsa and Bóta-Horváth 2012; Santiago and Roxas 2015), financial schemes for sectoral development (Dissanayake et al. 2013), nurturing intrapreneurship (Ikram et al. 2013; Otache 2017), and rural livelihood development (Mabaya et al. 2014; Panasyuk et al. 2014).

4.2.4 Organizational Studies

According to the entrepreneurship typologies used in this study, major organizational contexts include business, governmental agencies, nonprofit organizations, and international organizations (Wortman Jr 1987). There are no studies of agribusiness entrepreneurship in these contexts except for business subtheme where there were only eight such studies. A majority of the studies under the business context classification are more recent studies indicating a trend among the contemporary agribusiness researchers. Some major studies have focused on family farm (Lima et al. 2015) and farm-based businesses (Mirzaei et al. 2016) and cooperatives

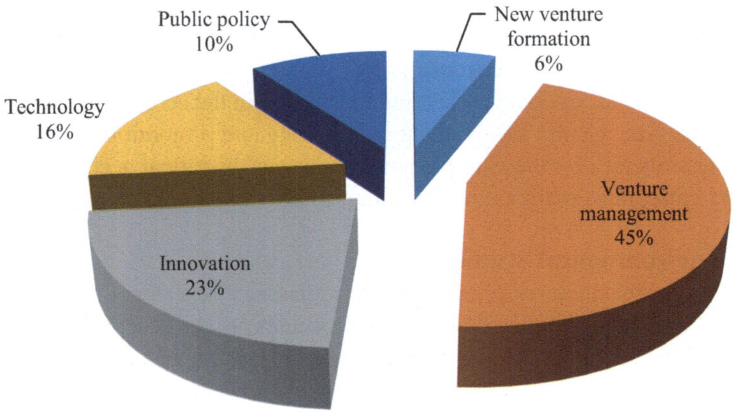

Fig. 4 Composition of theme areas under corporate entrepreneurship – functional studies

(Moreira et al. 2016; Siame 2016). Cooperatives have been identified as a specific business context which may create its own types of risks given its unique interactions (Moreira et al. 2016). Siame (2016) discusses the approach of creating, delivering, and capturing economic and social value through new generation cooperative business models to uplift the rural economic development. Furthermore, Rosairo and Potts (2016) have investigated into the factors affecting performance of farmer companies, and they have found that the performance of farmer companies is compromised by the absence of well-defined and regularly observed processes to develop and implement strategies and inadequate or inappropriate management skills.

4.2.5 Functional Studies

There is a greater diversity in functional studies, and 59% of the work is about corporate entrepreneurship, while the rest 41% is about individual entrepreneurship. Figure 4 portrays the composition of the study focuses under the corporate entrepreneurship theme.

Most of the corporate entrepreneurship studies have been related to the structural aspects. Specifically, a majority of the studies examined venture management (45%) and innovation (23%). In addition, 6% of scholarly work has looked into the overall behavior of the organization. The studies focusing on new venture formation address the issues such as factors determining joint-venture formation (Selassie and Hill 1993), effects of entrepreneurial quality (Darroch and McNaughton 2002), and challenges of starting agribusinesses (Bećirović et al. 2017). Areas such as decision support systems (Havlicek et al. 2003), organizational structure and culture (Kyriakopoulos et al. 2004; Martinez and Poole 2004), diversification strategy (Amanor-Boadu 2013; Bannikova et al. 2015; Hron et al. 2008), financial strategy (Gill et al. 2015; Liang et al. 2010), performance determinants (Rosario et al. 2011), marketing channels (Broderick et al. 2011), business training (Muniafu et al. 2013), networking (Yunita and Dhewanto 2015), business coaching (Smile et al. 2015), entrepreneurial orientation (Mahindarathne and Gunaratne 2015; Silveira-Martins and Vaz 2017), governance (Ma et al. 2017), and human capital accumulation (Udimal et al. 2017) have been the main focuses of studies under the venture management classification.

In innovation terms of corporate entrepreneurship theme, for instance, entrepreneurial innovations for agricultural mechanization (Mujeyi et al. 2015), innovation and Small and Medium Enterprises (SME) growth (Antonites and Haguma 2011; Sultan 2013), design-inspired innovation (Ekman et al. 2011), and intellectual property (Vieira et al. 2012) are key thematic areas popular among the agribusiness researchers.

Under the individual entrepreneurship, the major classification areas investigated include operations (11) and personal characteristics (17) and job characteristics (7). Rarely, a study has examined new venture formation and career aspects of individual entrepreneurship. In new venture formation terms, there are two Sri Lankan studies which uncover the entrepreneurial processes (Kodithuwakku and Rosa 2002) and entrepreneurial strategies (Nawaratne et al. 2009) adopted by rural farmers. Kodithuwakku and Rosa (2002) demonstrate the importance of entrepreneurial processes for the emergence of successful entrepreneurs in an unpromising context. Further, based on case study approach, they have uncovered that successful farmer entrepreneurs have been more creative and persistent in mobilizing resources instead of being more innovative in opportunity identification than those of unsuccessful villagers. Moreover, social networks and contacts of these successful farmer entrepreneurs have been a key determinant of their capital accumulation process. In another study Nawaratne et al. (2009) examined the role of entrepreneurship in accumulating socioeconomic wealth and found that successful farmer-fishermen are pluriactive compared to the unsuccessful counterparts who continued to have identical routine activities.

Another set of studies was related to the associations and external groups. These studies had mainly examined the issues such as the effects of industry associations on market sensing capabilities of farmers (Dentoni et al. 2014) and impact of strategic partnerships on "commercialization" of emerging farmers (Bitzer and Bijman 2014). Based on an inductive grounded theory, Dentoni et al. (2014) examined the effects of public R&D and market orientation on market sensing capability of small farmers. They uncovered that decentralizing the marketing decisions to individual farmers rather the industry associations as a way of enhancing the market sensing capabilities of small farmers. In another study, with a qualitative approach, Mardia, Iriani, and Kamaluddin (2017) found that the development of natural silk business through agribusiness has been supported by the community institutions such as farmers' groups, cooperatives, and entrepreneurs and government counselling and mentoring practices.

A significant number of studies had addressed the behavioral elements of individual entrepreneurship with the focus of personal characteristics (17) and job characteristics (7) but not that many studies around career aspects of individual entrepreneurs. The studies examining personal characteristics had mainly captured the issues such as female entrepreneurship (Adam et al. 2017; Attygalle et al. 2014;

Jackova et al. 2017; Okezie and Joshua 2016), entrepreneurial orientation of farmers (Mahindarathne and Gunaratne 2015), knowledge levels and entrepreneurial skills (Ratajczak 2016; Smith et al. 2015), determinants of entrepreneur's decision (Ariwodor and Agwu 2016; Rosmiza et al. 2015), determinants of farmer's innovation participation (Agwu et al. 2015), entrepreneurial intention (Bosompem et al. 2017; Higgins et al. 2018; Hosenally 2016; Lin et al. 2013), and risk attitude (Shadbolt and Olubode-Awosola 2016).

Another set of studies looked into the job characteristics of individual entrepreneurs. These studies have examined the approaches to develop the skills and competencies of entrepreneurship and thereby to develop the socioeconomic well-being and the rural and economic development in particular regions, in general. For instance, Adashie, Gyan-Kesse, Sarpong, and Boakye-Agyei (2016) had looked into the possibilities of developing entrepreneurs' capabilities through technology usage to knowledge sharing and developing engagement among the smallholder businesses. Further, Vidanapathirana, Hirimburegama, Hirimburegama, Nelka, and Kim (2015) examined the approaches of improving socioeconomic well-being of entrepreneurs through information and communications technology (ICT) education. The literature pertaining to the Sri Lankan context have focused on uncovering the nuances of personal characteristics of entrepreneurs linking to the entrepreneurship theoretical grounds. For instance, Galappaththi et al. (2017) have investigated the start-up motives of shrimp farmers and found that they are driven with both necessity and opportunity. Interestingly, this study reveals that these entrepreneurs face the environmental challenges with diverse set of responses based on existing capital types and willingness to collaborate. In another study, Hemachandra and Kodithuwakku (2010) found that poor farmer entrepreneurs tend to be more production oriented than becoming market-oriented and hence, they suggest to develop market-oriented mindsets among the poor farmers and thereby to make the socioeconomic development a success. There is a paucity of studies about external groups and associations and their effects on individual entrepreneurship except for Esham and Usami (2005).

There is a similarity between the main typologies focused by overall agribusiness research and the Sri Lankan literature. Functional studies have been the major typology (12) underpinning the Sri Lankan agribusiness entrepreneurship literature (e.g., Esham et al. 2006; Rosairo 2010), in a similar manner to the overall agribusiness research with entrepreneurship emphasis. Interestingly, a large majority of the functional studies have investigated about individual entrepreneurship (e.g., Galappaththi et al. 2017; Hemachandra and Kodithuwakku 2010) rather than the corporate entrepreneurship. Only four studies have concentrated on microenvironmental typology (e.g., Palmås and Lindberg 2013), while there is one theoretical study that looked into the social realities of Sri Lankan agri-entrepreneurs (e.g., Gamage et al. 2003). However, there is a great difference between these two groups in terms of the specific thematic areas of investigation.

5 The Way Forward for Sri Lankan Agribusiness Management Research

The major objective of this chapter was to uncover the existing status of agribusiness research with entrepreneurship emphasis with a special attention to Sri Lankan literature and present a research agenda for the field of agribusiness management. Therefore, based on the findings revealed in the previous section, the following section would outline the possible future research avenues which are important to consider by the Sri Lankan agribusiness researchers as well as general researchers in the field as a whole.

Shane and Venkataraman (2000) state that "for a field of science to have usefulness it must have a conceptual framework that explains and predicts a set of empirical phenomena that are not explained or predicted by conceptual frameworks already in existence of other fields of study." This is of significant relevance to the agribusiness entrepreneurship research given the paucity of theoretical studies. Future studies can study the entrepreneurial behaviors in agribusiness venture creation either in terms of corporate or individual entrepreneurship focus. For instance, new venture creation behaviors can be explained by theories of entrepreneurship such as effectuation (Sarasvathy 2009), causation (Shah and Tripsas 2007), and bricolage (Baker and Nelson 2005). Thus, future studies can apply these theoretical models as base models to reflect on specific agribusiness entrepreneurship models. As research matures, it is essential to map the field of agri-entrepreneurship and empirically test and establish the relationships among the key dimensions. This will provide a unifying framework for the future researchers to build on. In addition, a comprehensive review of research methods adopted in the agribusiness entrepreneurship research would be of greater potential to expand the boundaries of existing methodological domains of the field. Currently, there are no such studies conducted, although a vast majority of studies are apparently based on case studies. Hence, a critical review and a research agenda of the research methodological approaches would warrant uncovering the insights into the methodological advancement in the field.

Although there is a huge amount of studies addressing micro-environmental elements in agribusiness context, still, the future studies can specifically look into environmental dimensions such as munificence, hostility, and heterogeneity (Tsai et al. 1991) and how do they influence agribusiness entrepreneurship practices. Findings demonstrated that a significant number of studies have examined the environmental behaviors pertaining to particular industries and sectors. Although it is a promising approach to uncover the peculiarities in different sectors, researchers can still conduct comparative studies on the environmental elements and their influences on corporate and individual entrepreneurship. Further, future research could focus on whether there is a difference between the influences of environmental elements on corporate entrepreneurship and individual entrepreneurship. In addition, future studies should focus on analyzing the effects of international environment on agribusiness to uncover the entrepreneurship in foreign settings. Such an understanding would be of greater significance to agribusinesses engaging in foreign operations as it may guide the international operations of such businesses shifting the boundaries from theoretical implications to managerial or practical implications.

In terms of organizational studies, results revealed that there is a very limited focus on the alternative contexts of entrepreneurship in the existing agribusiness literature. The contextual aspects of entrepreneurship are dramatically shifting toward new forms such as governmental businesses, social businesses, social enterprises, and many other nonprofit structures. As a result, the emergence of hybrid forms of businesses could be seen as a trend in the contemporary world. For instance, social enterprises are a form of hybrid organizations where they try to achieve a social mission with a sustainable commercial mechanism (Doherty et al. 2014). Therefore, for instance, future studies can examine the social enterprise formation in agribusiness sector. In addition, such studies can be further extended to the analysis such as the potential impacts of social enterprises and nonprofit organizations on agribusiness entrepreneurship, differences in entrepreneurial processes in agribusiness social enterprises, and effects of hybridity on agribusiness performance. Studying the interface between social entrepreneurial aspects linking to agribusiness entrepreneurship would have a tremendous impact on expanding the boundaries of both mainstream entrepreneurship and agri-entrepreneurship uncovering the potential applicability of entrepreneurship theories and practices.

Although a large number of studies have focused on functional typology, there is a limited investigation on the public policy implementation and its relevance and impact on agribusiness formation and management. Further, an extensive amount of functional studies had looked into various managerial elements and their links to venture performance and success. Yet, there is a paucity of studies addressing how environment, strategy, and organizational aspects would collectively influence agribusiness performance. This is because a mere direct effect modeling of the relationships among the variables will not provide an adequate understanding of complex organizational mechanism, but instead a configurational approach (Harms et al. 2009; Wales 2016; Wiklund and Shepherd 2005). In addition, contingency theorists Drazin and Van de Ven (1985) argue that organizational processes need to be aligned with the environment in which they operate. Therefore, the joint consideration of strategy, organizational characteristics, and environmental elements by way of configurational approaches and contingency approaches would provide important insights into agri-entrepreneurship advancing the theoretical and methodological knowledge.

Especially, with reference to Sri Lankan literature, more quantitative and theory-driven studies are of timely importance. Although a methodological discussion is beyond the scope of this study, the initial investigations revealed that there is a handful of studies applying quantitative methods in their research designs (e.g., Mahindarathne and Gunaratne 2015; Wickramaratne et al. 2014). What is prominent is the qualitative case-based studies (Esham et al. 2006; Galappaththi et al. 2017). As well as theory building through case studies is important, theory testing through quantitative approaches is also of paramount importance if a field is to be developed and find its applications and boundaries. It is also necessary to uncover the applicability of major entrepreneurship theories and concepts in the Sri Lankan context to uncover the contextual differences. This would have significant implications for both theory and practice. Yet, there seems to be a very limited set of studies immersed in the mainstream entrepreneurship theories and concepts (Galappaththi et al. 2017; Kodithuwakku and Rosa 2002; Mahindarathne and Gunaratne 2015; Wickramaratne et al. 2014).

6 Conclusions

Agribusiness entrepreneurship literature is growing as an important research stream which offers tremendous opportunities to future scholars to utilize and integrate mainstream entrepreneurship theories and frameworks. Yet, at the moment, the literature is fragmented and missed numerous important areas of investigation. Thus, the review of agribusiness research with entrepreneurial emphasis using entrepreneurship typology framework has shed new lights on uncovering future potential avenues of agribusiness research. Specifically, the review findings uncovered the existing status of agribusiness research focus and offered the suggestions for future expansion of knowledge and understanding. By doing so, this study advances the understanding of potentially meaningful research opportunities that can enrich both theory and practice of agri-entrepreneurship. As with all studies, this study is not free from limitations. One of such limitations is that this study is restricted to peer-reviewed journal articles and conference papers. Therefore, future studies can incorporate books and reports also into the sample. This is because in the Sri Lankan context, for instance, there are some studies that stand as published reports which are not included in the sample of the current study. In addition, the search was performed only in few search engines and local journals which does not capture all the possible publication sources.

References

Adam RI, Osano P, Birika J, Amadi AAN, Bwisa H (2017) The situation of women in the agribusiness sector in Africa. Dev Pract 27(6):892–898. https://doi.org/10.1080/09614524.2017.1338670

Adashie R, Gyan-Kesse B, Sarpong G, Boakye-Agyei K (2016) The Kosmos innovation model-using integrative processes to redefine social investments for greater social impact in Ghana. Paper presented at the Society of Petroleum Engineers – SPE African health, safety, security and environment and social responsibility conference and exhibition 2016

Agwu NM, Agodi JE, Onwukwe FO, Iroh OA (2015) Determinants of agribusiness entrepreneurs' participation in innovations: a study of Abia state, Nigeria. Sci Pap Manag Econ Eng Agric Rural Dev 15(3):13–18

Aman M (2015) Agribusiness and value chain management students' attitude towards entrepreneurship in Haramaya University, Ethiopia: promises or practices of modularisation? Int J Manag Enterp Dev 14(1):1–10. https://doi.org/10.1504/IJMED.2015.069306

Amanor-Boadu V (2013) Diversification decisions in agriculture: the case of agritourism in Kansas. Int Food Agribus Manag Rev 16(2):57–74

Amanor-Boadu VR, Marletta P, Biere A (2009) Entrepreneurial supply chains and strategic collaboration: the case of Bagòss cheese in Bagolino, Italy. Int Food Agribus Manag Rev 12 (3):49–68

Anastasiadis F, Poole N (2015) Emergent supply chains in the agrifood sector: insights from a whole chain approach. Supply Chain Manag 20(4):353–368. https://doi.org/10.1108/SCM-08-2014-0259

Andri KB, Santosa P, Arifin Z (2011) An empirical study of supply chain and intensification program on Madura Tobacco industry in East Java. Int J Agric Res 6(1):58–66. https://doi.org/10.3923/ijar.2011.58.66

Antonites AJ, Haguma JJ (2011) Assessing the innovative nature of the agricultural based small businesses in Rwanda-the case study of the coffee industry. Afr J Agric Res 6(3):757–770

Ariwodor PC, Agwu NM (2016) Effect of entrepreneurial skills acquisition on the welfare of agribusiness households in Abia state, Nigeria. Sci Pap Manag Econ Eng Agric Rural Dev 16 (3):53–59

Attygalle K, Hirimuthugodage D, Madurawala S, Senaratne A, Wijesinha A, Edirisinghe C (2014) Female entrepreneurship and the role of business development services in promoting men entrepreneurs in Sri Lanka

Baker T, Nelson RE (2005) Creating something from nothing: resource construction through entrepreneurial bricolage. Adm Sci Q 50(3):329–366

Bannikova NV, Baydakov AN, Vaytsekhovskaya SS (2015) Identification of strategic alternatives in agribusiness. Mod Appl Sci 9(4):344–353. https://doi.org/10.5539/mas.v9n4p270

Banson KE, Nguyen NC, Bosch OJH, Nguyen TV (2013) A systems thinking approach to address the complexity of agribusiness for sustainable development in Africa. Paper presented at the 57th annual meeting of the International Society for the Systems Sciences, ISSS 2013: curating the conditions for a Thrivable Planet

Bećirović S, Plojović Š, Ujkanović E, Plojović S (2017) Challenges at starting an agribusiness in the hilly-mountainous regions of southwest Serbia. Econ Agric 64(4):1669–1686

Becvarova V (2008) Issues of competitiveness of the present agriculture. Agric Econ (Zemedelska Ekonomika) 54(9):399–405

Bitzer V, Bijman J (2014) Old oranges in new boxes? Strategic partnerships between emerging farmers and agribusinesses in South Africa. J South Afr Stud 40(1):167–183. https://doi.org/10.1080/03057070.2014.877647

Boehlje M (1999) Structural changes in the agricultural industries: how do we measure, analyze and understand them? Am J Agric Econ 81(1):1028–1041

Bosompem M, Dadzie SKN, Tandoh E (2017) Undergraduate students' willingness to start own agribusiness venture after graduation: a Ghanaian case. Contemp Issues Entrep Res 7(1):75–105

Broderick S, Wright V, Kristiansen P (2011) Cross-case analysis of producer-driven marketing channels in Australia. Br Food J 113(10):1217–1228. https://doi.org/10.1108/00070701111177656

Cook ML, Chaddad FR (2000) Agroindustrialization of the global agrifood economy: bridging development economics and agribusiness research. Agric Econ 23(1):207–218

Darroch J, McNaughton R (2002) Examining the link between knowledge management practices and types of innovation. J Intellect Cap 3(3):210–222

Dentoni D, English F, Schwarz D (2014) The impact of Public R&D on marketing and supply chains on small Farms' market sensing capability: evidence from the Australian seafood industry. Int Food Agribus Manag Rev 17(1):37–58

Dissanayake D, Udugama J, Jayasinghe-Mudalige U (2013) Development of an alternative microfinance scheme to finance entrepreneurship in tea smallholding sector: a success story. J Food Agric 3(1–2):31–40

Doherty B, Haugh H, Lyon F (2014) Social enterprises as hybrid organizations: a review and research agenda. Int J Manag Rev 16(4):417–436

Drazin R, Van de Ven AH (1985) Alternative forms of fit in contingency theory. Adm Sci Q 30 (4):514–539

Ekman S, Ekman A, Salunkhe U, Agarwal A (2011) Design inspired innovation for rural India. Paper presented at the ICED 11 – 18th international conference on engineering design – impacting society through engineering design

Esham M, Usami K (2005) Present state and issues of farmers-agribusiness linkages through farmer organization/group in Sri Lanka: a case study of hybrid maize. J Agric Dev Stud 43(1):137–142

Esham M, Usami K, Kobayashi H, Matsumura I (2006) An economic study of contract farming in the fruit and vegetable industry in Sri Lanka: a case study of gherkin (Cucumis Sativus) production. J Rural Probl 42(1):14–23

Farafonova NV (2013) Analysis of resource provision for agribusiness under market conditions. Actual Probl Econ 142(4):158–166

Galappaththi IM, Galappaththi EK, Kodithuwakku SS (2017) Can start-up motives influence social-ecological resilience in community-based entrepreneurship setting? Case of coastal shrimp farmers in Sri Lanka. Mar Policy 86:156–163

Gamage H, Cameron D, Woods E (2003) Developing a research framework for understanding the social realities, with special reference to Sri Lankan entrepreneurs. Paper presented at the 14th international farm management congress, Perth, Australia

Geraci VW (2000) The family wine-farm: vintibusiness style. Agric Hist 74(2):419–432

Gill A, Mand HS, Obradovich JD, Mathur N (2015) Factors affecting ethical sources of external debt financing for Indian agribusiness firms. Corp Ownersh Control 13(1CONT4):435–445

Goldberg R (1999) The business of agriceuticals. Nat Biotechnol 17(1):5–6

Harling KF (1995) Differing perspectives on agribusiness management. Agribusiness 11 (6):501–511

Harms R, Kraus S, Schwarz E (2009) The suitability of the configuration approach in entrepreneurship research. Entrep Reg Dev 21(1):25–49

Havlicek J, Havlicek Z, Vanek J, Silerova E (2003) Agris–a portal with decision support system for agribusiness. J Decis Syst 12(2):177–191. https://doi.org/10.3166/jds.12.177-191

Hemachandra D, Kodithuwakku S (2010) Business orientation among poor dry zone farmers in Sri Lanka. Sri Lankan J Agric Econ 8(1):31–49

Higgins LM, Schroeter C, Wright C (2018) Lighting the flame of entrepreneurship among agribusiness students. Int Food Agribus Manag Rev 21(1):121–132. https://doi.org/10.22434/IFAMR2016.0166

Holloway SS, Sebastiao H (2010) The role of business model innovation in the emergence of markets: a missing dimension of entrepreneurial strategy? J Strateg Innov Sustain 6(4):86–100

Hosenally N (2016) Videos promote young agricultural entrepreneurs. Appropr Technol 43 (3):36–38

Hron J, Stusek J, Arnost M, Huml J (2008) Diversification strategy in small and medium size agribusinesses in the Czech Republic – impulses for searching business opportunities. Agric Econ (Zemedelska Ekonomika) 54(11):505–509

Hussain MS, Baksh ME, Miah MAM (1996) Credit and input supply system for vegetable agribusiness: present status and policies. Asian Vegetable Research & Development Ctr, Taipei

Iglécias W (2007) Entrepreneurs in Brazilian agribusiness: collective action and forms of political activity. A case study of the sugar and cotton wars in World Trade Organization. Revista de Sociologia e Politica 1(28):75–97

Ikram S, Udugama J, Jayasinghe-Mudalige U (2013) Nurturing intrapreneurship: exploring the developments in plantation sector of Sri Lanka. J Food Agric 3(1–2)

Ioris AAR (2016) The politico-ecological economy of neoliberal agribusiness: displacement, financialisation and mystification. Area 48(1):84–91. https://doi.org/10.1111/area.12240

Jackova S, Kapsdorferova Z, Kadlecikova M (2017) Challenges and opportunities for rural Slovak women in agribusiness. In: Tomsik K (ed) Agrarian perspectives XXVI: competitiveness of European agriculture and food sectors. Czech University Life Sciences Prague, Prague 6, pp 123–130

Kassa BT, Haile AG, Essa JA (2011) Determinants of sheep prices in the highlands of northeastern Ethiopia: implication for sheep value chain development. Trop Anim Health Prod 43 (8):1525–1533. https://doi.org/10.1007/s11250-011-9837-x

Katon-Kovácsa J, Bóta-Horváth N (2012) Rural team-entrepreneurs: an answer to innovative, multi-disciplinary human capital education. Int Food Agribus Manag Rev 15(Special Issue A):93–97

Kilian B, Pratt L, Jones C, Villalobos A (2004) Can the private sector be competitive and contribute to development through sustainable agricultural business? A case study of coffee in Latin America. Int Food Agribus Manag Rev 7(3):21–45

Kodithuwakku SS, Rosa P (2002) The entrepreneurial process and economic success in a constrained environment. J Bus Ventur 17(5):431–465

Kristiansen S (2002) Competition, innovation and knowledge in Javanese rural business. Singap J Trop Geogr 23(1):52–69. https://doi.org/10.1111/1467-9493.00118

Kuhn ME, Darroch MAG, Ortmann GF, Graham DH (2000) Improving the provision of financial services to micro-entrepreneurs, emerging farmers and agribusiness: lessons from Kwazulu-Natal. Agrekon 39(1):68–81. https://doi.org/10.1080/03031853.2000.9523568

Kyriakopoulos K, Meulenberg M, Nilsson J (2004) The impact of cooperative structure and firm culture on market orientation and performance. Agribusiness 20(4):379–396. https://doi.org/10.1002/agr.20021

Lazzarini SG, Chaddad FR, Cook ML (2001) Integrating supply chain and network analyses: the study of net chains. J Chain Netw Sci 1(1):7–22

Liang LW, Shih KH, Chung YH (2010) Financing instruments and strategies of agribusiness: evidence from Taiwan. Afr J Bus Manag 4(3):320–332

Lima CC, Quintino SM, Andrade LMN, Dal Magro EF (2015) Empreender Na Gestão Agropecuária Da Amazônia: O Caso Das Agroindústrias Familiares Em Rondônia. Revista Metropolitana de Sustentabilidade 5(2):49–74. (ISSN 2318–3233)

Lin X, Carsrud A, Jagoda K, Shen W (2013) Determinants of entrepreneurial intentions: applying Western model to the Sri Lanka context. J Enterp Cult 21(02):153–174

Ma J, Gilmour B, Dang H (2017) Promise, problems and prospects: Agri-biotech governance in China, India and Japan. Chin Agric Econ Rev 9(3):453–475. https://doi.org/10.1108/CAER-02-2017-0028

Mabaya E, Jackson J, Ruethling G, Carter CM, Castle J (2014) Wild fruits of Africa: commercializing natural products to improve rural livelihoods in southern Africa. Int Food Agribus Manag Rev 17(Special issue B):69–74

Mahindarathne PP, Gunaratne L (2015) Entrepreneurial orientation of organic farmers: a case of organic vegetable farmers in the Badulla District of Sri Lanka. J Int Food Agribus Mark 27 (4):324–336

Mardia IN, Kamaluddin (2017) The agribusiness model of natural silk system in Soppeng regency. In: Dirawan GD (ed) Proceedings of the 2nd international conference on education, science, and technology, vol 149. Atlantis Press, Paris, pp 22–24

Martinez MG, Poole N (2004) Analysing linkages between strategy, performance, management structure and culture in the Spanish fresh produce industry. Int Food Agribus Manag Rev 7 (4):16–39

Micheels ET, Gow HR (2008) Market orientation, innovation and entrepreneurship: an empirical examination of the Illinois beef industry. Int Food Agribus Manag Rev 11(3):31–55

Mirzaei O, Micheels ET, Boecker A (2016) Product and marketing innovation in farm-based businesses: the role of entrepreneurial orientation and market orientation. Int Food Agribus Manag Rev 19(2):99–130

Moreira VR, Freier A, Veiga CP (2016) A review of concepts, strategies and techniques management of market risks in agriculture and cooperatives. Int Bus Manag 10(6):739–750. https://doi.org/10.3923/ibm.2016.739.750

Mujeyi K, Mutambara J, Siziba S, Sadomba WZ, Manyati TK (2015) Entrepreneurial innovations for agricultural mechanisation in Zimbabwe: evidence from an informal metal industry survey. Afr J Sci Technol Innov Dev 7(4):276–285. https://doi.org/10.1080/20421338.2015.1082367

Muniafu M, Wambalaba F, Wanyama W, Nduati G, Ndirangu D (2013) Using Oer as a tool for agribusiness management training for hard-to-reach rural farmer populations. J Asynchronous Learn Netw 17(2):21–30

Nawaratne YB, Kodithuwakku SS, Murray F, Little D (2009) Entrepreneurial strategies adopted by rural inland fishermen in the dry zone of Sri Lanka

Okezie CR, Joshua A (2016) Analysis of women entrepreneurship development in agribusiness sector in Benue state, Nigeria. Sci Pap Manag Econ Eng Agric Rural Dev 16(3):177–184

Osterwalder A, Pigneur Y (2010) Business model generation: a handbook for visionaries, game changers, and challengers. Wiley, Hoboken

Otache I (2017) Agripreneurship development: a strategy for revamping Nigeria's economy from recession. Afr J Econ Manag Stud 8(4):474–483. https://doi.org/10.1108/AJEMS-05-2017-0091

Palmås K, Lindberg J (2013) Livelihoods or ecopreneurship? Agro-economic experiments in Hambantota, Sri Lanka. J Enterp Commun People Places Glob Econ 7(2):125–135

Panasyuk MV, Safiullin LN, Gafurov IR, Safin AR (2014) Role of small businesses in the agricultural sector of the region. Mediterr J Soc Sci 5(28):56–59. https://doi.org/10.5901/mjss.2014.v5n28p56

Pitt C, Nelle S (2008) Applying a sectoral system of innovation (SSI) approach to the Australian red meat industry with implications for improving innovation and entrepreneurship in the Australian agrifood industry. Int Food Agribus Manag Rev 11(4):1–24

Pittaway L, Holt R, Broad J (2014) 6. Synthesising knowledge in entrepreneurship research–the role of systematic literature. In: Handbook of research on small business and entrepreneurship, vol 83. Edward Elgar, Cheltenham

Poláková J, Koláčková G, Tichá I (2015) Business model for Czech agribusiness. Sci Agric Bohem 46(3):128–136. https://doi.org/10.1515/sab-2015-0027

Ratajczak M (2016) Knowledge of the concept of corporate social responsibility in agribusiness enterprises (based on the example of the Sme sector in Malopolska). Manag Poland 20 (1):337–351. https://doi.org/10.1515/manment-2015-0043

Rosairo HS (2010) Factors affecting the performance of farmer companies in Sri Lanka. Lincoln University

Rosairo HR, Potts DJ (2016) A study on entrepreneurial attitudes of upcountry vegetable farmers in Sri Lanka. J Agribusiness Dev Emerg Econ 6(1):39–58

Rosario L, Sachiko N, Masanari S (2011) Japanese social enterprises: major contemporary issues and key challenges. Soc Enterp J 7(1):50–68. https://doi.org/10.1108/17508611111130158

Rosmiza MZ, Davies WP, Rosniza Aznie CR, Jabil MJ, Mazdi M, Wan Toren WY, Che Rosmawati CM (2015) Stagnation of rice straw agribusiness development in Malaysia: the entrepreneurs' perspectives. Mediterr J Soc Sci 6(4):523–530. https://doi.org/10.5901/mjss.2015.v6n4p523

Santiago A, Roxas F (2015) Reviving farming interest in the Philippines through agricultural entrepreneurship education. J Agric Food Syst Commun Dev 5(4):15–27. https://doi.org/10.5304/jafscd.2015.054.016

Santini C (2017) Ecopreneurship and ecopreneurs: limits, trends and characteristics. Sustainability (Switzerland) 9(4):492. https://doi.org/10.3390/su9040492

Sarasvathy SD (2009) Effectuation: elements of entrepreneurial expertise. Edward Elgar Publishing, Cheltenham

Sardana GSG, Dasanayaka SW (2008) Business facilitation through technology incubators in India, Pakistan, and Sri Lanka: a comparative perspective. Market Forces 4(2):18–29

Selassie HG, Hill RW (1993) Factors determining joint venture formation in the agribusiness sector of sub-Saharan African countries. J Int Food Agribus Mark 5(1):73–93. https://doi.org/10.1300/J047v05n01_05

Shadbolt NM, Olubode-Awosola F (2016) Resilience, risk and entrepreneurship. Int Food Agribus Manag Rev 19(2):33–52
Shafer SM, Smith HJ, Linder JC (2005) The power of business models. Bus Horiz 48(3):199–207
Shah SK, Tripsas M (2007) The accidental entrepreneur: the emergent and collective process of user entrepreneurship. Strateg Entrep J 1(1–2):123–140
Shane S, Venkataraman S (2000) The promise of entrepreneurship as a field of research. Acad Manag Rev 25(1):217–226
Siame M (2016) Social venturing and co-operative entrepreneurship business model (Svce-Bm) for growing Msmes in Zambia. In: Aaltio I, Eskelinen MT (eds) Proceedings of the 11th European conference on innovation and entrepreneurship. Acad Conferences Ltd., Nr Reading, pp 988–1001
Silveira-Martins E, Vaz CS (2017) Entrepreneurial orientation associated with performance: an analysis of agribusiness in Rs, Brazil. Agroalimentaria 23(44):93–105
Smile D, Mawuko D, Franklin O (2015) Business coaching and the development of agric-businesses in Africa. Paper presented at the Proceedings of the European Conference on Innovation and Entrepreneurship, ECIE
Smith JO, Garcia ZA, Suyundikov A, Kock T (2015) Measurable changes in pre-post test scores in Iraqi 4-H leader's knowledge of animal science production principles. J Youth Dev 10 (2):102–111
Sterns JA, Schweikhardt DB, Peterson HC (1998) Using case studies as an approach for conducting agribusiness research. Int Food Agribus Manag Rev 1(3):311–327
Sultan SS (2013) Innovation and Sme growth: the case of Palestine. Euromed Press, Marseille Cedex 9
Tsai WM-H, MacMillan IC, Low MB (1991) Effects of strategy and environment on corporate venture success in industrial markets. J Bus Ventur 6(1):9–28
Udimal TB, Jincai Z, Ayamba EC, Sarpong PB (2017) Human capital accumulation and its effect on agribusiness performance: the case of China. Environ Sci Pollut Res 24(27):22091–22101. https://doi.org/10.1007/s11356-017-9867-7
van der Have RP, Rubalcaba L (2016) Social innovation research: an emerging area of innovation studies? Res Policy 45(9):1923–1935. https://doi.org/10.1016/j.respol.2016.06.010
Vázquez FR (2016) Commitments for a diversified economy: the insertion of Mendoza grape (Argentina) in foreign markets (1907–1930). America Latina en la Historia Economica 23 (1):152–183. https://doi.org/10.18232/alhe.v23i1.67
Vidanapathirana NP, Hirimburegama K, Hirimburegama WK, Nelka SAP, Kim JH (2015) Socio economic development of farmers through ICT-based education and rural entrepreneurship development: case study in Sri Lanka. In: GomezChova L, LopezMartinez A, CandelTorres I (eds) Edulearn15: 7th international conference on education and new learning technologies. IATED-International Association of Technology Education & Development, Valenica, pp 4327–4333
Vieira ACP, Bolson EA, Neto RJ, Watanabe M, Yamaguchi CK (2012) Analysis of innovation management and intellectual property for the development of agribusiness. Espacios 33(10):12
Wales WJ (2016) Entrepreneurial orientation: a review and synthesis of promising research directions. Int Small Bus J 34(1):3–15
Wallace I (1997) Agricultural education at the crossroads: present dilemmas and possible options for the future in sub-Saharan Africa. Int J Educ Dev 17(1):27–39. https://doi.org/10.1016/S0738-0593(96)00036-3
Wickramaratne A, Kiminami A, Yagi H (2014) Entrepreneurial competencies and entrepreneurial orientation of tea manufacturing firms in Sri Lanka. Asian Soc Sci 10(18):50–62. https://doi.org/10.5539/ass.v10n18p50
Widmar DA, Gray AW, Gunderson M (2015) Hoover Seeds, Inc.: a period of transition. Am J Agric Econ 97(2):628–634. https://doi.org/10.1093/ajae/aau110
Wiklund J, Shepherd D (2005) Entrepreneurial orientation and small business performance: a configurational approach. J Bus Ventur 20(1):71–91

Wortman MS Jr (1987) Entrepreneurship: an integrating typology and evaluation of the empirical research in the field. J Manag 13(2):259–279

Yunita F, Dhewanto W (2015) Networking among small medium agribusiness enterprise and agribusiness community. Adv Sci Lett 21(6):1867–1870. https://doi.org/10.1166/asl.2015.6139

Zaridis AD, Mousiolis DT, Karamanis K, Rontogianni A (2015) The determinants of the market. The case of the Greek agribusiness sector. In: Kavoura A, Sakas DP, Tomaras P (eds) Proceedings of the 3rd international conference on strategic innovative marketing, vol 175. Elsevier Science, Amsterdam, pp 559–566

Compendium of Principles Applied and Technologies Developed for Managing Municipal Solid Wastes in Sri Lanka

B. F. A. Basnayake, R. T. K. Ariyawansha, and A. K. Karunarathna

1 Introduction

In the 1990s, the local authorities (LAs) began to grapple with the problems of an increase in waste generation due to improved living standards and unplanned and uncontrolled urbanization. The vacant lands in the urban fringe areas soon became unavailable for the purposes of processing and disposing of wastes because urban planning processes were inadequate to deal with increased demand. At the same time, these same urban populations became aware of the health hazards partly due to the discomfort from having to be associated with decomposing wastes in haphazard dumpsites. Therefore, the lack of space and the common attitude of 'not in my backyard' (NIMBY) syndrome exhibited by some parts of the population became obstacles for the safe disposal of wastes. We could, therefore, claim that the inadequacies of politicians, government officials and academic advisors combined with a lack of planned social development were the main causes for the mismanagement of wastes. It is an undeniable fact because we lost lives and paid a heavy price for the shortcomings and mistakes of the politicians and government officials. The dumpsite tenants were and still are some of the poorest and marginalized populations, still exploited and downtrodden and forgotten only to be referred to by one and all as the 'underworld'. It is clear that pollution levels contribute toward climate change, thus affecting agricultural production and sustainable development. We badly needed to transfer and develop technologies to prevent catastrophic consequences for all the population in any given area and especially the conditions

B. F. A. Basnayake (✉) · A. K. Karunarathna
Department of Agricultural Engineering, Faculty of Agriculture, University of Peradeniya, Peradeniya, Sri Lanka
e-mail: benb@pdn.ac.lk

R. T. K. Ariyawansha
Solid Waste Management Research Unit, Department of Agricultural Engineering, Faculty of Agriculture, University of Peradeniya, Peradeniya, Sri Lanka

R. P. De Silva et al. (eds.), *Agricultural Research for Sustainable Food Systems in Sri Lanka*, https://doi.org/10.1007/978-981-15-3673-1_22

that are faced by people living in and around dumpsites. Of major concern for us is the effect of despoliation of the environment and the safety and health of the populations around these dumpsites.

The populations living around these dumpsites with the support of the environmental lobbyists took several local authorities (LAs) to court for not managing these dumpsites. This resulted in the first research being done in 1996 on municipal solid waste (MSW) in keeping with the pledge to the Dehiwala-Mt. Lavinia Court by the Urban Development Authority (UDA). We commenced the research with our own funds. It was a small-scale composting system to determine the air requirements for maintaining aerobic conditions. The second was conducted to study the chimney effect on composting. Armed with the necessary scientific information, the Inclined Step Grate (ISG) composting system with chimneys was conceived as a robust design with relevant calculations on solar incidences, evapotranspiration rates, predictions on temperature fluctuations during day and night and computations on slope stability of decomposing wastes. A patent was obtained for the design (Basnayake 1998). Soon after, the UDA promoted the technology through the World Bank, Metropolitan Environmental Improvement Program (MEIP). For better or worse, the MEIP insisted on private sector participation. Fortunately, one of the plantation companies along with MEIP and UDA funded the pilot project which was established at the Meewathura Farm of the University of Peradeniya (UoP). Because of our efforts, we were selected to participate as the national research institute in the sub-project Sustainable Landfill Management in Asia of the project Asian Regional Research Program on Environmental Technologies (ARRPET) funded by the Swedish International Development Cooperation Agency (Sida) and coordinated by Asian Institute of Technology (AIT).

This review paper is an outcome of the research and development efforts dealing with the efforts in the characterization of raw as well as disposed wastes, including energy contents. It focuses on the basic principles developed in characterization of wastes based on degradability. These methodologies enabled us to develop appropriate technologies of converting waste to useful products. It describes an array of technologies that were developed ranging from aerobic composting (landfill pretreatment), anaerobic digestion, landfill bioreactors, sustainable landfilling and rehabilitation of dumpsites for the assistance of local authorities (LAs), particularly Kandy Municipal Council in managing the wastes. The considerations of socioeconomic aspects are discussed in relation to capacity mobilization, issues of waste management and economic value of wastes. It also outlines the salient points highlighted by the researchers of the Solid Waste Management Research Unit (SWMRU) in collaboration with government officials to formulate a policy framework required to implement the Comprehensive Integrated Solid Waste Management Plans (CISWMP) for the country sponsored and initiated by the UN-Habitat with the request of the Ministry of Mahaweli Development and Environment.

Table 1 Physical properties of waste (Basnayake et al. 2018)

Type of wastes	Karadiyana[1]	Colombo[2]	Kandy[3]	% MC	Size (mm)	Density (kg/m^3)
Combustibles % wet weight						
Biodegradable[a]	52.2	61.00	59.20	65	<50	600–700
Biodegradable[b]	13.53	11.00	18.20	30	>50	200–800
Coconut shells	13.53	5.00	6.00	15	100–450	350
Paper	4.81	5.00	2.31	4	Variable	350–500
Cardboard	1.28	1.50	3.20	2	Large	350
Textiles	1.74	2.04	0.50	2	Commingled	300
Plastic 1	8.15	6.00	4.70	3	Variable	50–800
Plastic 2	1.6	1.87	0.70	1	Variable	100–450
Nylon	0.73	0.85	0.28	1	Variable	50–800
Rubber	0.14	0.16	0.80	2	Variable	100–800
Non-combustibles						
PVC	0.13	0.15	0.05	1	Variable	1200–1500
Metal		2	0.82		Small	
C&D	1.03	1.20	1.57	12	Large	800–2400
Glass	0.94	2	1.60	1	Variable	100–1800
E-wastes	0.05	0.06	0.02	1	Small	Variable
Clinical	0.03	0.04	0.01	1	Small	100–350
Others	0.11	0.13	0.04	1	Small	Variable
Total	100.00	100.00	100.00			

Source: 1. Feasibility study Karadiyana (2011), 2. BETL report on Colombo (2001), 3. Menikpura et al. 2007a

Note: Some of the missing types of wastes were deduced by reducing the category of others
50% of the wastes are less than 50 mm particle size and 60% less than 70 mm, and all the other sizes need size reductions to 100 × 100 mm

[a]Short term, rapid decomposing volatile biodegradable

[b]Long term, lignified biomass of garden wastes, banana stems, etc.

2 Waste Characteristics

The production of materials and the point of waste generations determine the characteristics of wastes. It will depend on the social status, social values, income, urbanization, political will, availability and markets of commodities, capacity and capability of LAs (Madugeethika et al. 2002) and government officials, climatic conditions, geographical locations and compulsive consumerism (Mahees et al. 2010). Those are some of the key factors that dictate the type, quantity, quality of waste generations and collections. In view of changing scenarios of wastes, it was necessary to develop databases and information on waste characteristics that can be updated frequently. Over the years, as a prerequisite, we undertook many composition studies in different parts of the country and found out the quantum of actual generations and collections of wastes as given in Table 1 and Figs. 1 and 2. In recent

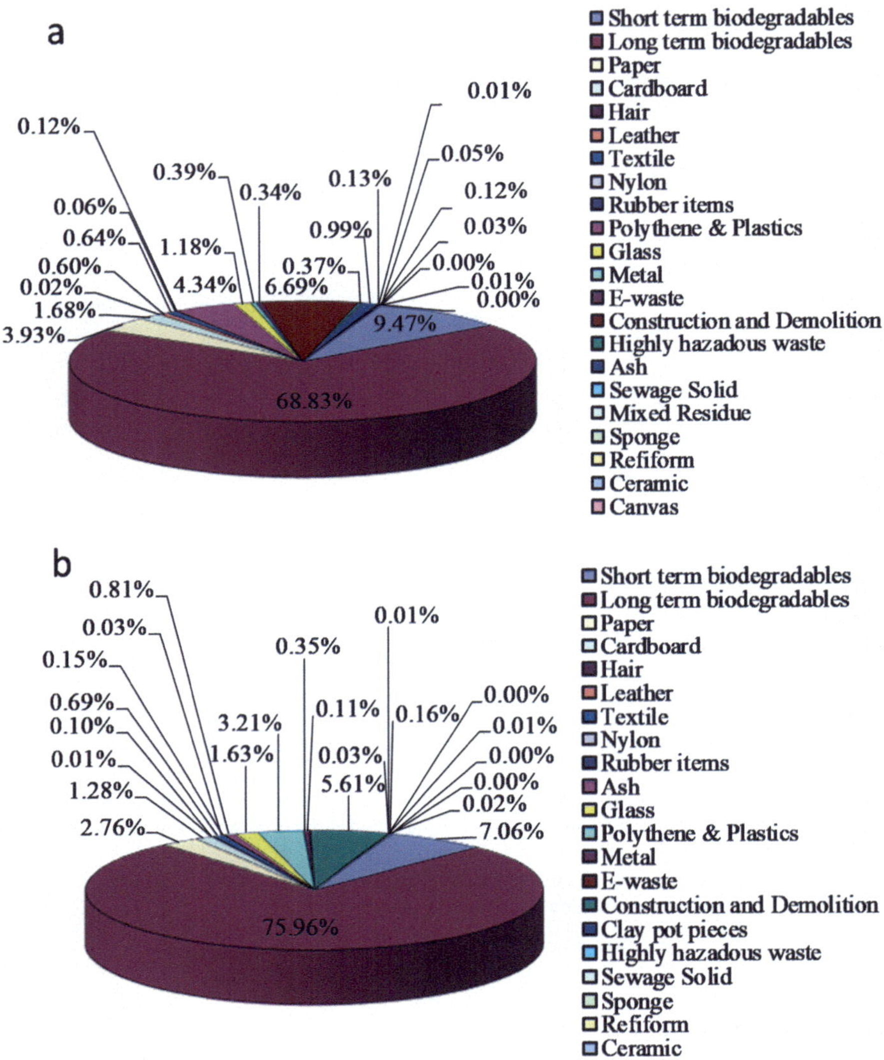

Fig. 1 Average waste composition of households: (**a**) Kalmunai MC; (**b**) Sammanthurai PS. (Ariyawansha et al. 2010b)

times, changes can be seen of the composition and the rate of increase of wastes is between 1% and 2%. The stability characteristics of wastes are important in filling wastes, behaviour in decomposition and soil moisture storage. Therefore, we wanted to find out: How high a landfill can be? Is the dumpsite stable? One of the fundamental principles governing stability is to determine the mechanics of sliding surfaces. It is directly related to type, size of particles and level of degradation of wastes. Therefore, the influence of particle size was included in the study to

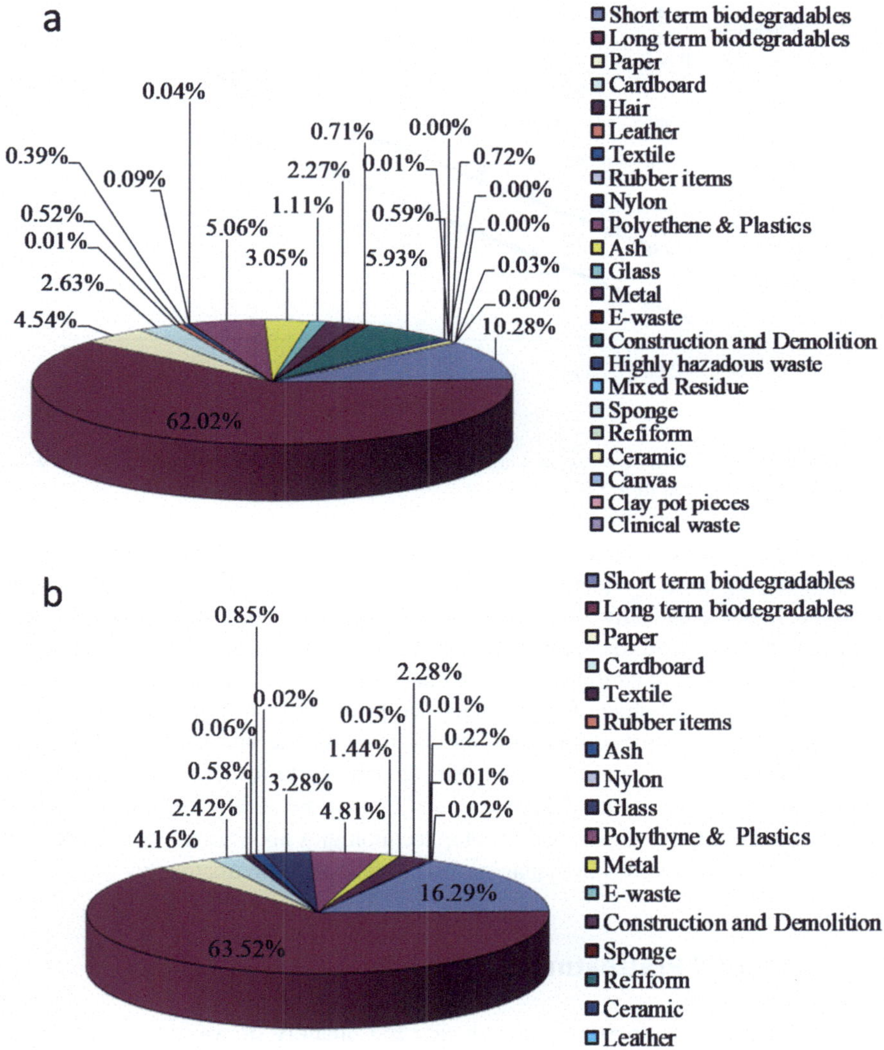

Fig. 2 Average waste composition of shops: (**a**) Kalmunai MC; (**b**) Sammanthurai PS. (Ariyawansha et al. 2010b)

determine the relationship of shear stress vs strain of wastes under increased loading until failure of raw wastes as given in Fig. 3. Raw wastes are much more prone to fail than decomposed wastes. We also found that the moisture content and degree of co-mingling of wastes influence the static and dynamic behaviour of wastes. The increase in moisture content and compaction of wastes restricts gas flow paths through wastes, such that the pressure losses increased linearly and square of the velocity of flow due to potential and kinetic energy losses, influenced by the porosity

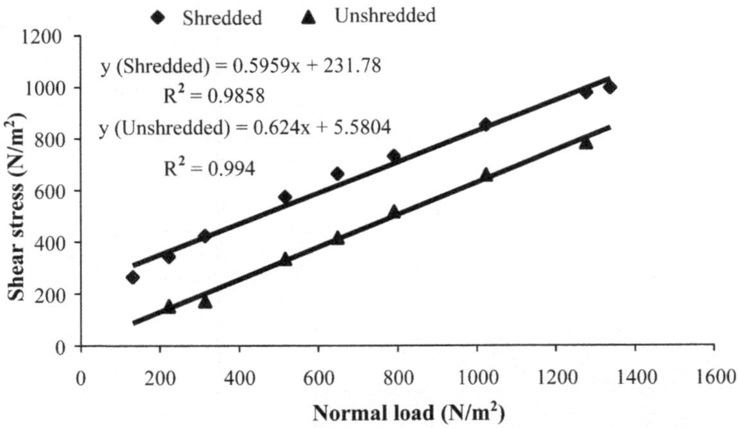

Fig. 3 Relationships of shear stress and normal stress for shredded and un-shredded wastes with respective angles derived from $\tau = c + \sigma \tan \varphi$

of the wastes. The settlement of wastes and leachate emissions were observed from simulations of three lysimeters on a long-term study, and those results were interpreted mathematically to predict the behaviour of dumpsites (Ariyawansha et al. 2010a) and landfills (Gunawardana et al. 2009). In the context of energy content in wastes, a scientific study led to the application of an important principle of the 'Hess law' in which energy of formation of different types of wastes was found. In addition, we modified an empirical formula that was used to deduce the energy content of agricultural wastes (Shafizadeh 1981) to determine the energy contents in municipal wastes (Menikpura and Basnayake 2009). The characteristics of wastes are needed to design and develop technologies in different waste sites, and one of the basic researches in managing wastes is landfill pretreatment.

3 Landfill Pretreatment

One of the ways of diverting and reducing the quantity of wastes ending up in landfills is to undertake landfill pretreatment (Gunawardana et al. 2009). The easiest methods of landfill pretreatment were developed for co-mingled wastes to reduce the quantity and organic loading of wastes on subsequent landfilling. The final recommendations were to pile wastes up to 3 m in height with or without supporting structures, however with adequate ventilation to reduce anaerobic pockets, thus preventing the generation of leachates. This practice was successfully applied by some of the LAs. Most of the LAs preferred extraction of compost from the wastes, making the system more environmentally friendly, and it was a recyclable resource. One of the best and versatile compost technologies that have been developed for sloping lands is the Inclined Step Grate (ISG) composting system (Basnayake and Ekanayake 2005; Ekanayake et al. 2005). It can process sorted or unsorted wastes at very high temperatures, producing high-quality 'thermophilic' compost. Two of the

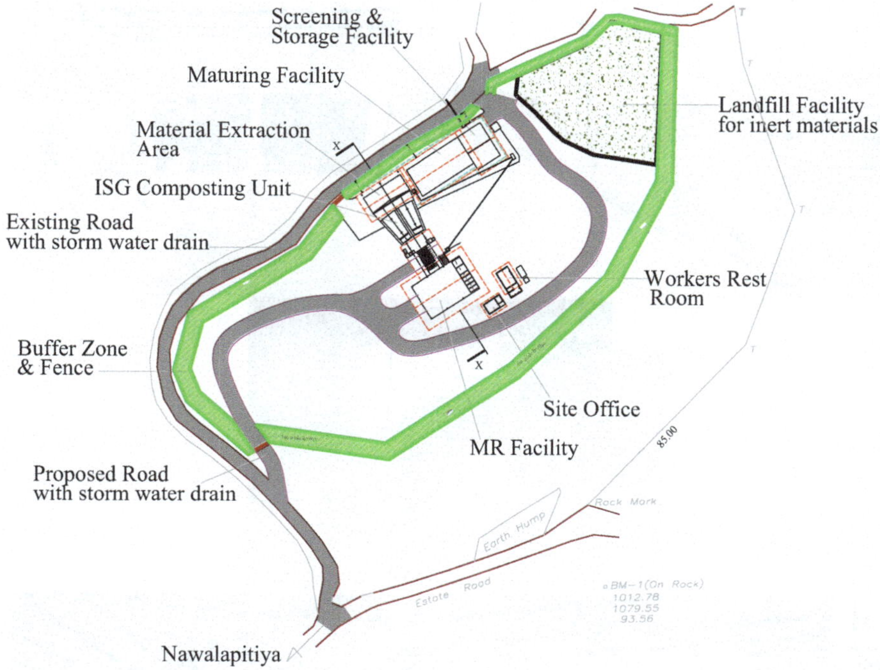

Fig. 4 Layout plan of the environmental preservation centre of Nawalapitiya

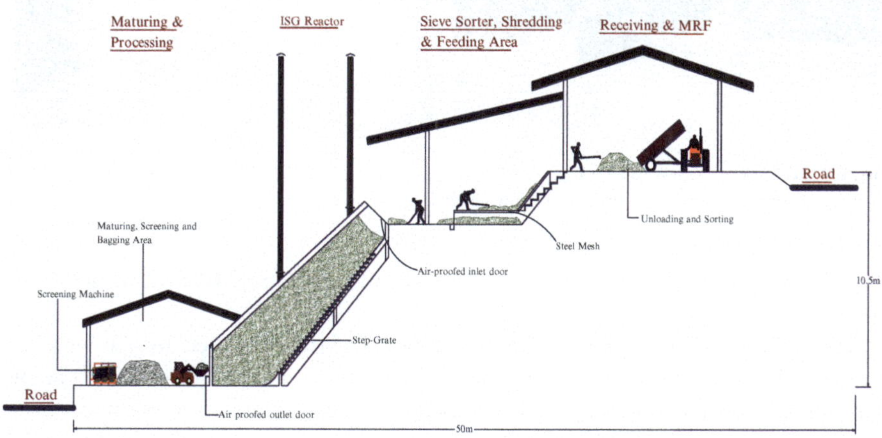

Fig. 5 Operational activities of material recovery facility and ISG unit

facilities still operate, and the one at Nawalapitiya (Figs. 4, 5 and 6) (Ariyawansha et al. 2010c) produces biochar-based fertilizer. Another complimenting technology was developed for flatlands, called 'Solar Aerated Chimney' (SAC) composting system. A system was installed at Batticaloa (EML Consultants Pvt. Ltd 2009)

Fig. 6 ISG unit at Nawalapitiya UC. (**a**) Overview, (**b**) inlet of ISG for waste, (**c**) Inclined Step Grates, (**d**) outlet of ISG and (**e**) maturation yard

(Fig. 7) and is still functioning but not at an optimum level (Ariyawansha et al. 2012).

These two systems operate with convectional currents caused by Archimedes principle. Even at optimum aeration, there will always be fibrous materials left in the first run. It is best then to recycle those residual organic materials to optimize these systems, thus reducing landfilling. By doing so, it increases the bulk density of the input waste loads and reduces as well as moderates moisture contents for optimum composting conditions. Most importantly, the nitrogen content is withheld within the increased specific surfaces of the recycled content from the decomposing materials like food wastes and grasses otherwise lost to the environment. Biofiltration systems are needed to curtail emissions of odorous gases, particularly from composting systems that process food wastes (Ariyawansha et al. 2010d; Menikpura et al. 2007b). Ideally, downdraft composting systems either passive like SAC or an active

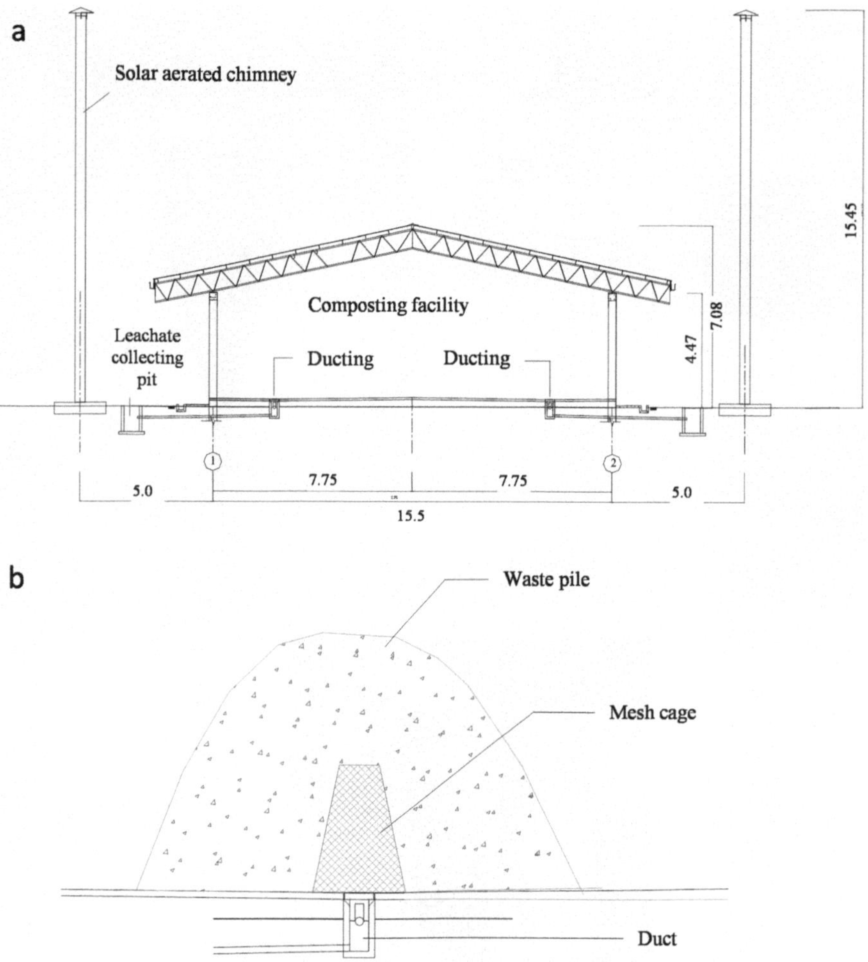

Fig. 7 Design of a passive aeration static pile composting system. (**a**) Sectional view. (**b**) Cone-shaped mesh cage inside the waste pile. (EML consultants Pvt Ltd 2009)

fan can incorporate what is known as 'wet and dry layered media' biofiltration system (Fig. 8) to reduce or eliminate odour pollution (Ariyawansha et al. 2010d; Ariyawansha et al. 2009). It captures nitrogenous and sulphurous compounds that eventually could be used as fertilizer since the media becomes live filtration that grows over time.

Strong odour is a major nuisance in transporting food and kitchen wastes. These wastes inevitably cause many problems to the public health and environment. The ambient pathogen levels increase, if the quality of the wastes has deteriorated. One other way is to introduce daily collection of wastes, which is costly and may not be the best option for a given location. Therefore, it is best to treat or pretreat the waste

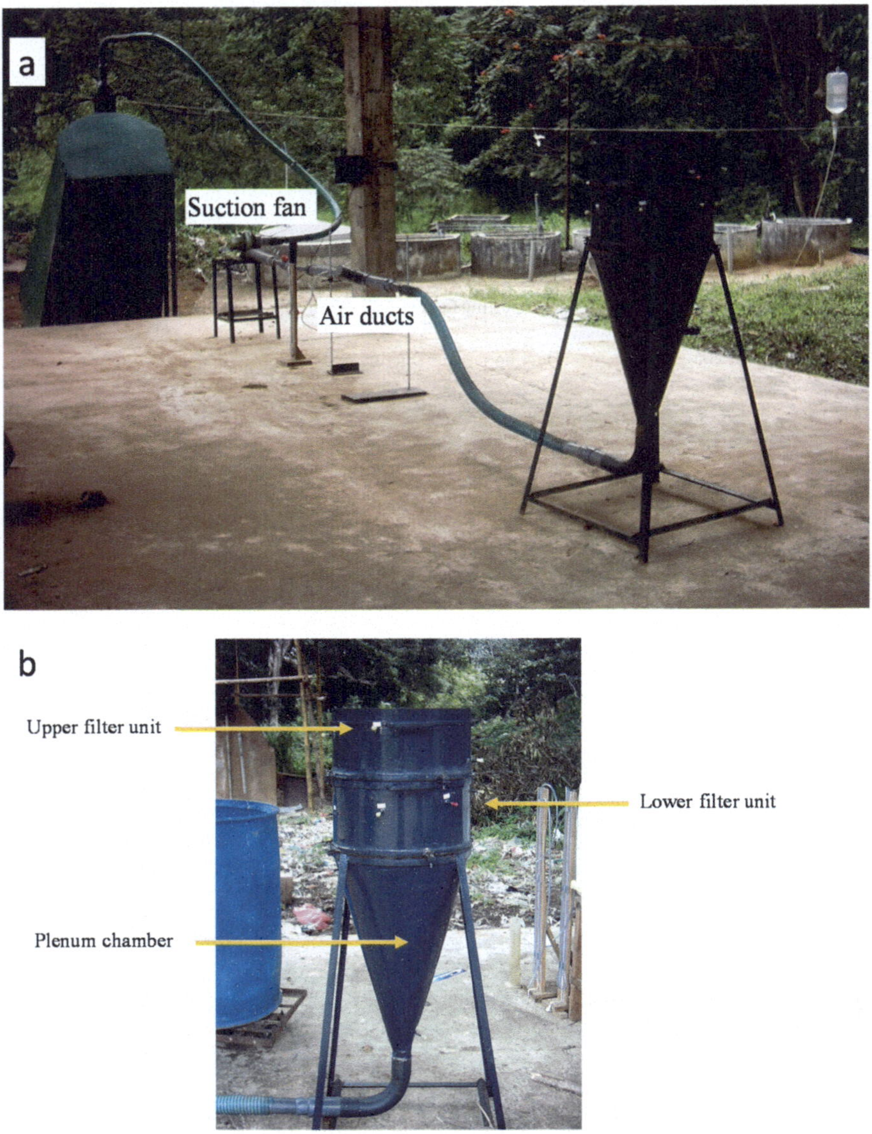

Fig. 8 Biofiltration unit. (**a**) Experimental setup. (**b**) Biofilter unit. (Ariyawansha et al. 2009)

prior to transporting the wastes. Home composting can be done as long as dry fibrous materials like garden wastes or old compost are incorporated to increase the bulk density and reduce the biodegradability of the wastes, thus preventing anoxic conditions promoting odorous gases. Also, it should be mandatory to incorporate a tall chimney into the composting unit, preferably with a biofiltration system.

Another option is to promote anaerobic digesters only for treating food wastes, thus compelling the separation of fibrous materials generated from the kitchen. In the attempt of improving existing and novel systems, a number of laboratory and prototype models were developed and tested. One that gave the best results was an upflow anaerobic sludge blanket (UASB) incorporated with a composite liner system that was developed by the UoP (Ariyawansha et al. 2018; Karunarathna et al. 2017). Although several interventions were needed to generate gas with the incorporation of phosphorous and lime at the start-up of the reactor, a first-order mathematical relationship was obtained for the reduction in total dissolved solids (TDS) over time (Ariyawansha et al. 2018). The productions of TDS occur in the initial hydrolysis and acidogenesis phases of reactions, increasing the concentration of ionic compounds that usually reduces the performance of classical UASB reactors. In this laboratory model, the composite biofilter liner system mineralized these inorganic compounds, thus reducing inhibiting reactions (Karunarathna et al. 2017). All of these factors were considered in upscaling the reactor for increasing the organic loading rate (OLR) (Ariyawansha et al. 2018). The new upscaled reactor (Bandara 2018) (Fig. 9) is undergoing a series of tests for optimizing and automation of the operational system. It is likely that the liner incorporated UASB reactors can be developed to convert food wastes to methane, carbon dioxide and trace gases. In fact, there are many technologies available to convert organic municipal wastes to biogas, but they are costly needing large investments. Instead, some LAs can rely on simple and proven technologies of landfill bioreactors to convert the wastes to landfill gas in landfill bioreactors.

4 Landfill Bioreactor

The technology of capturing and using methane emissions from landfills is an important issue both in terms of protecting the environment and gaining economic benefits from energy generations (Reinhart and Tounsend 1998). The process of extracting landfill gas (LFG) is accelerated by recycling the leachate that is sent over the wastes once the landfill is closed. In effect, it is a sanitary landfill that uses enhanced microbiological processes to transform and stabilize readily and moderately decomposable organic waste constituents within 5–10 years of bioreactor processes being implemented (Thivyatharsan et al. 2012). There could be several landfill bioreactors incorporated into a large landfill. The smallest of the units is termed 'biocell' (Reinhart and Tounsend 1998; Basnayake et al. 2007). The reasons for promoting smaller units are to reduce environmental impacts of prolonged active operational phase of landfills, such as odour nuisance, escape of methane and other landfill gases and operational costs of leachate treatment. Yet, the ionic strength of recycled leachate would increase when the organic contents of the wastes are converted to LFG. This increase in inorganic content creates inhibitive reactions, which slows down the gas generation rates, although the action of recycling leachate vastly contributes to accelerating the processes of anaerobic digestion. Another inhibitive compound that prevents methane gas generations, thus reducing the

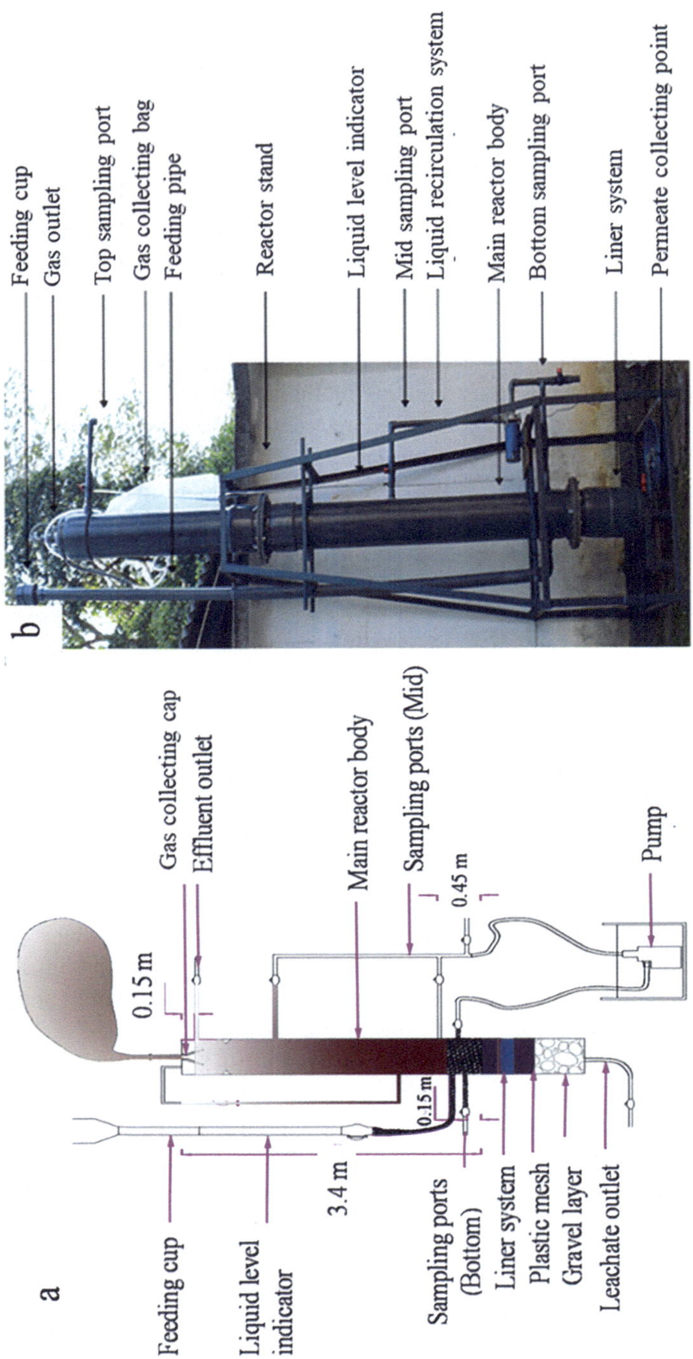

Fig. 9 Upflow anaerobic sludge blanket reactor. **(a)** A schematic diagram. **(b)** Constructed reactor. (Bandara 2018)

potential of carbon reduction and transformation reactions, is the formation of ammonia in the decomposition processes, such that the hydrogen released during the transformations is combined with the nitrogen instead of carbon. Because of these reasons, it was necessary to develop a technology to promote the natural equilibrium between ionic strength and pH, thus promoting enzyme activity ideal for methane generations.

4.1 Development of Liner System

Therefore, we were searching for a solution to reduce the high ionic content in the leachate oozing out of decomposing wastes. Our observation of the mineralization mechanism of natural ecosystems in peat bogs (marshy lands) that produces the methane was the evidence needed to develop a biofilter media that promotes vertical movement of leachate while mineralizing the inorganic materials in the media. The inorganic substances derived from decomposing vegetation are very fine, like clay, but with time colloidal formations of these substances turn to mineral sands. These clays and sands which originated from the same plant materials have dissimilar surfaces that promote higher cation exchange capacity with increased movement of dissolved substances between the surfaces. The microbial activity is very high between these surfaces. The replacement of sand with polythene increases the dissimilarity (electrical polarity) which is an added advantage compared to other materials like gravel, brick, or even charcoal. We decided to test a composite liner system of waste polythene and clay in the laboratory. The required permeability of 1×10^{-7} cm/s or below is needed through a liner system for compacted clay liners (Qian et al. 2002). It is the vertical movement needed for natural attenuation. In the case of our liner, it was found that a minimum thickness of 150 mm was required after compacting 150 mm of waste polythene between top and bottom layers of 50 mm thick clay. The experimental schematic layout at Sammanthurai and actual at Peradeniya (Gunarathna et al. 2007, 2010; Thivyatharsan et al. 2010, 2012) are shown in Figs. 10 and 11. The results were similar even at very high leachate heads (impounding of leachate above liner) as illustrated in Fig. 12. The permeability in both locations reduced with time and reached a steady state. In addition to this finding, interestingly, the permeability values under both conditions of saturated and unsaturated beneath the liner (simulated by a sand layer) fluctuated within a very narrow margin. Therefore, we can deduce that liner permeability will not increase or decrease very much when the water table fluctuates from a lower value to a higher one in the seasonal variations from dry to wet conditions. However, chemical diffusion can take place under saturated conditions. The findings can now be applied to landfills that could even be constructed on rehabilitated dumpsites since there are freely available waste polyethylene films. They can be used to reduce the permeability instead of costly high-density polyethylene (HDPE) liner systems that may fail due to differential settlements that could be expected in old dumpsites. Therefore, clay-polyethylene-clay liner can be recommended specially to build small landfills that are needed for small local authorities by rehabilitating existing open dumps and

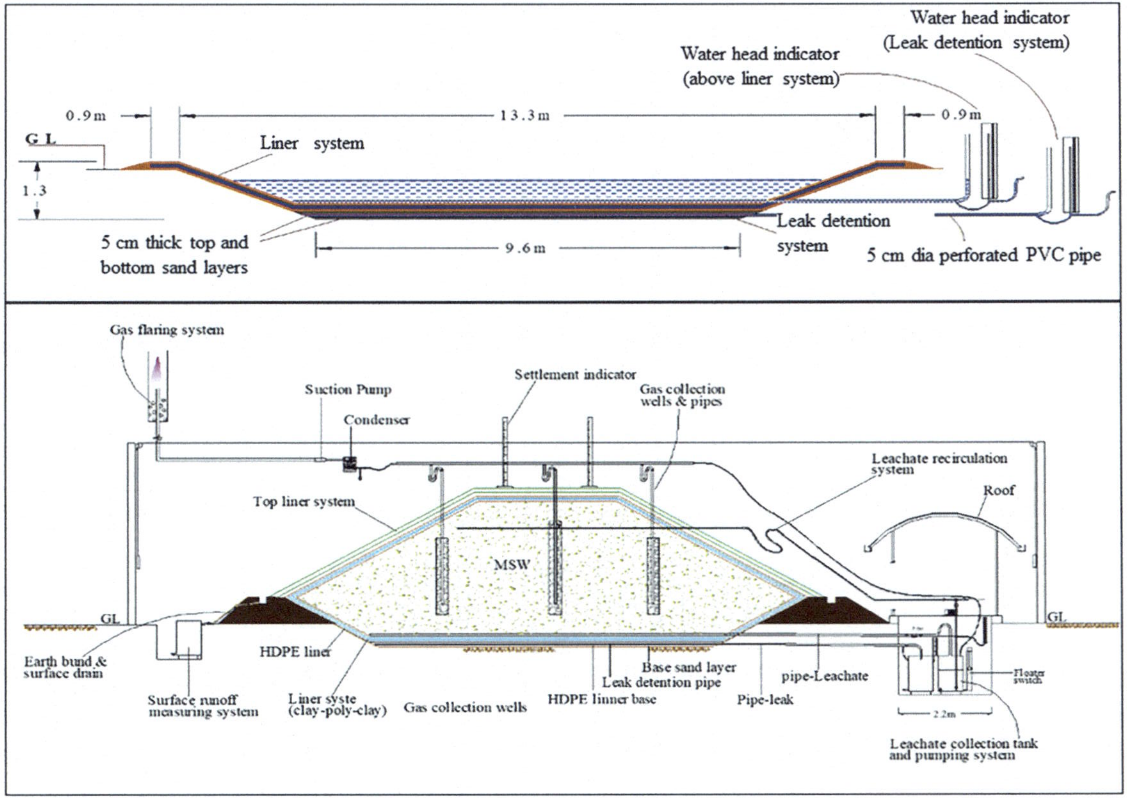

Fig. 10 A schematic diagram of landfill bioreactor test cell. (Thivyatharsan et al. 2010)

Fig. 11 Constructed landfill bioreactor test cells (**a**) at University of Peradeniya and (**b**) in Sammanthurai

transforming them into even landfill bioreactors. In the case of large-scale sanitary landfills, like Aruwakkalu landfill, which is in a sensitive ecosystem, they could have additional protection with the developed composite liner constructed above the HDPE liner system with hardly any differences in costs. In fact, it may be cheaper because the clay requirement for HDPE could be reduced. Also, biocells can be constructed to reduce pollution loads as described in Sect. 4. A Presidential Award was given for the Best Environmental Technology for the Patent (Basnayake 2011).

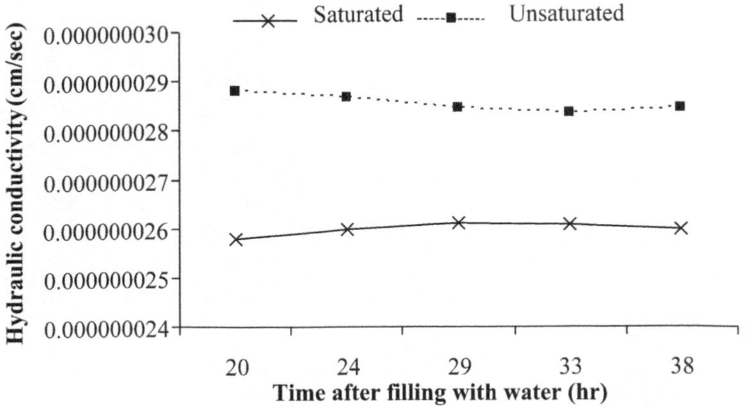

Fig. 12 Hydraulic conductivity of the field scale liner at a hydraulic head of 86.2 cm in saturated and unsaturated conditions. (Thivyatharsan et al. 2010)

4.2 Test Cells

Once we checked the integrity of the liner systems, we constructed the leachate collection systems and then landfilling operations were done in both locations. The composition and weights of the waste loads were obtained prior to landfill operations. The wastes were compacted in layers to attain a bulk density of 800 kg/m^3. Also, the gas extraction pipes with properly designed wells were installed while filling the wastes. The leachate recirculation pipes were installed before construction of the cover system. Clay and waste polyethylene were applied in the same way as the bottom liner system but at a lesser compaction. Unlike in the case of Peradeniya, at Sammanthurai, we decided to apply a sand layer above the clay-waste polyethylene-clay composite cover in order to drastically reduce evapotranspiration, especially during the dry period. Finally, the turf was placed by using grass swards taken from the surrounding area to complete the cap for the test cell.

In both systems, we found that the liner permeability was lesser after loading with wastes. There were differences in all of the parameters between the composite liner and high-density polyethylene (HDPE) liners used in landfills. The normal tendency is an increase in total solids (TS), volatile solids (VS), total suspended solids (TSS), volatile suspended solids (VSS), volatile fixed solids (VFS), ion concentrations, ammonia, chemical oxygen demand (COD) and biochemical oxygen demand (BOD) values in the leachate with time, since solubility increases inorganic compounds in the leachate. We believe that when these concentrations increase, particularly ion concentrations, pH and temperature influence the production of vastly different types of enzymes inhibiting favourable reactions to reduce the strength of the leachate, eventually making toxic conditions for the survival of large microbial populations to convert the available carbon sources to gases and water. Although it is possible to manipulate pH and temperature, the reduction of the ionic concentration is difficult. The very reason why we introduced the composite

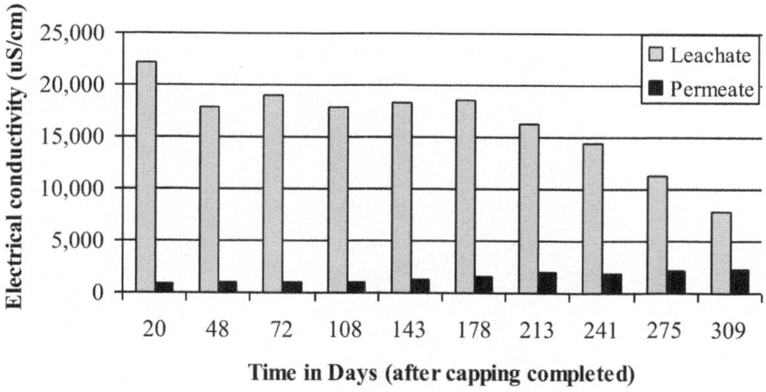

Fig. 13 Variations of electrical conductivity between leachate and permeate. (Thivyatharsan et al. 2012)

liner is to mineralize the inorganic compounds. Most of these inorganic compounds precipitate due to mineralization process by the microorganisms found in the interfaces of the biofilm liner. We can see the result in Fig. 13 when we compare electrical permeability results between leachate and permeate. There were large reductions of ionic strength in leachate contrary to the permeate values. The samples of permeate were obtained from the sand layer underneath the composite liner and above the impermeable barrier of HDPE liner that was installed for this purpose of monitoring natural attenuation taking place. As reported by many authors, under anaerobic conditions, ammonia tends to accumulate in the leachate and above 1500 mg/L, it becomes toxic, and the pH increases (Karthikeyan and Joseph 2014). In this test cell, at Sammanthurai, the values ranged between 15.46 and 62.5 mg/L, and ammonia concentrations in permeate were negligible. It seems a proportionately smaller quantity of ammonia had been oxidized to nitrate and a larger quantity was likely to have been directly converted to nitrogen gas through a possible path of anammox bacteria action (Shalini and Joseph 2017). All of the other parameters were then controlled to provide adequate microbial enzyme secretions to convert wastes to gases, particularly methane.

The top cover, too, certainly has influenced gas productions. Although it allowed water to enter the cell, it also prevented gas from escaping since the cover was saturated in most instances with heavy rainfall experienced throughout the experimentation period at Peradeniya. Irrigation of the cover was practiced at Sammanthurai. It also prevented the cracking of the surfaces. With this passive sealing, the gas extractions were 2.8 L/min, and it was augmented to 4.2 L/min by increasing suction pressure. The gas productions began very much earlier than reported (Mata-Alvarez 2003); perhaps it is the fastest rate so far for landfill bioreactors/biocells. However, the energy content of the gas was 32–37% of the input wastes. There were considerable losses from the large exposed area amounting to 27–32%. Moreover, the aeration operation before mining the test cell could have oxidized some of the carbon. Unfortunately, we were compelled to deactivate the

Fig. 14 Slicing through the landfill bioreactor profile. (Ecotech Lanka Ltd 2011)

cell before completion of the active phase in stabilizing the wastes. In fact, the reported values in most large landfills are between 15 and 20%. The remaining wastes contained 36% after excavation. The profile in Fig. 14 shows the residual wastes as well as a distinct profile of mineral accumulations in the liner and just below it.

5 Sustainable Landfills

We have just come across what is meant by a sustainable landfill. It is because once the gases diminish, the residual wastes can be extracted to allow raw waste to be disposed. The wastes consisting of organic and non-biodegradable wastes disposed in landfills can be used to process residual derived fuel (RDF) for energy and power generations. Even small landfills can be used for systematic disposal of wastes in different biocells and excavations of decomposed wastes, over and over again to produce gas and RDF for sale. The sand content and unwanted materials are separated and sorted to produce high energy content RDF. We obtained high heating values (HHV) between 20 and 24 MJ/kg. The lower heating value (LHV) will depend on moisture and hydrogen contents in RDF, and approximately 15–18 MJ/kg can be expected at equilibrium moisture contents in the wet zone. The energy of these RDF can fuel waste to energy plants.

6 Waste to Energy Systems

There are numerous types of power plants in the world. Most of them are incinerators rather than power generation systems. Although they produce electricity, the primary task is to incinerate as much as possible large quantities of wastes generated in highly urbanized cities. All of them are dependent on the tipping fee (disposal fee) for generating profits because tariff for electricity productions in those developed countries is so low that it is not economically feasible to make business sense in generating power. Therefore, the system efficiencies are within the range of 15–25%. Moreover, the operational costs are very high for cleaning flue gases consisting of acids, particulates in the form of fly ash, NOx and sulphur compounds. However, in the recent past, the RDF plants, particularly in Germany and France, reach 40% efficiency. The RDF is made from raw wastes after application of mechanical and biological treatment (MBT) technology, thus reducing all of the flue gas pollutants. We found that MBT technology is very expensive and operational costs are somewhat high, but less than compared to treating and capturing pollutants in raw waste combustion.

In our attempts to promote RDF thermal technology, we proposed the path of very low-cost landfill bioreactors to reduce the nitrogen, sulphur and chlorine compounds in the subsequent production of RDF and also, by cleaning the LFG before combusting it in a dual fuel system with RDF to increase the conversion to go beyond even 40% efficiency in a combined cycle system. It is even better to use a gasifier to generate high-quality syngas from RDF to drive a gas turbine and use the residual waste heat along with the cleaned LFG to fuel a boiler to produce steam to drive a generator for producing electricity (Ecotech Lanka Ltd 2011). Since these approaches are new, we decided to undertake a pilot scale project for Kandy Municipal Council (KMC) in Gohagoda.

6.1 Pilot Project at Gohagoda, Kandy

At the end of the second phase of the ARRPET project, it was decided to implement a pilot project at Gohagoda. Unfortunately, the pledged support for the third phase by the Swedish Government was withdrawn as an outcome from the Copenhagen summit 2009, because the benefiting countries were either technically capable and or economically sound to implement the Research and Development (R&D) findings. We took the challenge; since the commencement of ARRPET, we had trained researchers and were very familiar with the particular problems of the Gohagoda dumpsite. The aim of the pilot project was to rehabilitate the Gohagoda dumpsite and establish an integrated solid waste management project for the KMC. As a country, we were in the category of having the knowledge and know-how but not financially sound to undertake the project. At that time, although the country was going through several crises, we were hopeful in convincing the private sector to take the challenge. Our first attempt was to organize and conduct an investor forum at the Postgraduate Institute of Agriculture (PGIA) for the private sector, banks and government

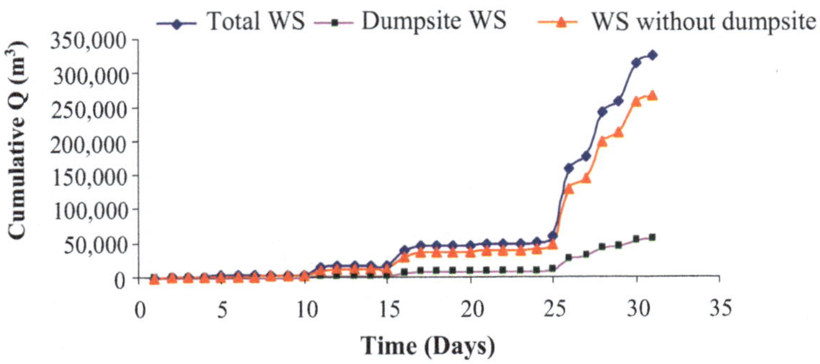

Fig. 15 A comparison of the cumulative discharges of the entire watershed (WS) and dumpsite catena (DC) and adjoining catena (AC). (Ecotech Lanka Ltd 2011)

agencies. After many meetings, a foreign company with a local partner decided to invest in the project, but they withdrew because of the burden of rehabilitating the dumpsite and relocating dumpsite inhabitants/population. Moreover, we did not have any of the approvals to make the project attractive for investors. Therefore, a company was incorporated, and the KMC in the absence of any assistance from the government still decided to support the project. A number of companies came to our rescue and funded some of the operations, particularly to obtain an environmental impact assessment (EIA) study approval.

6.1.1 EIA Study

Any EIA study is unique, but lessons can be learned from them. The site investigations were meticulous, although we faced many obstacles in assessing the dumpsite. Hydrological studies showed far reduced stream flows of dumpsite catena with a slightly reduced area by 10% in comparison to an adjoining one, of the same watershed. The quantities of water and gas emissions were significant, since the water holding capacity of the dump was much more than natural vegetation with urban dwellings and abundant paddy fields. In these two streams, the rates of base flows from the dump were far less compared to the adjoining catena and in both cases as expected base flows were higher in the night than daytime. In undertaking the water balance study of the dump, we introduced an index termed 'storage depletion' (SD):

$$\mathrm{SD} = \frac{\sum ET + \sum B_f}{\sum RF} \tag{1}$$

where *EF* is the rainfall, *ET* is the evapotranspiration and B_f is the base flow. The absorption capacity will depend on the maximum SD value for the duration considered. The moisture contents of disposed wastes were also considered in determining the water balance, and it was included in calculating SD index (Fig. 15). We found in the flood occurrence aspect of the study, it is imperative to allow inundation of the paddy fields; thus, filling and expansion of facility on that side of the catena should not be considered. The other important aspect of the study was on water quality surrounding the dumpsite. It was found that the BOD of the main stream flowing into the river from the dump during the wet season was 7500 mg/L (Menikpura et al. 2008) and it reached 32,000 mg/L in the puddle water in the dry season. The polluting parameters including pathogenic bacteria during the study period far exceeded the Central Environmental Authority (CEA) discharge standards. The comparative samples of the river water of up and downstream from the discharge point to the Mahaweli River showed the actual impact of the dump. We observed the catastrophic effects of washouts from the dumpsite, particularly during a high rainfall experienced on 25 December 2010. All of these pollutants end up in the Mahaweli irrigation systems.

Apart from polluted leachate, the ambient air quality of the surrounding area of the site shows considerable levels of pollution as given in Table 2. Estimated level of methane emission from Gohagoda dumpsite was 288 g/m^2/day. Also, according to our calculations, methane emissions from continuous dumping of waste is given in Fig. 16 based on IPCC model estimations (Menikpura et al. 2008). The study report indicates that both leachate and gas emissions have had an impact on fauna and flora. In fact, toxic levels of heavy metals during heavy runoff (washouts) periods have caused farmers to abandon the cultivation of paddy fields in the watershed (Wijesekara et al. 2014). Therefore, the livelihoods of neighbouring farmers were affected due to the dumping of wastes.

(a) The Socioeconomic Environment

The Google Images clearly show the settlement patterns within and around the dumpsite. The area is semi-urban to rural and can be considered a suburban area of the Kandy city. A survey was conducted to ascertain the socioeconomic status of establishing the project over a radius covering 500 m from the proposed facility. Some of the salient features were that a significant number of families (43 or 28.7% of the sample) work for the Kandy Municipality in various capacities, making the Kandy Municipality the largest single employer. Among them are those who are working at the dump as well. It was also interesting to note that 6.6% of the households work on site, meaning the dump, and they are totally dependent on the dump for income generations. These people and their families were highly exposed to the impacts of the dump, and it was imperative to relocate 53 homes at the cost of the project (Ecotech Lanka Ltd 2011). Most of these houses were in a dilapidated state.

(b) Impact Assessment of the Project

The Leopold matrix was prepared in relation to the identified project activities and environmental parameters. The effect of each project activity on each environmental parameter was evaluated. A numerical value was given to quantify the magnitude (size of impact) and importance (relative importance) separately, and finally by multiplying both values, a composite value was obtained. We will discuss

Table 2 Landfill gas concentration in several locations in and around Gohagoda dumpsite

Location	Landfill gas concentration								
	SO_2	NO_2	VOC	CO	CO_2	CH_4	Benzene	H_2S	NH_3
	(ppm)	(ppm)	(ppm)	(ppm)	(ppm)	(%)	(ppm)	(ppm)	(ppm)
Close to the leachate collection tank	<0.001	0.001	0.01	1	480	0.7	<0.01	0.02	<0.01
Close to the residential	0.001	0.001	0.01	1	460	0.7	<0.01	0.01	<0.01
Middle of the dumpsite	0.001	0.002	0.02	2	470	1	<0.01	0.02	<0.01

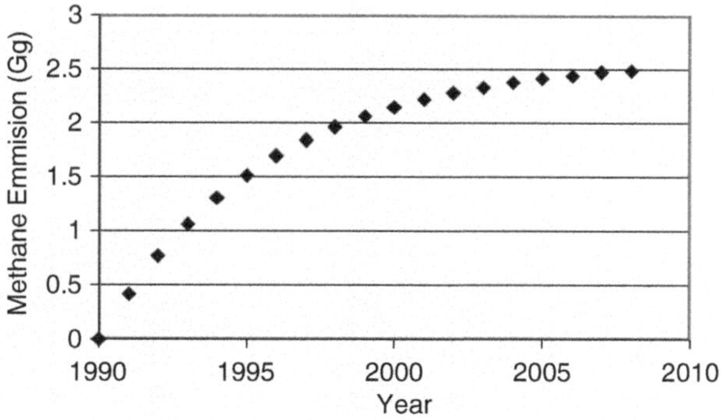

Fig. 16 Estimation of methane emissions from Gohagoda dumpsite with continuous dumping of wastes. (Menikpura et al. 2008)

Table 3 Stack emission levels from the proposed power plant

Parameter	Unit	Value
Flue gas emission	m^3/h	6000
NO	g/s	2.14
SO_2	g/s	0.66
CO	g/s	0.23
Particulate matter	g/s	0.166

some of the important and interesting issues. For example, it was necessary to apply an effective technology to capture methane gas from the dump. We found that it is best to install horizontal pipes below the surface of the dumpsite to capture 80–85% of the LFGs, thus reducing over 3200 kg/day out of 4000 kg/day. The stack height required for the proposed power plant was another important issue. It has to be high enough to prevent the deposition of pollutants on an undulating terrain (receptors) at higher concentrations than accepted air quality standards. It is also dependent on emission levels specified by the supplier and according to best practices in the latest gas cleaning systems and efficiency of moisture condensation. A summary of stack emissions is given in Table 3. These stack emissions should then meet ambient air quality standards under worse conditions of wind flow and direction. In fact, low wind speeds and unidirectional wind flow will concentrate more of the pollutants depending on the elevation of receptors than turbulent high winds that will take away and mix better with air. When we applied the modified Gaussian model (Nevers 1972), we were able to visualize and spot high to low deposition concentrations in specific locations as shown in Figs. 17 and 18. Table 4 gives permissible and predicted maximum concentrations. These values were obtained for a design height of 102 m to meet ambient air quality standards. Therefore, a height greater than 102 m will reduce the risk of polluting the hot spots shown in Figs. 17 and 18.

The most nerve-wracking aspect of the project was the possible collapse of wastes. It took us a considerable length of time and effort to develop scientifically and technologically sound low-cost technologies for stabilizing the embankments, collecting the leachate oozing out of the dump with interceptor drains and then treating it because the existing technologies were very expensive and not so reliable

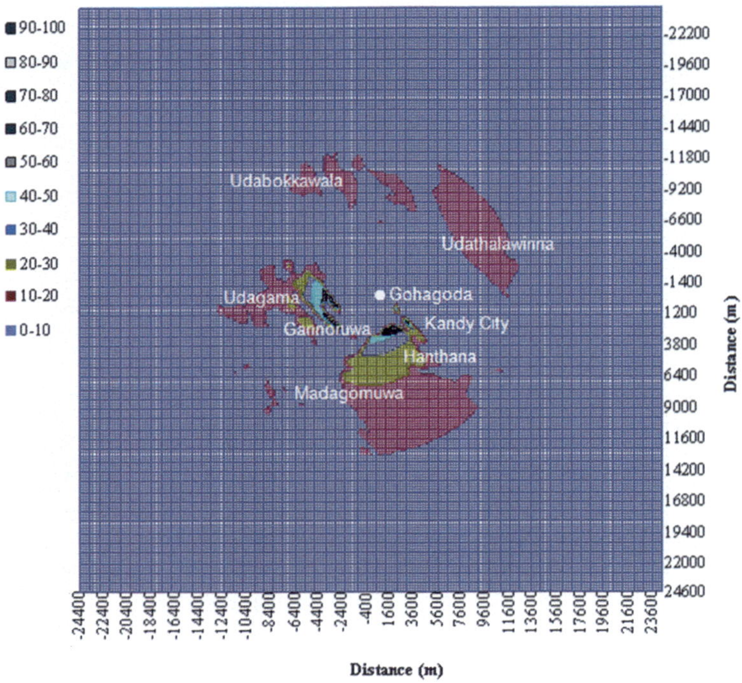

Fig. 17 Distribution of NO concentration over the study area. (Ecotech Lanka Ltd 2011)

in tropical climates. To this end, the study initiated by the National Institute of Fundamental Studies (NIFS) (Wijesekara et al. 2015) pinpointed the existing problems of perch leachate strata in the body of the dump. This allowed us to precisely dig and install 'French drains' and connect them to peripheral pipes, which were also intercepting the seeping leachate flows from the sides of the dump. Then the concept of a landfill bioreactor was used to develop a leachate treatment bioreactor (LTB) (Fig. 19) (Gnanakaran et al. 2017). Instead of raw wastes, old wastes were used to develop microbial films for treating the leachate. The effluent from the LTB was treated in an algal pond and finally in constructed wetlands. The EIA approval was given accepting the expected level of pollution from a dumpsite. In complying with the EIA approval, the dumpsite rehabilitation took precedence in the implementation programme of the pilot project.

6.1.2 Implementation of the Pilot Project (PP)

When we commenced activities in 2010, there were many small dumps in different locations, because KMC was unable to access the dump during wet weather (Fig. 20). Therefore, roads were constructed on the dump to bring in wastes from scattered small dumps while the raw wastes were disposed in the main dump in a systematic manner to prevent double handling. The slope stability calculations and on the ground, supervision of terracing and compaction were needed to ensure safe

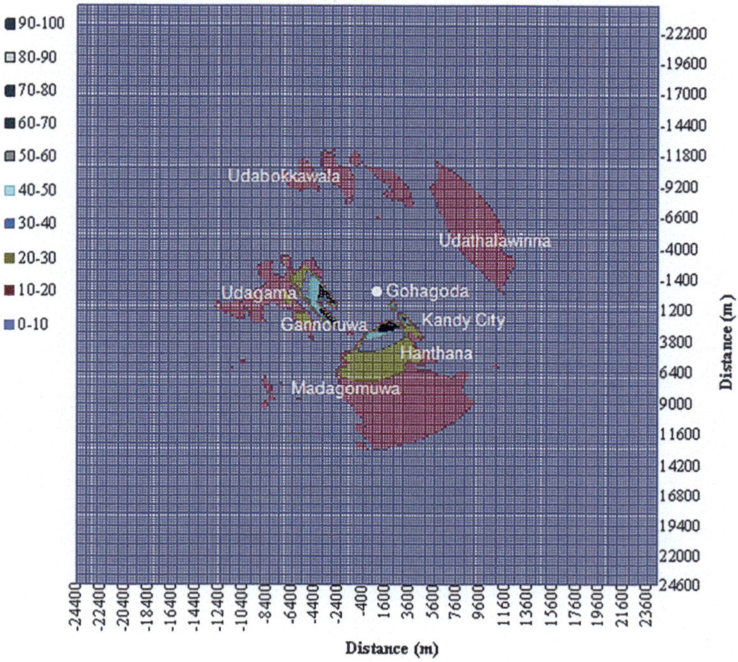

Fig. 18 Distribution of PM concentration over the study area. (Ecotech Lanka Ltd 2011)

Table 4 Permissible and predicted maximum concentration levels

Pollutant	Permissible level (μg/m^3)	Maximum concentration predicted (μg/m^3)
Carbon monoxide	58,000	11.7
Particulate matter	50	8.3
Nitrogen dioxide	100	93.5
Sulphur dioxide	80	32.3

levels of stability. Once the earthworks were completed, surface and sub-surface drains were constructed to prevent leachate contaminations. In the meantime, leachate collection pipes were installed, and treatment systems were constructed and made operational.

The results of the leachate treatment were good because all of the polluting parameters were approaching permissible CEA water quality discharge standards (Priyashantha et al. 2015). It was a 'tall order' for us to accomplish because none of the dumpsites can be rehabilitated to the level of a sanitary landfill. But we still continued to improve the LTB. After 2 years of operations, we decided to replace the decomposed wastes with a mixed media of waste rubber and decomposed wastes in layers. A laboratory unit of the LTB was tested. The results were very promising, and we were able to mathematically predict the reduction of polluting parameters (Gnanakaran et al. 2017) for well over 5 years or more. Once again, the LTB was

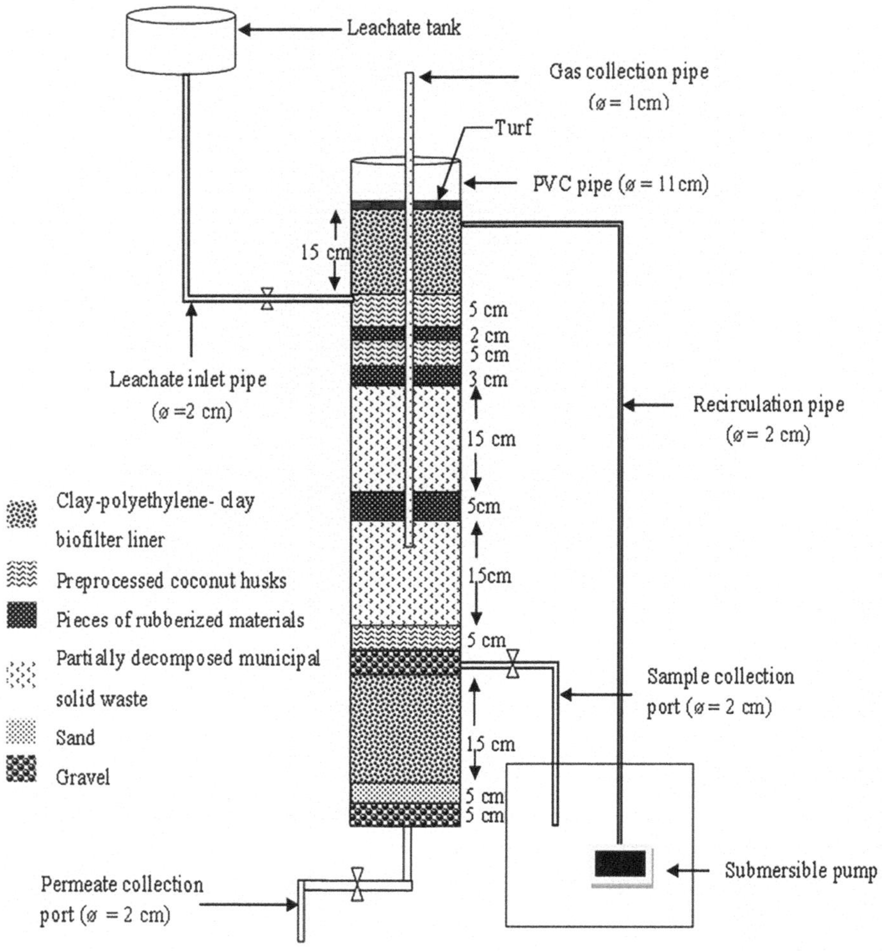

Fig. 19 A schematic diagram of the laboratory scale LTB. (Gnanakaran et al. 2017)

made operational. Although we were limited by a lack of funds, we launched the production of biochar and compost to divert the wastes from the dump. Also, the first stage of the landfill bioreactor was constructed with the composite liner system, and the remaining incoming wastes were disposed as described in the EIA study report (Ecotech Lanka Ltd 2011).

Investments for PP with Public-Private Partnership

The EIA approval and the valorization of the dumpsite were the key factors for assured investments because most of the investors, bankers, accountants and auditors refused to accept rehabilitated dumpsites as assets. In fact, one of them was a director of a Korean company. He said that his company almost went bankrupt

Fig. 20 Small dumps in different locations of the dumpsite

Table 5 Deduced resource utility value of the Gohagoda dumpsite (Senervirathne et al. 2013)

Description	Unit	Amount
Available stock of wastes in dumpsite	MT	250,000
Recovery of RDF (nominal)	%	60
Total quantity of RDF	MT	150,000
Sale price of RDF, equivalent to coal	US$/MT	90
Income from sale of RDF	US$	13,500,000
Cost of mining, processing and manufacture	US$/MT	46
Cost of RDF production	US$	6,923,077
Forecasted revenue from RDF	US$	6,576,923
Estimated cost of rehabilitation of dumpsite	US$	1,153,846
Estimated cost of relocation of houses	US$	823,077
Resource Utility Value (RUV)	US$	4,600,000

in trying to treat leachate and that 'without a tipping fee, it is suicidal'. Nevertheless, armed with a low cost and effective leachate treatment system, we resorted to calculate the Resource Utility Value (RUV) of the dumpsite (Senervirathne et al. 2013). The summary is given in Table 5. It should be noted that we have not included the value of LFG because it was supposedly to be used in the feasibility of power generations and it was an income generator, like RUV in the financial feasibility of the company. The feasibility studies showed that the total operating profit in converting waste to electricity is Rs. 24,000/Mg of raw wastes estimated over a period of the next 10 years (Basnayake et al. 2015). The return on investment (ROI) was less for generating 10 MW, but the internal rate of return (IRR) of 15% remained for smaller units; thus, a gradual increase in the capacity would have been the best option. Therefore, the proposed integrated solid waste management (ISWM) system had many investment scenarios, and one arm of a leading bank is an investment bank that offered the company 70% loan and 30% equity to install a plant to generate 2.5 MW of power while manufacturing biochar and compost. The excess wastes were to be disposed in landfill bioreactors for extracting gas and residual wastes. The intention of the investment was also to complete the rehabilitation of the dumpsite while relocating dumpsite inhabitants. Unfortunately, the majority shareholder of the company turned down the offer because he could not raise the capital without forgoing his shares. It was the same fate with a similar offer from the Asian Development Bank through a national leading bank for a higher capacity of minimum 5 MW at a lower interest. One must not get entangled with so-called investors and promoters because they are opportunists who are there to make a quick buck and get away from the responsibilities (Basnayake et al. 2015). In fact, politicians and government officials forced us to accept the investor and ignored the offers made by these reputed financial institutions until they get the best offer from a foreign company. As time passed, the EIA approval was cancelled.

7 Social and Economic Appraisals of MSW Management

All of these issues arose because the KMC and most other LAs were not given adequate directives and they were not willing and capable of initiating measures to overcome serious concerns with the local communities residing near dumpsites and recycling facilities. Moreover, not all generated wastes were collected because inadequate resources and proper disposal facilities created numerous environmental issues in many cities and towns. To date, some of these issues are not resolved. Therefore, at the end of 2015, UN-Habitat sponsored and initiated a study in eight of the provinces at the request and support of the Ministry of Mahaweli Development and Environment. The completion of the study is still pending, but a draft of the Comprehensive Integrated Solid Waste Management Plan (CISWMP) has been circulated among the expert committee, UN agencies and relevant ministry officials and UDA. It was also presented at the National Forum on Solid Waste Management on 23–24 June 2017. Furthermore, a manuscript was published in a journal publication (Basnayake et al. 2018). The study revealed that only 27.2% of the generated wastes are collected. In these eight provinces, the non-biodegradable, including plastic, is estimated to be 382–475 Mg/day, and approximately 232 Mg/day of plastics are burnt, emitting approximately 10–14 kg of polychlorinated dibenzo-p-dioxins (PCDDs) and dibenzofurans (PCDFs) every day. The calculation was based on emissions reported by Lemieux et al. 1999. The ignorance and attitude of the people have increased the dreadful habit of burning plastics. The respondents in most of the surveys conducted always blame the LAs, stating that collections are irregular, and they are compelled to burn wastes, including plastics. They also emphasize on a quality service being provided if they are to pay for it. Otherwise, the tendency is to burn the wastes without paying additional amounts to LAs. Few of the respondents are not willing to pay additional amounts because they already pay adequate taxes. When we explained the need to increase taxes in the way of a 'polluter pay' scheme with adequate measures of transparency, participation and the need to recycle the wastes, they are more amiable to become active participants in the proposed planning and implementation programmes.

After many discussions with LAs, regarding all of these issues and in view of minimizing wastes, the suggestion is to increase the fleet of vehicles for collections.

It will satisfy 50% of the gap between generations and present collections. It will be 39.7% of the total project costs. The construction of landfills will cost 23.1% with an allocation of 28.7% for administrative and management of all operations and services. The remaining 8.5% will be needed for R&D and education and training (E&T). We worked out that the polluter pay commitment is only Rs. 350 per capita per annum, considering only 50% of the population will be subject to this levy. The national CISWMP is based on applying 3R concept with the selection of applicable and suitable technologies of composting, anaerobic digestion and RDF fuelled power generation plants, allowing only 15% of non-hazardous inert wastes to be landfilled. These proposed methodologies will gradually reduce wastes ending up in landfills over the next 5 years. The LAs will be able to use the savings on landfilling on preferred technologies; some of them would need public-private partnerships (PPP). Therefore, the LAs with the communities will be empowered to develop plans at Provincial Council and LA levels; thus, the national plan will be amended accordingly. These plans will have eight sub-plans according to each subject as given in Table 6.

8 Way Forward

It is hoped that the recommendations in the UN-Habitat-sponsored study are implemented, particularly rehabilitation of dumpsites, formulation of 5-year CISWMPs at all levels with the incorporation of landfills, ensuring diversion of wastes by effective implementation of 3R in the planned strategy and establishment of Environmental Court/Tribunal (ECT) (Basnayake et al. 2018) for providing judicial services, thus empowerment of civil society groups. Accountability at all levels are included in the new policy on waste management, which will have a gazette notification shortly. The UN-Habitat study also contains a recommendation to set up an Evidenced-Based Informative and Interactive Database (EBIID) to enable accuracy in planning processes, monitoring and reporting on waste management. Furthermore, the suggestion is to engage regional and national universities to take the initiatives in supporting E&T and R&D activities with special emphasis on socioeconomic aspects to ensure sustainability of projects. In addition, the R&D activities should focus on electrically driven haulage vehicles to minimize transportation costs, thereby minimizing sound and air pollution. It is advisable to improve anaerobic digesters and develop a joint research programme with a technologically advanced country to develop small-scale combined cycle power generation units coupled with an absorption cycle refrigeration system to convert waste heat from a power generation plant to dry and wet ice. The latter system was proposed earlier (Basnayake 2009). It will increase refrigeration capacities in urban centres, thus reducing food wastes and contributing to preserving seeds. It is imperative to establish controlled environment of low-temperature seed banks for greater food security systems, which is badly needed in the country to combat climate change. Moreover, the power generations with MSW should be reduced with increased recycling of plastics, but those recyclables should then be replaced with agricultural

Table 6 Comprehensive integrated solid waste management plans (Basnayake et al. 2018)

Institutional development plan	Infrastructural development plan	Research and development (R&D) and education and training (E&T) plans	Social development plan	Environmental management plan	Resource allocation plan	Monitoring, risk assessment and disaster management plans	Regulatory and law enforcement plan
Coordinating body at LA, PC and national level	Selection of locations for facilities; landfill, 3R centres, transfer stations	Preparation of ISWM guidelines, LCA and feasibilities	Education programmes will be launched based on social research	Determination of N values for ensuring adequate capacities for collections	Calculations of THVs and financial feasibility studies	Determining mass balances of MSW as part of the monitoring programme	Establishment of ECT is a prerequisite for transparency, inclusiveness and accountability
An Institute of Material Cycles and Waste Management (IMCWM)	Upgrading and relocating of facilities.	IMCWM support of regional universities and institutes will conduct E&T programmes and workshops to teach preparation of feasibility studies	Subjects environment preservation and sustainable development	Selection of technologies design and implementation of ISWM system based on reducing landfilling	The plan is not to exceed the allocated land filling costs by diverting the wastes	LCA studies and social research for improving the systems. Monitoring frequencies on specified intervals	Rules and regulations between LAs should be consistent, and ECT is expected to ensure such cohesive understanding between LAs and stakeholder
CEA, NSWMSC and LAs	Ensuring facilities for workers and security of facilities					Risk assessment studies and frequent reporting	
Nomination and appointment of representative stakeholders in the ECT apart from experts		R&D institutions will undertake one or more listed fields of studies	Public will be informed and educated on grading system of LAs		Private sector participation for energy generations.	Disaster management plan: flooding, droughts, landslides, etc. Rapid removal, storage and recycling	ECT will request for frequent monitoring reports.
							Appointment of experts to ECT

wastes or dendro for maintaining the designed and desired power outputs from waste to energy plants, and it will also improve the efficiency of waste conversions to marketable products.

Acknowledgements The authors thank the government officials, particularly UDA, Kandy Municipal Council and Nawalapitiya Urban Council, UNOPS, academic and non-academic staff of the Department of Agricultural Engineering and PGIA for supporting the research and development efforts. Most of the investigations were supported by UDA and ARRPET research programme of Sida, coordinated by AIT and UNOPS. We wish to thank Aurora Basnayake for English corrections and valuable comments on comprehension.

References

Ariyawansha SWGRTK, Basnayake BFA, Dharmaweera WDNP, Sithamparanathan K (2009) Wet and dry layered humidified biofilter for destruction of gaseous contaminants from composting of municipal solid waste. In: 1st National symposium on natural resources management, Sabaragamuwa University, Sri Lanka 28–29 August 2009

Ariyawansha RTK, Basnayake BFA, Pathirana KPMN, Chandrasena ASH (2010a) Open dump simulation for estimation of pollution levels in wet tropical climates. J Trop Agric Res 21 (4):340–352

Ariyawansha RTK, Jayakody Lahiru N, Gunarathne HAYR, Senevirathne SADN, Kuruparan P, Chandrasekara WWCK, Thivyatharsan R, Basnayake BFA (2010b) Estimation of quantum of waste generation and solid waste characterization at Kalmunai Municipal Council (KMC) and Sammanthurai Pradeshiya Sabha (SPS). In: Third international symposium, Sabaragamuwa University of Sri Lanka on 26th–28th August 2010

Ariyawansha RTK, Chandrasena ASH, Senevirathne SADN, Basnayake BFA(2010c) Development of a sustainable Environmental Preservation Centre (EPC) at Nawalapitiya for urban solid waste management. In: Proceedings of international conference on sustainable build environment on 13th and 14th of December, 2010 at Earl's Regency Hotel, Kandy, Sri Lanka, pp 181–188

Ariyawansha RTK, Basnayake BFA, Kariyawasam KHPP, Jayasinghe IH(2010d) Applications of double layer compost biofilter for odor control in industrial and waste management facilities in Sri Lanka. In 9th Asia Pacific roundtable for sustainable consumption and production 10th–12th June 2010, Colombo Sri Lanka

Ariyawansha RTK, Basnayake BFA, Chandresena ASH, Senevirathne SADN (2012) Inferences drawn from converting a dumpsite quarry to a sustainable landfill in the East Coast of Sri Lanka. In: The 7th Asian Pacific landfill symposium sustainable solid waste management for a better life on 8th–11th October 2012. ISBN 978-602-18925-0-3, OSL 12-400–OSL12-405

Ariyawansha RTK, Basnayake BFA, Karunarathna AK (2018) Scaling-up of an up-flow anaerobic sludge blanket incorporated with a bio-filter liner system for treating kitchen wastes. Trop Agric Res 29(3):302–315

Bandara AMDKS (2018) Development of a scaled-up up-flow anaerobic sludge blanket reactor incorporated with a bio-filter liner system. Thesis. Department of Agricultural Engineering, Faculty of Agriculture, University of Peradeniya, Peradeniya, Sri Lanka

Basnayake BFA (1998) The invention of a sanitary composting process and equipment for urban solid waste management. Patent No. 10836

Basnayake BFA (2009) Business plan for operating an integrating solid waste management system at Science Park of south eastern university, Centre for Renewable Energy Sources (CRES), Postgraduate Institute of Agriculture, University of Peradeniya, Peradeniya

Basnayake BFA (2011) Composite clay-polyethylene biofilter liner and cover systems for enhancing landfill bioreactors and other bioremediation systems. Patent No: 15676

Basnayake BFA, Ekanayake KM (2005) Evaluation of different MSW landfill pretreatment (composting) systems in Sri Lanka. In: Proceedings Sardinia tenth international waste management and landfill symposium, S. Margherita di Pula, Cagliari, Italy; 3–7 October 2005

Basnayake BFA, Karunarathna AK, Chandresena ASH, Jayaweera MW (2007) An experimental approach for developing a sustainable landfill in Kaduwela, Sri Lanka. In: International conference proceedings on solid waste management, Chennai, India, 5–7 September 2007

Basnayake BFA, Ariyawansha RTK, Senevirathne SADN, Chandrasena ASH (2015) Valorization of municipal solid waste in Sri Lanka. Econ Rev (The Quarterly Economic Journal of People's Bank) 42(2):9–13

Basnayake BFA, Popuri S, Visvanathan C, Jayatilake A, Weerasoori I, Ariyawansha RTK (2018) Concerted initiative for planned management of municipal solid waste in target provinces in Sri Lanka. J Mater Cycles Waste Manage. https://doi.org/10.1007/s10163-018-0815-5

Ecotech Lanka Limited (2011) Environmental impact assessment study report: rehabilitation of Gohagoda dumpsite and establishment of an integrated solid waste management system for Kandy municipal council. 629/3, 8th Lane, Muththettugoda Road, Thalangama North, Battaramulla, Sri Lanka

Ekanayake KM, Basnayake BFA, Gunatilake SK, Chandrasena ASH (2005) Introduction of inclined step grade composting system for Asian countries (a case study in Tangalle area in Sri Lanka). In: Proceedings of the seventh international summer symposium, Japanese Society of Civil Engineers, July 30, 2005, Tokyo, Japan, pp 339–346

EML Consultants Pvt. Ltd (2009) Preparation of initial environmental examination report and detail design for the proposed composting plant and to improve the existing dumping site at Thirupperumthurai in Batticaloa district, 68, Davidson Road, Colombo 04, Sri Lanka

Gnanakaran M, Ariyawansha RTK, Basnayake BFA (2017) Simulation of lab-scale leachate treatment bioreactor with application of logistic growth equation for determining design and operational parameters. Int J Sci Eng Res 8(1):1061–1070

Gunarathna HAYR, Kapukotuwa ASB, Karunarathna AK, Basnayake BFA(2007) Construction and evaluation of a low cost, wasted polyethylene based liner for small landfills. In: Proceedings international conference on solid waste management, Chennai, India, 5–7 September, 2007

Gunarathna HAYR, Karunarathne SA, Basnayake BFA, Galagedara LW, Sangeetha T (2010) Evaluation of reactivity and inhibitions in developing municipal solid waste landfill bioreactors for tropical climatic conditions. Trop Agric Res 21(4):378–390

Gunawardana EGW, Shimada S, Basnayake BFA, Iwata T (2009) Influence of biological pre-treatment of municipal solid waste on landfill behaviour in Sri Lanka. Waste Manag 27 (5):456–462

Karthikeyan OP, Joseph K (2014) "Anammox" a novel process for nitrogen management in bioreactor landfills – a review. In: Centre for environmental studies. Anna University, Chennai

Karunarathna RHM, Ariyawansha RTK, Basnayake BFA (2017) Start-up of an up-flow anaerobic sludge blanket reactor built-in with a bottom liner of for food waste. In: Proceedings of 4th 3R international scientific conference on material cycles and waste management New Delhi, India, 8–10 March 2017

Lemieux PM, Lutes CC, Abbott JA, Aldous KM (1999) Emissions of polychlorinated dibenzo-p-dioxins and polychlorinated dibenzofurans from the open burning of household waste in barrels. Environ Sci Technol

Madugeethika JNK, Basnayake BFA, Nirosha K (2002) Development of an integrated solid waste management system for three local authorities in Sri Lanka through capacity mobilization. J Trop Agric

Mahees MTM, Sivayoganathan C, Basnayake BFA (2010) Food consumption, solid waste generation and water pollution in Pinga Oya Sub-catchment, Sri Lanka. In: 12th Royal asiatic international conference on Sri Lankan Studies

Mata-Alvarez J (2003) Biomethanization of the organic fraction of municipal solid wastes. IWA, London, pp 1–137

Menikpura SNM, Basnayake BFA (2009) New applications of 'Hess Law' and comparisons with models for determining calorific values of Municipal Solid Wastes (MSW) in Sri Lankan context. Renew Energy 34(6):1587–1594

Menikpura SNM, Basnayake BFA, Boyagoda PB, Kularathne IW (2007a) Application of waste to energy concept based on experimental and model predictions of calorific values for enhancing the environment of Kandy city. Trop Agric Res 19:389–400

Menikpura SNM, Ariyawansha SWGRTK, Basnayake BFA, Tharinda YGWGMC (2007b) Design and development of biofilters for controlling odor emissions from composting of municipal solid waste (MSW). In: Proceedings of international conference proceedings on Solid Waste Management, Chennai, India, 5–7 September 2007, pp 250–257

Menikpura SNM, Basnayake BFA, Pathirana KPMN, Senevirathne SADN (2008) Prediction of present pollution levels in Gohagoda dumpsite and remediation measures: Sri Lanka. In: The fifth Asian-Pacific landfill symposium, 22nd–24th October 2008, Sapporo, Hokkaido, Japan

Nevers ND (1972) Air pollution control engineering. Academic Press – Elsevier, USA

Priyashantha KAS, Ariyawansha RTK, Senevirathne SADN, Basnayake BFA, Chandrasena ASH (2015) Development and performance evaluation of the leachate treatment system at Gohagoda municipal solid waste disposal site. In: Proceedings of 6th international conference on structural engineering and construction management on 11th –14th December 2015, Kandy, Sri Lanka, pp 31–38

Qian X, Koerner RM, Gray DH (2002) Geotechnical aspects of landfill design and construction. Prentice-Hall, Upper Saddle River, pp 29–181

Reinhart DR, Tounsend TG (1998) Landfill bioreactor design and operation. Leusis Publishers, Boca Raton

Senervirathne SADN, Basnayake BFA, Chandrasena ASH, Ariyawansha RTK (2013) Municipal solid waste dumpsites; a burden or an obscure buried resource. In: Proceedings of the international conference on solid waste 2013 – innovation in technology and management, Hong Kong SAR, P.R. China, 5–9 May 2013, pp 718–721

Shafizadeh F (1981) Basic principles of direct combustion. In: Sofer SS, Zaborsky OR (eds) Biomass conversion processes for energy and fuels. PlenumPublishing Corporation, New York, pp 103–124

Shalini SS, Joseph K (2017) Combined Sharon and Anammox processes for ammoniacal nitrogen stabilisation in landfill bioreactors. Bioresour Technol. https://doi.org/10.1016/j.biortech.2017.10.077

Thivyatharsan R, Gunarathna HAYR, Basnayake BFA, Galagedara LW (2010) Determination of suitability and evaluation of a composite liner system for designing a landfill bioreactor on confined bedrock at Sammanthurai. Trop Agric Res 21(3):266–274

Thivyatharsan R, Gunarathna HAYR, Basnayake BFA, Kuruparan P (2012) Performance and suitability of a landfill bioreactor with low cost biofilm contained clay-waste polyethylene-clay composite liner system for tropical climates of Asian countries. Mater Cycles Waste Manage. https://doi.org/10.1007/s10163-011-0039

Wijesekara SSRMDHR, Mayakaduwa SS, Siriwardana AR, Silva N, Basnayake BFA, Kawamoto K, Vithanage M (2014) Fate and transport of pollutants through a municipal solid waste landfill leachate in Sri Lanka. Environ Earth Sci. https://doi.org/10.1007/s12665-014-3075-2

Wijesekara HR, De Silva SN, Wijesundara DT, De S, Basnayake BFA, Vithanage MS (2015) Leachate plume delineation and lithologic profiling using surface resistivity in an open municipal solid waste dumpsite, Sri Lanka. Environ Technol 36(23):2936–2943

Promising Modified Atmosphere Storage Methods to Protect Shelf-Stable Food Commodities in Sri Lanka

Batugahage Don Rohitha Prasantha

1 Introduction

Shelf-stable food grains after harvest are usually stored until the next harvesting season for sale and home consumption. Nearly 30–40% of the shelf-stable food commodities such as cereals, grain legumes, oil seeds, nuts, and spices harvested in Sri Lanka are kept by farmers for consumpon, seeds, and future sale for a period of 3–9 months (Adhikarinayake 2005; Prasantha et al. 2014a). This kind of on-farm storage provides about 70% of the total food requirements of the farm family, and it is a substantial contribution to the farm income during off-season. Nearly 50% of the paddy harvested in Sri Lanka is kept by farmers (Hafeel et al. 2008) for consumption, seeds, and future sale for a period of 6–12 months. Current statistics reveal that annual per capita consumption of rice in Sri Lanka is approximately 116 kg. Next to the rice, grain legumes play an important role in Sri Lankan diet. Grain legume is an inexpensive source of dietary protein supplement for more than 67% of Sri Lankans as an alternative to animal protein. Due to their high nutritional value, they are often being used in various dishes such as soup, porridge, curry, and some traditional confectioneries. Coconut mixed boiled mung bean and cowpea dishes are considered as one of the most popular breakfast diets among Sri Lankan next to the bakery products and rice/rice-based products. Spice crops such as nutmeg, cardamom, cinnamon, clove, vanilla, turmeric, ginger, pepper, and chili are some of the cash crops which are grown and processed by farmers solely for their income. Long-term storage of spices is also necessary to overcome the price fluctuations in market and respond to the demand in international markets. Most of the spices today are organically grown, and more than 80% of the products are exported in bulk to the EU, Japan, Mediterranean countries, and to the USA as organic products and sell at a

B. D. R. Prasantha (✉)
Department of Food Science & Technology, Faculty of Agriculture, University of Peradeniya, Peradeniya, Sri Lanka
e-mail: bdrp@pdn.ac.lk

R. P. De Silva et al. (eds.), *Agricultural Research for Sustainable Food Systems in Sri Lanka*, https://doi.org/10.1007/978-981-15-3673-1_23

price of $ 10–20 per kilogram. The value of spice export is equivalent to 1.5% of total export and 8% of agricultural export earning of Sri Lanka. Although high-quality cinnamon, nutmeg, clove, etc., are produced in Sri Lanka, postharvest protection systems that meet the international standard have not yet been developed. Therefore, occasionally export restrictions are imposed to local export shelf-stable commodities due to records of pest infestation (presence of dead or live insects), fungal infections, weed seeds, pesticide residues, etc. However, for a stable supply of shelf-stable foods, it is necessary to increase production and minimize postharvest losses during the handling process.

Shelf-stable stored products are characterized by having relatively low moisture content and include raw and semi-processed foods products. Grains, spices, nuts, and other shelf-stable agricultural commodities are commonly stored in woven polybags in Sri Lanka similar to many Asian and African countries. These bags are inexpensive and convenient but offer no protection to the stored commodities. Polybags are extensively used by farmers since it is convenient for transport, exchange, and storage under many conditions. Even though the storage duration may reach few months, shelf-stable food products such as grains, spices, nuts, herbs, and other dry foods stored in polybags are under constant threat of attack by many insect pests and fungi. Polybags do not provide any required safety for long-term storage of grains and other commodities, especially under humid climatic condition. Exact information of both quantitative and qualitative losses of storing food grains is extremely limited. Postharvest loss of on-farm storage of grains, spices, and nuts is high as about 15% due to improper storage techniques and lack of proper storage facilities in Sri Lanka (Sartaj and Ekanayake 1991). Poor storage conditions or practices cause rapid fluctuation of moisture withing the stored products due to the influence of environmental relative humidity (RH) and temperature. It has been estimated that on-farm storage loss of paddy in Sri Lanka is about 8% (out of 3.8 million MT per annum) due to poor storage (Adhikarinayake et al. 2006). Percentage of grain damage could be high as 60% in farm-level stored legumes (mung bean and cowpea) under polybags storage after 4–6 months (Prasadi 2014; Prasantha et al. 2014a). In an extreme situation, storage loss of legume grains (mung bean, cowpea, black gram, soya, etc.) due to damage of insects could be high as 98% (Prasantha et al. 2014b, 2018). Although the grains may have stored in suitable silos, postharvest losses may still occur due to uncleanness, physiological deterioration, spillage, and spoilage caused by insects and molds. Therefore, suitable modern technologies are necessary to achieve a significant reduction of postharvest losses of shelf-stable food commodities in Sri Lanka.

Insect pests and subsequent fungal infestations cause significant damage to the quality of grains during storage (Fig. 1) mainly in tropical countries due to high ambient temperature (>25 °C) and RH (>70%). Ambient temperature plays an important role on rapid multiplication of insects under tropical climatic condition. Fungal attack of stored food commodities generally occurs where drying has been inadequately conducted, or where large numbers of insect pests are present in the product. As a result of that, temperature may rise in the grains due to metabolic activities, and the stored product may easily get decomposition due to accumulation of high moisture and temperature. On the other hand, products are stored in damp

Fig. 1 Insect pest infestation in food grains: (**a**) mung bean infested by *Callosobruchus maculatus* (F.), (**b**) cowpea infested by *Callosobruchus chinensis* (L.), (**c**) secondary infestation of soybean by *Tribolium castaneum* (Herbst), and (**d**) red rice infested by *Sitotroga cerealella* (Olivier) (*NB: Soybean has primarily been infested by Callosobruchus chinensis* L.)

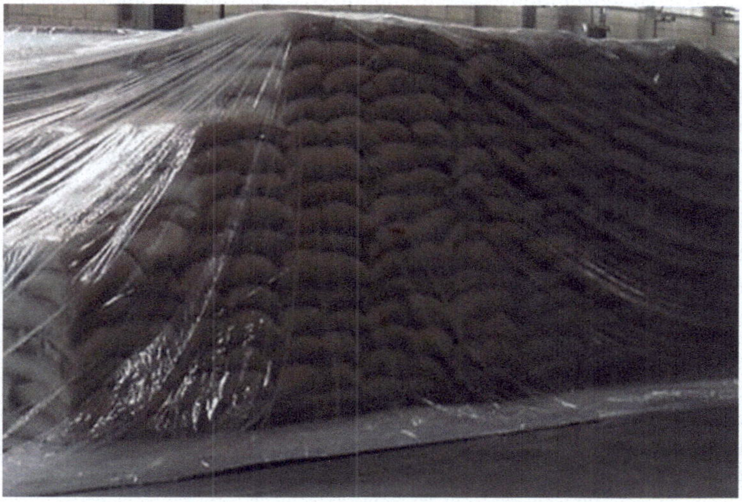

Fig. 2 Phosphine (PH_3) fumigation of shelf-stable food commodities using aluminum phosphide tablets

conditions could also absorb high moisture from the surrounding atmosphere. This tends to speed up undesirable chemical changes in the stored commodities such as discoloration, rancidity development, flavor changes, splitting, and in further formation of mycotoxins (Raj and Singaravadivel 1990; Seitz and Sauer 1996).

Today, a considerable amount of grains, spices, herbs, and their products are produced organically, but synthetic pesticides must not be applied. Even in conventional grain storage, the use of synthetic pesticides such as pirimiphos-methyl application and phosphine (PH_3) fumigation (Fig. 2) presents risk to workers, may develop resistance among insect pests, and ultimately, pesticide residues could cause a grave problem to the environment and health of end users. Pest control measures are costly when the stored products have heavily been infested by insects and their quality has been affected. Therefore, producers, especially farmers, apply many hazardous pesticides to protect their stored grains soon after harvesting and several times during storage. Phosphine fumigation is not recommended at the farm level because of the non-unavailability of skill and safety required in the application. As an alternative to the considerable storage loss, farmers sell their grains soon after harvesting at low price prior to the infestation by insects. However, during the off-season, price of grains such as mung bean and cowpea increases significantly in the retail market, ranging between $1.00 and 2.50 per kilogram. Long-term storage of grains and spices is also necessary to overcome the price fluctuations in the retail market for the benefit of consumers.

Because of limited shelf-life and restricted use of chemical preservatives, quality of perishable and semi-perishable food commodities deteriorates immediately after harvesting or processing mainly due to microbial growth. Compared to perishable foods, shelf-stable but nonperishable food commodities can be stored for a longer period of time. However, due to prevailing unfavorable intrinsic and extrinsic factors in the stored environment, quality and quantity of stored foods may be affected within a few months. Therefore, food industry is encouraged to search for an alternative nonchemical pest control technique for stored food preservation. Physical and low-risk control techniques such as controlled or modified atmospheric storage (CA/MA), radio frequency, microwave, and high temperature have advantages of being free of chemical residues, although there is a variation among insect species susceptibility to physical control technique (Fleurat-Lessard and Torc'h 2001; Fields and White 2002; Adler 2010). CA or MA storage systems are generally considered as promising alternative methods to pesticide application or chemical fumigation for storage systems under tropical conditions (Adler et al. 2000; Ferizli et al. 2001) for grains, nuts, dried fruits, spices, and dried herbs during the storage. Although the CA and MA techniques are more or less similar techniques, CA storage is a more sophisticated technique than MA storage. Therefore, in developing countries, MA-based technology can be used as a cost-effective storage solution at farm level. Generally, MA storage technique keeps the grain and other shelf-stable agricultural commodities safe from insect and fungal infestation while maintaining them at an acceptable germination potential (Villers et al. 2006; Navarro 2012). It is helping to protect farmers from seasonal fluctuations in commodity prices and availability of seeds as planting materials.

2 Modified and Controlled Atmospheric Storage

The principal mechanism of CA and MA technologies is to reduce the rate of respiration of product, insects, and microbial growth while retarding enzymatic activity by changing the gaseous environment around the stored food products. The generation of CA/MA is achieved by altering the normal ratio of atmospheric gases. A reduction in the oxygen (O_2) content $\leq 1\%$ and increasing carbon dioxide (CO_2) $\geq 5\%$ is required to kill all stages of insects and cease the growth of aerobic microorganisms and retard metabolic activities (Moreno-Martinez et al. 2000).

In MA storage, the gas composition of the storage environment is generally created mainly by the interaction of the commodity's respiration while permeation of respiratory gases through the packaging material. MA conditions can also be created by applying a vacuum and replacing the storage atmosphere with a desirable inert gas mixture such as CO_2 or nitrogen (N_2). A major difference between CA and MA storage is in the degree of gaseous composition control in the storage atmosphere. The CA implies a higher degree of control than the MA in maintaining specific levels of O_2, CO_2, and other gases such as ethylene (C_2H_4). Also, in MA storage, the composition of the atmosphere surrounding the product is generally created and maintained by the interaction of the commodity's respiration with the permeation of respiratory gases through the packaging material. Modified atmosphere conditions can also be established and adjusted by pulling a slight vacuum and replacing the package atmosphere with a desirable gas mixture, which can be further adjusted through the use of O_2 or CO_2. The development of both fungi and insects can be prevented throughout the storage period when $O_2 < 3\%$ in the MA storage atmosphere (Adhikarinayake 2005; Adhikarinayake et al. 2006; Prasantha et al. 2014a). In Sir Lanka process, cashew kernels are treated with CO_2 and nitrogen to prevent insect pest infestation (Surendra 1997) before packaging for export as MA treatment.

Although CA storage is more similar to MA storage, CA maintains specific levels of high CO_2 (> 1 and < 15%) and/or low $O_2 > 5\% < 10\%$ with N_2 and control C_2H_4 using scrubber throughout the storage period being the main difference between CA and MA techniques (Courtesy 1983; Wright and Kader 1997). In CA storage system, both temperature and relative humidity (RH) in the storage atmosphere are also regulated or controlled at optimum condition especially for the preservation of the quality of fruits and vegetables throughout the storage period (Coursey 1983). Air consists of approximately 78.08% N_2, 20.95% O_2, 0.93% Ar, 0.036% CO_2, and traces of several other gases that have no physiological significance. Sen et al. (2010) studied the effectiveness of high temperature CA for controlling storage pests and its effect on the quality of dried figs. They recommended CA as one of the best alternative methods for chemical fumigants to control storage pests of dehydrated fruits within relatively short time periods. The highest concentration of CO_2 (>40%) treatment is considered as an effective control technique of many insects without development of resistance and leaving no any residue in the stored food stuff (White and Jayas 1991). However, even at a high concentration of CO_2 at temperatures ≤ 25 °C, it takes relatively longer exposure times to kill the target insects than with PH_3 fumigation of same products. Despite the efficacy of CA used to control the stored pests of shelf-stable commodities, hermetic and vacuum are fairly inexpensive systems for local industries. On the other hand, operational cost is comparatively lower than CA and chemical fumigation. Therefore, this type of treatment is good for high-value organic shelf-stable commodities such as spices, nuts, and herbs which can earn high income in the international markets.

3 Hermetic Storage

Hermetic or airtight storage is an excellent MA technique for controlling insects in stored grain and is an alternative to chemical pesticides (Navarro 2006). Hermetic storage uses airtight containers to modify the storage atmosphere, obtaining a low O_2 and high CO_2 atmosphere after a few weeks of storage (Busta et al. 1980; Hafeel et al. 2008; Prasantha et al. 2014a, b). Alteration of the storage atmosphere is achieved through the respiratory metabolism of stored grains, insects, and microorganisms present in the storage environment, and the results are mortality of insects (Fig. 3), reduced reproductive capacity of insects, and inhibition of mycotoxin-producing fungi due to lack of oxygen and high carbon dioxide levels in the storage ecosystem (Caliboso and Sabio 1998). The development of both fungi and insects can be prevented when $O_2 < 10\%$ and $CO_2 > 5\%$ in the hermetic storage environment (container) within 30 days depending on the size of pest population (Moreno-Martinez et al. 2000; Adhikarinayake et al. 2006; Prasantha et al. 2014a). The other advantages are a reduction of grain metabolic activity, a reduction of fat oxidation, flavor retention, and maintaining better processing characteristics (Prasantha et al. 2014a, b). During exposure to comparatively high concentrations of CO_2, insects are killed under the stress of hypercarbia and/or hypoxia (Adler 1993). Murdock et al. (2012) reported that the cause of insect death was due to hypoxia and desiccation under hermetic storage system. In order to achieve significant mortality in the high CO_2 level, interstitial space air should be completely replaced from the commodities before inundation with CO_2. To achieve hermetic condition, products must be stored in relative low-cost plastics/PVC structures which can be hermetically sealed.

Israeli researchers (Donahaye et al. 1991) first investigated such possibility of storing bagged paddy (type "Nadu") under tropical outdoor conditions in Sri Lanka

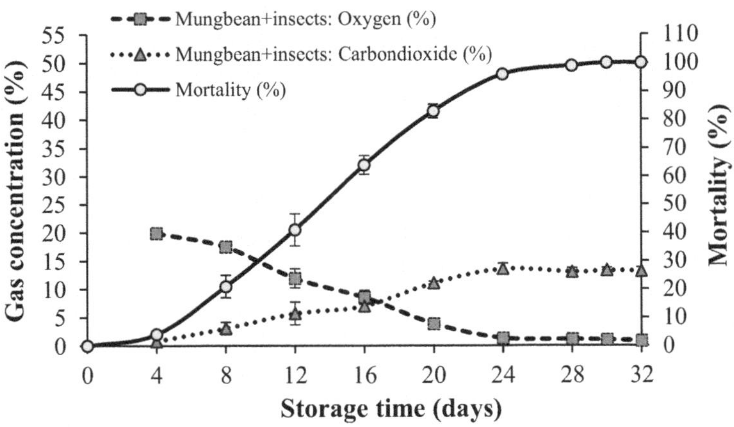

Fig. 3 Laboratory study of changes in gas composition and bean weevil (*Callosobruchus chinensis*) mortality of hermetically stored mung bean samples. (Source: Prasantha et al. 2014b)

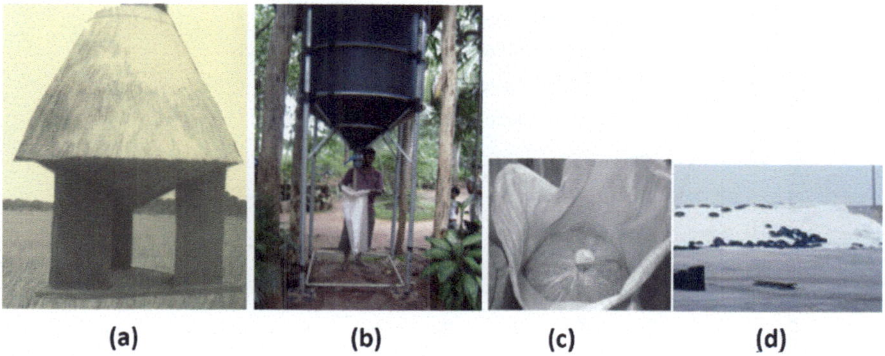

Fig. 4 Hermetic type grain storage structures developed in Sri Lanka; (**a**) airtight reinforced concrete paddy store bin, (**b**) grain storage PVC hermetic tank, (**c**) IRRI super bag, and (**d**) sealed flexible plastic liners. (Source: Donahaye et al. 1991; Adhikarinayake 2005; Hafeel et al. 2008; Prasantha et al. 2016)

using MA created in hermetically sealed plastic liners call storage cubes. They found 5% O_2 and 10–15% CO_2 of gas composition, 0.33–0.64% dry matter loss, and 1.24% loss of dry weight due to insect (*Tribolium castaneum*) and mold attack of 10–20 tons of bagged paddy after 6 months stored in storage cubes. Adhikarinayake et al. (2006, 2007) developed an airtight reinforced concrete bin (Fig. 4) to store paddy in Sri Lanka. In that trials, paddy (Bg 94-1) was stored inside the outdoor storage structure for 6 months while obtaining excellent hermetic conditions of low O_2 (2.7%) and high CO_2 (9%) due to initial development of insects such as *Ephestia cautella*, *Sitophilus* spp., and *Rhyzopertha dominica* (Adhikarinayake 2005). They have recorded better milling outturn, lower changes of moisture content, very low mass loss (<0.4%), and high head rice yield (36%) but poor germination rate of paddy after 6 months of storage in the hermetic bin.

Hermetic IRRI super bag is another alternative storage technique, similar to common polybag storage (Ben et al. 2006; Hafeel et al. 2008, 2011; Prasantha et al. 2014a). The O_2 content was dropped in the range of 7–13.8%, and CO_2 content rose in the range of 3–9% for infested and uninfested samples stored in the hermetic IRRI bag, respectively (Fig. 5). They also found that the hermetic storage of dried paddy had no adverse effect on milling yield, grain whiteness, and other quality parameters up to 4–9 months of storage (Table 1). Ambient storage (28–32 °C and RH 75–85%) is generally used for commercial storage of paddy, and two cropping seasons (9–12 months) are considered a long period for paddy storage under ambient conditions.

Prasadi (2014) and Prasantha et al. (2014a) found O_2 content dropped to 0.7% and CO_2 content rose to 8% within 28 days where freshly harvested mung bean and cowpea samples (obtained from farmers) were stored in 5 L hermetic PET bottles (Figs. 6 and 7). Although no live bean weevils were found in hermetic samples, significant change in the O_2 and CO_2 contents was detected in the hermetically stored samples, indicating that field bean sample may have already been infested by

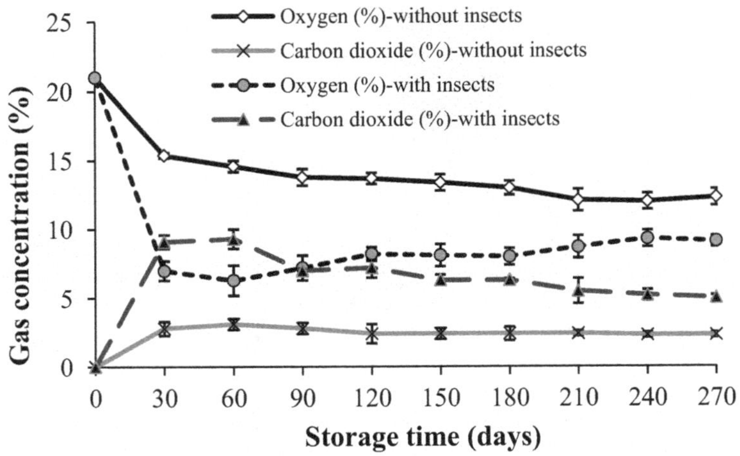

Fig. 5 Modification of gas compositions (%) of paddy with or without infestation of insects (*Sitophilus* spp.) stored in IRRI hermetic bags. (Source: Prasantha et al. 2014a)

either insect or fungi. These results indicate most of the fungal or insect infestation starts at the field level rather than during storage (Fig. 6).

However, grain quality deterioration is unavoidable under common storage conditions. Hard-to-cook (HTC) defect is another problem of legume beans associated with poor storage at high temperature and high humidity (Nasar-Abbas et al. 2008). HTC is characterized by extending cooking times for softening of grains (Liu and Bourne 1995), an increase in the grain hardness, and changing color of the seed coat (Fig. 8). These types of HTC grains need additional energy for cooking, which may affect nutritional qualities and acceptability by consumers (Deshpande et al. 1984). The extended cooking time and poor textural quality reduce the consumer preference and market value of legume beans. The other main problem is the loss of stored grain viability due to development of HTC characteristics and insect attack which hinders the percentage of seed germination. Therefore, an effective storage method should be able to prevent the growth of pest and molds, while maintaining the physicochemical and functional properties of grains. Shiga et al. (2009) reported that common beans stored at 30 °C and 75% RH for 8 months showed development of HTC characteristics and subsequently decrease in the protein and ash contents with seeds ageing, with changes in physical properties of the carbohydrates. According to our previous study, hermetic storage can prevent the development of HTC characteristics in terms of harness (N) development and cooking time (Fig. 9) and significantly reduce the postharvest loss of cowpea and mung bean (Prasadi 2014; Prasantha et al. 2014b).

Prasadi (2014) reported modifications of different protein fractions of mung bean and cowpea (SDS-PAGE gel electrophoresis method) samples after 6 months of storage in hermetic and control conditions. Although the exact reason for this protein change is not known, it may be correlated with the development of HTC characteristics. However, mung bean samples showed differences only in

Table 1 Physical and nutritional properties of cowpea, mung bean, and paddy samples stored 6–9 months in polybags (control) and hermetic structures under ambient conditions

	Storage condition		
Grain quality characteristics	Initial	Hermetic	Control
Cowpea and mung bean†			
Crude protein (%)	26.5–23.2	26.5–23.3	26.0–22.7
Crude fat (%)	2.5–0.9	2.5–0.9	2.5–1.1
Ash (%)	3.7–5.0	3.4–5.0	4.0–5.0
Bulk density (kg/m^3)	752.3–790.7	752.0–780.0	653.3–732.6
Antioxidant activity (%)*	27.0–25.0	50.0–35.0	63.0–32.0
Paddy: Bg 94-1†			
Total milling yield (%)	72.8	73.0	72.5
Head rice yield (%)	43.7	45.8	40.0
Thousand grain mass (g)	22.7	22.2	21.4
Paddy: Bg 352, Bg 300, Bg 358, and Bg 360**			
Total milling yield (%)	70.7–74.3	70.5–75.5	70.0–74.0
Head rice yield (%)	50.0–59.0	55.0–56.6	51.0–56.8
Thousand grain mass (g)	14.0–27.8	14.0–27.2	13.3–27.1
Bulk density (kg/m^3)	605.6–634.8	581.0–621.8	609.5–642.2
Amylose content (%)	26.2–28.7	30.0–32.3	31.1–32.7
Crude protein (%)	6.6–8.5	5.8–7.2	6.1–7.5
Crude fat (%)	2.4–3.0	2.0–2.1	1.6–2
Ash (%)	1.4–1.7	1.3–1.5	1.4–1.5
Free fatty acid (mg KOH/100 g)	1.4–1.7	2.5–2.6	4.0–4.1
Niacin (μg/g)	36.8–41.5	35.4–38.8	26.0–39.3

Source: Adhikarinayake (2005), Hafeel (2010, 2011), Prasadi (2014), Prasantha et al. (2014a)
†Storage of 6 months in PET bottles
*DPPH radical scavenging activity (%)
**Storage of 9 months in IRRI hermetic and polybags

the intensity of protein bands rather than different kDa protein bands. Although the exact reason for this protein change is not known, it may be correlated with the development of HTC characteristics.

However, due to practical difficulties and the inappropriateness of current technology such as structural weakness and handling, implementation of a hermetic storage system has not yet been well adopted among many farmers. Product feasibility, durability, and easy handling are important factors to popularize the hermetic storage method among local farmers. Therefore, modified PVC tank (Fig. 4) would be a good option to store shelf-stable food commodities in the future (Prasantha et al. 2016, 2018).

Field application of 1000-L-capacity PVC hermetic tanks showed better performance over control samples (non-hermetic PVC tank) of storage mung bean and cowpea (Prasantha et al. 2018). The O_2 content of hermetic samples stored in PVC tank was dropped to 13.4 ± 1.2%, and CO_2 content was increased up to 5 ± 0.7% (our unpublished data) within 6 months. After 6 months, grain germination

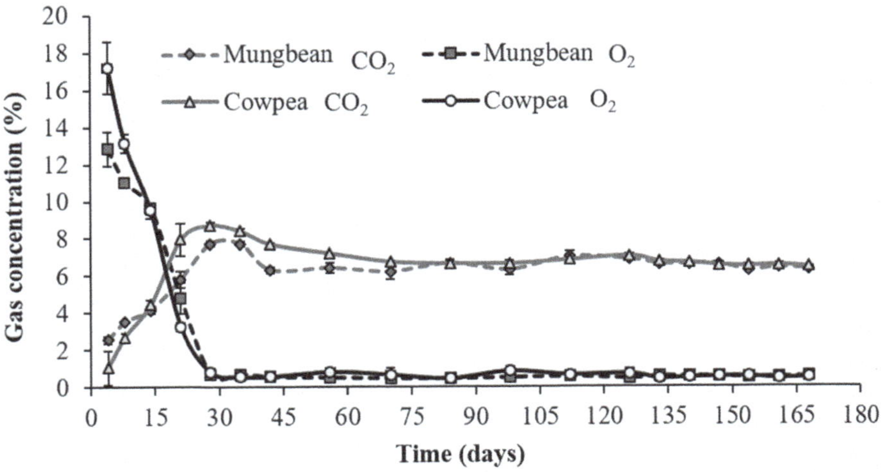

Fig. 6 Changes in gas composition of hermetically stored mung bean and cowpea samples obtained from farmers. (Source: Prasadi 2014)

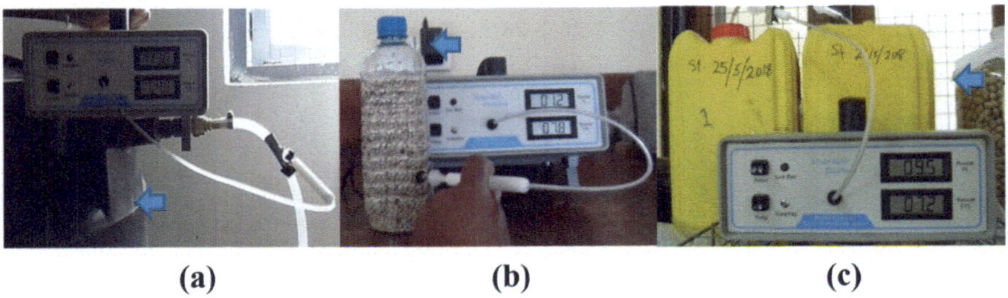

Fig. 7 Measuring gas compositions (O_2% and CO_2%) in hermetic experiments. (**a**) Mung bean stored in a 1000-L volume PVC hermetic tank, (**b**) cowpea stored in a 1-liter volume PET bottle, and (**c**) groundnut stored in a 25-L volume plastic cans

decreased from 95% to 92% in the hermetically stored samples in PVC tank, whereas it was decreased from 95% to 36% in the control tank sample due to grain damage. Percentage of grain damage in the hermetic sample was only 4.5 ± 1% compared to the heavy insect damage of the control samples (Table 2).

Hermetic storage itself reduces the need of insecticidal or fumigation treatments, with important economic and environmental advantages. It has been shown that hermetic storage is the most effective method to store grain in Sri Lanka due to its simplicity and low cost (Donahaye et al. 1991; Adhikarinayake et al. 2006, 2007; Hafeel et al. 2008; Prasantha et al. 2014a, b) similar to the use of hermetic storage in other countries (Caliboso and Sabio 1998; Villers et al. 2006; Quezada et al. 2006). Therefore, a hermetic storage system can be a safe and viable alternative to the conventional bag system for storage of stable-food commodities. They are capable of preserving the quality of dry food products while minimizing the weight loss.

Fig. 8 Color change of the mung bean seed coat during storage in polybag at ambient condition. (**a**) Freshly harvested mung bean and (**b**) HTC characteristic developed in mung bean sample after 6 months

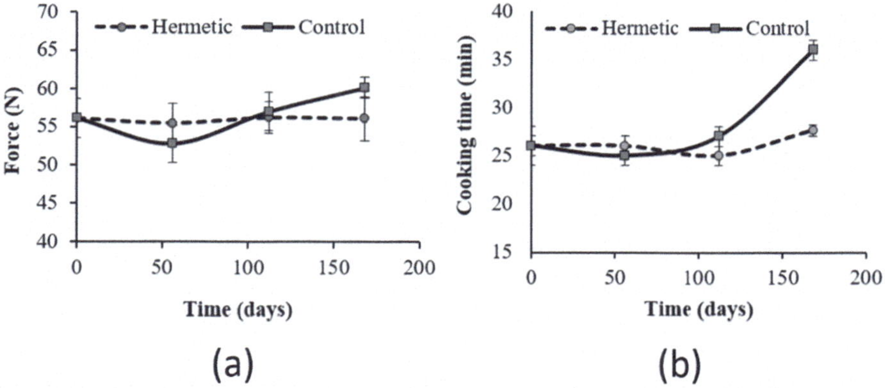

Fig. 9 Changes in HTC characteristic of mung bean stored in hermetic PET bottles. (**a**) Grain hardness and (**b**) cooking time. (Source: Prasantha et al. 2014b)

However, future research and development studies are needed to find out the effect of hermetic storage on quality preservation of various other shelf-stable food commodities during long-term storage. Feasible and easy handling is also important to popularize the hermetic storage technique among farmers. Further development of appropriate hermetic technique is therefore vital to promote hermetic storage among farmers and industries to reduce postharvest loss and to increase the food security in the country.

4 Low-Pressure-Modified Atmosphere Storage

Common storage pests such as the cosmopolitan species of Indian meal moth (*Plodia interpunctella* (Hübner)), Mediterranean flour moth (*Ephestia kuehniella* (Zeller)), Angoumois grain moth (*Sitotroga cerealella* (Olivier)), weevils (*Sitophilus*

Table 2 Average values of weight loss (%), grain damage (%), and germination (%) of mung bean, cowpea, and paddy samples stored in polybags (control) and hermetic structures under ambient conditions

Storage method	Storage (months)	Weight loss (%)		Grain damage (%)		Germination (%)	
		Control	Hermetic	Control	Hermetic	Control	Hermetic
	Mung bean						
PVC tank	0	0.00	0.00	0.00	0.00	93–97	93–97
	6	98.6	4.8–5.5	100[a]	4.0–5.5	33–37	90–93
	Cowpea and mung bean						
PET bottles	0	0.00	0.00	0.00	0.00	93–95	93–95
	6	100	4.0 ± 0.4	100[a]	3.1–4.2	46–58	87–88
	Paddy: Bg 94-1						
Concrete bin	0	0.00	0.00	0.00	0.00	85.0	85.0
	6	5.7[c]	2.2	3.3[a]	0.5	38	0
	Paddy: At 405, At 306, Bg 352, Bg 300, Bg 358 and Bg 360						
IRRI bag	0	0.00	0.00	0.00	0.00	88–90	88–90
	9	1.4–3.7[c]	0.46–1.4	7.8[d]	0.2	26–33	0–18
	Paddy: Nadu type						
Storage cubes[b]	0	0.00	0.00	–	–	–	–
	6	5.00	0.3–0.6[c]	–	–	–	–

Source: Donahaye et al. (1991), Adhikarinayake (2005), Hafeel (2010), Prasadi (2014), Prasantha et al. (2014b)
[a]Directly from field without fumigation
[b]Sealed PVC flexible liners
[c]Based on dry matter loss (%)
[d]Paddy samples have been fumigated prior to storage

spp.), red flour beetle (*Tribolium castaneum* (Herbst)), bean weevils (*Callosobruchus* spp.), lesser grain borer (*Rhyzopertha dominica*), and drugstore beetle (*Stegobium paniceum* L.) are some insect pest of stored products. These insects cause considerable economic loss to the food industry in the form of direct damage to the product, sanitary problems, control costs, and customer returns of the infested product and indirect damage to the machineries and other apparatus. Adult insects fly toward manufacturing foods, warehouse areas, and other food processing facilities for oviposition. Adults would later be present in final packaged products such as flour, pet food, biscuits, nuts, dried fruits, chocolate, powdered milk, and other foodstuffs (Riudavets et al. 2009; Noomhorm et al. 2009). Under some circumstances, larvae of moths ready for pupation penetrate through the packages to outside (Mullen and Pedersen 2000). Although the stored product moths are relatively more susceptible to MA/CA treatments, their eggs and diapausing larvae (i.e., *Plodia interpunctella*) are relatively tolerant to high CO_2 environment (Adler 1999; Adler et al. 2000; Johnson 2010).

Low-pressure vacuum storage (hypobaric storage) or packaging of foods in a low O_2 (i.e., vacuum, N_2 or CO_2 environment) content is considered as an effective MA technique for maintaining the shelf-life of many foods. It provides low oxygen environments in a comparatively short time through the use of vacuum pumps to evacuate the atmosphere within storage containers. It is an alternative storage technique for refrigeration and CA storage of shelf-stable and some perishable food products. The main function of the vacuum storage is low O_2 and/or comparatively high level of CO_2 development in a sealed container. The anaerobic environment of vacuum store prevents the growth of spoilage microorganisms and infestation of insects which are responsible for development of odor, flavor, and adverse textural changes. Although some physical, biochemical, and physiological changes were reported in shelf-stable foods at ambient temperature, except fresh fruits and vegetables, quality of other perishable food products did not get affected during storage at the refrigerated temperature under vacuum conditions. In order to get the effective protection of the stored food commodities under vacuum storage, O_2 should not drop <2% and CO_2 increase >30%. Vacuum storage is still not commercially applied to many perishable foods, but it is used as MA technique to preserve many shelf-stable package food products.

Vacuum may also be used in combination with high CO_2 or N_2. Although prior studies have shown that losses are reduced when rice is stored under MA conditions (Hafeel et al. 2008), few studies have evaluated the insecticidal and fungicidal effects of MA treatments on product quality parameters under field conditions. Johnson and Zettler (2009) and Johnson (2010) showed that vacuum treatment could effectively be used to control storage pests of tree nuts. Although it is considered as a very effective MA technique, RH in the vacuum chamber and moisture content of the product played a significant role during the vacuum treatments. As a result of low RH, stored foods are subjected to the excessive dehydration, but this may not be a problem of shelf-stable food commodities compared to the perishable foods. According to Mbata et al. (2005), 99% mortality of adult cowpea weevils (*Callosobruchus maculatus* F.) was achieved within an hour at low pressure (32.5 mmHg), but pupae and eggs were the most tolerant stages to low-pressure treatment. A recent study of Mortazavi and Ferizli (2014) found that total mortality of eggs, larvae, pupae, and adults of *Callosobruchus chinensis* can be obtained 144 h, 480 h, 432 h, and 192 h, respectively, after low pressure (88.8 mbar or 66.6 mmHg) treatment at 25 °C storage temperature. Adler et al. (2016) found that *Sitophilus granarius* (L.) can be successfully controlled if infested wheat sample was stored in a vacuum bag with low O_2 content.

Changes in quality of a commodity in storage could be predicted from knowledge of the storage temperature, moisture, atmosphere, and initial quality of the product. Common storage pests of spices such as the cigarette beetle (*Lasioderma serricorne* F.), drugstore beetle (*Stegobium paniceum* L.), and red flour beetle (*Tribolium castaneum* (Herbst)) cause considerable economic loss to spices, nuts, and herbal industries. The adult beetles damage the products, and the female insects lay many eggs during their lifetime. The larva hatching out from the egg feeds on the stored food products and also contaminates stored foods with frass, larval skins, and feces which later on help to harbor the fungal spores. The final instar larvae tunnel through the produce to pupate causing indirect damage (Riudavets et al. 2009; Noomhorm et al. 2009). Although stored product insects are relatively more susceptible to the

Table 3 Percentage mortality of adult drugstore beetle (*Stegobium paniceum* L.) exposed to low pressure (13.33 kPa) or control pressure (101.33 kPa) at room temperature (28 °C) and 75% relative humidity

Products	Exposure time (h)	Mortality (%) of adult insects		Vacuum oxygen content (%)
		Vacuum (40.0 kPa)	Control (101.32 kPa)	
Cashew nuts	6.0	10.3–12.0	0.00	2.8%
	24.0	32.7–40.0	0.00–1.00	3.2%
Nutmeg kernels	18.0	9.6–15.3	0.00	3.0%
	36.0	42.6–48.0	1.0–3.00	3.8%

MA/CA treatments, their eggs are relatively tolerant to high CO_2 environment (Adler 1999; Adler et al. 2000; Johnson 2010). Our previous study of vacuum treatment of cashew nut and nutmeg kernels (Prasantha 2020) indicated that vacuum treatment can effectively be used to control the adult drugstore beetle (*Stegobium paniceum* L.) as a test insect (Table 3). Another vacuum experiment was conducted in a 2-L vacuum desiccator (unpublished data). A vacuum of 13.33 kPa (100 mm Hg) was applied using 25 test insects (drugstore beetles) that infested cashew nut or nutmeg kernels. Drugstore beetles are feeding on many dry biological materials, including shelf-stable foods and non-food materials such as leather and animal feeds.

The vacuum storage is more effective if a considerable amount of O_2 could be replaced by adding N_2 into the vacuum atmosphere (Adler et al. 2016); otherwise, O_2 may increase in the vacuum during storage due to gradual release of available O_2 in the product (Table 3). Vacuum storage is important for long-term storage of sufficiently dry shelf-stable products, even if they were moderately infested by stored product pests. Low O_2 content and vacuum interfere the movement, feeding, mating, and respiration of many insects. Mortality rates of insects under hermetic or vacuum storage depend on the size of the insects and their developmental stages, CO_2 susceptibility, product type, temperature, and whether infestation occurs inside or outside of the treated commodity.

Information on the rate of change of stored commodity quality is necessary for cost–benefit analysis of the treatments. Therefore, information and understanding about moisture and gas diffusion, isothermal characteristics, and physicochemical and structural changes of the vacuum treated commodity is essential to establish proper control of stored product pests and obtain optimum quality of the product.

5 High-Pressure Carbon Dioxide Treatment as Modified Atmosphere Storage

In contrast to the conventional chemical fumigants and CA/MA treatments, high-pressure CO_2 treatment can be used as a preventive method to ensure pest-free food without leaving toxic chemical residues (Fig. 10). High-pressure CO_2 treatment extremely reduces the lethal exposure time of insect mortality in minutes to a few

Fig. 10 High-pressure carbon dioxide treatment chamber

hours (Nakakita and Kawashima 1994; Reichmuth and Wohlgemuth 1994; Prozell and Reichmuth 2001) compared to CA/MA treatments at ambient pressure. Fleurat-Lessard and Torc'h (2001) found that less than 1 h was sufficient to control all stages of *Tribolium castaneum*, *Dermestes maculatus* (de Geer), and *Plodia interpunctella* (Hübner) under high pressure CO_2 at 19 atm (=1.9252 MPa) treatment. Prozell and Reichmuth (2001) showed that at low temperature (10 °C) 15 bars (≈1.52 MPa) treatment, 100% mortality cannot be achieved within 10 h exposure time of some stored product pests. This indicated high-pressure CO_2 treatment may not be effective under low temperature. Eggs of insects and young larvae stages are more tolerant to CO_2 under pressures of 15–40 bars, especially if they are internal feeders (Adler et al. 2000). Fleurat-Lessard and Torc'h (2001) showed eggs of *D. maculatus* and *P. interpunctella* were preserved relatively unaffected under these conditions. Therefore, mortality rates of high-pressure exposure insects depend on the size of the insects and their developmental stages, CO_2 susceptibility, product type, temperature, and whether infestation occurs inside or outside of the treated commodity.

In general, product quality should not be affected by any treatment applied to the stored product. Under the high levels of CO_2 treatment at high pressure, gas molecules penetrate into the mass and are absorbed by the internal structure and bound water of grains and nuts (Cofie-Agblor et al. 1998). The possibility of CO_2 sorption by the product may play an important role in determining the amount of CO_2 required for treatment and maintain the required pressure level in the treatment chamber. Prasantha et al. (2012) examined the amount of CO_2 absorbed by inshell walnuts treated under high pressure with saturated CO_2 (98%) during 100 h (Fig. 11). The possibility of CO_2 sorption of the product may play an important

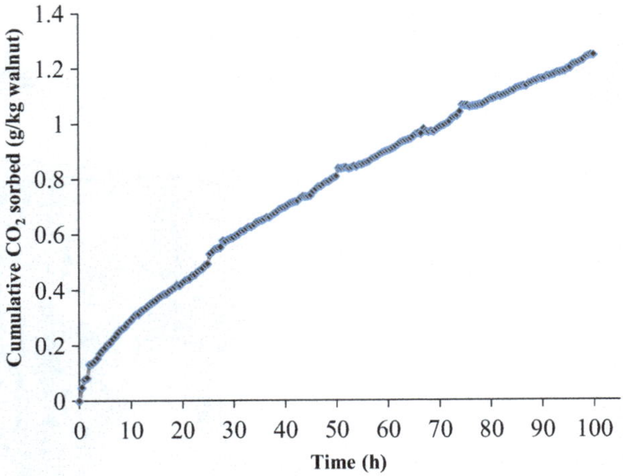

Fig. 11 CO_2 sorption by walnut at 8% (dry basis) moisture content and at 43 °C treatment temperature. (Source: Prasantha et al. 2012)

role in determining the amount of CO_2 required for treatment and may not be effective against target insects.

Noomhorm et al. (2009) reported that high-pressure CO_2 treatment produced more distinctive changes in the rice quality than low-pressure CO_2 treatment. Several high-pressure CO_2 trials have been conducted to determine the quality changes of treated cocoa beans and hazelnuts (Adler et al. 2000). Only the treated hazelnuts and almond showed a tendency to turn rancid earlier than untreated samples (Buranasompob et al. 2007). Tree nuts and nut flours are new product with potential use in baking and confectionery industries as a supplement or even as a replacement for wheat flour in the manufacture of bread, biscuits, and cakes and provide a source of natural flavoring in these foods. According to our previous study (Prasantha et al. 2012), rancidity development of inshell walnut and groundnut can significantly be controlled by nuts treated with high-pressure (250 kPa) saturated CO_2 (98%) and stored 10 days at 35 °C temperature compared to the control samples (Table 4).

Changes in the quality of a commodity in storage could be predicted by knowledge of storage temperature, moisture, storage atmosphere, and initial quality of the product. Information on the rate of quality change of a stored commodity under high pressure CO_2 is necessary for feasibility studies of the design pest control system. Therefore, information and understanding about the moisture, physical, biochemical, and structural changes of the high-pressure CO_2 treated commodity is essential to establish proper control of stored product pests and obtain optimum quality of the product. Despite the efficacy of MA storage treatments used to control the stored pests of shelf-stable commodities are substantial, effective high-pressure CO_2 treatment is a fairly expensive system for local industries. The setup requires expensive pressure chambers and safety control systems. However, operational cost is comparatively lower than during chemical fumigation. Therefore, this type of treatment is good for high-value organic commodities like spices, nuts, and herbs which can earn high income in the international markets.

Table 4 Mean free fatty acid (%) and peroxide values (meq/kg) of control and high-pressure carbon dioxide (98%) treated inshell walnut and groundnut stored at 35 °C

	Free fatty acid content (%)		Peroxide value (meq/kg)	
Storage time (day) at 35 °C	Control (101.3 kPa)	CO_2 treatment (250 kPa)	Control (101.3 kPa)	CO_2 treatment (250 kPa)
Inshell walnut				
Initial sample	0.21 ± 0.01	0.21 ± 0.01	0.17 ± 0.01	0.17 ± 0.01
0[a]	0.37 ± 0.02	0.27 ± 0.00	0.86 ± 0.21	0.21 ± 0.12
10	0.59 ± 0.13	0.38 ± 0.06	1.45 ± 0.33	0.30 ± 0.04
Inshell groundnut				
Initial sample	0.14 ± 0.01	0.14 ± 0.01	0.2 ± 0.03	0.20 ± 0.03
0[a]	0.36 ± 0.02	0.15 ± 0.00	0.25 ± 0.05	0.22 ± 0.04
10[b]	0.55 ± 0.11	0.25 ± 0.17	0.7 ± 0.03	0.25 ± 0.08

[a]Store 10 days under CO_2 and control treatments at 35 °C
[b]Soon after CO_2 treatments

References

Adhikarinayake TB (2005) Technical and economic feasibility of airtight ferrocement bin for on-farm storage of paddy in Sri Lanka. PhD thesis, Wageningen University, The Netherlands, Kingdom of the Netherlands

Adhikarinayake TB, Palipane KB, Müller J (2006) Quality change and mass loss of paddy during airtight storage in a ferro-cement bin in Sri Lanka. J Stored Prod Res 42:377–390

Adhikarinayake TB, Müller J, Oostdam JWM, Huisman W, Richard P (2007) An airtight paddy storage system for small-scale farmers in Sri Lanka. Agric Mechanization Asia Afr Lat Am 38:48–55

Adler C (1993) Zur Wirkungmodifizierter Atmosphären auf Vorratsschädlinge in Getreide am Beispiel des Kornkäfers *Sitophilus granarius* (L.) (Col., Curculionidae). (Effects of modified atmospheres on stored product pests in grain as represented by the granary weevil *Sitophilus granarius* (L.) (Col.: Curculionidae)). PhD thesis, Freie Universität Berlin, Shaker Verlag Aachen. isbn 3-86111-401-1

Adler C (1999) Efficacy of modified atmospheres against diapausing larvae of the Indian meal moth *Plodia interpunctella* (Hübner). In: Jin Z, Liang Q, Liang Y, Tan X, Guan L (eds) Proceedings of the 7th international working conference on stored product protection. Beijing, China, pp 685–691

Adler C (2010) Physical control of stored product insects. International European symposium on stored product protection "Stress on chemical products", Berlin. Deutschland Julius-Kühn-Archiv 429:33–35. https://www.openagrar.de/receive/openagrar_mods_00010594. Accessed 23 Sept 2018

Adler C, Corinth HG, Reichmuth C (2000) Modified atmospheres. In: Subramanyam B, Hagstrum DW (eds) Alternatives to pesticides in stored-product IPM. Kluwer Academic Publishers, Boston, pp 105–146

Adler CS, Ndomo-Moualeu AF, Begemann J, Münzing K (2016) Effect of vacuum storage of wheat (*Triticum aestivum*) grain on the granary weevil, Sitophilus granarius and wheat quality. In: Navarro S, Jayas DS, Alagusundaram K (eds) Proceedings of the 10th international conference on controlled atmosphere and fumigation in stored products. CAF Permanent Committee Secretariat, Winnipeg, pp 287–290

Ben DC, Liem PV, Dao NT, Gummert M, Rickman JF (2006) Effect of hermetic storage in the super bag on seed quality and milled rice quality of different varieties in Bac Lieu, Vietnam. Agric Eng 31:55–56

Buranasompob A, Tang J, Powers JR, Reyes J, Clark S, Swanson BG (2007) Lipoxygenase activity in walnuts and almonds. LWT-Food Sci Technol 40:893–899

Busta FF, Smith LB, Christensen CM (1980) Microbiology of controlled atmosphere storage-an overview. In: Shejbal J (ed) Controlled atmosphere storage of grains. An international symposium at Rome-Italy, Development in agricultural engineering. Elsevier, New York, pp 121–132

Caliboso FM, Sabio GC (1998) Hermetic storage of grains in the tropics. In: Nawa Y, Takagi H, Oguchi A (eds) Post harvest technology in Asia: a step forwards a stable supply of food products, The 5th JIRCAS international symposium Japan, pp 59–720

Cofie-Agblor R, Muir WE, Jayas DS, White NDG (1998) Carbon dioxide sorption by grains and canola at two CO_2 concentrations. J Stored Prod Res 34:159–170

Coursey DG (1983) Postharvest losses in perishable foods of the developing world. In: Lieberman M (ed) Post-harvest physiology and crop preservation. Plenum Press, New York, pp 485–514

Deshpande SS, Sathe SK, Salunkhe DK (1984) Interrelationships between certain physical and chemical properties of dry bean (*Phaseolus vulgaris* L.). Plant Foods Hum Nutr 34:53–65

Donahaye E, Navarro S, Ziv A, Blauschild Y, Weerasinghe D (1991) Storage of paddy hermetically sealed plastic liners in Sri Lanka. Trop Sci 31:109–124

Ferizli AG, Navarro S, Donahaye JE, Rindner M, Azriel A (2001) Airtight granary for use by subsistence farmers. In: Dhnahaye EJ, Navarro S, Leesch JG (eds) Proceedings of international conference of controlled atmosphere and fumigation in stored products, pp 37–43

Fields PC, White ND (2002) Alternative to methyl bromide treatment for stored product and quarantine insects. Annu Rev Entomol 47:331–359

Fleurat-Lessard F, Torc'h JL (2001) Control of insects in postharvest: high temperature and inert atmospheres. In: Vincent C, Panneton B, Fleurat-Lessard F (eds) Physical control methods in plant protection. Springer, Berlin, pp 107–127

Hafeel RH (2010) Effect of storage methods on grain quality of different rice varieties in Sri Lanka. M. Phil dissertation, Postgraduate Institute of Agriculture, University of Peradeniya, Sri Lanka

Hafeel RF, Prasantha BDR, Dissanayake DMN (2008) Effect of hermetic-storage on milling characteristics of six different varieties of paddy. Trop Agric Res 6:102–114

Hafeel RF, Prasantha BDR, Dissanayke DMN, Edirisinghe EMRD (2011) Hermetic storage method, a promising alternative for rice storage. Ann Sri Lanka Dep Agric 13:59–70

Johnson JA (2010) Effect of relative humidity and product moisture on response of diapausing and non-diapausing Indian mealmoth (Lepidoptera: Pyralidae) larvae to low pressure treatment. J Econ Entomol 103:612–618

Johnson JA, Zettler JL (2009) Response of postharvest tree nut lepidopteran pests to vacuum treatments. J Econ Entomol 102:2003–2010

Liu K, Bourne MC (1995) Cellular, biological, and physicochemical basis for the hard-to-cook defect in legume seeds. Crit Rev Food Sci Nutr 35:263–298

Mbata NG, Johnson M, Phillips TW, Payton M (2005) Mortality of life stages of cowpea weevil (Coleoptera: Bruchidae) exposed to low pressure at different temperatures. J Econ Entomol 98:1070–1075

Moreno-Martinez E, Jiménez S, Vázquez ME (2000) Effect of (*Sitophilus zeamais*) and (*Aspergillus chevalieri*) on the oxygen level in maize stored hermetically. J Stored Prod Res 36:25–36

Mortazavi H, Ferizli AG (2014) Effectiveness of low pressure to control azuki bean weevil, Callosobruchus chinensis (L.) (Col: Bruchidae). In: Arthur FH, Kengkanpanich R, Chayaprasert W, Suthisut D (eds) Proceedings of the 11th international working conference on stored product protection, Thailand. http://spiru.cgahr.ksu.edu/proj/iwcspp/iwcspp11.html. Accessed 23 Sept 2018

Mullen MA, Pedersen JR (2000) Sanitation and exclusion. In: Subramanyam B, Hagstrum DW (eds) Alternatives to pesticides in stored-product IPM. Kluwer Academic Publishers, Boston, pp 29–72

Murdock LL, Margam V, Baoua I, Balfe S, Shade RE (2012) Death by desiccation: effects of hermetic storage on cowpea bruchids. J Stored Prod 49:166–170

Nakakita H, Kawashima K (1994) A new method to control stored product insects using carbon dioxide with high pressure followed by sudden pressure loss. In: Highley E, Wright EJ, Banks HJ, Champ BR (eds) Proceedings of 6th international working conference on stored-product protection. CAB International, Canberra/Wallingford/Oxon, pp 126–129

Nasar-Abbas SM, Plummer JA, Siddique KH, White P, Harris D, Dods K (2008) Cooking quality of faba bean after storage at high temperature and the role of lignins and other phenolics in bean hardening. LWT-Food Sci Technol 41:1260–1267

Navarro S (2006) Modified atmospheres for the control of stored product insects and mites. In: Heaps JW (ed) Insect management for food storage and processing. AACC International, St. Paul, pp 105–145

Navarro S (2012) The use of modified and controlled atmospheres for the disinfestations of stored products. J Pest Sci 85:301–322

Noomhorm A, Sirisoontaralak P, Uraichuen J, Ahmad I (2009) Effects of pressurized carbon dioxide on controlling *Sitophilus zeamais* (Coleoptera: Curculionidae) and the quality of milled rice. J Stored Prod Res 45:201–205

Prasadi VPN (2014) Effect of Hermetic storage on physico-chemical and functional properties of grain legumes. M. Phil dissertation, Postgraduate Institute of Agriculture, University of Peradeniya, Sri Lanka

Prasantha BDR, Obenland D, Walse S, Johnson J (2012) High temperature CO2 treatments for inshell walnuts. International research conference on methyl bromide alternatives and emissions reductions. In: Obenauf G (ed) Proceedings of the methyl bromide alternatives outreach, Orlando, pp 37.1–37.4. http://mbao.org/2012/Proceedings/37JohnsonJ.pdf. Accessed 23 Sept 2018

Prasantha BDR, Hafeel RF, Wimalasiri KMS, Pathirana UPD (2014a) End-use quality characteristics of hermetically stored paddy. J Stored Prod Res 59:158–166

Prasantha BDR, Prasadi VPN, Wimalasiri KMS (2014b) Effect of hermetic storage on end-use quality of mungbean. In: Arthur FH, Kengkanpanich R, Chayaprasert W, Suthisut D (eds) Proceedings of the 11th international working conference on stored product protection, Chiang Mai, pp 373–384. http://spiru.cgahr.ksu.edu/proj/iwcspp/iwcspp11.html. Accessed 23 Sept 2018

Prasantha BDR, Kumarasinghe HKMS, Emitiyagoda GAMS (2016) Hermetic type grain storage PVC tank. Patent number LK/1/2/11775. National Intellectual Property Office Sri Lanka

Prasantha BDR, Kumarasinha KMH, Emitiyagoda GAMS (2018) Storage of mung bean in Hermetic PVC Tank. In: Adler CS et al (eds) Proceedings of the 12th international working conference on stored product protection (IWCSPP) in Berlin, Germany, October 7–11, 2018. Julius-Kuhn-Archiv 463(1):441–447

Prasantha BDR (2020) Effects of low-pressure storage of food commodities on the mortality of adult stored product insects. Sri Lankan J Biol 5(1):8

Prozell S, Reichmuth C (2001) Carbon dioxide under high pressure for stored-product protection in temperate climates. In: Donahaye EJ, Navarro S, Leesch JG (eds) Proceedings of international conference of controlled atmosphere and fumigation in stored products. Executive Printing Services, Fresno, pp 719–725

Quezada MY, Moreno J, Vazquez ME, Mendoza M, Mendez-Albores A, Moreno-Martınez E (2006) Hermetic storage system preventing the proliferation of *Prostephanus truncatus* Horn and storage fungi in maize with different moisture contents. Postharvest Biol Technol 39:321–326

Raj SA, Singaravadivel K (1990) Biodeterioration in rice (*Oryza sativa* L.) due to low, medium and high moisture. Int Biodeterior 27:237–248

Reichmuth C, Wohlgemuth R (1994) Carbon dioxide under high pressure of 15 bar and 20 bar to control the eggs of the Indian meal moth *Plodia interpunctella* (Hübner) (Lepidoptera: Pyralidae) as the most tolerant stage at 25°C. In: Highley E, Wright EJ, Banks HJ, Champ BR (eds) Proceedings of 6th international working conference on stored-product protection. CAB International, Wallingford/Oxon, pp 163–172

Riudavets J, Castane CR, Alomar O, Pons MJ, Gabarra R (2009) Modified atmosphere packaging (MAP) as an alternative measure for controlling ten pests that attack processed food products. J Stored Prod Res 45:91–96

Sartaj IZ, Ekanayake S (1991) Postharvest losses. Trop Agric Res 3:115–132

Seitz M, Sauer DB (1996) Volatile compounds and odors in grain sorghum infested with common storage insects. Cereal Chem 73:744–750

Sen F, Meyvaci KB, Turanli F, Aksoy U (2010) Effects of short-term controlled atmosphere treatment at elevated temperature on dried fig fruit. J Stored Prod Res 46:28–33

Shiga TM, Cordenunsi RB, Lajolo FM (2009) Effect of cooking on non-starch polysaccharides of hard-to-cook beans. Carbohydr Polym 76:100–109

Surendra GBB (1997) Integrated production practices in cashew in Sri Lanka. Expert consultation on integrated production practices in cashew in Asia. FAO, Bangkok.. http://www.fao.org/docrep/005/ac451e/ac451e00.htm#Contents. Accessed 23 Sept 2018

Villers P, de Bruin T, Navarro S (2006) Development and applications of the hermetic storage technology. In: Lorini BB, Beckel H, Deckers D, Sundfeld E, dos Santos JP, Biagi JD, Celaro JC, Faroni LRDA, de Bortolini LF, Sartori MR, Elias MC, RNC G, da Fonseca RG, Scussel VM (eds) Proceedings of the V.M. 9th international working conference on stored product protection. Brazilian Post-harvest Association – ABRAPOS, Passo Fundo, RS, Brazil, Campinas/São Paulo, pp 719–729

White NDG, Jayas DS (1991) Factors affecting the deterioration of stored flaxseed including the potential of insect infestation. Can J Plant Sci 71:327–335

Wright KP, Kader AA (1997) Effect of controlled-atmosphere storage on the quality and carotenoid content of sliced persimmons and peaches. Postharvest Biol Technol 10:89–97

The manufacturer's authorised representative in the EU is Springer Nature Customer Service Centre GmbH, Europaplatz 3, 69115 Heidelberg, Germany. If you have any concerns regarding our products, please contact ProductSafety@springernature.com

Printed and bound by CPI Group (UK) Ltd, Croydon, CR0 4YY
30/09/2026
02225508-0019